ELECTRICAL S·Y·S·T·E·M·S TECHNOLOGY

·CIRCUITS·
·MOTORS·
·GENERATORS·
·CONTROLS·

ELECTRICAL S·Y·S·T·E·M·S TECHNOLOGY

·CIRCUITS·
·MOTORS·
·GENERATORS·
·CONTROLS·

314 191259

Walter L. Bartkiw, B.A., MSc.
Technical Director
Governor Simcoe Secondary School
St. Catharines, Ontario

Kenny T. Sookhoo, B.Sc., P. Eng.
Assistant Technical Director
Governor Simcoe Secondary School
St. Catharines, Ontario

McGraw-Hill Ryerson Limited

Toronto   Montréal   New York   Auckland   Bogotá   Cairo   Guatemala
Hamburg   Lisbon   London   Madrid   Mexico   New Delhi
Panama   Paris   San Juan   São Paulo   Singapore   Sydney   Tokyo

ELECTRICAL SYSTEMS TECHNOLOGY:
CIRCUITS, MOTORS, GENERATORS, CONTROLS

ISBN 0-07-548798-5

2 3 4 5 6 7 8 9 0   D   5 4 3 2 1 0 9 8

Printed and bound in Canada

Canadian Cataloguing in Publication Data

Bartkiw, Walter L.
    Electrical systems technology

Includes index.
ISBN 0-07-548798-5

1. Electric engineering. I. Sookhoo, Kenny T.
II. Title.

TK146.B37 1985   621.3   C85-098667-2

ACKNOWLEDGMENTS

AEG - Telefunken Corporation
Allen-Bradley Canada Ltd.
Amprobe Instruments
B & K Precision
Bodine Electric Co.
Canadian General Electric Co. Ltd.
Century Electric, Inc.
John Dubiel
Ferranti-Packard Transformers Ltd.
Franklin Electric
Reliance Electric Limited
Square D Company Canada Ltd.
Steelman Electric Manufacturing Co. Ltd.
Triplett Corporation

Enzy Stienstra, Chairman of Science and Technology,
Cawthra Park Secondary School, Mississauga, Ontario.
Author of Chapters 14 and 16.

This book is dedicated to the
authors' parents, John and
Anastasia Bartkiw, and Seuradge
and Radhica Sookhoo.

CHENAUX GENERATING STATION
Chenaux Generating Station is located on the Ottawa River 13 kilometres north of Renfrew. It was placed in service by Ontario Hydro in 1950. Three dams were necessary to close the river here—Limerick Dam and Portage du Fort Dam, and a small auxiliary dam on the Quebec side—with a total length of 1220 metres. The plant has a capacity of 122 400 kilowatts from eight units.

Courtesy: Ontario Hydro

PREFACE

The purpose of writing this text was to provide a single comprehensive book which covers the curriculum topics in secondary schools, vocational-technical schools, technical institutes, community colleges, and in military and industrial training programs that prepare students for careers in electrical technology.

The authors have attempted to present the material in a concise form, beginning with fundamentals of applied electricity, basic electric circuits, motors, generators, transformers, measuring devices, conventional, solid state, and computer control.

The final chapter on industrial and computer electronics will, we hope, make the student more aware of the rapidly changing industrial world. It introduces the student to important electronic components and logic devices that are utilized in control and computer systems.

Both the technical student wishing to specialize in electricity and the student seeking a single electrical credit will find this book a storehouse of relative material. A summary of important points and a set of questions have been included at the end of each chapter. This will serve as a useful review, and also enhance the reader's understanding of the chapter just studied.

The instructor does not have to follow the same sequence of chapters as laid out by the authors. The content readily lends itself to any particular course outline.

In accordance with the International System of Units, SI units and symbols are used.

We have listed below a list of occupations to be found in the electrical field.

We would like to express our sincere thanks to the technical students of Governor Simcoe Secondary School, who were the first to field test many parts of the initial manuscript. Thank you, to the secretarial staff of Governor Simcoe Secondary School for their invaluable assistance; to the editors Don Lipsett and Jen-

nifer Joiner; to the many companies who supplied us with valuable pictures and product information. A special note of appreciation to Enzy Stienstra for his critical review of the complete manuscript. And last but not least, a sincere thanks to our wives Noreen Bartkiw and Basdaye Sookhoo for their encouragement, personal sacrifice, and also their help in the typing of the manuscript.

Walter L. Bartkiw
Kenny T. Sookhoo

ELECTRICAL SAFETY

Safety is everyone's responsibility, not only personal safety, but also a consideration for the safety of others. It is important that one be aware of the danger that exists in the operation of electrical equipment and even more important to know the safety rules and practices in order to eliminate potentially dangerous conditions.

An electrical shock is the result of an individual coming in contact with a "live" conductor, be it from an electrical circuit or improperly grounded equipment. The amount of current that may pass through the body may vary, depending on the resistance of the human body. Dry skin may produce high resistance of several hundred thousand ohms, but if the voltage is great enough, an electric current will pass through. When the skin is wet, the body resistance may drop to a thousand ohms or less, allowing more current to pass through the body, causing greater damage. Under this condition, a low voltage of 120 volts or less may produce a current that may cause a dangerous shock that could be fatal.

A current of approximately 10 to 15 mA may cause a painful shock, but not prevent voluntary muscular action. However, a current of 15 to 20 mA may cause permanent damage and loss of muscular control in adjacent muscles. Currents of 20 to 50 mA produce great pain and severe muscular contraction, making it difficult to breathe, but currents greater than 50 mA may paralyze the heart and respiratory muscles, causing death. Apply artificial respiration immediately if breathing has stopped.

The best way to avoid electrical shock is to practise safe work habits. The following safety precautions should be followed.

- Avoid horseplay—consider the result of such acts.
- Report unsafe conditions, equipment immediately.
- Wear proper clothing, according to shop standards. Do not wear loose clothing. Remove rings, wristwatches, bracelets, etc., when working on electrical apparatus.
- Do not work on energized circuits; never assume that an electric circuit is dead. Check it with a voltmeter or neon lamp, then lock it out and tag it.
- Do not touch electrical apparatus when hands are wet, or while standing on a wet floor.
- Use tools that are insulated (moulded-type) when working on exposed electrical apparatus.
- Do not ground yourself when working on electrical apparatus. Always stand on dry wood on a concrete floor when working with power tools.
- Make sure that the electrical power tools-

apparatus is properly grounded by a three-prong plug. The safest approach is to use a ground fault interrupter.
- Do not work alone with high voltage circuits; have someone present at all times to render assistance in an emergency. Use proper protective gear such as rubber gloves, shoes, rubber mats. Never use two hands when working on live circuits.
- When working with batteries, use rubber gloves, aprons, and boots, as well as proper eyeshield, for the prevention of acid burns. When a battery is being charged, explosive hydrogen gas is produced; thus care should be taken to prevent short-circuits or sparks. Follow manufacturers' instructions.
- Warning signs and guards should be posted to alert personnel of high-voltage areas.
- Use electrical tools and appliances that bear the stamp of the Canadian Standards Association (CSA), or the Underwriter's Laboratories (UL) in the case of the United States. These labels mean that the device has been tested and meets the safety standard of the designated association.

Finally, practise safety whether at home or at work. Accidents are always caused by negligence, or lack of understanding of the proper safety procedures. Remember the safety slogan "Accidents don't just happen; they are caused."

LIST OF OCCUPATIONS IN THE ELECTRICAL FIELD

Electrician	Electrical Inspector
Electrical Appliance Repair	Electrical Medical Equipment
Design and Development Engineer	Railway Locomotive
Circuit Design Engineer	Oil Exploration
Electrical Construction	Automotive Electrical Wiring and Repair
Electrical Research Technologist	Industrial Electrician
Power Transmission	Heating and Cooling Industry
Power Generation Technician	Electrical-Meter Installer
Traffic Circuit Engineer	Electrical-Testing Technician
Nuclear Operations Technologist	Electrical Discharge Machine Tender
Robotics	Electric Cranes Technician
Computer-Related Technologies	Electric Toy Repairs
Mining Electrician	Food Processing Equipment Repair
Marine Electrician	Traffic Control Systems
Instrumentation Technician	Electrician, Communications Equipment
Aircraft Industry	Electrical Engineer
Electrical Repair	

CONTENTS

Preface

1 FUNDAMENTALS

1

Energy / Elements and Compounds / The Atom / Law of Charges / Electron Shell Arrangements / Conductors, Insulators, and Semi-Conductors / Electric Charge / Current Flow / Multiple and Submultiple Units of Current / Potential Energy Difference / Converting Multiple and Submultiple Units of Voltage / Converting Multiple and Submultiple Units of Resistance / Factors Affecting the Resistance of a Conductor / Temperature Coefficient of Resistance / Conductor Materials / Conductance / Ohm's Law / Measuring Current / Measuring Voltage / Measuring Resistance / Measurement of Electric Power

2 BATTERIES

18

What Is a Battery? / How It All Began / Voltaic Cell / Polarization / Internal Resistance / Dry Cells / Local Action / Alkaline-Maganese Cell / Silver Oxide Cell / Magnesium Cell / The Lithium Cell / Multiple-Cell Energy Sources / Secondary Cells / Lead-Acid Cells / Battery Operation / Safety / Precautions

3 DIRECT CURRENT CIRCUITS

36

Series Circuits / Ohm's Law / Parallel Circuits—Voltage / Parallel Circuits—Current / Parallel Circuits—Resistance / Series-Parallel Circuits / Work, Power, and Energy / Terminal Voltages and Power Losses / Power Transfer / Complex Circuits

4 MAGNETISM

58

Magnetic and Nonmagnetic Substances / Theory of Magnetism / Magnetic Circuits / Consequent Poles / Electromagnetism / Magnetic Circuits / Magnetic Flux (ϕ) / Flux Density (β) / Magnetizing Force or Magnetic Intensity (H) / Reluctance (R) / Permeability (μ) / Magnetization (β-H) Curves / Magnetic Circuit Problems / Hysteresis / Electromagnetic Coils in Series and Parallel

5 ELECTROMAGNET INDUCTION

78

Induction / Requirements for Electromagnetic Induction / Conditions That Affect Amount of Induced Voltage / Faraday's Law of Electromagnetic Induction / Left-Hand Generator Rule / Lenz's Law of Electromagnetic Induction / Magnetic and Electromagnetic Applications

6 DC ELECTRICAL MEASURING INSTRUMENTS

87

The Galvanometer / The Basic Ammeter / Increasing the Range of the Ammeter / Computing the Shunt Resistance / Multirange Voltmeters /

Series Multiplier Arrangements / Voltmeter / Sensitivity (Ohms-Per-Volt) / Loading Effect of Voltmeters / The Ohmmeter / Series Ohmmeter / Multiple Ohmmeter / Ranges / The Shunt Ohmmeter / The Wheatstone Bridge / The Slide Wire Bridge / The Megger (Megohmmeter) Insulation Tester

7 DC GENERATORS 109

Introduction / Generating an AC Sine Wave / The Simple DC Generator / Practical DC Generators / Generator Field Exitation / The Magnetization Curve / Generator Internal Voltage Losses / Correcting Armature Reaction / Voltage Regulation—Load Characteristics / The Separately Excited Generator / The Self-Excited Shunt Generator / Generator Power Losses—Efficiency

8 DC MOTORS 136

Introduction / Basic Motor Principle / Torque and Rotary Motion / Communication / Lap and Wave Windings / Calculation of Torque / Torque Measurement / Armature Reaction / Interpoles / Counter Voltage (V_C) in a Motor Factors Affecting Motor Speed / Speed Regulation / Classification of DC Motors / Motor Ratings / Motor Losses / Efficiency / The Shunt DC Motor / Series DC Motors / Compound DC Motors

9 DC MOTOR CONTROL 155

DC Motor Starters / Motor-Starting Requirements / Manual Starters / Manual Starting Rheostats for DC Series Motors / Combination Manual Starters and Speed Controllers / Face-plate Controller / Drum Controllers / Introduction of Automatic Motor Control / Electrical Diagrams and Symbols / Typical Control Circuit / Magnetic Contactors / Blowout Coil / Overload Protective Devices / Thermistor Overload Relay / Automatic Reversing / Anti-Plugging Relay / Reduced Voltage DC Starters / Dynamic Breaking / Ward-Leonard System of Speed Control / Solid-State Motor Control

10 ALTERNATING CURRENT FUNDAMENTALS 182

Alternate Current Versus Direct Current / Alternating Voltage / Basic Trigonometric Functions / Voltage and Current Values / Voltage and Current Phase Relationships / Phasors (Vectors) and Addition of Phasors / AC Circuits with Only Resistive Loads

11 INDUCTANCE AND REDUCTIVE REACTANCE 202

Inductance / Inductive Reactance (X_L) / Phase Relationship in a Purely Inductive Circuit / Power Taken by a Pure Inductance / The Effective Resistance of a Coil / The Q-Factor of a Coil / R-L Circuit Impedance / Apparent Power (Volt-Ampere) / Power Factor and Phase Angle

12 CAPACITORS AND CAPACITIVE REACTANCE 219

The Capacitor / Capacitance and Energy Storage / The Dialectric of a Capacitor / Direct Current R-C Circuits / Capacitors in AC Circuits / Phase Relationship in a Purely Capacitive Circuit / Power in a Capacitor (Reactive Power) / R-C Series Circuit

13 SINGLE-PHASE *R-L-C* CIRCUITS 240

Series R-L-C Circuits / Circuit Power / Parallel R-L-C Circuits / Power Factor Correction / Series Parallel Circuits

14 TRANSFORMERS 261

Transformers / Transformer Construction / Transformer Polarity / Principles of Transformer Operation / Transformer Losses and Efficiency / Instrument Transformers / Autotransformers / Tap-Changing Transformers / Voltage Regulating Transformers (Regulators) / Three-Phase Transformers

15 THREE-PHASE (POLYPHASE) CIRCUITS 285

Introduction to Three-Phase Circuits / Phase Rotation / The Wye or Star Connection / The Delta Connection / Power Factor of a Three-Phase System / Power in a Three-Phase System / Three-Phase Power Measurement / Three-Phase Transformer Connections

16 AC GENERATORS (ALTERNATORS) 314

Construction of AC Generators / Rotor Field Discharge Circuit / Ventilation and Cooling / Induced Voltages in an AC Generator / Voltage Regulation / Parallel Operation of AC Generators / Hunting of Alternating Current Generators / AC Generator Losses and Efficiency

17 THREE-PHASE INDUCTION MOTORS 333

Squirrel-Cage Induction Motor / Motor Power Factor / Types of Induction Motor—Code Identification / Speed Control of Squirrel-Cage Motors / The Wound-Rotor Induction Motor

18 SYNCHRONOUS MOTORS AND SELF-SYNCHRONOUS DEVICES 356

Three-Phase Synchronous Motors / Construction, Operation, Starting / Self-Synchronous Devices / Small Single-Phase Synchronous Motors

19 SINGLE-PHASE AC MOTORS 376

The Single-Phase Induction Motor Principle / Producing a Rotating Field from Two 90° Out-of-Phase Fields / Split-Phase Inductive Motors / The Shaded-Pole

Motor / Universal Motors / The Repulsion-Start Induction-Run Motor / Repulsion-Induction Motors

20 AC MEASURING INSTRUMENTS 390

Rectifier-Type Meter / Electromagnetic-Type Meters / Thermocouple Meters / Electrodynamometer Movement / Power Measurements / Energy Measurements in AC Circuits / Power-Factor Meters / The Clamp-On Volt-Ammeter / Multitesters / Frequency Meters

21 AC MOTOR CONTROL 406

Motor Controllers / Contactors / Relays / Reduced-Voltage Starting / Solid State Starter / Reversing Controllers / Interlocks / Sequence Control / Jogging / Braking Induction Motors / Multispeed Motors / Wound-Rotor Motor Control / Wound-Motor Motors / Synchronous Motor Control / Static Motor Control / Inverters / Variable Frequency Speed Control / Programmable Controllers

22 INDUSTRIAL AND COMPUTER ELECTRONICS 433

Semiconductor Devices / PN Junction, Zener Diode, LED, Thermistor, SCR, Junction Transistors, FET, MOSFET, IC / Rectifier Circuits / Analog Versus Digital Systems / Digital Principles and Devices / Op-Amp, Differential Amp, Binary System, Logic Devices, Flip-Flop / Microcomputer Systems Interfacing

Appendix A Structure of Electron Shells 466
Appendix B Colour Coding of Resistors 470
Appendix C The J operator 471
Appendix D Natural Trigonometric Functions 475

FUNDAMENTALS

ENERGY

One need only look around to realize the importance of energy. For example, the sun has sent energy to earth for millions of years. A large part of this energy is absorbed by plants, and much of it is dissipated in the form of heat. Electricity is the study of energy provided by the flow of electrons in conductors and devices that form electric circuits. *Energy* is the ability to do work. It exists in two forms: potential and kinetic.

Potential energy is energy that is stored — for example, a water dam holding back a large reservoir of water that is waiting to be released, a car battery stores electrical energy that is waiting to be released when a load is connected across the terminals.

A moving body contains *kinetic energy* — such as a baseball in flight, or water released by a dam that flows through a turbine. In the example of the car battery, once the circuit is connected to the terminals, the resulting electron flow causes a lamp to light up (work being done).

Energy exists in many different forms: *mechanical energy* — a man driving a nail into a piece of wood; *chemical energy* — sugar dropped into a cup of coffee that dissolves quickly because of a chemical reaction; *light energy* — a laser beam burning a hole through a piece of steel; *heat energy* — in the ancient example of rubbing one stick against another, the two sticks are set on fire; *solar energy* — sunlight striking a solar cell will produce electricity. These examples are a few of the many forms of energy that exist.

Work is done when one form of energy is converted into other forms of energy. Our ability to convert one form of energy into another has made possible great scientific progress.

Electrical energy is one of the fundamental forms of energy used in the world today. Electricity in motion will produce light, heat, and will energize motors, appliances, TV sets, computers and many other electrical devices. Since electricity is generated in one place but is required in another, the study of electricity involves generation, transmission, controls, and all the devices that use this energy to accomplish useful work. To understand what this energy is, we must investigate the structure of all matter.

ELEMENTS AND COMPOUNDS

What is matter? *Matter* is anything that we can feel, see, or smell. To express it in another form, matter is anything that has mass and occupies space. It can be in the form of a solid, a liquid, or a gas. Wood, rubber, and cement are forms of solid matter; water, paint, and oil are examples of forms of liquid matter, while oxygen, helium, and hydrogen are examples of forms of gaseous matter.

All matter is made up of basic materials that are called *elements*. An element is a pure substance. Oxygen, copper, gold, carbon and mercury are a few examples of elements. There are 106 known elements. Ninety-two of these elements occur in nature (natural elements), and the other 14 are man-made (artificial elements). Appendix A shows a complete list of natural and artificial elements known today. The last three artificial elements have yet to be named.

As we look around us, it becomes obvious that there are more types of matter than there are elements. Matter such as water, salt, and rubber are actually called *compounds* because they are made up of more than one element. The smallest particle into which a given compound may be divided without destroying its chemical properties is called a *molecule*.

THE ATOM

The smallest particle to which an element can be reduced and still retain the properties of that element is called an *atom*. Molecules are made up of atoms which are bound together. The water molecule is really a combination of two hydrogen (H) atoms and one oxygen (O) atom. That is why the chemical formula for water is H_2O.

The atom is made up of microscopic particles: the electron, the proton, and the neutron. These particles have different characteristics. However, as far as is known, all electrons are alike, all protons are alike, and all neutrons are alike. Figure 1-1 shows how electrons, protons, and neutrons are arranged to form an atom.

The atomic model shown in Fig. 1-1 is known as the Bohr model, named after the scientist who developed it. Niels Bohr (1885–1962), a Danish physicist studied the nature of the

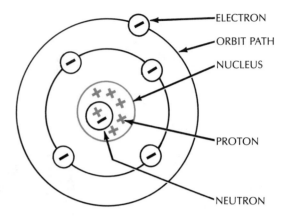

Figure 1-1 The structure of the atom.

nucleus, electrons, and their orbits.

The structure of the atom, Fig. 1-1, is similar to our own solar system. In the centre of the atom, there is a large mass called the nucleus, and it may be compared to the sun of our solar system. The nucleus is made up of protons and neutrons, which are about equal in mass. The proton carries a positive charge, while the neutron is electrically neutral. The number of neutrons that the nucleus contains will depend on the type of atom; also, the number of protons and neutrons in the nucleus will determine the mass of the atom. The proton, however, is 1840 times as heavy as the electron that rotates in an orbital path around the nucleus. Electrons carry a negative charge.

The attraction between a proton and an electron in an atom is due to the fact that the two particles exert an electrostatic force on one another, producing electrostatic lines of force between the protons and electrons within the atom. These electrostatic lines of force are responsible for keeping the electrons orbiting in their shells, or orbits. Within atoms, there are both gravitational forces and electrostatic forces. The electrostatic interaction between a proton and an electron is 2.3 times 10^{39} stronger than the gravitational forces between them.

Atoms of different elements differ from each other in the number of electrons they

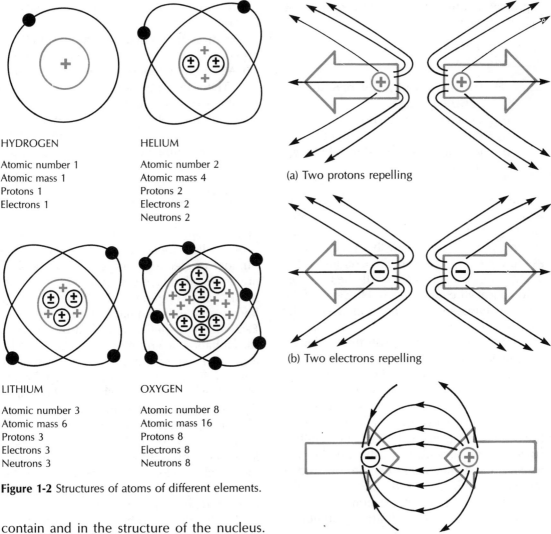

HYDROGEN

Atomic number 1
Atomic mass 1
Protons 1
Electrons 1

HELIUM

Atomic number 2
Atomic mass 4
Protons 2
Electrons 2
Neutrons 2

(a) Two protons repelling

(b) Two electrons repelling

LITHIUM

Atomic number 3
Atomic mass 6
Protons 3
Electrons 3
Neutrons 3

OXYGEN

Atomic number 8
Atomic mass 16
Protons 8
Electrons 8
Neutrons 8

Figure 1-2 Structures of atoms of different elements.

contain and in the structure of the nucleus. Figure 1-2 shows the basic structure of a hydrogen, helium, lithium, and oxygen atom.

(c) Electron and proton attracting

Figure 1-3 Behaviour of like and unlike charges.

LAW OF CHARGES

In an atom, the negative charge of an electron is equal but opposite to the positive charge of a proton. The lines of force associated with these charges produce an electrostatic field which, in turn, will cause an attracting or repelling force. The law of electrical charges states that "like charges repel each other and unlike charges attract each other." Figure 1-3(a), (b),

and (c) shows the behaviour of like and unlike charges.

In Fig. 1-3(a) we see two positive charges repelling each other, while Fig. 1-3(b) illustrates two electrons repelling each other, and Fig. 1-3(c) shows how the fields of an electron and proton interconnect, causing the two charges to attract. The law of charges can be expressed mathematically in the form of the following equation:

$$F = \frac{q_1 \times q_2}{d^2} \qquad (1\text{-}1)$$

where F = the force (of attraction or repulsion) between charges

q_1 = the charge on one body
q_2 = the charge on the second body
d^2 = the square of the distance between the bodies

This law was discovered by Charles Coulomb, a French scientist (1736–1806). By substituting arbitrary values in Eq. (1-1) for q_1, q_2, and d^2, we can see how the force changes as the values are changed. For example, if the charges are doubled (increased), the force will automatically be quadrupled. However, if the distance between the bodies is increased, the force will decrease.

The total positive charge on the nucleus is determined by the number of protons in the nucleus. This number, called the **atomic number** of the element, is what distinguishes one element from another. For example, the atomic number of hydrogen is 1 because it has but one proton in the nucleus. Helium has two protons in its nucleus; its atomic number is 2 (see Fig. 1-2). Lithium has three protons and its atomic number is 3, and so on up to the heaviest natural element, uranium, which has 92 protons in its nucleus; its atomic number is 92.

The total number of protons and neutrons in the nucleus determines the atomic mass. Prior to 1960, the atomic mass of an atom was denoted by the number resulting from the addition of the number of protons and the number of neutrons. For example, the oxygen atom has 8 protons and 8 neutrons resulting in an atomic mass of 16.

However, at an international meeting held in 1960, it was decided that the definition of atomic mass should be changed.

The atomic mass of an atom is now expressed in *unified atomic mass units (μ)*, where $1\,\mu = 1.660 \times 10^{-27}$ kg.

The mass of one hydrogen atom is 1.007 8 μ and one aluminum atom is 26.98 μ. The symbol A is used to denote the atomic mass number. The value for A is the nearest whole number expressed in the unit μ. For example, the actual mass of aluminum is 26.98 μ, but the value of A is expressed as A = 27 μ (rounded to the nearest whole number).

ELECTRON SHELL ARRANGEMENTS

Electrons spin around the nucleus in specific orbits called *shells*. Each atom has a fixed number of electrons and shells. These shells are arranged in layers as in the rings of a tree trunk that has been cut. The atoms of all the known elements can have up to seven shells. Each electron shell is identified by a letter in the alphabet (or a number), and each shell has a certain maximum number of electrons it can contain. Figure 1-4 shows the arrangement and identification of each shell, while the table in Fig. 1-4 shows the distribution of electrons. The shell closest to the nucleus 1(K) is the lowest energy level; 2(L) is the next higher level; the third is 3(M), and so on. The outermost shell is called the valence shell; the electrons in that shell occupy the highest energy levels and are called *valence electrons*.

The electrical and chemical properties of an atom are determined by the actions and energy content of the valence electrons. The number of electrons in a valence shell gives an indication of the atom's ability to lose or gain extra electrons. An atom that has a number of electrons in the valence shell close to its full complement will gain electrons quite easily trying to complete its shell and achieve equilibrium. Conversely, an atom which has a few valence electrons, compared to the required number, will tend to lose valence electrons quite easily. In addition to orbiting about the nucleus, the electron spins on its own axis.

SHELL	MAXIMUM NUMBER OF ELECTRONS
1ST SHELL (K)	2
2ND SHELL (L)	8
3RD SHELL (M)	18
4TH SHELL (N)	32
5TH SHELL (O)	18
6TH SHELL (P)	10
7TH SHELL (Q)	2

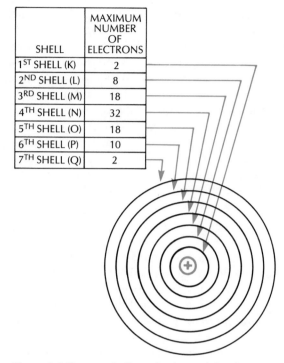

Figure 1-4 Electron shell numbering and distribution of electrons in each shell.

This action, illustrated in Fig. 1-5, will be studied in Chapter Four, on magnetism.

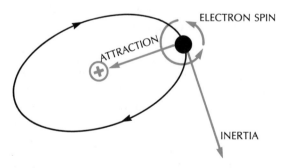

Figure 1-5 Interaction of forces in an orbiting electron.

THE ION

Atoms are affected by external influences such as heat, light, or electric fields. Any one of these forces can upset the balanced state of the atom by removing or adding electrons. When this occurs, the number of positive and negative charges are no longer equal; thus the atom will have either a net negative or positive charge. An atom that has gained or lost an electron is called an *ion*. The process of changing an atom to an ion is called *ionization*. Figure 1-6(a) shows a neutral atom with four electrons and four protons.

Figure 1-6(b) shows a positive ion that has one more proton than electron. Figure 1-6(c) shows a negative ion that contains more electrons than protons.

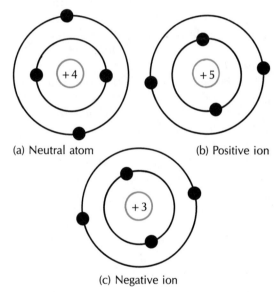

(a) Neutral atom   (b) Positive ion

(c) Negative ion

Figure 1-6 Ions.

CONDUCTORS, INSULATORS, AND SEMI-CONDUCTORS

Materials which have valence electrons that can be easily released are called *conductors*. In a good conductor, the valence electrons have much freedom to move. Two important factors that affect the amount of freedom of the valence electron are:

1. The number of "empty valence spaces." For example, in the copper atom, the valence shell

contains one electron out of a possible 32. Therefore, there are 31 "empty spaces." The silicon atom contains 4 valence electrons and so there are 14 "empty valence spaces." The copper atom contains a larger number of "empty valence spaces" than the silicon atom; thus, copper is a better conductor.

2. The second factor which affects the freedom of the valence electrons of an atom is the distance separating these electrons from the nucleus. The greater the distance, the weaker the attracting force holding the electron to the nucleus. For example, the valence electrons for copper are located in the fourth shell and for silicon they are in the third shell. Therefore, copper valence electrons are farther away from the nucleus than are silicon valence electrons. Thus, copper valence electrons are freer.

Substances such as rubber, porcelain, and phosphorous have very few "empty valence spaces." Since these valence electrons have little freedom, these substances are extremely poor conductors and so are called "insulators."

Elements such as silicon and germanium are neither good conductors nor good insulators—thus, they are called semi-conductors. All semi-conductor elements have four valence electrons that act as a semi-closed shell.

Instead of trying to define conductors, semi-conductors, and insulators in terms of the number of empty valence spaces, we can state it in the following manner. Elements whose atoms have less than four valence electrons are good conductors. (The fewer valence electrons, the better the conductor.) Those that have exactly four valence electrons are semi-conductors and those that have more than four valence electrons are good insulators. A list of elements and their electron shells is shown in Appendix A.

ELECTRIC CHARGE

As was shown earlier, there is a movement of electrons between atoms. Some atoms gain electrons, while others lose electrons. When a transfer of electrons takes place, the normal distribution of positive and negative charges no longer exists and the object now has an electric charge (quantity symbol Q). The unit of measure for the magnitude of an electric charge is the *coulomb*. The symbol for coulomb is C. An object that has a charge of one coulomb has either gained (negative charge) or lost (positive charge) 6 240 000 000 000 000 000 electrons. This may be read as 6 quintillion, 240 quadrillion, and is usually written in terms of scientific notation as 6.24×10^{18}.

CURRENT FLOW

As we have seen earlier, electrons orbiting around the nucleus of an atom are attracted to the nucleus. The farther away from the nucleus, the smaller the force that bonds them to the atom. The valence electrons of a conductor such as copper or aluminum, which are loosely held, are able to drift from one atom to another in all directions. These electrons are called *free electrons*. This type of drift occurs in all conductors, but has little practical use. Figure 1-7 illustrates the random movement of the free electrons in a copper conductor.

In order to produce an electric current, the free electrons must be influenced to flow in the same direction through the conductor. This can be achieved by placing electrical charges on each end of the conductor, a negative charge at one end, and a positive charge at the other end. (See Fig. 1-8.) Between these charged ends of the conductor, there exists a

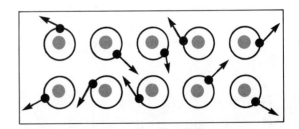

Figure 1-7 Random movement of free electrons in a conductor.

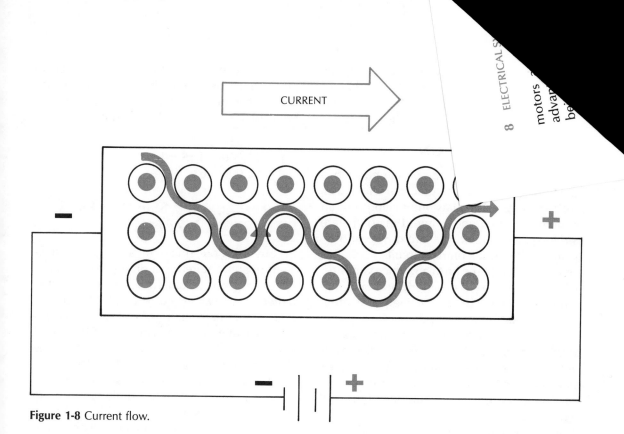

CURRENT

Figure 1-8 Current flow.

difference of potential energy.

If a copper wire were connected to a potential difference, as shown in Fig. 1-8, the free electrons in the copper wire would move in the same direction from the negatively charged end to the positively charged end. The negatively charged end repels electrons into the conductor and the positively charged end attracts electrons.

The forced movement of electrons constitutes a current flow quantity symbol, I. Electric current is measured by the number of electrons that pass by a point in a given time. A current flow of one electron is an extremely small unit. The greater the number of electrons flowing per second, the larger the current flow. A current flow of one coulomb per second through a conductor is called an *ampere*. Thus, the ampere is a unit of measure, and the symbol for ampere is A. In 1822, André Marie Ampère distinguished the difference between electromotive force and electric cur-

rent. (An electromotive force is the electrical potential that causes electric current to flow in a circuit.) The instrument used to measure current flow is the *ammeter*.

DIRECT CURRENT (dc)

Direct current is produced by a generator or a battery. Direct current flows in one direction only, from negative to positive, and does not vary over time. Direct current was widely used for many years before alternating current was known. Direct current is used in automotive vehicles and electronic and electrical circuits.

ALTERNATING CURRENT (ac)

Alternating current is also produced by a generator (called an alternator). Alternating current flows in one direction for a fixed period of time and then reverses itself to flow in the opposite direction for the same period of time. Alternating current is universally used in

and electric control circuits. It has
...tages that dc does not have: that of
...ing able to be "stepped up" or "stepped
down" and transmitted for long distances.

SCIENTIFIC NOTATION

Although the ampere is the basic unit of current flow, it is not always a practical unit to use, since currents as small as one millionth of an ampere are often encountered in solid state and integrated circuits that are now predominant in electrical control circuits. Subunits such as milliamperes (mA) and microamperes (μA) equal to one thousandth and one-millionth of an ampere, respectively, are often employed (Tables 1-1 and 1-2).

In the field of electricity, it is often necessary to use very large numbers in the millions as well as numbers less than one, such as one-millionth. These numbers may be written in their basic units, which would take time and greatly increase the chance of error.

An abbreviated method called "scientific notation," or simply "power of ten," expresses the numbers in a convenient form as powers of ten. To express a quantity in a scientific notation, move the decimal point until there is

one significant digit to the left of the decimal place, and then multiply the result by the appropriate power of ten to make the number equal to its basic value. For example:

$$328 = 3.28 \times 10^2$$
$$825\,000 = 8.25 \times 10^5$$
$$0.006 = 6 \times 10^{-3}$$

MULTIPLE AND SUBMULTIPLE UNITS OF CURRENT

In converting one unit of measure to an equivalent quantity stated in a different prefix, it is necessary to establish the relationship between the values. For example, to convert 0.125 to mA, (large units to small units), Table 1-2 shows that the prefix milli = 0.001, and 1 A = 1000 mA. The answer will be one thousand time greater than 0.125. Therefore, when changing large units to small units, the answer is made larger by multiplication. To make the conversion, simply multiply 0.125 by 1000. This can be done conveniently by moving the decimal point in 0.125 three places to the right. Hence, 0.125 A = 125 mA.

To convert 325.2 μA to A (small units to large units), the prefix micro = 0.000 000 1; and

Table 1-1 Metric prefixes

PREFIX	SYMBOL	MULTIPLIER	EXPONENT FORM
exa	E	1 000 000 000 000 000 000	10^{18}
pera	P	1 000 000 000 000 000	10^{15}
tera	T	1 000 000 000 000	10^{12}
giga	G	1 000 000 000	10^9
mega	M	1 000 000	10^6
kilo	k	1 000	10^3
hecto	h	100	10^2
deca	da	10	10
deci	d	0.1	10^{-1}
centi	c	0.01	10^{-2}
milli	m	0.001	10^{-3}
micro	μ	0.000 001	10^{-6}
nano	n	0.000 000 001	10^{-9}
pico	p	0.000 000 000 001	10^{-12}
femto	f	0.000 000 000 000 001	10^{-15}
atto	a	0.000 000 000 000 000 001	10^{-18}

Table 1-2 Changing units of current

TO CONVERT FROM	TO	MOVE DECIMAL POINT	EXAMPLE
A	mA	3 places to the right	0.125 A = 125 mA
A	μA	6 places to the right	2 A = 2 000 000 μA
mA	A	3 places to the left	47 mA = 0.047 A
μA	A	6 places to the left	325.2 μA = 0.000 325 2 A
mA	A	3 places to the left	517 mA = 0.517 A
A	mA	3 places to the right	702 A = 702 000 mA

1 A = 1 000 000 μA. The answer is made smaller by division. To convert 325.2 μA to A, the number 325.2 is divided by 1 000 000. This can be done conveniently by moving the decimal point in 325.2 six places to the left. Hence, 325.2 μA = 0.000 325 2 A.

POTENTIAL ENERGY DIFFERENCE

Electrical charges can be produced by a transfer of electrons from one object to another. When electrons leave one point, they produce a positive charge at that point; they show up at another point to produce a negative charge. Thus a *potential energy difference* is created between the two points. These charges can be produced by:

1. *Friction* (Fig. 1-9) charged objects are produced when two materials such as silk and a glass rod are rubbed together.

2. *Chemical Action* (Fig. 1-10) chemicals combining with certain metals initiate a chemical reaction that will transfer electrons which, in turn, produces a potential energy difference.

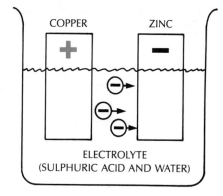

Figure 1-10 Voltage produced by chemical action.

3. *Pressure* (Fig. 1-11) a force applied to Rochelle salts, ceramics, or piezo-crystals will force the electrons out of orbit from one side of the material to accumulate on the other side, thus building up an electric charge.

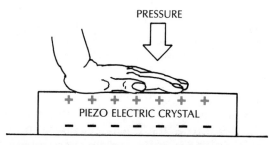

Figure 1-11 Voltage produced by pressure.

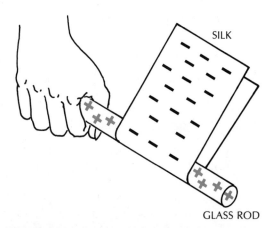

Figure 1-9 Voltage produced by friction.

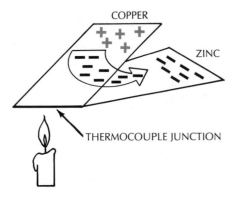

Figure 1-12 Voltage produced by heat.

4. *Heat* (Fig. 1-12) in a thermocouple (two dissimilar metals joined together), heat energy applied to the junction point will cause a transfer of electrons to take place. One metal readily gives up electrons and the other metal accepts the electrons, thus producing a potential energy difference.

5. *Light* (Fig. 1-13) when light strikes certain materials such as potassium, sodium, cesium, lithium, selenium, germanium, cadmium, and lead sulfide, these materials release electrons, thus producing a potential energy difference.

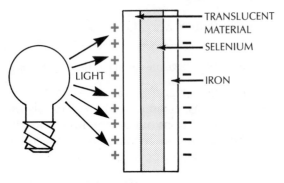

Figure 1-13 Voltage produced by light.
Light striking the selenium releases electrons to the iron, causing the outside plates to build up opposite charges.

6. *Magnetism* (Fig. 1-14) when a conductor is passed through a magnetic field, the force of the field causes the valence electrons in the conductor to move in one direction, produc-

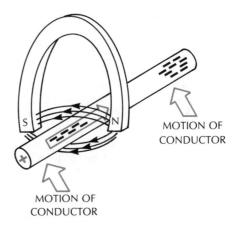

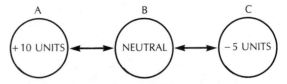

Figure 1-14 Voltage produced by magnetism.

ing a potential energy difference between the ends of the conductor.

The potential energy of charged bodies depends upon the amount of charge. The greater the concentration of electrons, the greater the force produced. The difference of potential energy between two charged objects is the electromotive force or electrical pressure, and is commonly called *voltage*, (quantity symbol V). The unit of measure is the volt, and the symbol for volt is V.

A potential difference of one volt exists between two points in a circuit if a charge of one coulomb gives up one joule of energy when it flows from one point to the other. The instrument that measures potential difference is called the *voltmeter*. Figure 1-15 illustrates the concept of potential energy.

Figure 1-15 Potential energy difference.

In Fig. 1-15, point A has a charge of +10 units and point B is neutral. Thus, the difference in potential energy from B to A is +10 units. However, going from point A to point B, the potential difference is −10 units. Likewise, going from point A to point C, the potential energy difference is −15 units, while going from point C to point B, the potential energy difference would be +5 units. When the difference in potential energy is a fraction of a volt, then millivolts (mV) and microvolts (μV) are used for measuring electromotive force. These small voltage values are encountered in solid state circuits, while large values of voltages such as a kilovolt (kV), equal to one thousand volts, are often employed in electrical circuits. Voltages larger than one kilovolt are not used except in special applications of electrical equipment such as generation and transmission equipment.

CONVERTING MULTIPLE AND SUBMULTIPLE UNITS OF VOLTAGE

Units of voltage measurement are changed in the same way that units of current are changed, by moving the decimal place either to the left or to the right.

RESISTANCE

All materials offer some opposition to the flow of current. This is a physical property that is determined by the manner in which the material is constructed. This opposition to current flow is called *resistance* (quality symbol is *R*). The unit of measure is the *ohm*, named after its discoverer, German physicist Georg Simon Ohm (1787–1854). The symbol for the ohm is the Greek letter (Ω) omega. A circuit has a resistance of one ohm, when one volt produces a current of one ampere.

Resistance is a property that ordinarily does not change, even though the voltage in a circuit may vary. Electrical devices with different values of resistance are connected in a circuit to control the flow of current. Since resistance provides opposition to the flow of current, it can be said that *the current in a circuit is inversely proportional to the resistance in a circuit*. As the resistance in a circuit is increased, the current decreases, or when the resistance is reduced, the current increases. (This statement will be combined with additional details to form Ohm's law, which will appear later in this chapter.)

Most of the time resistance values may be expressed in ohms. But, as we found in measuring volts and amperes, the basic unit is not always the most practical unit.

CONVERTING MULTIPLE AND SUBMULTIPLE UNITS OF RESISTANCE

Units of resistance are changed in the same manner as units of current or voltage, by moving the decimal point either to the left or to the right.

RESISTORS

Resistors can be classified into two groups, wire-wound and composition. Wire-wound resistors are used in circuits with large currents, or circuits that require a high degree of resistance accuracy, as in test equipment applications. Composition resistors are constructed with powdered carbon. High values of resistance can be constructed at a low cost. These resistors cannot be made to a close tolerance, as can wire-wound resistors. Composition resistors are used in circuits with small currents, where close tolerances are not required.

RESISTOR COLOUR CODE

Resistance values of resistors can be quickly determined by means of a simple colour code. Appendix B explains the colour coding of resistors.

FACTORS AFFECTING THE RESISTANCE OF A CONDUCTOR

The resistance of a conductor depends not only on the material of which it is made, but

Table 1-3 Specific resistance and temperature coefficient of common conductors for solid wire 1 mm in diameter and 1 m long at 20°C

MATERIAL	SPECIFIC RESISTANCE OHMS/METER	TEMPERATURE COEFFICIENT PER DEGREE C PER OHM AT 20°C
silver	0.016	+ 0.003 8
copper	0.017	0.003 93
gold	0.023	0.003 4
aluminium	0.028	0.003 9
tungsten	0.056	0.004 5
brass	0.070	0.001 0
nickel	0.086	0.006 0
iron	0.096	0.005 5
steel	0.165	0.002 0
lead	0.218	0.003 9
manganin	0.435	0.000 03
konstantin	0.485	0.000 008
mercury	0.952	0.000 7
nichrome	1.115	0.000 16
zinc	24.0	0.004 0
carbon	36 300.0	– 0.000 5

also on its length, cross-sectional area, and temperature. The resistance of a material 1 m in length having a cross-sectional area of 1 mm² at 20°C is a constant for that material. This property is called *resistivity* or specific resistance. For example, the resistivity of annealed copper is 0.017. Table 1-3 lists the specific resistance and temperature coefficient of common conductors.

The resistance of a conductor is directly proportional to its length. A 300-m copper wire with 1-mm² cross-sectional area and at 20°C will have a resistance of $300 \times 0.017 = 5.1\Omega$.

In a conductor with a large cross-sectional area, the number of free electrons is greater than in a wire of small cross-sectional area. Thus, the greater the diameter of conductor, the lower its resistance. If the diameter of a wire is doubled, its resistance is reduced by half.

The resistance of a conductor at the standard temperature or 20°C may be calculated using the following formula:

$$R = \frac{\rho\ell}{a} \qquad (1\text{-}2)$$

where R = resistance in ohms
ℓ = conductor length in metres
a = cross-sectional area in mm²
ρ = resistivity of conductor in ohm metres at 20°C

Example 1-1. Find the resistance of a copper conductor 5 km in length and 4 mm in diameter.

Solution:

$$R = \frac{\rho\ell}{a}$$

$$R = \frac{0.017 \times 5000}{3.14 \times 2^2}$$

$$R = \frac{0.017 \times 5000}{12.56}$$

$$R = \frac{85}{12.56}$$

$$R = 6.76 \ \Omega$$

TEMPERATURE COEFFICIENT OF RESISTANCE

Most materials experience an increase in resistance when the temperature is increased. For example, a tungsten filament lamp has a much lower resistance when it is cold than it has when it is red hot; this is due to vibration of the atoms. The increasing temperature causes the atoms to vibrate at a faster rate, continually changing the spacing between the atoms and making it more difficult for the free electrons to pass through. Therefore, the resistance of a conductor increases with the rise in temperature. For example, in Table 1-3, the resistance of a copper conductor changes by a percentage of 0.003 93 for every degree change in temperature above or below 20°C. If the resistance increases as the temperature increases, such materials are said to have a positive temperature coefficient (PTC). Materials whose resistance decreases as the temperature increases have a negative temperature coefficient (NTC). Carbon is one substance that has a negative temperature coefficient. The carbon atoms' rhythmic vibrations tend to assist rather than obstruct the movement of the free electron.

The relationship between temperature and resistance may be expressed mathematically as follows:

$$R_2 = R_1[1 + \alpha(t_2 - t_1)]$$

where R_2 = resistance of conductor at t_2
  R_1 = resistance of conductor 20°C (1-3)
  α = temperature coefficient of conductor at 20°C
  t_2 = new level of temperature

Example 1-2: If a copper wire has a resistance of 15 Ω at 20°C, calculate its resistance at 90°C:

where $R_2 = R_1[1 + \alpha(t_2 - t_1)]$
  $R_2 = 15[1 + 0.003\ 93(90\text{-}20)]$
  $R_2 = 15(1 + 0.275)$
  $R_2 = 19.13\ \Omega$

CONDUCTOR MATERIALS

Motor windings, transformers, generators, transmission lines, and most electrical devices are constructed of electrical conductors made of copper or aluminium. A good conductor must not only have a low resistance, but it must be light weight, have a high tensile strength, and be economically priced. Materials such as gold and silver make good conductors as well, but are very costly.

Transmission lines must have a high tensile strength and at the same time light weight; therefore, a conductor used for this purpose must have an inner core made of steel, with a copper or aluminium outer layer to give it the light weight.

CONDUCTANCE

There are times when it is more convenient to consider how well a circuit is able to conduct current instead of how much resistance it offers. *Conductance* is the ease with which a circuit conducts current. It is the opposite of resistance. The less resistance, the higher the conductance. The symbol for conductance is the letter G. The unit of measure is the siemens (S), named after its inventor Ernst von Siemens. The formula for conductance is:

$$\text{Conductance} = \frac{1}{\text{Resistance}} \quad (1\text{-}4)$$

or

$$G = \frac{1}{R}$$

Example 1-3: Find the conductance for a resistance of 5 Ω.

Solution:

$$G = \frac{1}{R}$$

$$G = \frac{1}{5}\ S$$

or $G = 0.2\ S$

OHM'S LAW

Every circuit has three basic electrical properties: voltage, current, and resistance, which are related to each other. Dr. George Simon Ohm discovered the basic relationships between voltage, current, and resistance.

These fundamental relationships are:

1. When the voltage in a circuit is increased by a certain percentage but the resistance remains the same, the current flow in the circuit will increase by the same percentage.

2. When the voltage in the circuit remains the same but the resistance is increased by a certain percentage, the current flow in the circuit will decrease by the same percentage.

When the above two relationships are combined, we have the most basic law for electric circuits, which is called Ohm's Law. It states:

The current flowing in an electric circuit is directly proportional to the voltage and inversely proportional to the resistance.

Ohm's law expressed in the form of a mathematical relationship is:

$$\text{Current} = \frac{\text{voltage}}{\text{resistance}} \qquad (1\text{-}5)$$

or

$$\text{Amperes} = \frac{\text{volts}}{\text{ohms}}$$

or

$$I = \frac{V}{R}$$

where I is the current in amperes (A)
V is the voltage in volts (V)
R is the resistance in ohms (Ω)

Ohm's law, therefore, can be used in calculating circuit current, where the values of voltage and resistance are known.

Example 1-4: If the voltage in an electric circuit is 100 V, and the resistance is 20 Ω, what is the current flow?

$$I = \frac{V}{R}$$
$$I = \frac{100}{20}$$
$$I = 5\text{ A}$$

From the basic equation of Ohm's law, $I = \frac{V}{R}$, we can derive the formula for:

voltage $\qquad V = I \times R$

and $\qquad$ resistance $\qquad R = \frac{V}{I}$

Example 1-5: In an electric circuit, a current of 1.5 A flows through a resistance of 20 Ω. Find the value of the voltage applied to the circuit.

$$V = I \times R$$
$$V = 1.5 \times 20$$
$$V = 30\text{ V}$$

Example 1-6: What value of resistance is needed in a circuit that has an applied voltage of 8 V to produce a current flow of 0.004 A?

$$R = \frac{V}{I}$$
$$R = \frac{8}{0.004}$$
$$R = 2000\ \Omega$$

MEASURING CURRENT

Since an electric current is measured in amperes, the meter used for measuring current flow is called an *ammeter*. An ammeter measures the number of coulombs per second (amperes) flowing through the circuit. Figure 1-16 shows a simple circuit with an ammeter connected in the circuit.

To measure the current flow through the lamp, the ammeter must be connected in the circuit so that all the current that flows through the lamp must also flow through the meter. This is called a *series* connection. The current flows from the negative terminal of the source, enters the negative terminal of the meter, and flows out the positive terminal of the meter.

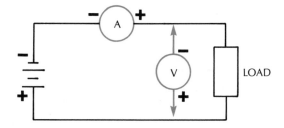

Figure 1-16 Measuring current and voltage.

From Ohm's law, we know that an increase in circuit resistance will cause the current flowing through the circuit to decrease. Therefore, the ammeter, which is connected directly into the circuit, must have a very low resistance so as not to alter the flow of current.

MEASURING VOLTAGE

Voltage is measured in volts, and the meter used for measuring potential energy difference is called a *voltmeter*. Figure 1-17 shows a voltmeter connected in a circuit, measuring the voltage across the load.

Note that the voltmeter forms a second path for current flow when connected to the circuit. Thus, the resistance of the voltmeter must be appreciably large so as not to draw too much current from the circuit and alter the actual circuit conditions. The voltmeter must be connected with the proper polarity in the circuit. The negative terminal of the voltmeter must be connected to the negative side of the supply. Similarly, the positive terminal of the voltmeter must be connected to the positive side of the supply.

MEASURING RESISTANCE

The instrument used for measuring resistance is called an *ohmmeter*. Unlike the two previ-

ous meters described, the ohmmeter contains its own internal power supply. To ensure accuracy and the safe use of this instrument, the voltage in the electric circuit being metered must be disconnected. The ohmmeter must be properly zeroed before measuring the resistance.

The discussion of meters in this chapter is merely to review the connections and uses of commonly used electrical measuring instruments. The design and construction of these instruments will be dealt with in greater detail in Chapter Six.

MEASUREMENT OF ELECTRIC POWER

Electrical power can be measured by an instrument called a *wattmeter*. The wattmeter (containing four terminals) is basically a voltmeter and an ammeter combined. The ammeter terminals are connected in series and the voltmeter terminals are connected in parallel with the circuit in which the power is being measured. Figure 1-17 shows how the wattmeter is connected in the circuit. Electrical power and the measurement of electric power will be studied in greater detail in Chapter Three, and the wattmeter in Chapter Six.

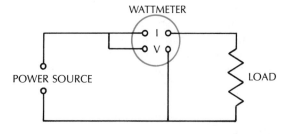

Figure 1-17 Measuring electrical power.

SUMMARY OF IMPORTANT POINTS

1. Energy is the ability to do work.
2. Matter is anything that has mass and occupies space.
3. All matter is made up of basic materials that are called elements.
4. Compounds are made up of two or more elements.
5. The atom is the smallest particle to which an element can be reduced.
6. The smallest particle into which a given compound may be divided without destroying its chemical properties is called a molecule.
7. The atom on one element differs from the atom of another element in the number of electrons, protons, and neutrons.
8. Law of charges: like charges repel and unlike charges attract.
9. An atom that has gained or lost an electron is called an ion.
10. Conductors have many free electrons, while insulators have few free electrons.
11. Free electrons moving in the same direction produce a current flow.
12. Potential difference can be created by the following: friction, chemical action, pressure, heat, light, magnetism.
13. Resistance is the opposition to current flow.
14. Ohm's Law: The current in a circuit is inversely proportional to the resistance and directly proportional to the voltage.
15. Factors affecting resistance: material, cross-sectional area, and temperature.
16. Conductance is the ease with which a circuit conducts current.
17. An ammeter is connected in series with the load.
18. A voltmeter is connected in parallel with the load.
19. Do not connect an ohmmeter in a circuit when the power is on.

REVIEW

1. Define the following:
   (a) element
   (b) molecule
   (c) compound
   (d) difference of potential energy
   (e) energy
2. Name six methods whereby a voltage may be produced.
3. State and explain Ohm's Law.
4. List the four factors that affect the resistance of a conductor.
5. What is meant by a coulomb of charge?
6. What is the difference between a conductor and an insulator?
7. Define what is meant by the term "semiconductor."
8. Describe what is meant by "potential energy difference" and give its unit of measurement.
9. Draw a diagram showing how you would connect an ammeter and voltmeter in a simple electric circuit with one load.
10. What are the units of measurement for:
    (a) current
    (b) voltage
    (c) resistance
11. List twelve metric prefixes along with their symbol and mathematical equivalents.
12. Define the following terms:
    (a) resistivity
    (b) potential energy
    (c) kinetic energy
13. What unit of measurement is used to define the cross-sectional area of a conductor?
14. Describe the basic structure of an atom.
15. If the resistance in a circuit is doubled and we wanted to maintain the current flow to its original value, what would have to be done to the source voltage?
16. If the resistance in a circuit were reduced by one-half, what will happen to the current if the supply voltage is not changed?

17. Why does a tungsten lamp draw more current at room temperature than at its operating temperature?
18. Explain how an ion is formed.
19. Define the term conductance and give the unit of measure and symbol.
20. What is matter? Give examples.
21. Describe the nature of an electric current.
22. Explain what is meant by the terms, (a) positive temperature coefficient, (b) negative temperature coefficient.

PROBLEMS

1. Convert the following units:
   (a) 54 mA   = _____ A
   (b) 5000 V   = _____ kV
   (c) 0.3 MΩ   = _____ Ω
   (d) 260 000 Ω   = _____ kΩ
   (e) 5563 μA   = _____ mA
   (f) 735 mA   = _____ A
   (g) 138 V   = _____ mV
   (h) 3.2 kV   = _____ V
   (i) 75 mV   = _____ kV
   (j) 8723 MV   = _____ kV
2. Find the cross-sectional area in mm² of the following conductors:
   (a) diameter = 4 mm
   (b) diameter = 2.5 mm
3. Find the diameter in millimeters of a wire that has an area of 225 mm².
4. Find the resistance of a copper conductor with a resistivity of 21.2×10^{-9} Ω/m. Its cross-sectional area is 115 mm² and its length is 1000 m.
5. A 5 Ω resistance is connected across a 15-V source. What is the value of the current flowing in the circuit?
6. An ammeter reading is 0.36 A when a lamp is connected across 110 V. Find the resistance of the lamp.

7. Find the voltage in a circuit if a resistance of 65 kΩ draws a current of 82 mA.
8. Calculate the missing value below:

	V	I	R
(a)	5 mV	___ A	2 Ω
(b)	___ V	12 mA	1.5 MΩ
(c)	22 kV	50 A	___ Ω
(d)	17 V	___ A	40 Ω
(e)	___ V	2.6 A	3.3 kΩ
(f)	27 mV	2.2 A	___ Ω
(g)	110 V	___ A	3 kΩ

9. A relay coil having a resistance of 2000 Ω operates an electric switch. In order to operate the relay, a current of 0.22 A is required. What voltage must be applied to the relay?
10. Calculate the resistance of an aluminium conductor which has a length of 1000 m and a diameter of 1.5 mm.
11. If the resistance of a copper wire is 300 Ω at 20°C, calculate the resistance of the wire when it is heated to 80°C.
12. A conductor has a resistivity of 0.027 42 Ωm, the cross-sectioned area is 120 mm², and the length is 1000 m. Find its resistance.
13. If a conductor of 500 m has a resistance of 2 Ω, what is the resistance of the same conductor that is 800 m in length?
14. How many metres of copper wire 1.4783 mm in diameter are required in a coil which is to have a resistance of 26.3344 Ω?
15. A copper wire has a resistance of 15 Ω at 20°C. Calculate its resistance at 90°C.
16. If constantin has a resistance of 9000 Ω at 20°C, what is its resistance when the temperature rises to 70°C?

CHAPTER
TWO

BATTERIES

WHAT IS A BATTERY?

The battery is a storage device for electricity, one of the most widely used sources of energy today. It is a self-contained energy converting device to which no other source of energy can lay claim. It converts chemical energy and makes available electrical energy as needed. When several cells are connected electrically (usually in series), they form a battery. The battery is a source of dc voltage that can be used for portable equipment, such as flashlights, automobiles, photographic cameras, radios and television, measuring instruments, tape recorders, calculators, hearing aids, paging devices, watches, heart pacemakers, and toys of all kinds. It is the most versatile energy source available.

The large generators that produce high voltages, in energy produced as opposed to energy consumed, are only about 40 percent efficient. But in an energy cell, nearly 90 percent of the chemical energy stored in the cell is converted to electrical energy. Today, in the midst of a fuel crisis that appears to be longterm, the full potential of energy cells, and of the utilization of natural chemical forces, is only now beginning to be realized. The development of transistors and microscopic circuits has opened up a whole new portable energy low-voltage world.

Batteries are classified into two categories: primary and secondary. The primary cell converts chemical energy into electrical energy by consuming the chemicals within the cell to initiate the action. The secondary cell also converts chemical energy into electrical energy, but it may be recharged when its chemical energy supply becomes depleted. By sending a current in a reverse direction through the cell, the chemical action is reversed, which restores the electrolyte and plates to their original chemical state.

Primary batteries are called dry cells. They are made out of a chemical paste and are used where a limited current is required. The secondary cell uses liquid chemicals and is called the wet cell. These are used where a heavy current is required.

HOW IT ALL BEGAN

The production of a voltage by a chemical means was brought about through the efforts of two Italian scientists, Luigi Galvani and Alessandro Volta.

In 1792, Luigi Galvani, a medical doctor, had removed dissected frogs' legs from a salt solution and then suspended them by means of a copper wire. He discovered that every time he touched the legs with an iron scalpel, they twitched. Galvani realized that electricity was produced but thought the frogs' legs produced the electricity.

In 1800, Volta disputed Galvani's electrical theory, and claimed that electricity was produced as a result of the chemical action between the copper wire, the iron scalpel, and the salt solution. Using the information he gained from this discovery, he built the first electrical battery, called the *voltaic* pile.

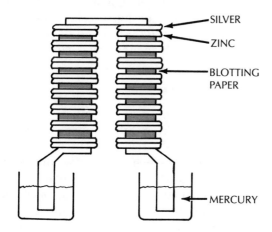

Figure 2-1 The voltaic pile.

Over the years, many combinations of metals, metal oxides, and chemicals were assembled. In the process, zinc, which had been used by Volta, was proven to be one of the best anode metals to use in the cell. In 1860, Georges Leclanché of France developed a wet cell which was the forerunner of the carbon cell. At the turn of the century, the rechargeable lead-acid battery was developed. To meet new application requirements, the nickel-cadmium, mercury-oxide, alkaline, and silver-zinc batteries were then developed. This is where the state-of-the-art remained for approximately 25 years. In 1970, the first long-life, primary lithium cell was developed. During this decade, other inventions have surfaced, such as the aluminum air battery and the reduction-oxidation chemical principle battery.

VOLTAIC CELL

A voltaic cell, which is a primary cell, consists of two dissimilar metals immersed in a chemical solution so that they do not touch each other. A chemical reaction takes place where chemical energy is converted into electrical energy. The two metal plates are called *electrodes* and the chemical solution is called the *electrolyte*.

The amount of voltage produced depends entirely on the composition of the two plates and the electrolyte. The distance between the plates, size of the plates, or the amount of electrolyte have no effect on the voltage of a battery. The electrolyte causes one electrode to lose electrons, making it positive, while developing a surplus of electrons or negative charge on the other electrode. The voltage of the cell is the difference in potential between the two electrodes.

The basic voltaic cell shown in Fig. 2-2 is made up of a positive copper (Cu) and negative zinc (Zn) electrodes immersed in an electrolyte of dilute sulphuric acid and water (H_2SO_4). The electrolyte breaks down into two positively charged atoms of hydrogen (H^{+2})

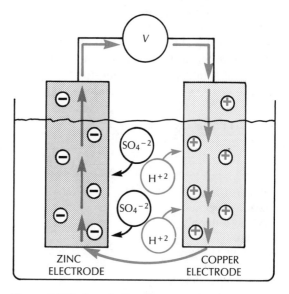

Figure 2-2 Basic voltaic cell.

and one sulphate ion which has two negative charges (SO_4^{-2}).

When the zinc electrode is immersed in the electrolyte, the sulphate ions (SO_4^{-2}) react with the zinc atoms to form molecules of zinc sulphate ($ZnSO_4$), which leaves an excess number of its electrons on the zinc electrode, giving it a negative charge. The positively charged hydrogen atoms drift to the copper plate, making it positive. The copper electrode becomes positively charged because the positive charges neutralize some of the free electrons in the copper electrode. Thus, the copper electrode will develop a deficiency of electrons, or a positive charge. Since the charges on the two electrodes are of opposite polarities, a difference in potential is produced. This voltage is approximately 1.1 V.

This chemical action takes place for a very short period, until the potential difference of 1.1 V is reached. The cell is then in a state of equilibrium and the chemical action ceases, as long as there is an open circuit between the copper and zinc electrodes.

If the copper electrode is connected externally through a load to the zinc electrode, electrons will flow from the negative zinc electrode through the external connection to the positive copper electrode. This electric current alters the chemical equilibrium between the electrolyte and the charged electrode. The chemical action will now continue to replenish the charges on the electrodes in order to maintain this equilibrium. This chemical action and the resulting current will continue until the zinc electrode is completely dissolved. Once dissolved, the cell is exhausted and will cease to produce a voltage. The chemical action in this cell is nonreversible; thus, this cell cannot be recharged.

The life of this cell will depend on the magnitude of the current and whether the cell is used intermittently or under continuous drain.

The voltage developed in any cell depends on the materials used for the electrodes and electrolyte. Table 2-1 lists some of the common primary cells and their characteristics.

POLARIZATION

When the hydrogen ions combine with the free electrons from the copper electrode, hydrogen gas is also released. This gas collects on the copper electrode. These hydrogen bubbles are nonconducting and increase the internal resistance of the cell by allowing fewer hydrogen atoms to reach the copper electrode. The hydrogen ions then collect around these bubbles and repel other hydrogen ions away from the copper electrode. This action, called *polarization*, increases the internal resistance of the cell and results in a decrease in the output voltage.

To remove the gas bubbles from the copper electrode, a depolarizing agent (potassium chromate) which has a surplus of oxygen in it is

Table 2-1 Primary cell comparison

	CARBON-ZINC	MANGANESE-ALKALINE	MERCURY-ZINC	SILVER-OXIDE	MAGNESIUM-CARBON	LITHIUM-CARBON
Nominal voltage	1.5 V	1.5 V	1.35 V	1.5 V	1.6 V	2.95 V
Load voltage	1.2 V	1.2 V	1.15 V	1.4 V	1.5 V	2.82 V
Wh/kg (energy density)	35	42	56	30	40	150
Shelf life storage temp. 22°C	1-2 years	2-3 years	3-4 years	3-4 years	10 years	10 years
Low temperature % of capacity at 40°C	20%	40%	40%	40%	50%	60%

added to the electrolyte. This depolarizing agent reacts with the hydrogen gas bubbles to prevent polarization. The reaction produces water which mixes with the electrolyte.

INTERNAL RESISTANCE

Any electrical conductor presents some resistance to the flow of electrons. All cells have an internal resistance made up of the cross-sectional area of the electrodes, the distance between electrodes, temperature, and the resistance of the electrodes and electrolyte. This internal resistance reduces the terminal voltage of the cell when current is flowing through a circuit. The internal resistance may be used to indicate the condition of the cell. When the cell is new and the chemical action is at its highest point, the internal resistance is low. As the cell is being utilized, the chemical action causes the cell to exhaust itself, resulting in an increase in the internal resistance. The internal voltage drop increases, resulting in a decrease in terminal voltage.

The internal resistance (R_i) may be determined by obtaining two voltage measurements, the open-circuit voltage (V_o) and closed-circuit voltage (V_c). In the open-circuit measurement, the terminal voltage (V_o) is measured with a high ohms-per-volt voltmeter so that it draws a negligible current from the cell. The closed-circuit voltage (V_c) is measured with a low resistance (R_L) of about 15 Ω placed across the terminals. The difference between the two measurements may be attributed to the internal resistance of the circuit. The internal voltage drop (V_i) may be calculated by the formula,

$$V_i = V_o - V_c \tag{2-1}$$

Thus, if an open-circuit measurement (V_o) was 1.5 V and the closed-circuit measurement (V_c) was 1.48 V, then the voltage loss (V_i) attributed to the internal resistance would be 1.50 − 1.48, which is 0.02 V. To find the value of the battery's internal resistance, it is neces-

sary to calculate the current flow (I_c) under load.

$$I_c = \frac{V_c}{R_L}$$

$$I_c = \frac{1.48}{15}$$

$$I_c = 0.099 \ A$$

The internal resistance (R_i) may now be calculated by using the value $I_c = 0.098$ A and the internal voltage drop $V_i = 0.02$ V.

$$R_i = \frac{V_i}{I_c}$$

$$R_i = \frac{0.02}{0.099}$$

$$R_i = 0.202 \ \Omega$$

DRY CELLS

The name "dry cell" is really a "misnomer" because the dry cell is not really dry internally. The electrolyte is in the form of a moist or semi-solid composition. These dry batteries are completely sealed to prevent leakage, and some types are rechargeable.

The most common dry cells manufactured today are the zinc-carbon, zinc-mercuric oxide, silver-oxide, alkaline, and the zinc-manganese dioxide cells. The magnesium and lithium cells are new arrivals which show considerable promise.

The basic principles by which dry cells produce electricity apply to all popular cells.

THE ZINC-CARBON CELL (CZn)

The zinc-carbon (Léclanché) cell (Fig. 2-3) is the most widely used dry cell because of its low cost and reliable performance. It is available in a variety of sizes, shapes and voltages, the nominal voltage being 1.5 V.

It has several advantages over the simple voltaic cell, in that the electrolyte is in the form of a paste composition, overcoming the hazard of spillage. A depolarizing agent is mixed in with the electrolyte, and the zinc container

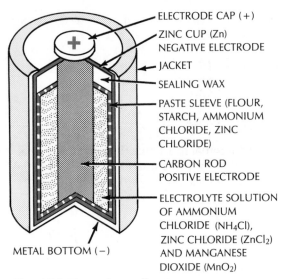

ELECTRODE CAP (+)

ZINC CUP (Zn)
NEGATIVE ELECTRODE

JACKET

SEALING WAX

PASTE SLEEVE (FLOUR, STARCH, AMMONIUM CHLORIDE, ZINC CHLORIDE)

CARBON ROD POSITIVE ELECTRODE

ELECTROLYTE SOLUTION OF AMMONIUM CHLORIDE (NH_4Cl), ZINC CHLORIDE ($ZnCl_2$) AND MANGANESE DIOXIDE (MnO_2)

METAL BOTTOM (−)

Figure 2-3 Zinc-carbon cell.

permits a larger area of interaction with the electrolyte. The electrolyte is made up of ammonium chloride (NH_4Cl), zinc chloride ($ZnCl_2$) and manganese dioxide (MnO_2) as the depolarizing agent.

The zinc cup serves a dual function; it contains the electrolyte, the depolarizing agent, and acts as the negative electrode of the cell. The inside of the zinc container is lined with a paste-like sleeve made of flour, starch, and ammonium chloride. This sleeve prevents carbon and other impurities from acting directly on the zinc electrode.

A carbon rod is inserted in the electrolyte, which serves as the positive electrode. The top of the cell is sealed with wax to prevent evaporation.

The chemical action of the dry cell is basically the same as the voltaic cell. The electrolyte decomposes into positive ammonium ions (NH_4^+) and negative chloride ions (Cl^-). The negative chloride reacts with the zinc cup, causing it to dissolve. The zinc atoms release electrons and produce positive zinc ions (Zn^+) that enter the electrolyte while the electrons build up a negative charge on the zinc cup. The positive zinc ions (Zn^+) repel the positive

hydrogen and ammonium ions to the carbon electrode, where they attract the free electrons from the carbon rod and produce a deficiency of electrons or positive charge on the carbon electrode. A voltage of 1.5 V builds up between the carbon rod and the zinc container. When a load is connected between the terminals, current flows from the zinc electrode, through the load, to the carbon electrode. If during discharge the zinc cup does not dissolve uniformly as current flows through the load, corrosion becomes excessive at certain spots, creating perforations. The electrolyte will leak out corroding materials outside the cell.

As in the voltaic cell, the chemical action results in a layer of hydrogen deposited on the carbon rod electrode. This process is called *polarization*. A mixture of carbon powder, manganese dioxide, and zinc chloride act as a depolarizing agent.

LOCAL ACTION

Impurities such as iron and carbon found in the zinc electrode material react chemically with the electrolyte to form small cells. These cells produce small electric currents around the zinc electrode even when there is no load connected to the dry cell. This is referred to as *local action*, and it causes the zinc container to dissolve and develop perforations, allowing the electrolyte to leak out. To prevent local action the zinc container is coated with mercury, which isolates the iron and carbon impurities from the electrolyte. This process of coating the zinc with mercury is known as *amalgamation*.

The zinc-carbon cell is highly efficient in converting the chemical energy into electrical energy as long as the discharge of the cell takes place at a low rate. The voltage discharge of a zinc-carbon cell is compared to other dry cells in Fig. 2-4.

The heavy-duty zinc-carbon, or zinc-chloride, cell is an improved version of the zinc-carbon cell; while similar in construction, the

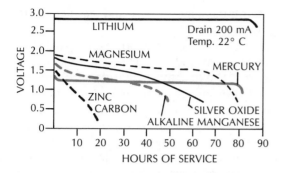

Figure 2-4 Primary cell (size D) voltage discharge.

major difference is that the electrolyte in the heavy-duty cell is essentially zinc chloride. This results in better performance characteristics, and the cell becomes drier during discharge, thus reducing the likelihood of perforations.

Both the zinc-carbon and zinc-chloride cells are recommended for intermittent service or medium-drain service, such as portable radios, flashlights, etc.

ALKALINE-MANGANESE CELL

The alkaline-manganese cell is commonly referred to as an *alkaline* cell, Fig. 2-5.

The alkaline cell is similar to the zinc-carbon

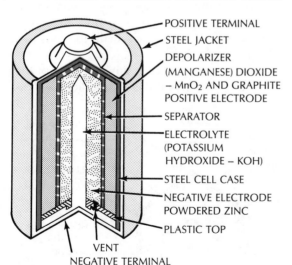

Figure 2-5 Alkaline-manganese cell.

cell in some respects. Both cells have negative electrodes made out of zinc. However, the positive electrode of the alkaline cell is made of manganese dioxide. (Manganese dioxide is a pure oxide that has a high oxygen content per unit of volume, which causes the chemical activity to increase substantially.) The basic difference between the cells, however, is in their construction and the composition of the electrolyte. In the alkaline cell, the negative electrode is inserted from the bottom in the centre of the cell and the positive electrode is the steel case. The negative electrode is fastened to the bottom of the container to form the negative terminal of the cell. The top terminal is fastened to the positive electrode (steel case) and serves as the positive terminal. Alkaline cells produce an output potential of 1.5 V (Table 2-1). The separator, negative and positive electrodes, are all soaked with electrolyte to produce maximum amount of energy with minimum internal resistance.

The alkaline cell produces a greater quantity of electrochemical energy than could possibly be obtained in a zinc-carbon cell of the same size. It has a longer shelf life and excellent low-temperature performance characteristics.

In the zinc-carbon cell, heavy current drains at continuous usage cause the cell to drop to a low value of efficiency. The alkaline cell is able to produce a continuous heavy current drain at high levels of efficiency. The voltage discharge characteristic of the alkaline manganese primary cell is shown in Fig. 2-4. Thus, alkaline cells are ideal for powering heavy drain applications, as found in motion picture cameras, radios, tape recorders, electric shavers, etc. They can be used to replace the zinc-carbon cell, but should not be mixed with them.

MERCURY CELL

The mercury cell is a miniaturized device that requires a low-rate continuous-drain application, such as in hearing aids, watches, heart pacers, and calculators. The mercury cell is

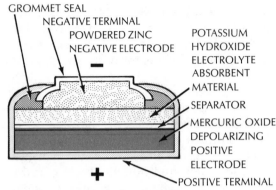

Figure 2-6(a) Mercury button cell.

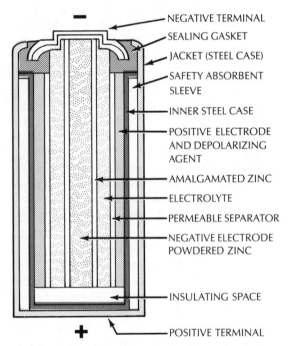

Figure 2-6(b) Cylindrical mercury cell.

produced in two different designs, the "button" or cylindrical shapes, shown in Fig. 2-6(a) and 2-6(b).

This cell has a high energy density, stable voltage (Fig. 2-4), and a long shelf life. It is unaffected by high temperatures, but does not perform well at low temperatures. It has very low internal resistance. Mercury cells produce an output potential of 1.35 V, in contrast to the

1.5 V of the zinc-carbon and alkaline cells (Table 2-1).

Mercury cells use mercuric oxide as a depolarizing agent and as the positive electrode material. The negative electrode is an alloy of powdered amalgamated zinc. A solution of potassium hydroxide saturated with zinc oxide acts as the electrolyte. A separator prevents the movement of solid particles in the cell from one electrode to another, thereby increasing the shelf life of the cell.

The position of the positive and negative terminals in a mercury cell is reversed from that of the carbon-zinc cell.

SILVER OXIDE CELL

The silver oxide cell will produce 1.5 V (Table 2-1), while the mercury cell will operate at 1.3 V. It is used in applications having low current drains, such as watches, calculators, and hearing aids. It uses silver oxide both as a depolarizing agent and positive electrode. The negative electrode is made up of zinc and the electrolyte is highly alkaline. Figure 2-7 shows the construction of a silver oxide cell.

Both electrodes are fitted into a short cylindrical can with the proper insulators and seals. The top of the container makes contact with the negative electrode. The bottom of the container makes contact with the positive electrode.

The discharge curve is similar to that of the alkaline cell. While it has a good energy den-

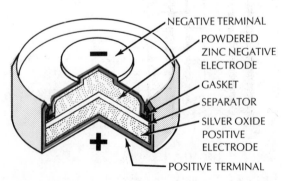

Figure 2-7 Silver oxide cell.

sity rating and a good high rate discharge characteristic (Fig. 2-4), it does have a low capacity, which restricts it to low drain applications.

The silver oxide cell can sustain severe abuse, such as short circuiting or exposure to high temperatures and humidity. It has an excellent shelf life and can deliver 90 percent of its nominal rating after one year of storage at 22°C.

Due to the higher cost of silver, silver cells are more expensive than existing cells of similar size.

Button cells are not interchangeable regardless of the similarities in their sizes and shapes, since each cell has a different drain characteristic.

MAGNESIUM CELL

Extremes in temperature and moisture have long been hazardous to dry cells. The magnesium cell has a high degree of reliability, stability, and performance capability under adverse conditions. It has a five-year shelf life, with excellent high drain and low temperature capabilities. Figure 2-8 shows a cross-sectional view of the magnesium cell.

A highly conductive carbon structure, formed in the shape of a cup, serves as the cell container. An integral centre rod is incorporated to reduce current paths. Liquids and gases cannot pass through the cup nor can the electrolyte corrode it.

The cell consists of the carbon cup, a cylindrical magnesium negative electrode, and paper separators. A mix consisting of manganese dioxide and magnesium bromide forms the electrolyte. The absorbent paper separator isolates the electrodes, but allows ionic conduction (movement of ions through the electrolyte) when the cell is under load. The electrolyte makes contact with the inside and outside surfaces of the negative electrode, the centre rod, and the inside surfaces of the cup. This design provides large carbon and negative electrode surface areas.

All magnesium cells have a voltage delay of a few seconds up to a few minutes in reaching operating voltages. This is the result of a protective coating which forms on the magnesium negative electrode and provides the excellent shelf life that characterizes these cells. The magnesium cell offers high rate and low temperature characteristics. The energy available from the magnesium cell varies with the applied load and temperature when discharged continuously. Figure 2-4 compares the discharge curves of the magnesium cell with other primary cells. The magnesium discharge curve is fairly flat, delivering between 1.4 and 1.6 V. Wide ranges in temperature have virtually no appreciable effect on the capacity of magnesium cells.

Magnesium cells are used in emergency transmitters, emergency lighting, and are an integral part of a standby weapons system.

THE LITHIUM CELL

The lithium cell is a relative newcomer. It was developed to serve a need that could not be met by other batteries. It is one of the most important breakthroughs in battery technology since the development of the zinc-carbon cell.

Lithium makes a far more reactive electrode

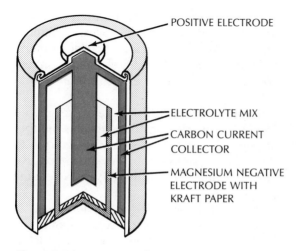

POSITIVE ELECTRODE

ELECTROLYTE MIX

CARBON CURRENT COLLECTOR

MAGNESIUM NEGATIVE ELECTRODE WITH KRAFT PAPER

Figure 2-8 Magnesium cell.

material than zinc, potassium, or sodium. Because of this, lithium cells develop a high voltage (3.04 V), Table 2-1.

Lithium is the lightest of all metals. In applications where weight and size are important factors, lithium cells get first consideration. The metal lithium offers the greatest energy density of any primary cell. Table 2-1 gives the comparison of energy densities of the various primary cells. Lithium cells offer nearly three times the energy density of mercury, four times that of alkaline cells, and six times that of carbon-zinc cells.

Lithium cells are produced in two basic designs, the cylindrical cell, Fig. 2-9(a), and the button cell, Fig. 2-9(b).

Because of its low internal resistance, the lithium cell can deliver a high rate of current flow for intermittent and continuous loads. It is capable of operation at temperatures from −55 to +75°C. Conventional cell performance deteriorates drastically at very low and very high temperatures. The voltage discharge characteristic is shown in Fig. 2-4.

Due to the fact that the cylindrical cell is hermetically sealed, preventing leakage, a ten-year shelf life at room temperature is possible with minimal capacity loss. During long periods of storage, a passive layer is formed over the lithium negative electrode. This layer prevents further reaction or loss of capacity. The problem of negative electrode corrosion, as it occurs in other conventional cells, is thus eliminated. This passive layer may cause a delay in the voltage buildup if high loads are applied after an extended storage period. However, it dissipates quickly and does not affect subsequent discharge.

The lithium cell consists of a lithium negative electrode. A separator, located between the negative and positive electrodes, prevents internal short circuits, and at the same time allows ions to pass freely between the electrodes. The positive electrode is made up of highly porous carbon. The electrolyte may be composed of lithium sulphur dioxide or lith-

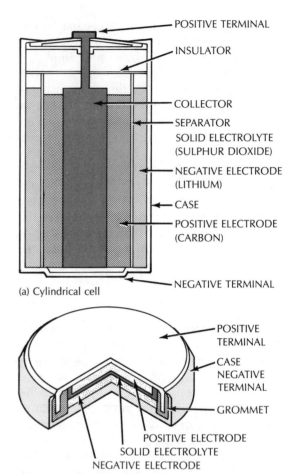

(a) Cylindrical cell

(b) Button cell

Figure 2-9 Lithium cells.

ium aluminium tetrachloride. Both solutions provide good stability, reliability, maximum shelf life, and safety. A current collector in the cylindrical cell provides an electrical connection between the positive electrode and terminal.

Because lithium cells contain pressurized contents, they are constructed with a number of safety features to prevent damage due to shorting. A safety-vent prevents excessive pressure buildup should the cell become overheated.

Lithium cells are used in a wide range of applications as a power source, some of which are as follows: telecommunications, instrumentation, business machines, process control, medical electronics, distress equipment, etc.

MULTIPLE-CELL ENERGY SOURCES

In the beginning of this chapter, it was shown that the term "battery" was used to mean a combination of cells. In most applications, the voltage and current that can be supplied by a single cell may not be enough to operate a device. The device may require either a higher voltage or higher current, and in some cases both. Thus, energy cells may be connected in several ways: series, parallel, or even series-parallel combinations. Cells connected in series provide a higher voltage, as shown in Fig. 2-10. Two 1.5-V cells connected in series provide a total voltage of 3 V.

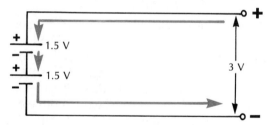

Figure 2-10 Cells in series.

The polarities all have to be in the same direction. If they are not, then the voltages will subtract. From Ohm's law, it was seen that current in a series circuit is the same; thus, the same current flows through all the cells in series-connected cells.

To increase the current rating, cells have to

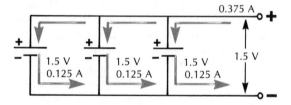

Figure 2-11 Cells in parallel.

be connected in parallel. Each cell supplies its own current and the sum of the currents from each cell will be the total current rating, (Fig. 2-11). The cells should be connected with the same polarity, otherwise the cells will supply current to each other (the higher voltage cell will supply current to the lower voltage cell).

Connecting cells in a series-parallel combination will provide a power source that will have both a higher voltage and current rating. By connecting four 3-V cells in a series-parallel combination, we can produce a high current 6-V energy source. (See Fig. 2-12.)

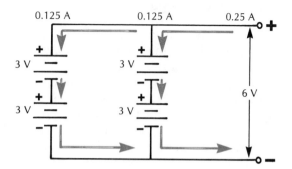

Figure 2-12 Cells in series-parallel combination.

SECONDARY CELLS

The secondary cell makes use of the energy of the charging current to build up energy-containing chemicals, which, in turn, produce electricity by chemical reaction. Thus, the secondary cell stores the energy until it is used (discharged), then it can be restored (recharged) to its original state by forcing an electric current from some other source through the cell in the opposite direction to that of discharge.

The most popular secondary cells are the lead-acid cells which produce a voltage of 2.2 V, and the nickel-cadmium with a voltage of 1.2 V. Other systems based on silver-zinc, silver-cadmium, and nickel-iron, which serve very specialized needs, will not be covered in this book.

LEAD-ACID CELLS

Since its development, the lead-acid battery has been the most widely used of all rechargeable storage battery systems. Although it has drawbacks in its large mass and size and the hazards of the sulphuric acid electrolyte, it also provides many advantages. It produces the highest voltage per cell of any commercial system, and the lowest cost per watt-hour of capacity. The lead-acid cell can be charged and discharged many times and possesses a relatively linear discharge curve.

CONSTRUCTION

The cut-away diagram of a lead-acid battery (Fig. 2-13) shows the component parts and how they are assembled. Two groups of coated lead plates are immersed in a dilute solution of sulphuric acid, which is the electrolyte. One group of plates forms the positive electrode and the other forms the negative electrode. Separators, made of wood, rubber, or glass, are used to prevent the plates from coming in contact with each other. These

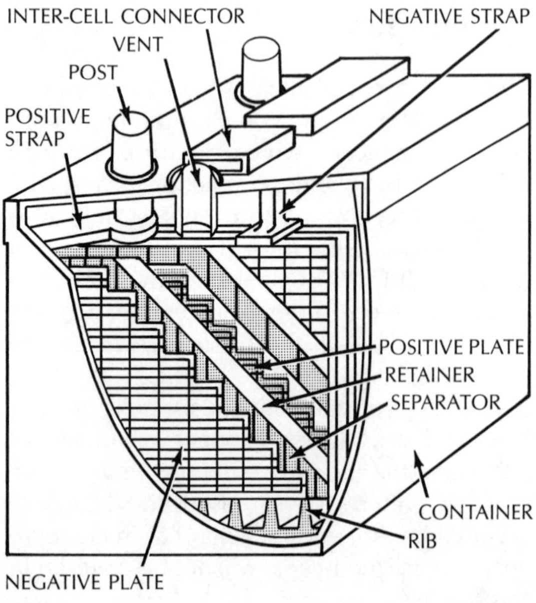

Figure 2-13 Construction of a lead-acid battery.

separators have a vertical groove to permit free circulation of the electrolyte around the positive plate.

Each cell container has a filler cap to permit the addition of distilled water that may be lost through evaporation. The filler cap has a vent hole that permits the gases to leave the cell while it is being charged.

At the bottom of the battery there is a ribbed section where the sediments are deposited.

The container and cell covers of portable lead-acid batteries are usually of hard rubber; whereas, for stationary batteries, the container is made of glass or plastic and the cover of hard rubber.

Each cell produces a voltage of 2 V. A lead-acid storage battery with three cells in series is rated at 6 V, while one with six cells in series is rated at 12 V.

BATTERY OPERATION

The uncharged cell with electrolyte in it contains two electrodes, both made of lead sulphate ($PbSO_4$). This uncharged cell cannot produce any electrical energy because the electrodes are not made of dissimilar metals. Before the lead-acid cell will work, both electrodes must be dissimilar metals so that the electrolyte will react with the electrodes. The voltage developed in a cell depends upon the types of metals and electrolyte used.

In the charged condition (Fig. 2-14(a)), the positive plate is made up of lead peroxide (PbO_2). The negative plate is made up of a different form of lead oxide called spongy lead (Pb). The electrolyte contains sulphuric acid (H_2SO_4); therefore, the relative density is at its maximum value. The sponge lead and lead peroxide are, in reality, different metals and they react in the battery in the same manner as unlike metals would react. When the positive and negative plates are submerged in the electrolyte, a solution of sulphuric acid and water, (H_2SO_4) (Fig. 2-14(b)), a voltage is produced and current will flow if the circuit is completed.

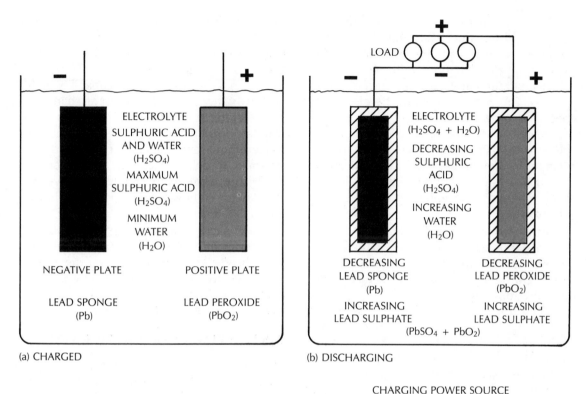

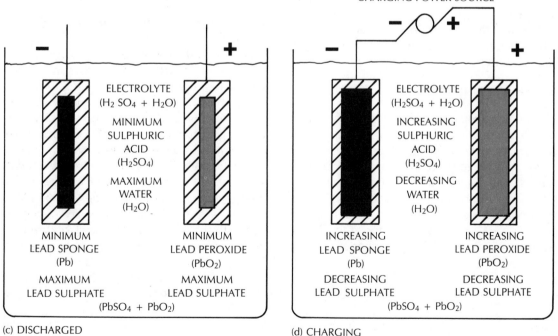

Figure 2-14 Chemical action of lead-acid cell.

During discharge (Fig. 2-14(b)), the sulphuric acid (H_2SO_4) reacts with the positive and negative plates and these are changed to lead sulphate ($PbSO_4$). Both have been changed to the same metal, so current ceases to flow. The water content of the electrolyte becomes progressively higher (acid content decreases). Thus, the relative density of the electrolyte will gradually decrease during discharge. In the completely discharged condition (Fig. 2-14(c)), there is minimum sulphuric acid (H_2SO_4) and maximum water (H_2O) in the electrolyte.

To recharge a battery, direct current from an external source must be applied to the battery terminals in the opposite direction to the current flow when the battery is discharging (Fig. 2-14(d)).

When the battery is being charged, a reverse action takes place. The acid that was being held in the sulphated plate material is returned to the electrolyte, thus raising its relative density. The discharged plates are changed back to their dissimilar forms. The positive plate is again pure lead peroxide (PbO_2) and the negative plate is pure lead (Pb). The chemical formula for a storage cell is shown below.

RATING OF BATTERY

Capacity is a term used to denote the ability of a fully charged battery to deliver a quantity of electricity over a specified period of time. This rating is expressed in *ampere-hours* ($A \cdot h$), and it can be obtained by multiplying the discharge in amperes by the discharge time in hours. A battery rated as $50 \, A \cdot h$ could deliver 5 A continuously for ten hours ($5 \times 10 = 50 \, A \cdot h$). This same battery could

also deliver 10 A for five hours ($10 \times 5 = 50 \, A \cdot h$), or even one ampere for 50 hours ($1 \times 50 = 50 \, A \cdot h$). As can be seen from the above examples, these ampere-hour ratings are not true, because the capability of the battery is affected by the rate of discharge.

Therefore, the battery industry has set a standard for rating automotive batteries. The ampere-hour rating of a battery is found from its ability to deliver a current for 20 hours at 27°C. Thus, a battery that is rated to deliver 5 A steadily for 20 hours at 27°C is said to have a $100 \, A \cdot h$ rating.

The capacity of a battery is affected by the following factors: the amount of material in the cell, the thickness of the plates, the rate of discharge, temperature, the quantity and concentration of the electrolyte, plate spacing, age and life history of the battery.

TESTING OF A LEAD-ACID BATTERY

The relative density of the electrolyte is measured with a hydrometer (Fig. 2-15). The hydrometer employs a calibrated float in a glass tube, into which a sample of the electrolyte can be drawn.

The strength of the electrolyte depends on the state of charge that the cell is in. As was seen, when the cell is fully charged, the electrolyte has a high content of sulphuric acid and a low content of water. When the cell is discharged, the reverse condition occurs: the sulphuric acid content is low and the water content is high. These changes are reflected in the mass of the electrolyte and can be measured in terms of its *relative density*. Relative density is the ratio of the density of a substance to that of water. (The mass of the electrolyte is compared with the mass of equal

$$\text{DISCHARGING} \longrightarrow$$
$$PbO_2 \quad + \quad Pb \quad + \quad 2H_2SO_4 \quad = \quad 2PbSO_4 \quad + \quad 2H_2O$$

$$\begin{array}{lllllll}
\text{Lead} & & & + & \text{Sulphuric} & = & \text{Lead} & & \\
\text{Peroxide} & + & \text{Lead} & & \text{Acid} & & \text{Sulphate} & + & \text{Water}
\end{array}$$
$$\longleftarrow \text{CHARGING}$$

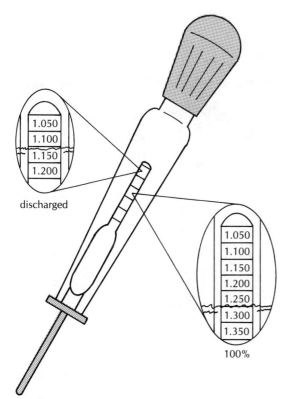

1.050
1.100
1.150
1.200

discharged

1.050
1.100
1.150
1.200
1.250
1.300
1.350

100%

Figure 2-15 The hydrometer.

volume of water at the same temperature.) Sulphuric acid has a greater density than water. Therefore, an electrolyte with a high acid content will have a higher density than an electrolyte with a lower acid content. The relative density of the electrolyte is an excellent indicator of the state of charge of a battery. Most automobile batteries will show a relative density of between 1.260 and 1.280 when fully charged and a reading of 1.160 when discharged.

BATTERY CHARGING

There are two basic methods that are used to charge batteries (both methods use dc current). They are the *normal charge* and the *fast* or *high-rate charge*. The normal charge provides a low rate of current flow and takes the longest time. The fast charge has a higher current flow, takes considerably less time, but batteries that are not in very good condition

can become damaged. This is because the chemical action is accelerated, causing the water to boil and the electrodes to deteriorate.

When a battery is being charged, hydrogen gas is released at the negative plate and oxygen gas at the positive plate. No smoking or electrical sparks should be permitted near the battery because of the possibility of an explosion.

MAINTENANCE OF LEAD-ACID BATTERIES

During normal operation, some water is lost from the electrolyte by evaporation. The level of the electrolyte should be kept above the top of the plates. Distilled water should be added at regular intervals. Sulphuric acid should not be added to the battery.

A battery that is allowed to remain at a low state of charge for excessive periods of time becomes "sulphated" and will not give normal service. This is a condition where a hard coating that forms on the negative electrodes during discharge becomes permanently set and cannot be removed in the normal manner by recharging. If a battery is to be stored for a period of time, check the level of the electrolyte and add water if necessary. Give the battery a full charge. Store the battery in the coolest and driest location available. Repeat the above procedure at three to four week intervals.

NICKEL-CADMIUM BATTERIES

The nickel-cadmium battery has been in use for more than 50 years. It is extremely reliable and may be recharged many times. It has a relatively constant discharge potential, a long life expectancy (over 20 years), and is maintenance-free.

The active material in the positive electrode is nickel hydrate. Cadmium is the active material of the negative electrode, and the electrolyte is a water solution of potassium hydroxide (KOH).

In a fully charged nickel-cadmium battery,

the nickel hydrate of the positive plate is highly oxidized, while the active material of the negative plate is metallic cadmium sponge. When the battery is discharged, the positive plate active material is reduced to a lower oxide, while the metallic cadmium sponge of the negative plate is oxidized.

Thus, the reaction consists mainly of a transfer of oxygen ions from one set of plates to the other, the electrolyte acting as transfer agent for the oxygen. As the electrolyte takes no part in the chemical reaction, its relative density does not change materially during charge or discharge. The average operating voltage of the cell under normal discharge conditions is about 1.2 V.

Nickel-cadmium cells are available in either sealed or vented types. It is the sealed type that is the most popular because it is maintenance-free and can be used in any position without danger of electrolyte leakage.

During the charge cycle, nickel-cadmium batteries generate gas. Oxygen is generated at the positive electrode and hydrogen is formed at the negative electrode. These gases are vented through a valve at the top of the cell (Fig. 2-16). In the sealed-type cell (Fig. 2-16(b)), the hydrogen gas is suppressed at the negative electrode. The oxygen formed at the positive electrode reacts with the cadmium of the negative electrode. This basic process is used to regulate gas pressure during overcharge.

The nickel-cadmium cell is available in three basic designs; the coiled-electrode sintered-plate cylindrical cell (Fig. 2-16(a)), the button cell (Fig. 2-16(b)), and the pocket plate (Fig. 2-16(c)).

Nickel-cadmium batteries have excellent low-temperature performance, light mass, high energy density and are long-operating. They can be used as direct replacements for primary cells in nearly all devices. They are used in many cordless electrical devices in consumer, industrial, commercial and scientific applications.

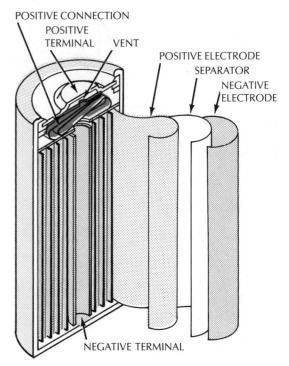

Figure 2-16(a) Cylindrical (coiled-electrode) cell.

One disadvantage of the nickel-cadmium cell is that it only has moderate charge retention. About one-quarter of the charge is lost after storage for a month at room temperature. Thus, after a lengthy storage period, nickel-cadmium batteries must be recharged before using. Another disadvantage is the high initial cost of the nickel-cadmium cell as compared to other cells. However, over a long period of

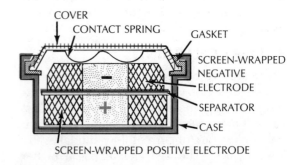

Figure 2-16(b) Nickel-cadmium button cell.

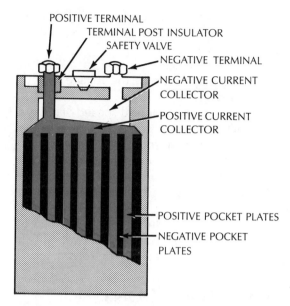

POSITIVE TERMINAL
TERMINAL POST INSULATOR
SAFETY VALVE
NEGATIVE TERMINAL
NEGATIVE CURRENT COLLECTOR
POSITIVE CURRENT COLLECTOR
POSITIVE POCKET PLATES
NEGATIVE POCKET PLATES

Figure 2-16(c) Pocket plate rectangular cell.

time the cost per hour of use is less.

The electrolyte used in nickel-cadmium batteries (potassium hydroxide) is highly corrosive. The same safety precautions should be utilized as with sulphuric acid.

SAFETY PRECAUTIONS

Cells can be damaged through improper use or handling. For example, if cells were wrapped in aluminum foil during storage, they could be short-circuited. Storage of primary cells at high temperatures can cause gas to be produced within the cell. If there is no proper venting protection, this would cause bulging in the cell. Cells should be stored in a dry place at normal room temperatures.

Discharging a cell at a higher rate than its current rating, or shorting the cell, can produce excessive internal heat, which will produce gas.

All cells connected in series should be replaced at the same time. Mixing old and new cells will not only reduce the overall service life, but will cause leakage.

Cells should not be disposed of in a fire, as they may explode. Attempting to recharge a nonrechargeable cell will cause the cell to rupture.

The electrolytes in secondary cells are corrosive (alkaline) or acidic (sulphuric acid) and can cause severe burns. If acid comes into contact with the skin, the affected area should be washed as soon as possible with water and covered with a petroleum jelly.

If the alkaline electrolyte from a nickel-cadmium battery is spilled on the skin, wash the affected area with water, vinegar, or a boric acid solution.

SUMMARY OF IMPORTANT POINTS

1. A cell is an electrochemical device which converts chemical energy into electrical energy.
2. Alessandro Volta made the first chemical cell with two different electrodes, zinc, and copper, immersed in a solution of acid and water, called the electrolyte.
3. The chemical action that takes place causes one electrode to produce a positive charge and the other electrode a negative charge. A potential energy difference is produced between the electrodes.
4. The voltage produced by a cell is determined by the metals used for the electrodes and the chemical reaction; the current capacity is determined by the size.
5. The primary cell converts chemical energy into electrical energy and it cannot be recharged. It consumes the chemicals as it produces electrical energy.
6. The secondary cell has to be charged with electrical energy before it can produce electrical energy. The secondary cell can be recharged. For this reason it is called a storage cell.
7. The zinc-carbon cell is the most common

type of primary cell. The positive electrode is made of carbon and the negative electrode consists of the zinc cup.

8. The zinc container of the zinc-carbon cell is amalgamated to reduce local action.

9. Heavy-duty zinc-carbon cells are made of leakproof steel jackets.

10. Alkaline cells use alkaline electrolytes, produce the same voltage as zinc-carbon cells, require less attention, and have a longer life.

11. The mercury cell is a primary cell that is basically alkaline. It has a long life, with a constant voltage of 1.35 V.

12. Silver oxide-alkaline-zinc batteries are used in miniature power sources. They provide a low current and constant higher voltage than other similar size batteries. They offer a long service life and have good low temperature characteristics.

13. Under adverse conditions, the magnesium cell performs with a high degree of reliability. It has a long shelf life, with excellent high drain and low temperature capabilities.

14. The lithium cell has a nominal cell voltage of 2.9 V and energy density nearly three times that of mercury and four times that of alkaline cells. It has extremely long shelf life.

15. The nickel-cadmium cell is a dry cell that is rechargeable. It has an output voltage of 1.25 V, and is used in many cordless applications.

16. The lead-acid cell is the most common type of secondary cell that can be recharged. The chemical process is reversed by passing a reverse current through the cell.

17. The relative density of a lead-acid cell is measured with a hydrometer.

18. Acid should never be added to any secondary cell, only distilled water.

19. When two or more cells are connected in series, the individual voltages add. When two cells are connected with opposing polarities, the voltages subtract.

20. Two cells connected in parallel increase the current capacity, but the voltage remains the same.

21. Connecting cells in a series-parallel combination will increase the voltage and current capacity.

22. Cells should never be disposed of in a fire.

23. Cells should be stored in a dry place at room temperature.

24. Alkaline electrolytes are corrosive and acidic and can cause severe burns.

25. Skin areas affected by an alkaline electrolyte should be washed with water, vinegar, or boric acid solution. Areas affected by acidic electrolyte should be washed with water and covered with petroleum jelly.

REVIEW QUESTIONS

1. Define the term "battery."
2. Describe the difference between a primary and secondary battery.
3. What are the essential parts of a cell?
4. Describe the difference between a wet cell and a dry cell.
5. Describe how a wet cell develops a voltage.
6. List the factors that affect the internal resistance of a cell.
7. Name four types of dry cells.
8. Distinguish between local action and polarization in a primary cell.
9. Define the term "amalgamation."
10. Draw a diagram of a common zinc-carbon cell and label the various parts.
11. If a white substance appears on a dry cell, what does it probably indicate?
12. Define (a) battery capacity
           (b) shelf life
13. How does an alkaline primary cell differ from a zinc-carbon cell?
14. Name two types of secondary cells.
15. Make a simplified diagram of a lead-acid cell. Label and identify all parts.
16. What are the electrodes of a lead-acid cell composed of when discharged?
17. Does an uncharged lead-acid cell meet

the requirements of a battery? Explain.

18. Explain fully the process in charging and discharging a lead-acid cell.
19. How is the relative density of the electrolyte related to the charge of a lead-acid cell?
20. What effect is increased in a lead-acid battery by an excessive rate of charge or discharge?
21. What is periodically added to the electrolyte of a battery?
22. What is the purpose of vent holes in secondary cells?
23. Name the active materials in the nickel-cadmium cell?
24. List the advantages of the nickel-cadmium cell.
25. List two basic methods used for charging secondary cells. Explain the advantage and disadvantage of each method.
26. State three advantages and three disadvantages of lead-acid cells.
27. Why is it good practice to keep storage batteries fully charged at all times?
28. List some of the safety precautions that should be taken when handling and charging lead-acid cells.
29. Should acid or water be added to a storage battery, and where should the level of the liquid be kept?
30. When not in use, storage batteries can be stored; explain the procedure to be followed.

CHAPTER
THREE

DIRECT CURRENT CIRCUITS

The analysis and applications of the rules governing series, parallel, and complex direct current circuits are discussed in this chapter. This discussion includes power, energy, and line loss as related to circuit properties.

The series circuit is the simplest electric circuit to analyze. In this circuit, devices are connected end-to-end to form a single continuous path.

The parallel-circuit connection is very common in residential wiring systems. In this type of circuit, there is more than one current path. In a parallel circuit, the devices are connected to form independent or separate current branches.

Almost all practical electric circuits are combinations of series and parallel connections. These series-parallel circuit combinations are called *complex circuits*. In this circuit, some devices are connected in series and others are connected in parallel. The rules for solving these types of circuits are to a large extent an expansion of the rules governing series and parallel circuits. Therefore, a sound understanding of series and parallel properties is important. Also, many of the concepts discussed in this chapter will be used in other chapters to analyze new electrical systems and circuit devices.

SERIES CIRCUITS

RESISTANCE

It was shown in Chapter One that if the temperature, cross-sectional area, and resistivity of a conductor remain constant, its resistance changes only if its length does. Therefore, if a circuit's length is increased, its resistance will increase. Connecting resistances in series has the same effect as increasing the length of the circuit, because there is only one current path in this circuit. In fact, the resistance of the total circuit is equal to the sum of the resistances connected in the series. In Fig. 3-1, the total resistance is:

$$R_T = R_1 + R_2 + R_3 \qquad (3\text{-}1)$$

CURRENT

In a series circuit, there is only one current path. As a result, the same amount of current must flow through every part of a series circuit. Therefore in Fig. 3-1, the current leaving the negative terminal of the source will flow through R_1, R_2, R_3, and back to the positive terminal of the source. This can be checked experimentally by connecting ammeters at various points in the circuit.

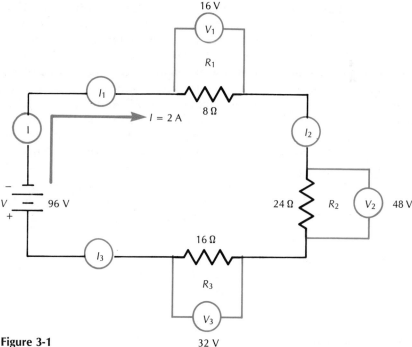

Figure 3-1

The current rule for a series circuit may be expressed mathematically as:

$$I_T = I_1 = I_2 = I_3 = \ldots \qquad (3\text{-}2)$$

The value of the current flow in a series circuit is dependent on the source voltage and the total circuit resistance. As long as this voltage and resistance remain constant, the current flow will be constant. Changing either the voltage or the resistance will change the current value to a new level. The rule, however, remains that whatever current leaves the negative source terminal stays the same throughout the series and enters the positive source terminal.

VOLTAGE

The source voltage is the electrical pressure that is responsible for causing a current flow through a closed circuit. If the circuit resistance were increased, the source voltage would have to be increased in order to maintain the same level of current flow. Since the same current flows in every part of a series circuit, a larger load resistance in the series circuit would require a larger part of the source voltage than would a smaller load resistance. Therefore, in Fig. 3-1, since the series resistance R_2 is three times larger than R_1, the voltage V_2 will be three times larger than V_1 in order to cause the same current flow.

The voltage across a resistor is called a voltage drop. Since the entire source voltage is used to cause the current flow, it follows that the sum of the voltage drops in a series circuit must equal the source voltage. This is known as Kirchhoff's Voltage Law, and may be expressed mathematically as:

$$V_T = V_1 + V_2 + V_3 + \ldots \qquad (3\text{-}3)$$

This property can be verified by measuring the source and voltage drops, as in Fig. 3-1.

$$\begin{aligned} V_T &= V_1 + V_2 + V_3 \\ 96\ V &= 16\ V + 48\ V + 32\ V \end{aligned} \qquad (3\text{-}4)$$

Voltage equations such as equations (3-3) and (3-4) can be written as:

$$(V_1 + V_2 + V_3) - V_T = 0 \qquad (3-5)$$

The voltage equation in this form is often referred to as a *voltage loop equation*. In a series circuit, there is only one closed loop so there will be only one loop equation.

The loop equation for any closed loop, as in Fig. 3-2, can be derived as follows. First, assign the polarities of the voltages in the loop.

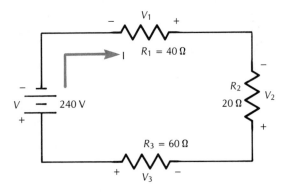

Figure 3-2

Next, go through the loop, starting at one point and, in either a clockwise or counterclockwise direction, proceed to each voltage drop, giving each voltage drop the sign of the polarity first seen for that voltage drop, until the entire loop is traversed. Finally, set the equation equal to zero.

$$V_T - V_1 - V_2 - V_3 = 0 \qquad (3-6)$$

OHM'S LAW

This law was discussed in Chapter One. It establishes the relationship between current (I), voltage (V), and resistance (R) in a circuit. The three forms of this law are repeated below because of their importance in analyzing electrical systems. The student will be using them repeatedly when working with electricity.

$$V = IR \qquad I = \frac{V}{R} \qquad R = \frac{V}{I} \qquad (3-7)$$

Ohm's law must be properly applied to avoid errors in calculations. The law may be applied to an entire circuit or to any part of a circuit. However, it is important that when used for an entire circuit the values of the voltage, current, and resistance must be those of the entire circuit. When the law is used for only a part of a circuit, the values of the voltage, current, and resistance must be only those for that particular part of the circuit.

Example 3-1: In the circuit given in Fig. 3-2, find the total circuit resistance, circuit current, and the voltage drop across each resistor.

Solution:

$$R_T = R_1 + R_2 + R_3$$
$$R_T = 40\ \Omega + 20\ \Omega + 60\ \Omega = 120\ \Omega$$

$$I = \frac{V_T}{R_T} = \frac{240\ V}{120\ \Omega} = 2\ A$$

$$V_1 = I \times R_1 = 2\ A \times 40\ \Omega = 80\ V$$
$$V_2 = I \times R_2 = 2\ A \times 20\ \Omega = 40\ V$$
$$V_3 = I \times R_3 = 2\ A \times 60\ \Omega = 120\ V$$

Example 3-2: In the circuit given in Fig. 3-3, determine the circuit current, the resistances R_2 and R_3, and the source voltage.

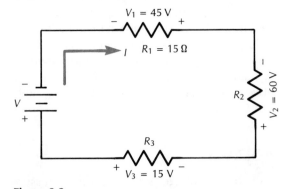

Figure 3-3

Solution:

$$I = \frac{V_1}{R_1} = \frac{45\ V}{15\ \Omega} = 3\ A$$

$$R_2 = \frac{V_2}{I} = \frac{60 \text{ V}}{3 \text{ A}} = 20 \text{ }\Omega$$

$$R_3 = \frac{V_3}{I} = \frac{15 \text{ V}}{3 \text{ A}} = 5 \text{ }\Omega$$

$$V_T = V_1 + V_2 + V_3$$
$$V_T = 45 \text{ V} + 60 \text{ V} + 15 \text{ V}$$
$$V_T = 120 \text{ V}$$

Example 3-3: Write the voltage loop equation for Figs. 3-2 and 3-3, and verify the statement that the sum of the voltages in a closed loop is equal to zero.

Solution:

from Fig. 3-2,
$$V_T - V_1 - V_2 - V_3 = 0 \text{ V}$$
$$240 \text{ V} - 80 \text{ V} - 40 \text{ V} - 120 \text{ V} = 0 \text{ V}$$

from Fig. 3-3,
$$V_T - V_1 - V_2 - V_3 = 0 \text{ V}$$
$$120 \text{ V} - 45 \text{ V} - 60 \text{ V} - 15 \text{ V} = 0 \text{ V}$$

Both equations satisfy the statement that the sum of the voltages in a closed loop is equal to zero.

PARALLEL CIRCUITS

VOLTAGE

In a parallel circuit, there is more than one current path. These paths are often called *parallel branches*. In Fig. 3-4, there are three parallel branches connected across the source voltage. It is obvious from this connection that each of the branch voltages is equal to the source voltage.

This may be represented mathematically as:

$$V_T = V_1 = V_2 = V_3 \qquad (3\text{-}8)$$

In Fig. 3-4, there are six different closed loops. The voltage loop equations are:

$$V_T - V_1 = 0 \text{ V} \qquad (3\text{-}9)$$
$$V_T - V_2 = 0 \text{ V}$$
$$V_T - V_3 = 0 \text{ V} \quad V_1 - V_2 = 0 \text{ V}$$
$$V_1 - V_3 = 0 \text{ V}$$
$$V_2 - V_2 = 0 \text{ V}$$

A close comparison of these loop equations will again indicate that in a parallel circuit the voltages across the parallel branches are equal to each other. This property of parallel circuits is expressed mathematically as:

$$V_T = V_1 = V_2 = V_3 = \ldots \qquad (3\text{-}10)$$

PARALLEL CIRCUITS

CURRENT

In an electric circuit, the same number of electrons leaving one terminal of the source must return to the other source terminal. In Fig. 3-5, the current flow from the source to point A is the total circuit current I_T.

At point A, this current divides into I_1 and I_A, and at point B, the current I_A is further divided into I_2 and I_3. These two current divisions at point A and point B may be stated mathematically as:

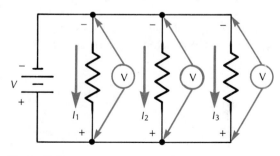

Figure 3-4

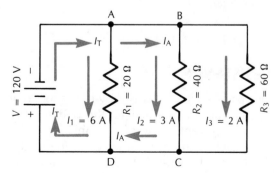

Figure 3-5

$$I_T = I_1 + I_A \quad \text{and} \quad I_A = I_2 + I_3 \qquad (3\text{-}11)$$

At point C, the currents I_1 and I_2 recombine to produce I_A, and at point D, the currents I_A and I_1 recombine to again produce the total current I_T, which is returned to the positive source terminal. The mathematical expression for the currents at each of these two points, C and D, is the same as given in Eq. 3-11.

Points such as A, B, C, and D in an electric circuit where the current either divides or the currents come together are called *electrical nodes* or *junctions*. Two important circuit properties may be summarized from what we have just seen.

First, the total current entering a node is equal to the total current leaving the node. This is referred to as Kirchhoff's Current Law.

Second, the source current is equal to the sum of the branch currents.

It was shown that for a parallel circuit, the same voltage exists across each of the parallel branches. Therefore, using Ohm's law, the branch currents in Fig. 3-5 may be found by using the equations:

$$I_1 = \frac{V_T}{R_1} \qquad I_2 = \frac{V_T}{R_2} \qquad I_3 = \frac{V_T}{R_3} \qquad (3\text{-}12)$$

From these equations, it should be noted that current to the parallel branches will divide in accordance to their resistance. That is, the branch with the largest resistance will allow the smallest current flow through it. This is illustrated in Fig. 3-5. In this circuit R_3 is three times larger than R_1, so the current I_3 will be three times smaller than I_1.

PARALLEL CIRCUIT

RESISTANCE

In a parallel circuit it was shown that the total circuit current is equal to the sum of the branch currents. Therefore, obviously the total circuit current is larger than any of the single branch currents. As a result, by applying Ohm's law for resistance, $R_T = \frac{V_T}{I_T}$, it can be

reasoned that the total resistance of a parallel circuit is smaller than the resistance of any of the parallel branches. Also, if the number of parallel branches is increased, the total circuit resistance will decrease.

The mathematical expression for calculating the total resistance of a parallel circuit may be found as follows. From Kirchhoff's current law,

$$I_T = I_1 + I_2 + I_3 + \ldots$$

and applying Ohm's law for current,

$$\frac{V_T}{R_T} = \frac{V_T}{R_1} + \frac{V_T}{R_2} + \frac{V_T}{R_3} + \ldots,$$

therefore,

$$\frac{1}{R_T} = \frac{1}{R_1} + \frac{1}{R_2} + \frac{1}{R_3} + \ldots \qquad (3\text{-}13)$$

This method of determining total resistance is sometimes called the *reciprocal method*. Note that it involves first finding $\frac{1}{R_T}$. This must then be inverted to give the value for R_T. For example, in Fig. 3-5,

$$\frac{1}{R_T} = \frac{1}{R_1} + \frac{1}{R_2} + \frac{1}{R_3}$$

$$\frac{1}{R_T} = \frac{1}{20} + \frac{1}{40} + \frac{1}{60}$$

$$\frac{1}{R_T} = \frac{6}{120} + \frac{3}{120} + \frac{2}{120}$$

$$\frac{1}{R_T} = \frac{11}{120} \text{ S (S = siemens)}$$

$$\therefore \frac{R_T}{1} = \frac{120}{11} \ \Omega, \ R_T = 10.91 \ \Omega$$

If there are only two resistances in parallel, the reciprocal resistance expression reduces to:

$$\frac{1}{R_T} = \frac{1}{R_1} + \frac{1}{R_2}$$

$$\frac{1}{R_T} = \frac{R_1 + R_2}{R_1 \times R_2}$$

$$\therefore R_T = \frac{R_1 \times R_2}{R_1 + R_2} \qquad (3\text{-}14)$$

This resistance equation is referred to as the *product over sum* expression, and is often simpler to use whenever there are only two resistances in parallel.

Example 3-4: If in the circuit shown in Fig. 3-5, $R_1 = 20\ \Omega$, $R_2 = 30\ \Omega$, $R_3 = 60\ \Omega$, and $V = 120\ V$, determine the circuit resistance R_T, and the various circuit currents.

$$\frac{1}{R_T} = \frac{1}{R_1} + \frac{1}{R_2} + \frac{1}{R_3}$$

$$\frac{1}{R_T} = \frac{1}{20} + \frac{1}{30} + \frac{1}{60}$$

$$\frac{1}{R_T} = \frac{6}{60}\ S$$

$$\therefore R_T = \frac{60}{6} = 10\ \Omega$$

$$I_T = \frac{V_T}{R_T}$$

$$I_T = \frac{120\ V}{10\ \Omega} = 12\ A$$

$$I_1 = \frac{V_T}{R_1}$$

$$I_1 = \frac{120\ V}{20\ \Omega} = 6\ A$$

$$I_A = I_T - I_1$$
$$I_A = 12\ A - 6\ A$$
$$I_A = 6\ A$$

$$I_2 = \frac{V_T}{R_2}$$

$$I_2 = \frac{120\ V}{30\ \Omega} = 4\ A$$

$$I_3 = \frac{V_T}{R_3}$$

$$I_3 = \frac{120\ V}{60\ \Omega} = 2\ A$$

Example 3-5: In the circuit shown in Fig. 3-6, determine the source voltage V, the circuit current I_T, and the total resistance R_T.

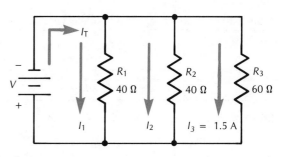

Figure 3-6

$$V_T = V_3$$
$$V_T = I_3 \times R_3$$
$$V_T = 1.5\ A \times 60\ \Omega$$
$$V_T = 90\ V$$

$$I_1 = \frac{V_T}{R_1}$$

$$I_1 = \frac{90\ V}{20\ \Omega} = 4.5\ A$$

$$I_2 = \frac{V_T}{R_2}$$

$$I_2 = \frac{90\ V}{40\ \Omega} = 2.25\ A$$

$$I_T = I_1 + I_2 + I_3$$
$$I_T = 4.5\ A + 2.25\ A + 1.5\ A$$
$$I_T = 8.25\ A$$

$$R_T = \frac{V_T}{I_T}$$

$$R_T = \frac{90\ V}{8.25\ A} = 10.91\ \Omega$$

SERIES-PARALLEL CIRCUITS

Series-parallel circuits are actually simple combinations of series and parallel-circuit connections. The rules governing these circuits are the same as those developed for series circuits and for parallel circuits. A summary of these rules is listed in Table 3-1. The student is encouraged to review these circuit laws carefully, as they will be helpful in the analysis of many electric circuits. Before analyzing more complex electric circuits, some other circuit concepts will have to be studied.

Table 3-1 Summary of series and parallel circuit properties

	SERIES CIRCUIT	PARALLEL CIRCUITS
Current	$I_T = I_1 = I_2 = I_3$	$I_T = I_1 + I_2 = I_3$
Resistance	$R_T = R_1 + R_2 + R_3$	$\dfrac{1}{R_T} = \dfrac{1}{R_1} + \dfrac{1}{R_2} + \dfrac{1}{R_3}$
Voltage	$V_T = V_1 + V_2 + V_3$	$V_T = V_1 = V_2 = V_3$
	The total resistance is greater than any one of the resistances in the circuit. When resistance is added, the circuit current decreases.	The total resistance is smaller than any one of the resistances in the circuit. When resistance is added, the circuit current increases.

The total voltage less the sum of the voltage drops in a closed loop is equal to zero.

The total current entering a node is equal to the current leaving the node.

Example 3-6: In the circuit shown in Fig. 3-7(a), determine the total resistance, the circuit and branch currents, and the load voltages.

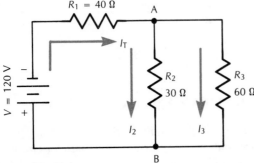

Figure 3-7(a)

Some of these will be found later at the end of the chapter.

Solution: The current flow through a circuit is often a useful guide for identifying which of the circuit components are in series and which

are in parallel. In the series connection, the same current will flow through each component, while in the parallel connection, the current divides into the parallel branches at one node to recombine at another node.

In this circuit, the current divides into I_2 and I_3 at node A and these recombine at node B. Therefore, the branches R_2 and R_3 are connected in parallel and their combined resistance is:

$$R_A = \frac{R_2 \times R_3}{R_2 + R_3}$$

$$R_A = \frac{30 \times 60}{30 + 60}$$

$$R_A = \frac{1800}{90}$$

$$R_A = 20\ \Omega$$

This resistance R_A can be used to replace the parallel connection from node A to node B as shown in Fig. 3-6(b). Obviously, R_1 and R_A are connected in series as the same current flows through each of them.

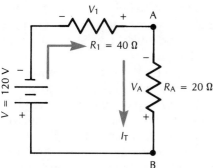

Figure 3-7(b)

Therefore, the total circuit resistance is:

$$R_T = R_1 + R_A$$
$$R_T = 40 \ \Omega + 20 \ \Omega$$
$$R_T = 60 \ \Omega$$

The total circuit current can now be calculated using Ohm's law:

$$I_T = \frac{V_T}{R_T} = \frac{120 \ V}{60 \ \Omega} = 2 \ A$$

Similarly, using Ohm's Law, the voltage drops V_1 and V_A can be determined:

$$V_1 = I_T \times R_1 = 2 \ A \times 40 \ \Omega = 80 \ V$$
$$V_A = I_T \times R_A = 2 \ A \times 20 \ \Omega = 40 \ V$$

Remember the properties of parallel circuits. The voltage between the nodes A and B (the voltage drop V_A) is the same as the voltage across the parallel branches connected between these two nodes. Therefore, the voltage drops V_2 and V_3 are:

$$V_A = V_2 = V_3 = 40 \ V$$

Using Ohm's law again, the branch currents I_2 and I_3 can now be determined:

$$I_2 = \frac{V_2}{R_2} = \frac{40 \ V}{30 \ \Omega} = 1.33 \ A$$
$$I_3 = \frac{V_3}{R_3} = \frac{40 \ V}{60 \ \Omega} = 0.67 \ A$$

As a check, the student can verify that the current entering node A is equal to the current

leaving the node:

$$I_T = I_2 + I_3$$
$$2 \ A = 1.33 \ A + 0.67 \ A$$
$$2 \ A = 2 \ A$$

A similar equation may be written for node B. Finally, the student can check that the sum of the voltages in a closed loop is equal to zero. There are three closed loops in Fig. 3-7(a). First, the polarities of the voltages are assigned. Next, going through each loop in a clockwise direction, the equations are:

loop

(1) $V_T - V_1 - V_2 = 0 \ V$
120 V – 80 V – 40 V = 0 V

loop

(2) $V_T - V_1 - V_3 = 0 \ V$
120 V – 80 V – 40 V = 0 V

loop

(3) $V_2 - V_3 = 0 \ V$
40 V – 40 V = 0 V

The usefulness of these two circuit properties on electrical nodes and closed loops is illustrated in the next example.

Example 3-7: In the circuit given in Fig. 3-8, determine the unknown current, the circuit voltages, and the total circuit resistance.

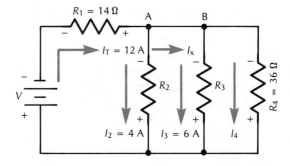

Figure 3-8

Solution:

At node A: $I_T = I_2 + I_x$
$I_x = 12 \ A - 4 \ A = 8 \ A$
At node B: $I_x = I_3 + I_4$
$I_4 = 8 \ A - 6 \ A = 2 \ A$

Applying Ohm's law:

$$V_4 = I_4 \times R_4$$
$$V_4 = 2 \text{ A} \times 36 \ \Omega$$
$$V_4 = 72 \text{ V}$$

From the voltage law for parallel circuits:

$$V_2 = V_3 = V_4 = 72 \text{ V}$$

Applying Ohm's law:

$$R_2 = \frac{V_2}{I_2} = \frac{72 \text{ V}}{4 \text{ A}} = 18 \ \Omega$$
$$R_3 = \frac{V_3}{I_3} = \frac{72 \text{ V}}{6 \text{ A}} = 12 \ \Omega$$

and

$$V_1 = I_T \times R_1$$
$$V_1 = 12 \text{ A} \times 14 \ \Omega$$
$$V_1 = 168 \text{ V}$$

Writing the loop equation:

$$V_T - V_1 - V_2 = 0 \text{ V}$$
$$240 \text{ V} - 168 \text{ V} - 72 \text{ V} = 0 \text{ V}$$

Applying Ohm's law:

$$R_T = \frac{V_T}{I_T} = \frac{240 \text{ V}}{12 \text{ A}} = 20 \ \Omega$$

WORK, POWER, AND ENERGY

Work is defined as the product of a force and the distance through which the force acts. It is measured in joules (J); force is measured in newtons (N), and distance in metres (m). Work is done when a force overcomes resistance.

Energy is the ability to do work; it is expended whenever work is done. Thus, the units of work and energy are the same. A fundamental physical law of energy is that it cannot be created or destroyed; it can only be transformed from one form into another. Work involves this change in energy.

The energy expended or used in a system is generally not all converted into useful work, as some of the energy is always expended in a nonuseful form. This nonuseful change in

energy represents the system's energy loss. The ratio of the useful output power to the total input power is called *efficiency*. It is thus evident that a system's efficiency should be kept as high as possible in order to minimize the system's losses.

Power is defined as the rate of doing work or as the rate at which energy is expended, and it is measured in watts (W). Power involves the quantity time (*t*). For example, if an electric motor can do a certain amount of work in six minutes and a second motor can do the same work in three minutes, then the second motor, which can do the work twice as fast, is said to have twice the power as the first electric motor.

POWER AND ENERGY CALCULATION AND MEASUREMENT

The equations which can be used to determine the power in an electric circuit are:

$$P = VI \qquad P = I^2R \qquad \text{and } P = \frac{V^2}{R} \qquad (3\text{-}15)$$

where P = power in watts (W)
V = voltage in volts (V)
I = current in amperes (A)
R = resistance in ohms (Ω)

In a series circuit, the total circuit power may be determined by using the total circuit values in the above equation (3-15). Or the total power may be found by adding the power of the individual loads that are connected in series.

The total power for parallel and other circuit connections is determined in the same way as for series circuits. For total circuit power, use total circuit values with the equations (3-15) or simply add up the power for each of the loads in the circuit.

Example 3-8: Determine the total circuit power for the circuit shown in Fig. 3-9.

$$R_T = R_1 + \left(\frac{R_2 \times R_3}{R_2 + R_3} \right)$$

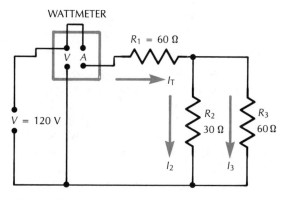

WATTMETER

$R_1 = 60\,\Omega$

$V = 120\,V$

R_2 30 Ω

R_3 60 Ω

I_T

I_2

I_3

Figure 3-9

$$R_T = 60 + \left(\frac{60 \times 30}{90}\right)$$

$$R_T = 80\,\Omega$$

$$I_T = \frac{V_T}{R_T}$$

$$I_T = \frac{120\,V}{80\,\Omega}$$

$$I_T = 1.5\,A$$

$$V_1 = I_T \times R_1$$
$$V_1 = 1.5\,A \times 60\,\Omega$$
$$V_1 = 90\,V$$

$$V_2 = V_3$$
$$V_2 = V_T - V_1$$
$$V_2 = 120\,V - 90\,V$$
$$V_2 = 30\,V$$

$$I_2 = \frac{V_2}{R_2}$$

$$I_2 = \frac{30\,V}{30\,\Omega}$$

$$I_2 = 1\,A$$

$$I_3 = \frac{V_3}{R_3}$$

$$I_3 = \frac{30\,V}{60\,\Omega}$$

$$I_3 = 0.5\,A$$

Total power

$$P_T = (I_T)^2 R_T$$
$$P_T = (1.5)^2 \times 80$$
$$P_T = 180\,W$$

or

$$P_T = P_1 + P_2 + P_3$$
$$P_T = (I_1)^2 R_1 + (I_2)^2 R_2 + (I_3)^2 R_3$$
$$P_T = 1.5^2 \times 60 + 1^2 \times 30 + 0.5^2 \times 60$$
$$P_T = 135 + 30 + 15$$
$$P_T = 180\,W$$

The power in a circuit may also be measured directly using a *wattmeter*. The actual construction of this instrument will be dealt with in a later chapter on meters. Briefly, the wattmeter contains two coils. One coil, called the voltage coil, is connected across the circuit's voltage, and the second coil, called the current coil, is connected in series with the circuit's load. The wattmeter therefore contains a total of four terminals. The method of connecting this instrument in a circuit is illustrated in Fig. 3-10. In this connection, the wattmeter will indicate total circuit power.

The electrical energy used by a system is found by multiplying its power by the length of time the system is functioning.

$$\text{Energy} = \text{Power} \times \text{Time} \qquad (3\text{-}16)$$

If power is in watts and time is in hours, the unit for energy will be the watthour (W·h). A larger and more commonly used unit of energy is the kilowatthour (kW·h).

$$1\,kW \cdot h = 1000\,W \cdot h$$

Example 3-9: Three 100-W lamps are connected in parallel across a 120-V supply. If all three lamps operate for 12 h, determine the total energy converted in kW·h.

$$\text{Total power} = 300\,W$$
$$\text{Energy used} = P \times t$$
$$= 300\,W \times 12\,h$$
$$= 3600\,W \cdot h$$
$$= 3.6\,kW \cdot h$$

The electrical energy converted by a system may also be directly measured with the use of a watthour-meter. This meter is in standard use by public utilities to measure the energy supplied to commercial and residential electrical systems. The energy supplied to a customer during a certain period is found by taking the difference between the present and previous meter readings. This difference in meter reading is used in calculating the customer's electric energy bill for that period.

TERMINAL VOLTAGES AND POWER LOSSES

All voltage sources and transmission lines have resistance. When a current flows through a circuit, this resistance can cause a significant drop in voltage. The terminal voltage available to the load can be much less than the actual generated source voltage. The resistance in a power supply is due to its internal parts and is called internal resistance. For electrical power supplies, such as a generator, this internal resistance is generally constant. For a battery, the resistance increases as the battery ages. The resistance in transmission lines is dependent on length, wire size, temperature, and wire resistivity. The voltage drop due to line resistance is called line voltage drop and is a particular concern of hydroelectric utilities which transmit electrical energy over long distances. Both the internal resistance of a source and the line resistance in transmission are considered to be connected in series with the actual source voltage. The following examples will illustrate the voltage drops caused by these resistances.

Example 3-10: In the circuit shown in Fig. 3-10, the internal source resistance, R_i is 4 Ω and the

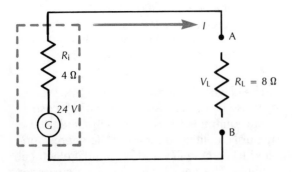

Figure 3-10

load resistance, R_L, is 8 Ω. Determine the terminal voltage V_L if the actual generated voltage is 24 V.

Solution: If the load R_L is removed, no circuit current will flow. Therefore, the internal resistance voltage drop V_i will be zero and so the voltage across the points A and B, V_{AB}, will equal 24 V, the actual source voltage. This voltage with the load removed is often referred to as the *open circuit voltage*. When the load is connected, a current of 2 A will flow in the circuit.

$$R_T = 4 \; \Omega + 8 \; \Omega = 12 \; \Omega$$
$$I = \frac{24 \text{ V}}{12 \; \Omega} = 2 \text{ A}$$
$$V_i = 2 \text{ A} \times 4 \; \Omega = 8 \text{ V}$$
$$V_L = 2 \text{ A} \times 8 \; \Omega = 16 \text{ V}$$

The internal resistance voltage drop will now be 8 V and the load or terminal voltage will be reduced to 16 V.

Careful examination of the circuit will indicate that this internal voltage drop is directly proportional to the circuit current. If the load resistance is decreased, the circuit current will increase, the voltage drop will increase, and the terminal voltage will decrease. If the load resistance is increased, the voltage drop will decrease, and the terminal voltage will increase.

The internal source voltage drop is important because it affects the terminal voltage. However, in many electric circuits, no mention is made of internal resistance. In these circuits, the student may assume that the internal resistance has already been considered or that it is so small as to be insignificant.

Example 3-11: In the circuit given in Fig. 3-11, the resistance of each of the two distribution lines is indicated as a single resistance, R_D, and each is equal to 3 Ω. Determine the terminal voltage, V_L, when the load resistance is 24 Ω.

Solution: If the load resistance is removed, no circuit current will flow. The line voltage drop

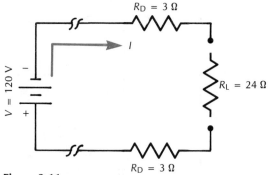

Figure 3-11

will be zero, and the terminal voltage (open circuit voltage) will be equal to the source voltage. When the load resistance is connected, a current of 4 A will flow in the circuit and the voltage distribution will be:

$$R_T = 3\ \Omega + 3\ \Omega + 24\ \Omega = 30\ \Omega$$
$$I = \frac{120\ V}{30\ \Omega} = 4\ A$$
$$V_D = 4\ A \times 3\ \Omega = 12\ V$$
$$V_L = 4\ A \times 24\ \Omega = 96\ V$$

The total line voltage drop is 24 V and the terminal voltage is reduced to 96 V. It should be evident that the line voltage drop, like the internal source resistance voltage drop, is directly proportional to the line current. Therefore, an increase in load resistance will decrease the circuit current, which will cause the line voltage drop to decrease and the terminal voltage to increase. A decrease in load resistance will produce the opposite effects; the circuit current will increase, which will cause an increase in the line voltage drop and a resulting decrease in the terminal voltage.

The internal resistance in a circuit's power supply and the line resistance in the distribution line of the circuit also cause an energy loss in the system. This energy is dissipated, or lost, in the form of heat and occurs whenever a current flows in the circuit. In fact, this energy loss varies as the square of the current and is often referred to as the $I^2 \times R$ power loss.

In power systems where power must be transmitted over long distances, the voltage is stepped-up at the point of transmission and then stepped-down at the point of distribution. The step-up and step-down of the line voltage is accomplished with the use of transformers. When the voltage at which power is transmitted is increased, the line current is reduced by the same factor by which the voltage is increased, because $P = V \times I$. By reducing the line current, the transmission line power losses are greatly reduced. These advantages are illustrated by the following example.

Example 3-12: A power transmission system has a total line resistance of 2.4 Ω and supplies power to a 72-kW load. Determine the line $I^2 \times R$ power losses when the load is supplied at (a) 2400 V and (b) 4800 V

(a) $I = \dfrac{P}{V} = \dfrac{72\ 000\ W}{2400\ V} = 30\ A$

Total line loss $= I^2R = (30)^2\ A \times 2.4\ \Omega$
$$= 2160\ W$$

(b) $I = \dfrac{P}{V} = \dfrac{72\ 000\ W}{4800\ V} = 15\ A$

Total line loss $= I^2R = (15)^2\ A \times 2.4\ \Omega =$ 546 W. This example shows that when the voltage supplied to the load is increased from 2400 V to 4800 V, the line current decreases from 30 A to 15 A, and the line power losses decrease from 2160 W to 540 W.

POWER TRANSFER

It is often desirable to transfer a maximum amount of energy per unit of time from one point to another. In a series circuit, the maximum power can be transferred from the power supply to a load resistance only when the load resistance is of a certain value. The load resistance necessary for maximum power transfer is illustrated by the following example.

Example 3-13: In the circuit shown in Fig. 3-12, find an equation for relating the power transfer to the load resistance. Use this equation to

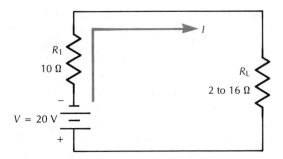

Figure 3-12

determine the power transfer for various values of load resistance. Tabulate the results and plot the graph for power transfer versus load resistance.

Solution:

$$R_T = R_1 + R_L$$
$$I = \frac{V}{R_T} = \frac{V}{R_1 + R_L}$$

Power transfer (P_L) to load $= I^2 R_L$

$$P_L = \left[\frac{V}{R_1 + R_L}\right]^2 \times R_L \qquad (3\text{-}17)$$

The power transfer for the load resistance ($R_L = 2\Omega$) is:

$$P_L = \left[\frac{20}{10 + 2}\right]^2 A \times 2\,\Omega = 5.56\ W$$

The power transfer for other load resistances can be found in a similar way. The results are tabulated in Table 3-2.

Table 3-2 Power transfer relating to values in load resistance

LOAD RESISTANCE R_L	POWER TRANSFER P_L
2 Ω	5.56 W
4	8.16
6	9.36
8	9.88
10	10.00
12	9.90
14	9.70
16	9.18

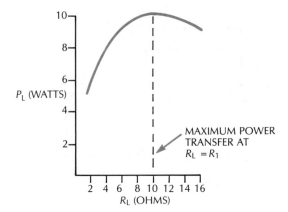

Figure 3-13

The graph for power transfer versus load resistance is shown in Fig. 3-13. It is obvious from this graph that as the load resistance is increased, the power transfer increases up to a maximum and then begins to decrease. Maximum power is transferred when R_L is 10 Ω. The circuit in Fig. 3-12 indicates that the other circuit resistance R_1 is also 10 Ω. Therefore, for maximum power transfer to a load resistance, the load resistance must equal the resistance of the rest of the circuit.

COMPLEX CIRCUITS

Some complex circuits can be analyzed readily by first reducing them to simple circuits through the process of combining resistance values. This is illustrated in Fig. 3-14.

Solution: In Fig. 3-14(a) the parallel combination of R_1 and R_2 can be reduced to its single equivalent resistance.

$$R_A = \frac{6 \times 3}{6 + 3} = 2\ \Omega$$

Similarly, the single equivalent for R_3 and R_4 is:

$$R_B = \frac{20 \times 30}{20 + 30} = 12\ \Omega$$

and the single equivalent for R_8, R_9, and R_{10} is:

$$\frac{1}{R_C} = \frac{1}{24} + \frac{1}{8} + \frac{1}{12}$$

$$\frac{1}{R_C} = \frac{1 + 3 + 2}{24}$$

$$\frac{1}{R_C} = \frac{6}{24} \ S$$

$$R_C = \frac{24}{6}$$

$$R_C = 4 \ \Omega$$

The equivalent resistance R_A, R_B, and R_C are now substituted into the circuit as shown in Fig. 3-14(b). The circuit can now be further simplified by the equivalents R_D and R_E shown in Fig. 3-14(c).

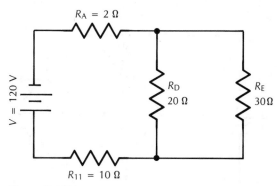

Figure 3-14(c)

$$R_D = R_7 + R_C$$
$$R_D = 16 \ \Omega + 4 \ \Omega$$
$$R_D = 20 \ \Omega$$

$$R_E = R_B + R_5 + R_6$$
$$R_E = 12 \ \Omega + 6 \ \Omega + 12 \ \Omega$$
$$R_E = 30 \ \Omega$$

From Fig. 3-14(c), the total circuit resistance can now be readily computed.

$$R_T = R_A + R_{11} + \frac{R_D \times R_E}{R_D + R_E}$$

$$R_T = 2 \ \Omega + 10 \ \Omega + \frac{30 \times 20}{30 + 20} \ \Omega$$

$$R_T = 24 \ \Omega$$

Many complex circuits cannot be solved by the conventional methods covered so far. The next two examples illustrate two such circuits and the additional steps which must be used in their solutions.

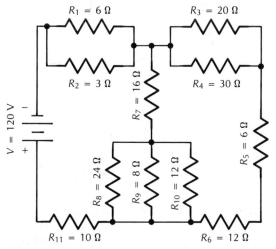

Figure 3-14(a)

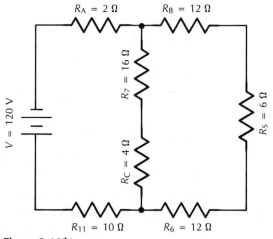

Figure 3-14(b)

Example 3-15: Fig. 3-15(a) shows a complex circuit in which resistor R_3 forms a bridge from one side of R_1 to the opposite side of R_5. This type of circuit is called a bridge circuit. The solution for this circuit involves the conversion of the delta (Δ) circuit formed by R_1, R_2, and R_3 into a Wye (Y) equivalent circuit. Equations are available for use in converting from a delta (Δ) circuit to a Wye (Y) circuit and from a Wye circuit into a delta circuit. Figure 3-16 illustrates these conversions.

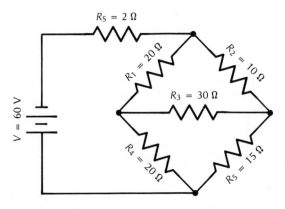

Figure 3-15(a)

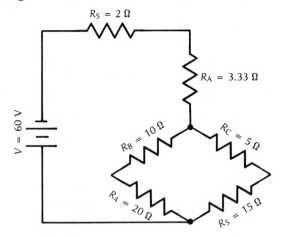

Figure 3-15(b)

To convert from delta to Wye:

$$R_A = \frac{R_1 \times R_2}{R_1 + R_2 + R_3}$$

$$R_B = \frac{R_1 \times R_3}{R_1 + R_2 + R_3}$$

$$R_C = \frac{R_2 \times R_3}{R_1 + R_2 + R_3} \qquad (3\text{-}18)$$

To convert from Wye to delta:

$$R_1 = \frac{R_A R_B + R_B R_C + R_C R_A}{R_C}$$

$$R_2 = \frac{R_A R_B + R_B R_C + R_C R_A}{R_B}$$

$$R_3 = \frac{R_A R_B + R_B R_C + R_C R_A}{R_A} \qquad (3\text{-}19)$$

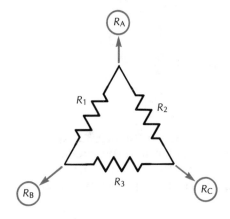

Figure 3-16(a) Delta (Δ) or Pi (π) circuit.

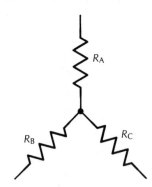

Figure 3-16(b) Wye (Y) or Tee (T) circuit.

Equation 3-19 can now be used to simplify the circuit shown in Fig. 3-15(a) to that shown in Fig. 3-15(b).

$$R_A = \frac{R_1 \times R_2}{R_1 + R_2 + R_3}$$

$$= \frac{20 \times 10}{20 + 10 + 30} = 3.33 \ \Omega$$

$$R_B = \frac{R_1 \times R_3}{R_1 + R_2 + R_3}$$

$$= \frac{20 \times 30}{20 + 10 + 30} = 10 \ \Omega$$

$$R_C = \frac{R_2 \times R_3}{R_1 + R_2 + R_3} = \frac{10 \times 30}{20 + 10 + 30} = 5 \ \Omega$$

The procedure for determining the total resistance of this circuit is now simply that of a series-parallel arrangement and the remaining solution is left as an exercise for the student.

Multiple source circuits as shown in Fig. 3-17(a) can be analyzed in different ways. Three solutions are described. The first will use the familiar loop and node equations; the second will illustrate the use of the Superposition Theorem and the third will describe Thévenin's equivalent circuit principle.

Example 3-16 Solution by Loop and Node Equation: When the direction of current flow in a circuit is unknown, the direction may be assumed during an analysis of the circuit. If the solution produces a positive answer, the assumed current direction is correct; if the answer obtained is negative, this indicates that the current is flowing in the opposite direction. However, the magnitude of the current obtained is correct. In Fig. 3-17(a), the current direction is assumed to flow as shown on the diagram.

The current equation for node A is:
$$I_L = I_1 + I_2 \tag{1}$$
Since there are three unknowns in this equation, three equations are required to find these unknowns. The two other equations which may be written are voltage loop equations, which are then expressed in current form, using Ohm's law.

loop 1
$$V_A - V_1 - V_L = 0$$
$$100 - 10\,I_1 - 20\,I_L = 0 \tag{2}$$
loop 2
$$V_B - V_2 - V_L = 0$$
$$80 - 30\,I_2 - 20\,I_L = 0 \tag{3}$$
By the process of substitution, these three equations can be reduced to a single simple equation.

Substitute (1) into (2)
$$100 - 10\,I_1 - 20(I_1 + I_2) = 0$$
$$100 - 30\,I_1 - 20\,I_2 = 0 \tag{2A}$$

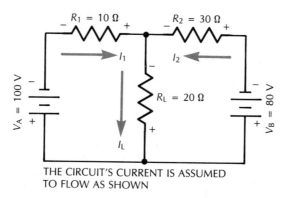

THE CIRCUIT'S CURRENT IS ASSUMED
TO FLOW AS SHOWN

Figure 3-17(a) Solution #1.

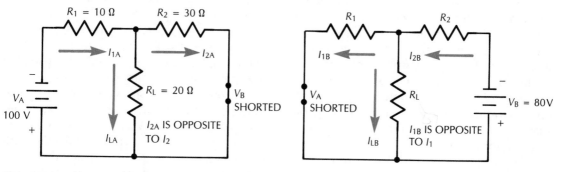

Figure 3-17 (b) (c) Solution #2.

Substitute (1) into (3)

$$80 - 30\,I_2 - 20(I_1 + I_2) = 0$$
$$80 - 20\,I_1 - 50\,I_2 = 0 \qquad (3A)$$

Rearrange (2A)

$$I_2 = \frac{1}{20}(100 - 30\,I_1) \qquad (2B)$$

Substitute (2B) into (3A)

$$80 - 20\,I_1 - \frac{50}{20}(100 - 30\,I_1) = 0$$
$$80 - 20\,I_1 - 250 + 75\,I_1 = 0$$
$$55\,I_1 = 170$$
$$I_1 = 3.09 \text{ A}$$

From (2B)

$$I_2 = \frac{1}{20}(100 - 30 \times 3.09) = 0.36 \text{ A}$$

From (1)

$$I_L = I_1 + I_2$$
$$I_L = 3.09 \text{ A} + 0.36 \text{ A}$$
$$I_L = 3.45 \text{ A}$$

Second Solution using the Superposition Theorem:

The proof of this theorem is left for more advanced study in network analysis. However, the principle described is simple enough that it may be effectively applied in the solution of many circuit networks having two or more voltage sources. The principle of superposition says that one finds the currents produced by each voltage source and then adds these together. This is permissible only if every element in the network is linear; that is, if all the resistances, inductances, and capacitances are constant. (Transistors and iron-core coils are typical nonlinear elements that may prevent application of the superposition theorem.)

The circuit shown in Fig. 3-17(a) contains two voltage sources and only linear elements and therefore, the superposition principle may be correctly applied. The circuit is redrawn for convenience.

Determine currents with V_B set to be zero (Fig. 3-17(b)) (V_B is short-circuited).

$$R_{A\,in} = 10 + \frac{30 \times 20}{50} = 22 \ \Omega$$

$$I_{1A} = \frac{V_A}{R_{Ain}} = \frac{100}{22} = 4.54 \text{ A}$$

$$I_{LA} = \frac{30}{50} \times 4.54 = 2.72 \text{ A}$$

$$I_{2A} = -I_{1A} - I_{LA} = -1.82 \text{ A}$$

(Note: this current is negative since it flows opposite to the direction assumed in the diagram.)

Now determine the currents with V_A set to be zero. (Fig. 3-17(d))

$$I_{2B} = \frac{V_B}{R_{B\,in}} = \frac{80}{36.67} = 2.18 \text{ A}$$

$$I_{LB} = \frac{10}{30} \times 2.18 = 0.73 \text{ A}$$

$$I_{1B} = -I_{2B} - I_{LB} = 1.45 \text{ A}$$

I_{1B} is negative for the same reason as I_{2A}

Finally, add the individual currents.

$$I_1 = I_{1A} + I_{1B} = 4.54 - 1.45 = 3.09 \text{ A}$$
$$I_2 = I_{2A} + I_{2B} = 1.82 + 2.18 = 0.36 \text{ A}$$
$$I_L = I_{LA} + I_{LB} = 2.72 + 0.73 = 3.45 \text{ A}$$

The current values for I_1, I_2, and I_L are identical to those values obtained in the first solution using loop and node equations.

Third Solution using Thévenin's Theorem:

Like the superposition theorem, the proof of Thévenin's theorem is also left for more advanced study in network analysis. However, the principle described by this theorem is relatively simple, easy to apply, and often useful in the analysis of many circuit networks.

Thévenin's theorem says that any live network can be represented by a single constant-voltage source, V_θ, in series with a single impedance, Z_θ. Thus, Thévenin's equivalent circuit can now be used to replace the active network. V_θ is the voltage which may be measured across the output terminals of the active network when all the sources within the network are operating normally. Z_θ is the impedance which may be measured across these two terminals when all the voltage sources have been reduced to zero (short-circuited). In this example, all the elements are purely resistive so the impedance Z_θ would also be purely resistive and may be represented by the single resistance, R_θ.

In Fig. 3-17 the circuit currents are required. Thévenin's principle will be used to find the current I_L and then loop equations will be used to determine the remaining two circuit currents. Since I_L is to be determined using Thévenin's principle, R_L would be considered an external circuit to which the rest of the circuit (active circuit) is connected. This is illustrated in Fig. 3-18. Thévenin's equivalent circuit is found for the active circuit.

From Fig. 3-18(a):

$$V_A - V_1 - V_2 - V_B = 0 \text{ V}$$
$$100 - 10I - 30I - 80 = 0 \text{ V}$$
$$-40I = -20$$
$$I = 0.5 \text{ A}$$
$$V_\theta = V_{ab} = V_B + V_2 = 80 + (0.5) \times 30 = 95 \text{ V}$$

From Fig. 3-18(b):

$$R_\theta = \frac{R_1 \times R_2}{R_1 + R_2} = \frac{10 \times 30}{10 + 30} = 7.5 \text{ }\Omega$$

From Fig. 3-18(c), which shows the Thévenin's equivalent circuit connected to the external circuit R_L:

$$I_L = \frac{V_\theta}{R_\theta + R_L} = \frac{95}{7.5 + 20} = 3.45 \text{ A}$$

From Fig. 3-17:

$$V_A - V_1 - V_L = 0$$
$$100 - 10I_1 - 20(3.45) = 0$$
$$I_1 = \frac{31}{10} = 3.10 \text{ A}$$

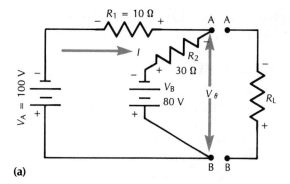

(a)

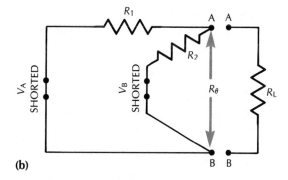

(b)

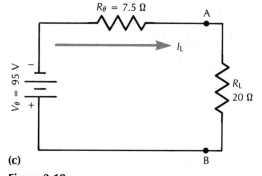

(c)

Figure 3-18

$$V_B - V_2 - V_L = 0$$
$$80 - 30I_2 - 20(3.45) = 0$$
$$I_2 = \frac{11}{30} = 0.36 \text{ A}$$

The results obtained for I_1, I_2, and I_L agree with those of the two previous solutions.

SUMMARY OF IMPORTANT POINTS

SERIES CIRCUITS

1. A series connection has only one current path. Therefore, the current is constant through this path.
2. The total resistance is equal to the sum of the resistances connected in series.

$$R_T = R_1 + R_2 + R_3 + \ldots$$

3. The source voltage is equal to the sum of voltage drops across the load resistances in a series circuit.

$$V_T = V_1 + V_2 + V_3 + \ldots$$

The above voltage law may be restated as the algebraic sum of the voltages in a closed loop is equal to zero.

4. Forms of Ohm's law are: $V = IR$, $I = \dfrac{V}{R}$, and $R = \dfrac{V}{I}$

PARALLEL CIRCUITS

5. In a parallel circuit, there are more than one independent current path.
6. The total current is equal to the sum of the parallel branch currents.

$$I_T = I_1 + I_2 + I_3 + \ldots$$

7. A point where the current splits up into the parallel branches or where the branch currents come together is called an electrical node or junction.
8. The algebraic sum of the currents at a node is equal to zero.
9. The voltage across parallel branches is the same.
10. The total resistance decreases as the number of parallel branches is increased.

$$\frac{1}{R_T} = \frac{1}{R_1} + \frac{1}{R_2} + \frac{1}{R_3} + \ldots$$

11. For only two resistances in parallel,

$$R_T = \frac{R_1 \times R_2}{R_1 + R_2}$$

WORK POWER ENERGY

12. The circuit power in watts may be determined from the equations:

$$P = VI \text{ or } P = I^2 R \text{ or } P = \frac{V^2}{R}$$

13. Electrical energy is given by the equation

Energy = Power × Time

14. Some energy units are

kW · h = 1000 W · h
1 W/s = 1 joule (J)

TERMINAL VOLTAGE AND POWER LOSSES

15. The resistance in transmission lines causes both a voltage and a power loss. The internal resistance present in electrical devices will cause similar losses.
16. A circuit's voltage losses is directly proportional to the circuit's current. The energy loss varies as the square of the current.
17. To reduce the losses during the transmission of energy over long distances, the voltage is stepped-up at the point of transmission. The line current will be reduced by the same factor the voltage is increased.

POWER TRANSFER

18. For maximum power transfer to a load, the load resistance must equal the resistance of the rest of the circuit.

COMPLEX CIRCUITS

20. To convert a delta-connected resistive circuit to a Wye-equivalent resistive circuit:

$$R_A = \frac{R_1 \times R_2}{R_1 + R_2 + R_3}$$

$$R_B = \frac{R_1 \times R_3}{R_1 + R_2 + R_3}$$

$$R_C = \frac{R_2 \times R_3}{R_1 + R_2 + R_3}$$

R_1, R_2, and R_3 are the resistances of the delta circuit and R_A, R_B, and R_C are those for the Wye circuit. As an example, the vertex of R_1 and R_2 gives the direction of R_A.

21. To convert from Wye circuit to a delta circuit:

$$R_1 = \frac{R_x}{R_B}$$

$$R_2 = \frac{R_x}{R_C}$$

$$R_3 = \frac{R_x}{R_A}$$

where $R_x = R_A R_B + R_B R_C + R_C R_A$

22. Two theorems which are sometimes useful in analyzing certain complex circuits are the superposition theorem and Thévenin's equivalent circuit principle.

REVIEW QUESTIONS

1. In the circuit given in Fig. 3-2, if $R_1 = 100\ \Omega$, $R_2 = 75\ \Omega$, $R_3 = 150\ \Omega$, and $I = 0.2$ A, what are the load voltages V_1, V_2, V_3 and the source voltage V_T?

2. In the circuit given in Fig. 3-2, if $R_1 = 2\ k\Omega$, $R_2 = 4.7\ k\Omega$, $R_3 = 1.5\ k\Omega$ and $V = 492$ V, determine the load voltages V_1, V_2, and V_3.

3. In the circuit given in Fig. 3-3, if $V = 240$ V, $V_1 = 60$ V, $V_2 = 80$ V, and $R_3 = 5\ k\Omega$, determine the load voltage V_3, the circuit current (I_T), and the total circuit resistance (R_T).

4. A series circuit has three resistors R_1, R_2, and R_3 connected across a voltage source V_T. If the total circuit resistance $R_T = 6\ k\Omega$, the load resistances $R_1 = 1\ k\Omega$, $R_2 = 2\ k\Omega$ and the voltage drop (V_3) across the load R_3 is 90 V, determine the following: (a) R_3, (b) circuit current (I_T) (c) voltage drops V_1 and V_2, and (d) the source voltage V_T.

5. In the circuit given in Fig. 3-6, if $R_1 = 4\ k\Omega$, $R_2 = 6\ k\Omega$, $R_3 = 9\ k\Omega$, and $V = 180$ V, determine the total circuit resistance and the source current.

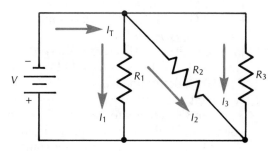

Figure 3-19

6. If in the circuit described in Q. 6, the load R_1 is shorted out, describe the effects on: (a) total circuit resistance (b) load resistances R_2 and R_3, (c) circuit current, (d) load voltage drops V_2 and V_3. (Note: assume the source voltage remains the same as in Q. 5.)

7. In the circuit shown in Fig. 3-19, if $R_1 = 60\ \Omega$, $R_2 = 30\ \Omega$, $R_3 = 40\ \Omega$, and $I_3 = 3$ A, determine the branch currents I_1 and I_2 and the source current I_T.

8. In the circuit shown in Fig. 3-19, if $I_T = 6.5$ A, $I_1 = 2$ A, $I_2 = 3$ A, and $R_3 = 160\ \Omega$, determine the source voltage (V_T) and the total circuit resistance (R_T).

9. In the circuit shown in Fig. 3-19, if $R_1 = 2\ k\Omega$, $R_2 = 6\ k\Omega$, $R_3 = 8\ k\Omega$, $R_4 = 8\ k\Omega$, and $V = 220$ V, determine the total circuit resistance (R_T), the source current (I_T), and the branch current I_2.

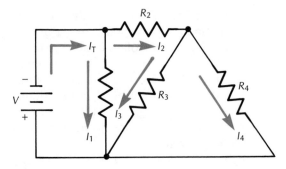

Figure 3-20

10. In the circuit shown in Fig. 3-20, if $R_1 = 25\ \Omega$, $R_2 = 15\ \Omega$, $R_3 = 60\ \Omega$, $R_4 = 30\ \Omega$, and $I_4 = 3$ A, determine the currents I_3, I_2, I_1, and I_T.

11. In the circuit shown in Fig. 3-20, if $I_T = 5$ A, $I_1 = 2$ A, $I_3 = 1.6$ A, $R_2 = 15\ \Omega$, and $R_4 = 50\ \Omega$, determine R_3, R_1, and R_T.

12. A circuit has three parallel resistors R_1, R_2 and R_3 connected across a source voltage V_T. If the total circuit resistance $R_T = 2$ kΩ, the branch resistances $R_1 = 6$ kΩ, $R_2 = 12$ kΩ, and the branch current I_3 through the branch resistance R_3 is 30 mA, determine the following: (a) branch resistance R_3, (b) source voltage V_T, (c) branch current I_1 and I_2, and (d) total circuit current I_T.

13. If in the circuit described in Q. 12, the branch resistance R_1 is open circuited, describe the effects on the following: (Note: assume the branch current I_3 remains the same.) (a) Total circuit resistance, (b) branch current I_2, (c) the circuit current I_T, and (d) the source voltage V_T.

14. Give four important statements about energy.

15. A dc motor is connected across 115-V lines and draws 40 A. What is the power input to the motor?

16. Three resistances, 20 Ω, 17 Ω, and 35 Ω are connected in series with a 208-V source. Determine the power in each of the loads and the total circuit power.

17. If in the circuit shown in Fig. 3-19,

$R_1 = 40\ \Omega$, $R_2 = 60\ \Omega$, $R_3 = 90\ \Omega$, and $V = 480$ V, (a) what is the total circuit power and (b) the energy converted by the circuit in 5 h?

18. Briefly, how does the construction of the watthour meter differ from the wattmeter?

19. A generator with an internal resistance of 2 Ω supplies 5 A to an external load the resistance of which (R_L) is 20 Ω. Determine (a) the load voltage, (b) the no-load voltage, (c) the load resistance required for maximum power transfer, and (d) the maximum power transfer to the load in (c).

20. A generator supplies a distant load, $R_L = 60\ \Omega$, with the rated load voltage, $V_L = 220$ V. If the resistance of each line is 0.2 Ω, determine (a) the generator's output terminal voltage, (V_G), and (b) the total line power losses.

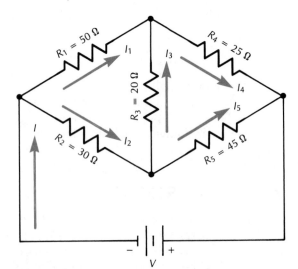

Figure 3-21

21. In the circuit shown in Fig. 3-21, replace the delta circuit formed by R_1, R_2, and R_3 with its Wye equivalent circuit and then determine the total resistance of the entire circuit.

22. Replace the delta circuit formed by R_3, R_4, and R_5 shown in Fig. 3-21 with its Wye equivalent circuit and then deter-

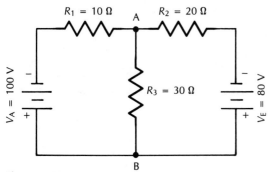

Figure 3-22

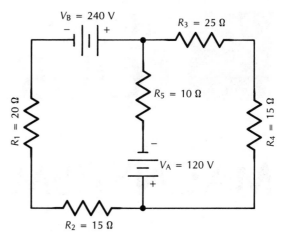

Figure 3-23

mine the resistance of the entire circuit. Compare your result to that for Q. 21.

23. Using the calculations developed for Q. 22, determine the currents I_1, I_2, and I_3 if the source voltage is equal to 60 V.

24. Using loop and node equations, determine the current flow through the load resistor R_3, shown in Fig. 3-22.

25. Determine the current flow through R_3 in Fig. 3-22 using the superposition principle.

26. Determine the current flow through R_3 in Fig. 3-22 using Thévenin's theorem.

27. Determine the current flow through R_4 in Fig. 3-23 using (a) loop and node equations, (b) the superposition principle and (c) Thévenin's theorem.

CHAPTER
FOUR

MAGNETISM

Magnetism and electricity are interrelated, to the point that the one cannot be studied by itself without having to involve the other. These two fundamental invisible forces are found in all electrical and electronic devices. Their applications range from a simple door chime used in a home to generators, motors, transformers used in factories, railway locomotives, etc. Electronic circuits used in television, satellites, computers, all make extensive use of magnetic force.

NATURAL MAGNETISM

The phenomenon of magnetism was discovered some 2000 years ago. It was found that a certain iron ore had the ability to attract small pieces of iron. This mineral magnetite (or lodestone, as it became to be called), was named after the people known as Magnetites who lived in the locality of Asia Minor. Magnetite is a natural magnet which has no value at present, since magnets with great force can be produced artificially.

Although a magnetic field is invisible, evidence of its force can be seen when certain metals can attract other metals without coming into contact.

ARTIFICIAL MAGNETS

An artificial magnet can be made by stroking a piece of steel with a lodestone or some other magnet. Thus, the magnetized steel is called an artificial magnet. Some materials such as soft iron can be easily magnetized, but they lose their magnetism very quickly after the magnetizing force has been removed. Any material that can be easily magnetized and that quickly loses its magnetism is said to have high *permeability*. Permability refers to the ease with which a material conducts magnetic force and the symbol for permeability is the Greek letter μ (mu).

Reluctance is the opposition offered by a substance to the flow of magnetic force, similar to resistance in electricity. Thus, soft iron that has high permeability also has low reluctance. Hard steel, on the other hand, is difficult to magnetize, therefore has lower permeability and higher reluctance. Thus, reluctance is the opposite of permeability. After steel has become magnetized, it will retain the magnetism indefinitely. Hard steel then is said to have high retentivity. The symbol for reluctance is $\mathcal{R}$.

The amount of magnetism retained by any substance being magnetized, after the magnetizing force has been removed, is called *residual magnetism*.

Magnets may be classified into two general groups, natural and artificial, and these in turn can be broken down into permanent or temporary magnets. Lodestone, hard steel, and certain alloys of nickel and cobalt, when magnetized, can be classified as permanent magnets. Temporary magnets are magnets that are produced by electric currents. They are easily magnetized and lose their magnetism when the electric current is turned off. These temporary magnets do retain a very small amount of residual magnetism, which plays a very important part in the operation of a generator, as will be shown in Chapter Seven. Stronger artificial magnets can be produced by electrical means. Today, magnets are classified by the type of material (alloy) used, as *Alnico* (an alloy of aluminum, nickel and cobalt) and *Cunife* (an alloy of copper (Cu) nickel (Ni) and iron (Fe)).

When a straight bar of steel is magnetized, it is called a bar magnet, and has its own magnetic field. The magnetic field of a magnet is the invisible magnetic influence which the magnet sets up around itself. The magnetic field is not distributed uniformly over the entire surface of the magnet. The magnetic field is strongest at the ends and weakest in the middle of the magnet. This can easily be observed by dipping a bar magnet in iron filings. The ends of the bar will attract the greatest number of iron filings, while only a few will be attracted to the centre of the magnet. The areas at the end of the magnet where the magnetic effects are the greatest are called the *poles* of the magnet. These poles are named after the direction in which they point when the magnet is suspended. The end that points toward the north is called the *North Pole* and the end that points toward the south is the *South Pole*.

strongly attracted by a magnetic field, and include iron, steel, nickel, cobalt, and alloys such as alnico.

Materials which are attracted only slightly by a strong magnetic field are classified as *paramagnetic* substances. These include aluminum, platinum, chromium, sodium, and oxygen. This magnetic force is about a million times smaller than that of iron.

Diamagnetic materials are slightly repelled by magnetic fields. These include bismuth, copper, zinc, mercury, silver, gold, glass, and water.

Materials that cannot be magnetized are classified as *nonmagnetic* substances. Wood, rubber, paper, wax, and plastics are a few examples.

The characteristic which determines if a substance is nonmagnetic, ferromagnetic, paramagnetic, or diamagnetic is its permeability. The *permeability* of a vacuum is assumed to be one and it is used as the standard to which all other substances are compared. Ferromagnetic materials have values of permeability from 50 to 5000. Paramagnetic substances have a permeability slightly greater than 1, while diamagnetic materials have a permeability rating of less than 1.

A recently developed magnet (a ceramic magnet, that is nonmetallic) has ferromagnetic properties with very high permeability. These *ceramic* magnets are made from powdered metal oxides called *ferrites*. Ceramic magnets are used extensively as ferrite cores, in coils for radio frequency transformers, and in various electronic circuits.

MAGNETIC AND NONMAGNETIC SUBSTANCES

When classifying magnetic materials, a distinction must be made between magnetic and nonmagnetic substances.

Materials that readily respond to magnetic fields (can be strongly magnetized) are called *ferromagnetic* substances. These materials are

THEORY OF MAGNETISM

As with the basic theories of electricity, magnetic theory starts with the electron. In Chapter One, we found that the electron has a negative charge. This charge produces an electric field that comes into the electron from all directions (Fig. 4-1(a)).

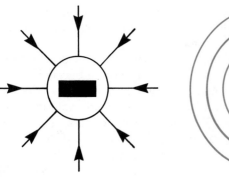 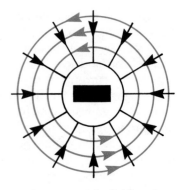

(a) Electric field (b) Magnetic field around electron (c) Electromagnetic field

Figure 4-1 Theory of magnetism.

However, an electron that is spinning as it rotates in its own orbit produces a magnetic field which exists in concentric circles around the electron (Fig. 4-1(b)). The polarity of this magnetic field is determined by the direction the electron is spinning.

Figure 4-1(c) shows the electrostatic lines of force and the magnetic lines of force acting at right angles to each other. The combined field forms an electromagnetic field around each electron.

It is possible to have some electrons spin in a clockwise direction and others in a counter-clockwise direction. In *nonmagnetic materials* there are an equal number of electrons spinning in a clockwise and counter-clockwise direction in each orbit. The magnetic fields cancel each other, resulting in no magnetic field. That is why nonmagnetic materials cannot be magnetized.

In *magnetic materials*, the number of clockwise and counter-clockwise spinning electrons are not equal; therefore, the magnetic fields do not cancel. The uncancelled spinning electrons (magnetic dipoles) arrange themselves into groups called *domains*. In an unmagnetized material such as iron, these domains (magnetic fields) are haphazardly arranged (Fig. 4-2(a)), cancelling each other. Therefore, no magnetic field can be detected in the iron.

However, when the iron is subjected to an external magnetic field, the lines of force from the external magnetic field travel through the iron causing the domains to align themselves

 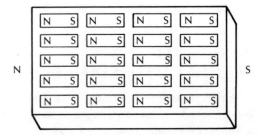

(a) ELECTRON DIPOLES (DOMAINS) NOT ALIGNED (b) ELECTRON DIPOLES (DOMAINS) ALIGNED

Figure 4-2 Alignment of electrons.

in the same direction (Fig. 4-2(b)). The iron is then magnetized.

When the external magnetizing field is removed from the iron, the domain rearrange themselves to the haphazard pattern of Fig.4-2(a).

THE MAGNETIC FIELD

Every magnet is surrounded by a space where its magnetic effects are present. This is called the *magnetic field*. The magnetic field is made up of magnetic lines of force, or *magnetic flux*. While the lines are invisible, a simple method of making the magnetic field visible is to place a piece of glass or paper over a bar magnet and then sprinkle iron filings on the glass (paper). The iron filings are affected by the magnetic field, and align themselves in a pattern following the lines of flux.

The greatest number of iron filings are attracted at the poles, indicating that the force of the magnetic field is greatest near the poles. As the distance from the poles increases, the field becomes weaker and fewer iron filings appear. Thus, the magnetic field may be characterized as a force whose intensity diminishes as the distance from the poles increases. Figure 4-3 shows lines of force which represent the magnetic field around a bar magnet. Arrowheads placed on each line of force show

that they enter the magnet at the south pole and leave the magnet at the north pole. Within the magnet, the lines of force flow from south to north.

In actuality the magnetic field completely surrounds a magnet, and does not exist only in a single plane. It extends in all directions around a bar magnet.

Lines of force have certain properties or characteristics:

1. Magnetic lines of force are continuous and form closed loops.
2. Flux lines have direction or polarity.
3. Flux lines never cross one another.
4. Lines of force follow the path of least resistance.

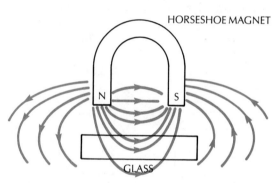

(a) NONMAGNETIC MATERIAL

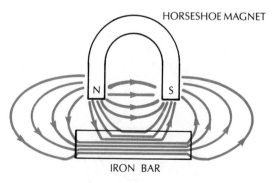

(b) MAGNETIC MATERIAL

Figure 4-4 Lines of force for magnetic and nonmagnetic material in a magnetic field.

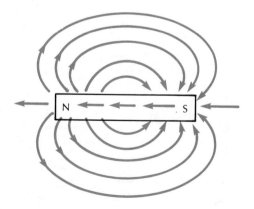

Figure 4-3 Lines of force around a bar magnet.

5. Magnetic lines of force that have the same direction tend to repel each other.

When nonmagnetic materials such as wood, glass, or rubber are placed in a magnetic field, the magnetic field lines are unaffected (Fig. 4-4(a)). The flux lines remain the same. But when a magnetic substance such as iron or alnico is placed in the magnetic field, the field lines become concentrated (Fig. 4-4B).

LAWS OF MAGNETISM

When two magnets are brought close together, whether its two like poles or two unlike poles, their magnetic fields will interact. One of the characteristics of a magnetic line is that it will not cross another line, and this fact determines how the fields will interact with each other. If the lines of force are acting in the same direction, they will join forces as they approach each other. This is why unlike poles attract (Fig. 4-5a).

If the lines are acting in the opposite direction, they will oppose each other. They apply a force against each other. Thus, like poles repel (Fig. 4-5).

As shown, the first rule of magnetism states that like poles repel and unlike poles attract. The second rule states that the force between two magnetic poles is directly proportional to the product of the strength of the two poles and inversely proportional to the square of the distance between the two poles. This law is expressed by the following mathematical equation;

$$F = \frac{m_1 \cdot m_2}{d^2} \tag{4-1}$$

where F = the attracting or repelling force between the two magnetic poles
$m_1 \cdot m_2$ = strength of the poles
d = distance between the poles

MAGNETIC INDUCTION

If a piece of soft iron is placed in the field of a magnet (but not in contact with the magnet), the iron will become magnetized. The lines of force emanating from the permanent magnet aligns electrons of the iron bar's atoms in the same direction instead of the random pattern found in unmagnetized iron (Fig. 4-6). The magnetized iron bar develops its own mag-

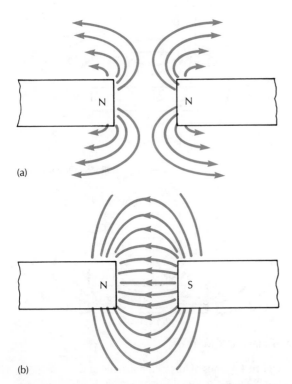

(a)

(b)

Figure 4-5 Magnetic fields between (a) like poles, (b) unlike poles.

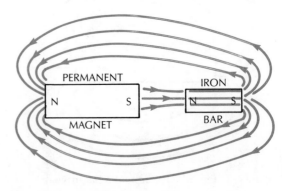

Figure 4-6 Magnetization by induction.

netic poles. The induced poles have opposite polarity from the poles of the permanent magnet. The fact that a magnetic field extends outward from a magnet is the basis for inductive effects in electric and electronic circuits. This will be covered in greater detail later in this chapter and in Chapter Five.

MAGNETIC SHIELDING

Watches, compasses, and other precision instruments are adversely affected by magnetic fields. These instruments may be protected by enclosing them in soft-iron magnetic shields (Fig. 4-7). The soft iron with a high permeability directs the magnetic lines around the object to be shielded.

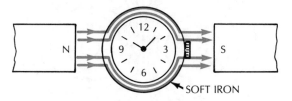

Figure 4-7 Magnetic shield protecting a watch.

CONSEQUENT POLES

It is possible to magnetize a substance that will have two similar poles at the ends, and two like poles in the centre of the magnet. The two like poles in the interior are called *consequent poles* (Fig. 4-8).

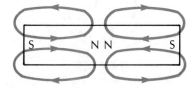

Figure 4-8 Consequent poles.

ELECTROMAGNETISM

In the year 1819, a Danish physicist, Hans Christian Oersted, discovered that electrons flowing through a conductor will set up a magnetic field around the conductor. This mag-

netic field, similar to the one produced by a permanent magnet, is known as *an electromagnetic field*. The magnetic field around the conductor forms a series of loops (concentric circles) around the conductor (Fig. 4-9). The more current that flows through the conductor, the greater the number of lines of flux that will be produced.

The direction of these circular lines of force can be determined by moving a compass around the conductor. The compass needle will align itself in the direction of the flux lines (Fig. 4-9). As with an ordinary magnet, the magnetic field intensity is greatest near the wire, and decreases with distance.

- MAGNETIC FIELD AROUND CONDUCTOR
- END VIEW
- DOT REPRESENTS HEAD OF ARROW

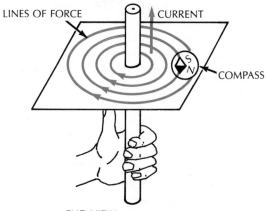

- END VIEW
- CROSS REPRESENTS TAIL OF AN ARROW
- MAGNETIC FIELD AROUND CONDUCTOR

Figure 4-9 Determining the direction of circular lines of force using a compass. The magnetic field around a conductor.

LEFT-HAND CURRENT RULE

The diagram in Fig. 4-9 shows that there is a relationship between the direction of electron flow and the direction of the flux lines. The direction of the flux lines can be determined if the direction of electron flow is known, or vice versa. The rule for determining this is called the left-hand rule for a straight conductor. It is illustrated in Fig. 4-9. The rule states:

> If you grasp a conductor in your left hand and wrap your fingers around the wire with your thumb pointing in the direction of electron flow, your fingers will point in the same direction as the magnetic flux.

Figure 4-9 also shows the magnetic field around a conductor and the action of the magnetic field around the conductor when viewed from either end of the conductor. If two such conductors were brought close together, the magnetic fields around both conductors would then interact. If the currents flowing in both conductors are in opposite direction, their magnetic fields will oppose each other (Fig. 4-10) and will cause repulsion. However, if the currents in both conductors are flowing in the same direction (Fig. 4-10(a)), their magnetic fields will link together and be located around both conductors, and the conductors will be attracted to each other. Thus, it can be stated that parallel currents flowing in the same direction produce an attracting force; while parallel currents flowing in opposite directions produce a repelling force.

The magnetic field produced by current flowing in a straight conductor is of little practical use. It does have direction but no polarity, and the magnetic field has little strength. By changing the shape of the wire to form a turn, the magnetic characteristics can be greatly improved (Fig. 4-11). All the circular flux lines enter on one side of the turn and leave on the other side. The flux lines are brought closer together and are compressed in the centre of the turn to create a strong magnetic field. In

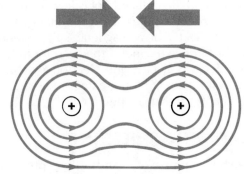

(a) currents flowing in the same direction,

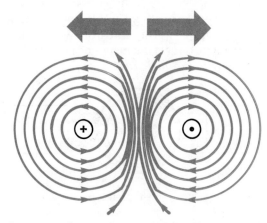

(b) currents flowing in opposite directions

Figure 4-10 Interaction of electromagnetic fields in parallel conductors.

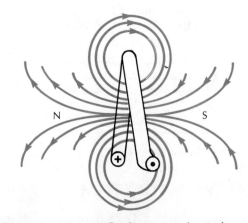

Figure 4-11 Magnetic flux lines around a single turn of wire.

addition, the turn has polarity; the north pole is on the side where the flux lines leave, and the south pole on the side where the flux lines enter.

If more turns are wound in the same direction, a coil is formed (Fig. 4-12). The magnetic flux around each turn will add to make a stronger magnetic field. By increasing the number of turns, the magnetic field becomes even stronger. If the turns are tightly compressed, the strength of the magnetic field increases. In addition, the more current that flows through the coil, the more flux lines there will be and the strength of the magnetic field will increase further. This coil has now become an electromagnet because the magnetic field results from the *movement* of electron energy.

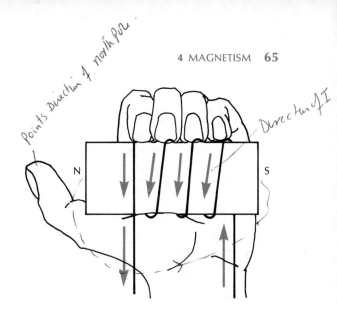

Figure 4-13 Left-hand rule for coils.

at right angles to the coil. The thumb then points in the direction of the north pole.

Thus, the direction of current flow can be determined if the north pole is known, or vice versa.

The strength of the magnetic field may be greatly increased by placing an iron core inside the coil. When the magnetic field enters the iron, it forces the electrons to align themselves so that the magnetic energy of the four uncancelled electrons in each atom is *added* to the current's magnetic field. Soft iron is used as a core material because of its high permeability (it is easily magnetized and demagnetized). Hard steel would become permanently magnetized.

MAGNETIC CIRCUITS

Many electrical and electronic devices such as meters, transformers, relays, generators, motors, microphones, reed switches, etc., depend on electromagnetic forces for their operation.

A magnetic circuit is a path (closed-loop) of low reluctance to the flow of flux lines produced by the magnetic force. There are three basic types of simple magnetic circuits: air, iron core, and iron core with an air gap (Fig. 4-14).

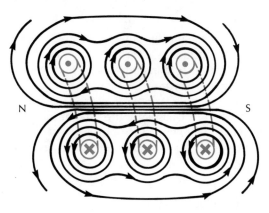

Figure 4-12 A cross-sectional view of a current-carrying coil and its electromagnetic field.

To reverse the polarity of an electromagnet, either the current flow needs to be reversed or the wire wound in the opposite direction.

A left-hand rule may be used to determine the direction of the magnetic field surrounding a coil. This left-hand rule for a coil (Fig. 4-13) is stated as follows:

Grasp the coil with your left hand, with the fingers pointing in the direction of current flow around the coil and extend the thumb

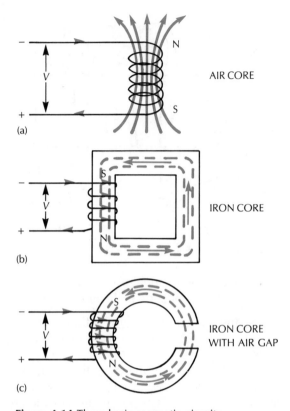

AIR CORE

(a)

IRON CORE

(b)

IRON CORE
WITH AIR GAP

(c)

Figure 4-14 Three basic magnetic circuits.

In order to understand magnetic circuits, a comparison is made between electric and magnetic circuits. Current flowing in an electric circuit can be compared to the flow of magnetic lines of force. In an electric circuit there is an electromotive force (emf), while in a magnetic circuit, a *magnetomotive force* (mmf), or magnetizing force, causes magnetic lines throughout the circuit. Sometimes the term magnetic field intensity is used as another name for magnetizing force.

The opposition to current flow in an electric circuit is *resistance*. In the magnetic circuit, the opposition to the flux lines is known as *reluctance*. With the exception of magnetic materials, most other nonmagnetic materials have high reluctance. The relationship between voltage, current, and resistance in an electric circuit is very similar to the relationship between magnetomotive force, flux, and reluctance in the magnetic circuit. Figure 4-15 compares the electric and magnetic circuit parameters.

MAGNETOMOTIVE FORCE ($\mathcal{F}$)

In order to compare the magnetic strength of different coils, the magnetomotive force (mmf) of the electromagnet has to be calculated. The magnetomotive force of a coil is directly proportional to the current flowing in the coil and the number of turns of wire per unit length. The number of turns of wire is multiplied by the number of amperes of current flowing in the coil. The unit of measure is the *ampere-turn* ($A \cdot t$). This is shown by the following equation:

$$\mathcal{F} = 4\pi NI$$
$$\text{or} \quad \mathcal{F} = 12.6 \ NI \quad \text{(Eq. 4-2)}$$

where $\mathcal{F}$ = magnetomotive force
N = number of turns in the coil
I = current in amperes
4π = a constant

Example 4-1: Calculate the magnetomotive force that will be produced by an electromagnet having a 20-turn coil and 5 A of current flowing through it.

$$\mathcal{F} = 12.6 \ NI$$
$$\mathcal{F} = 12.6 \times 20 \times 5$$
$$\mathcal{F} = 1260 \ A \cdot t$$

MAGNETIC FLUX (ϕ)

Magnetic flux refers to the total number of flux lines generated in a magnetic circuit by the magnetomotive force. Its symbol is the Greek letter ϕ (phi).

The flux in a magnetic circuit may be compared to the flow of current in the electric circuit. The unit of magnetic flux is the weber (Wb). One weber (1 Wb) equals 100 000 000 (10^8) lines of magnetic force. The weber is named after a German physicist, Wilhelm Weber (1804–1890).

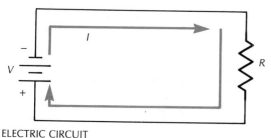

ELECTRIC CIRCUIT

MAGNETIC CIRCUIT

Figure 4-15 Comparison of electric and magnetic circuit parameters.

ELECTRIC CIRCUIT	MAGNETIC CIRCUIT
1. Electric Current (I) current flows in a complete circuit	Magnetic Flux (ϕ) magnetic flux forms complete loop
2. Electromotive force (emf) or Voltage (V) produces current flow	Magnetomotive force ($\mathcal{F}$) produces flux lines
3. Resistance (R) opposition to current flow	Reluctance ($\mathcal{R}$) opposition to flux lines
4. Ohm's Law $I = \dfrac{V}{R}$	Magnetic Law $\phi = \dfrac{\mathcal{F}}{\mathcal{R}}$

5. Factors affecting resistance
   —length
   —cross-sectional area
   —material
	Factors affecting reluctance
	—length
	—cross-sectional area
	—material

CALCULATING MAGNETIC FLUX

Example 4-2: A magnetic circuit has a magnetomotive force of 525 A·t applied to it, the reluctance of the circuit is 700 000 A·t/Wb. Calculate the total flux produced in the circuit.

Given $\mathcal{F}$ = 525 A·t
$\mathcal{R}$ = 700 000 A·t/Wb

Thus

$$\phi = \frac{\mathcal{F}}{\mathcal{R}}$$

$$\phi = \frac{525}{700\,000}$$

ϕ = 0.000 75 Wb or (75 000 flux lines)

RELUCTANCE ($\mathcal{R}$)

Reluctance is the opposition to flux lines offered by the magnetic circuit and is similar to resistance in the electric circuit. The amount of reluctance depends on the type of material (air, iron, nickel) and the length and cross-sectional area of the circuit.

The reluctance of magnetic materials is very low, while nonmagnetic materials have a high reluctance. The unit of measure for reluctance is the ampere-turn/weber (A·t/Wb).

$$\mathcal{R} = \frac{\ell}{\mu a} \tag{4-3}$$

where ℓ = length in metres
μ = permeability of the material
a = cross-sectional area (m²)

or

$$\mathcal{R} = \frac{\mathcal{F}}{\phi} \text{ (Magnetic law)}$$

For calculating reluctance, refer to Ex. 4-2.

FLUX DENSITY (β)

Flux density is defined as the number of flux lines passing through a cross-sectional area of

1 m². Since flux is measured in webers and area in square metres, thus flux density is webers per square metre (Wb/m²). However, one weber per square metre is called a tesla (T) named after a Yugoslavian scientist, Nikola Tesla (1857–1943). Thus, the unit of flux density is the tesla (T). The symbol for flux density is β. The flux density formula may be expressed as follows:

$$\beta = \frac{\phi}{a} \qquad (4\text{-}4)$$

where β = flux density (flux lines per square metre)

ϕ = total number of flux lines in the magnetic circuit

a = the cross-sectional area in square metre

CALCULATING FLUX DENSITY

Example 4-3: Find the flux density when 500 Wb lines of force pass through an area whose dimensions are 10 m long by 2 m wide.

$$\beta = \frac{\phi}{a}$$

$$\beta = \frac{500}{10 \times 2}$$

$$\beta = \frac{500}{20}$$

$$\beta = 25 \text{ T}$$

MAGNETIZING FORCE OR MAGNETIC INTENSITY (*H*)

The magnetizing force is simply the magnetomotive force acting on each metre length of material in the magnetic circuit. This unit of magnetic intensity is useful in showing that a longer piece of magnetic material will require a larger magnetomotive force to produce the same flux than a shorter length of the same substance.

The symbol for magnetic intensity is the letter H, and this quantity can be found by

using the following equation:

$$H = \frac{\mathscr{F}}{\ell} \qquad (4\text{-}5)$$

where H = magnetizing force

$\mathscr{F}$ = magnetomotive force

ℓ = length in metres

The magnetizing force is measured in ampere-turns per metre (A·t/m).

CALCULATING MAGNETIZING FORCE

Example 4-4: Find the magnetizing force of a magnetic circuit that is 0.5 m long and has an applied magnetomotive force of 1250 A·t.

$$H = \frac{\mathscr{F}}{\ell}$$

$$H = \frac{1250}{0.5}$$

$$H = 2500 \text{ A·t/m}$$

PERMEABILITY (μ)

Like conductance $\left(G = \frac{1}{R}\right)$, the inverse of reluctance is permeability.

It is an indication of the ability of a magnetic material to conduct lines of force. Any material that is easily magnetized (concentrates magnetic lines of force) has a high permeability.

The symbol for permeability is the Greek letter μ (mu). Air is used as the standard and is assumed to have a permeability of 10^{-7}, the permeability of all nonmagnetic materials is about the same. The permeability of a magnetic material such as iron may be 100 to 9000 times greater. It will be shown later that the permeability of a circuit is dependent not only on the medium, but also on the operating conditions.

The formula for permeability is:

$$\mu = \frac{\beta}{H} \qquad (4\text{-}6)$$

where β = flux density in tesla
H = magnetizing force in ampere-turns per metre

CALCULATING PERMEABILITY

Example 4-5: Calculate the permeability of a magnetic circuit of a piece of steel that has an applied magnetizing force of 1000 A·t/m, which produces a flux density of 1.4 T.

$$\mu = \frac{\beta}{H}$$

$$\mu = \frac{1.4}{1000}$$

$$\mu = 0.001\ 4\ \frac{\text{Wb/m}^2}{\text{A·t/m}}$$

MAGNETIZATION (β-H) CURVES

Magnetic materials behave quite differently from one another when under the influence of a magnetomotive force. The permeability differs from one substance to another. Also, the permeability of one substance will change as the magnetizing force is changed. For example, increasing the magnetizing force ten times for a particular material will not necessarily increase the flux density ten times.

Therefore, the permeability of a material must be determined experimentally from the existing operating conditions. However, many commonly used materials have been tested and graphs plotted to show how they behave toward the passage of magnetic lines of force. It may also be necessary to see how many lines of force flow as the magnetizing force is changed. Figure 4-16 shows the relation of the number of flux lines (β) generated in a magnetic circuit made out of cast iron, when the magnetizing force (H) is changed.

By observing the curve, it can be seen that the flux density increases rapidly as the magnetizing force increases. However, the flux density increases at a slower rate at the knee of the curve until a point called the saturation

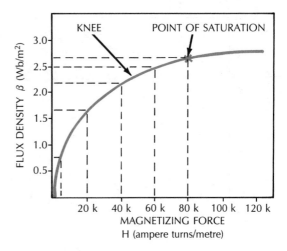

Figure 4-16 Magnetization curve for cast iron.

point is reached. Here, the magnetization curve begins to flatten out, and it becomes impossible to increase the flux density, regardless of the amount of magnetizing force used.

As a material is being magnetized, the electrons start to align themselves with the magnetizing force. There will be a point where all electrons in the magnetic material become aligned. At this point, increasing the magnetic force will not produce additional flux lines. This is known as the point of *saturation*.

The permeability of a magnetic material can be found by using the magnetization curve for different values of magnetizing force and substituting these values in the formula $\mu = \frac{\beta}{H}$.

The same magnetization curve cannot be used for all magnetic materials. Different types of steel and alloys will have different characteristics. (See Fig. 4-17 on page 70.)

MAGNETIC CIRCUIT PROBLEMS

Example 4-6: In Fig. 4-18, the coil contains 100 turns of wire and the core is made of wood. The mean (cross-sectional) diameter of the wood core is 0.01 m and of the ring 0.05 m. Find the value of $\mathcal{F}$ and I required to produce a flux flow of 10^{-6} Wb.

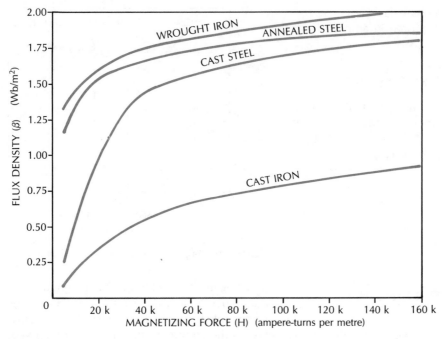

Figure 4-17 Magnetization curves for four magnetic materials.

Solution: The permeability for a nonmagnetic material such as wood, etc., is about constant and equal to 10^{-7}.

The wood core is nonmagnetic, thus $\mu = 10^{-7}$.

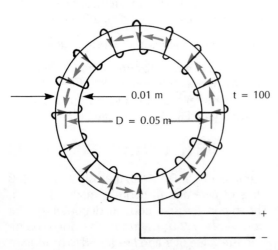

Figure 4-18 Circular magnetic circuit.

(a) Mean circumference

$$C = \pi D$$
$$C = 3.14 \times 0.05$$
$$C = 0.157 \text{ m}$$

(b) Cross-sectional area of wooden core

$$a = \frac{\pi D^2}{4}$$
$$a = \frac{3.14 \times 0.01^2}{4}$$
$$a = 7.85 \times 10^{-5} \text{m}^2$$

(c) Flux density

$$\beta = \frac{\phi}{a}$$
$$\beta = \frac{10^{-6}}{7.85 \times 10^{-5}}$$
$$\beta = 1.27 \times 10^{-2} \text{ Wb/m}^2$$

(d) Magnetizing force

$$H = \frac{\beta}{\mu}$$

$$H = \frac{1.27 \times 10^{-2}}{10^{-7}}$$
$$H = 1.27 \times 10^5 \, \text{A} \cdot \text{t/m}$$

(e) Magnetomotive force

$$\mathcal{F} = H \times \ell \; (\ell = \text{mean circumference } C)$$
$$\mathcal{F} = 1.27 \times 10^5 \times 0.157$$
$$\mathcal{F} = 1.99 \times 10^4 \, \text{A} \cdot \text{t}$$

but

$$\mathcal{F} = 12.6 \, NI$$
$$I = \frac{\mathcal{F}}{12.6 \, N}$$
$$I = \frac{1.99 \times 10^4}{12.6 \times 100}$$
$$I = 1.579 \, \text{A}$$

Example 4-7: Find the flux, permeability, and reluctance for the circuit shown in Fig. 4-19. The magnetic circuit is made of cast iron, with 75 turns of wire, and a current flow of 15 A.

(a) $\mathcal{F} = 12.6 \, NI$
$$\mathcal{F} = 12.6 \times 75 \times 15$$
$$\mathcal{F} = 14 \, 175 \, \text{A} \cdot \text{t}$$

(b) $H = \dfrac{\mathcal{F}}{\ell}$
$$H = \frac{14 \, 175}{0.44}$$
$$H = 32 \, 215 \, \text{A} \cdot \text{t/m}$$

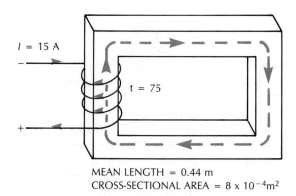

MEAN LENGTH = 0.44 m
CROSS-SECTIONAL AREA = $8 \times 10^{-4} \text{m}^2$

Figure 4-19 Rectangular magnetic circuit.

(c) Since the core is a magnetic material and the permeability depends upon the operating conditions, look up the β-H curve for cast iron. When $H = 32 \, 215$ At/m, therefore $\beta = 0.5$ Wb/m².

(d) $\phi = \beta \times a$
$$\phi = 0.5 \times 8 \times 10^{-4}$$
$$\phi = 4 \times 10^{-4} \, \text{Wb}$$

(e) $\mu = \dfrac{\beta}{H}$
$$\mu = \frac{0.5}{32 \, 215}$$
$$\mu = 0.000 \, 015 \, \frac{\text{Wb/m}^2}{\text{A} \cdot \text{t/m}}$$

or $\mu = 2 \times 10^{-5}$ at this particular operating point

(f) $\mathcal{R} = \dfrac{\ell}{\mu a}$
$$\mathcal{R} = \frac{0.44}{2 \times 10^{-5} \times 8 \times 10^{-4}}$$
$$\mathcal{R} = \frac{0.44}{16 \times 10^{-9}}$$
$$\mathcal{R} = \frac{44 \times 10^7}{16}$$
$$\mathcal{R} = 3 \times 10^7 \, \text{A} \cdot \text{t/Wb}$$

Example 4-8: In Fig. 4-20 the magnetic circuit is made up of two materials that provide different reluctances, a cast steel path and an air gap. Find (a) mmf, (b) total series reluctance.

Part I *Iron Path*:

(a) $\mathcal{F} = 12.6 \, NI$
$$\mathcal{F} = 12.6 \times 300 \times 10$$
$$\mathcal{F} = 37 \, 800 \, \text{A} \cdot \text{t}$$

(b) $H = \dfrac{\mathcal{F}}{\ell}$
$$H = \frac{37 \, 800}{0.5}$$
$$H = 75 \, 600 \, \text{A} \cdot \text{t/m}$$

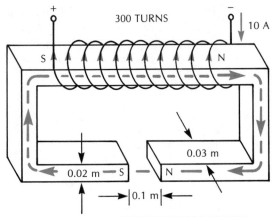

- MEAN LENGTH OF MAGNETIC PATH IN CAST STEEL
  0.5 m
- MEAN LENGTH OF AIR GAP IS 0.1 m
- CROSS SECTIONAL AREA = 0.02 × 0.03
  = 0.000 6 m²

Figure 4-20 Magnet with an air gap.

(c) From the β-H curve for cast steel, 75 600 A·t/m will produce 2.6 Wb/m².

Thus, $\phi = \beta \times a$
$$\phi = 2.6 \times 0.0006$$
$$\phi = 1.56 \times 10^{-3} \text{ Wb}$$

(d) $\mu = \dfrac{\beta}{H}$

$$\mu = \dfrac{2.6}{75\ 600}$$
$$\mu = 3.4 \times 10^{-5}$$
or
$$\mu = 340 \times 10^{-7} \dfrac{\text{Wb/m}^2}{\text{A}\cdot\text{t/m}}$$

(e) $\mathcal{R} = \dfrac{\ell}{\mu a}$

$$\mathcal{R} = \dfrac{0.5}{3.4 \times 10^{-5} \times 0.0006}$$
$$\mathcal{R} = 2.45 \times 10^7 \text{ A}\cdot\text{t/Wb}$$

Part II *Air Gap*: (permeability for air; $\mu = 10^{-7}$)

$$\mathcal{R} = \dfrac{\ell}{\mu a}$$

$$\mathcal{R} = \dfrac{0.1}{1 \times 10^{-7} \times 0.0006}$$
$$\mathcal{R} = \dfrac{0.1}{6 \times 10^{-11}}$$
$$\mathcal{R} = 0.0167 \times 10^{11} \text{ A}\cdot\text{t/Wb}$$
or
$$\mathcal{R} = 167 \times 10^7 \text{ A}\cdot\text{t/Wb}$$

Total Reluctance Path (cast steel plus air gap)

$$\mathcal{R} = 2.45 \times 10^7 + 167 \times 10^7$$
$$\mathcal{R} = 169.45 \times 10^7 \text{ A}\cdot\text{t/Wb}$$

HYSTERESIS

It was seen earlier that the permeability of a magnetic material changed when the magnetic field intensity (magnetizing force) varied, and as saturation is approached, permeability decreases in value.

The current flow in many machines, such as transformers, motors, and generators is continuously changing in intensity and direction. Therefore, the flux in the cores of such machines will also change continuously. In practice, however, the flux change does not follow exactly the change in magnetizing force, but instead lags behind because of the core's retentivity or opposition to magnetic change. This lag in magnetic flux behind the magnetizing force that produces the flux lines is called *hysteresis*. Hysteresis is a Greek word that means "to lag."

To obtain a clear picture of how magnetic materials behave in a magnetic field, the flux density must be plotted for magnetic fields in both the positive and negative directions. Figure 4-21 shows the β-H graph, where the field is first applied in one direction and then in the reverse polarity.

When a magnetizing force is applied to a magnetic substance and is gradually increased in a positive direction, the magnetic flux will increase in value initially and level off after the point of saturation is reached (A). If the magnetizing force is decreased from this value back to zero, the magnetic flux decreases

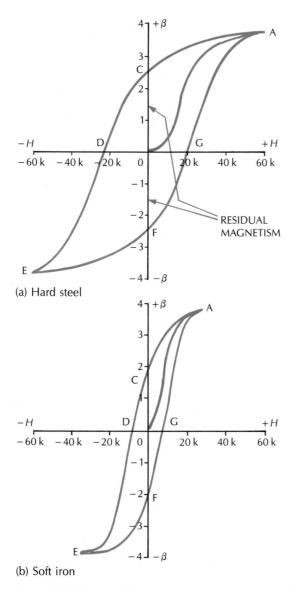

(a) Hard steel

(b) Soft iron

Figure 4-21 Hysteresis loop.

the magnetizing force has been removed. This property is called *retentivity*. Note that in Fig. 4-21(a), hard steel is capable of retaining a larger quantity of flux lines than the soft iron in Fig. 4-21(b). If the magnetizing force (*H*) is applied in a negative direction OD, the magnetization will be reduced to zero. This is the amount of magnetizing force that is required to reduce the residual magnetism to zero. The force, OD, is called the *coercive force*.

Additional negative magnetizing force will cause the magnetic material to reach saturation in the negative polarity at point E. Reducing the magnetizing force in a positive direction (towards O) will reduce the number of flux lines to point F. When the magnetizing force reaches point O, OF shows the amount of residual magnetism in the negative polarity. A further positive magnetizing force will complete the curve through point G and to A to complete what is called the *hysteresis loop*. This loop ACDEFGA is the result of the inability of the magnetic flux to follow changes in the magnetizing force. From the above explanation, it is apparent that the electrons in the magnetic material do not return to their original orientation when the magnetizing force is decreased. The property of lagging behind the magnetizing force is known as *hysteresis*. The *coercive force* required to return the electrons to their original orientation represents energy that is wasted or lost. The amount of energy lost varies with different magnetic materials. These losses also vary with the frequency the magnetizing force is changed. If the magnetizing force is reversed rapidly, the energy loss is high, and if the magnetizing force is reversed slowly, the energy loss may be negligible.

Residual magnetism can be both beneficial and a hindrance, depending upon its application. In generators, residual magnetism is beneficial in attaining a build-up of voltage; this will be discussed in greater detail in Chapter Seven, DC Generators. In meter movements, residual magnetism may result in an inaccurate reading.

along the line AC and not along the line of AO. When the magnetizing force is reduced to a value of zero, a considerable number of lines of force CO remain in the magnetic material. This value of flux CO is called residual magnetism. It also illustrates the ability of the magnetic material in retaining its magnetism after

ELECTROMAGNETIC COILS IN SERIES AND PARALLEL

In some electric circuits, coils may be connected in series or in parallel. Their combined magnetic effects depend on the flux direction from each winding. Figure 4-22 shows electromagnetic coils connected in series-aiding. The same current flows through both coils (series circuit), and both coils form magnetic fields at either end that aid each other. To produce an electromagnetic field so that the circuit is series-opposing, the winding in one leg may be reversed, or the current direction through the coil may be reversed.

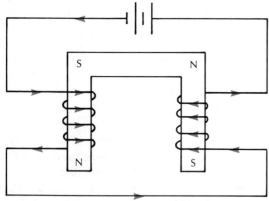

Figure 4-22 Electromagnetic coils, series-aiding.

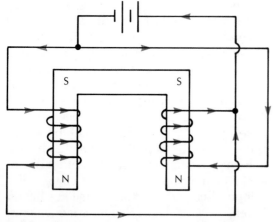

Figure 4-23 Electromagnetic coils, parallel-opposing.

In a parallel circuit, each coil draws its own current. Figure 4-23 shows electromagnetic coils connected in parallel-opposing. Here the coils form opposite poles at the same end.

SUMMARY OF IMPORTANT POINTS

1. Magnetism is used to produce electricity and electricity in turn produces magnetism.
2. Transformers, relays, motors, and generators rely on magnetic properties for their operation.
3. Atoms of ferromagnetic materials combine by sharing their valence electrons in a certain manner.
4. A substance that has its electrons aligned is said to be magnetized.
5. A magnet is a substance that has the property of attracting magnetic materials.
6. Magnetism is considered to be imaginary lines called flux lines.
7. A magnet has north and south polarities. It is at these poles that the attracting forces are greatest.
8. Magnetic lines of force travel from north to south in the external circuit. They form closed loops. The magnetic field is the space which is made up of flux lines. Lines of force do not cross each other.
9. The greater the concentration of flux lines, the stronger the magnetic field. Lines of force acting in the same direction attract (that is why unlike poles attract), while lines of force acting in opposite directions repel (that is why like poles repel). The forces of a magnet are stronger as the distance from the magnet is decreased, and get weaker as the distance becomes greater.
10. Residual magnetism is the magnetism retained in a magnetic material after the magnetizing force has been removed.

11. Permeability is the ability to conduct lines of flux.
12. Reluctance is opposition to magnetic flux.
13. Retentivity is the ability to retain magnetism.
14. Saturation occurs when all domains have been aligned.
15. Electron flow through a conductor will produce a magnetic field around the conductor.
16. Left-hand rule for a straight conductor: grasp the conductor in your left hand so that the thumb points in the direction of the current flow; the fingers will indicate the direction of the magnetic field. The greater the electron current, the stronger the magnetic field.
17. An electromagnet can be formed by winding an insulated conductor around an iron core and passing current through the coil.
18. The strength of the magnetic field of an electromagnet is determined by the number of turns of the coil and the amount of current flow.
19. Magnetizing force $(\mathcal{F}) = 4\pi NI$.
20. Left-hand rule for coils: grasp the coil in your left hand with the fingers pointing in the direction of current flow; the thumb will point in the direction of the north pole.
21. Parallel conductors with currents flowing in the same direction will attract each other.
22. Parallel conductors with currents flowing in opposite directions will repel each other.
23. Flux density (β) is the number of flux lines concentrated in a given area. It is expressed in tesla, which is webers per square metre.
24. Magnetic circuit field intensity (H) is measured in ampere turns per metre.
25. Hysteresis is the lagging behind of the electrons in aligning themselves, as the magnetizing force changes direction.

26.

Quantity	Symbol	Unit of Measure
Flux	ϕ	Weber (Wb)
Flux Density	β	Wb/m² = tesla (T)
Reluctance	$\mathcal{R}$	A·t/Wb
Permeability	μ	$\dfrac{\beta}{H}$
Magnetic Field Intensity	H	A·t/m

REVIEW QUESTIONS

1. Define
   (a) residual magnetism
   (b) retentivity
   (c) reluctance
2. Write the equation for Ohm's law as applied to magnetic circuits.
3. Explain the difference between paramagnetic, ferromagnetic, and diamagnetic materials.
4. What is the relationship between permeability and reluctance?
5. What factors in a magnetic circuit affect the reluctance?
6. Explain the difference between flux and flux density.
7. List the factors that affect the strength of any electromagnet.
8. Define
   (a) permeability
   (b) saturation
   (c) hysteresis
9. Give the magnetic symbol for each of the following:
   (a) flux
   (b) flux density
   (c) magnetizing force
   (d) permeability
10. Illustrate and fully label the hysteresis effect on soft iron. Give a brief description for this magnetic behaviour.
11. Why does an electromagnet have a considerably greater force than a coil without an iron core?

12. Define
    (a) magnetic circuit
    (b) magnetomotive force
13. Explain why permeability for a particular material varies at different operating points on the magnetization curve.
14. Explain what is meant by ferrites.
15. What are the adverse effects of hysteresis in electrical equipment?
16. Draw a typical magnetization curve, label fully, and mark the point of saturation.
17. Name the desirable magnetic qualities of steel necessary for making an electromagnet and a permanent magnet.
18. State the left-hand rule for a coil.
19. State the differences between temporary and permanent magnets.
20. List the two factors that affect the force of attraction or repulsion between two magnetic poles.

PROBLEMS

1. Calculate the mmf produced by an electromagnet having 250 turns of wire and 0.9 A of current flowing through them.
2. Find the flux density of a magnetic material that has a permeability of 0.005 $\dfrac{Wb/m^2}{A \cdot t/m}$, with a magnetic intensity of 835 $\dfrac{A \cdot t}{m}$
3. Find the current flowing through a coil that has 60 turns of wire and produces a mmf of 300 A·t.
4. Find the reluctance in Q. 3 if the flux is 30 Wb.
5. A coil wound on an iron ring has 100 turns. The mean diameter of the iron ring is 0.1 m. The cross-sectional area of the iron ring is 0.000 9 m², and the permeability is 1.5×10^{-5}. Find:
    (a) the mmf required to produce 1500 flux lines.
    (b) the current required to generate the magnetic field.

6. Calculate the flux density in a magnetic circuit that has a cross-sectional area of 0.000 3 m² and total flux of 2700 lines.
7. Find the reluctance in a magnetic circuit that has a magnetomotive force of 20 A·t which generates a total flux of 8000 lines.
8. Find the current in a magnetic circuit with 30 turns of wire having a cross-sectional area of 0.36 m², with a length of 0.03 m, which generates a flux density of 5 T. The permeability of the magnetic material is $2 \times 10^{-5} \dfrac{Wb/m^2}{A \cdot t/m}$
9. A solenoid with an iron core of 0.3 m long, has a permeability of 3×10^{-5}. The coil has 500 turns of wire and a current flow of 0.6 A. Calculate the (a) mmf (b) H (c) β.

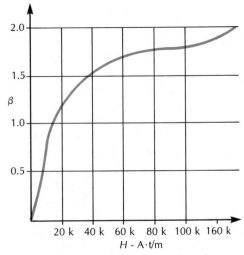

Fig. 4-24 β-H curve for laminated steel.

10. A magnetic circuit made of laminated steel has 500 turns of wire and a current flow of 12 A. The mean length of the circuit is 0.5 m and its cross-sectional area 8×10^{-4} m². Calculate the following: (a) mmf, (b) H, (c) β, (d) μ, (e) $\mathcal{R}$, (f) total flux (ϕ).
11. Find the magnetomotive force required to

produce 0.5 Wb of flux through an air gap that is 0.002 m in length and has a cross-sectional area of 0.000 064 m².

12. A cast-iron ring has a mean diameter of 0.017 m and a cross-sectional area of 0.000 036 m². Find the current that must flow through 500 turns of wire wound on the ring to produce 1.5 Wb.

CHAPTER FIVE

ELECTROMAGNETIC INDUCTION

INDUCTION

Induction may be defined as the affect of one body on another without any physical contact between them. In an earlier chapter, you saw that a charged body will induce a charge in a neutral body when the charged body is simply brought close to the neutral body. The electrostatic field, which is set up by a charged body around itself, is responsible for causing the induced charge in the neutral body without the two bodies actually touching. Induction produced by an electrostatic field is called *electrostatic induction*.

A second kind of induction is called *magnetic induction*. It is common experience that when a magnet is brought close to an unmagnetized piece of soft iron, the two objects will attract, as when the unlike poles of two magnets are brought close together. What actually happens when the magnet is brought close to the piece of soft iron is that the magnetic field which surrounds a magnet causes some of the magnetic domains in the soft iron to become aligned. The soft iron is thus made a weak magnet and the two magnets will now attract. The magnetism which is set up in the soft iron

by the magnetic field of the nearby magnet is said to be induced, and the process is called magnetic induction. The principle of magnetic

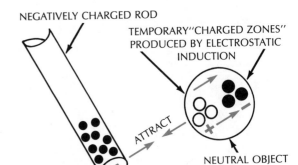

Figure 5-1(a) Electrostatic induction.

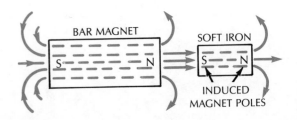

Figure 5-1(b) Magnetic induction.

induction is used in many devices, such as in magnetic switches, relays, and magnetic latches.

The third, and perhaps the most important type of induction, is called *electromagnetic induction*. This principle was first demonstrated by Faraday in 1831. This important discovery led to the development of generators, alternators, transformers, the telephone, and numerous other electrical devices. It is the basis on which all of these devices operate. It can be shown experimentally that a voltage develops between the ends of a conductor when the conductor is placed in a *varying* magnetic field.

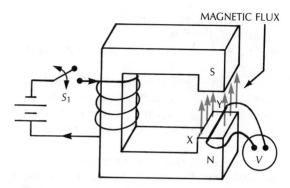

Figure 5-2 Producing voltage by electromagnetic induction.

THE REQUIREMENTS FOR ELECTROMAGNETIC INDUCTION

It was shown previously that when a current flows through a coil, as in Fig. 5-2, an electromagnet is produced. If the ends of a conductor are connected to a sensitive voltmeter, and the conductor is moved in and out of the magnetic field set up by the electromagnet, the voltmeter will be seen to deflect in one direction when the conductor is moved into the magnetic field, and in the opposite direction when the conductor is moved out of the magnetic field. If the magnet is now moved instead of the conductor, the voltmeter will again be seen to deflect in one direction as the magnet is moved towards the conductor and in the opposite direction as the magnet is removed. However, if the conductor were to be held stationary in a stationary and fixed magnetic field, no meter deflection will be observed. Now, if the conductor is again held stationary but the electromagnet is turned **on** and **off** with a switch, the voltmeter will again be seen to deflect in one direction at the instant the switch is turned **on**, and in the opposite direction at the instant the switch is turned **off**.

The foregoing discussion may be summarized by saying that when a varying magnetic field cuts through a conductor, a voltage is produced between the ends of the conductor. This voltage is produced because the moving magnetostatic force adds energy to the electrons of the conductor's atoms, causing them to become excited and move to one end of the conductor. The varying magnetic force and hence the voltage produced between the ends of the conductor will result from (a) moving the conductor so that it will cut the lines of force, (b) moving the magnet so that the conductor is cut by the flux lines, or (c) varying the magnetic flux, as when the switch in Fig. 5-2 is turned **on** and **off**. The voltage developed between the ends of the conductor in a varying magnetic field is often referred to as an *induced voltage*, and the current flow which is set up in the conductor is called an *induced current*. The process of inducing a voltage in a conductor is called electromagnetic induction.

CONDITIONS THAT AFFECT THE AMOUNT OF INDUCED VOLTAGE

The amount of voltage that is induced in a conductor by a varying magnetic field depends on four factors:
1. The strength of the magnetic field
2. The length of the conductor in the field

3. The speed at which the conductor cuts through the flux
4. The angle at which the conductor cuts the flux

When the field strength is increased, the conductor will cut more lines of force per second for a given conductor speed, and so the voltage induced in the conductor will increase.

If the conductor is wound into a coil of several turns, the effective conductor now in the magnetic field is increased, and so the induced voltage will increase. Each turn of the coil now has a voltage induced in it, and since the turns are in series, the total induced voltage of the coil is equal to the sum of the voltages induced in the turns. The induced voltage increases directly with the increase in the number of turns of the coil.

When the speed at which the conductor cuts through the lines of force is increased, the induced voltage also increases. Increasing the conductor's speed has a similar effect to increasing the field strength. An increase in speed will increase the number of flux lines cut by the conductor per second. The induced voltage is directly proportional to the speed at which the lines of force are cut.

The angle at which the conductor cuts the

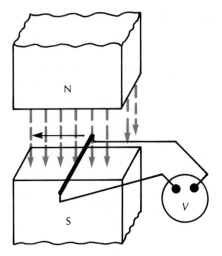

Figure 5-3(b) When a conductor is moved at 90° to flux, maximum induction occurs.

field is also important. It can be shown experimentally that maximum voltage is induced when the conductor cuts the flux at 90°, and less voltage is induced when the angle between the flux and direction of conductor's motion is less than 90°. At right angles, the conductor will cut through the maximum amount of flux and maximum induction will occur. If the conductor is moved parallel to the lines of force, it will not cut any flux lines and so the voltage induced will be zero.

FARADAY'S LAW OF ELECTROMAGNETIC INDUCTION

All of the above factors for increasing the amount of induced voltage refer to methods used to increase the amount of flux cut by the conductor per second. A basic law of electromagnetic induction is called *Faraday's Law* and it states: "The voltage induced in a conductor is directly proportional to the rate at which the conductor cuts the magnetic lines of force." In other words, the more lines of force being cut per second, the higher the induced voltage will be.

It was seen that when the flux is varied in a coil, a voltage is induced in the coil. When the

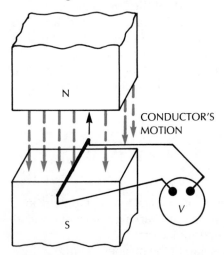

Figure 5-3(a) When a conductor is moved parallel to flux it produces minimum induction.

flux cutting a one-turn coil is varied at the rate of one weber (10^8 lines of force) per second, the voltage induced in the single turn coil is one volt.

Example 5-1: The flux in a 100-turn coil is varied at the rate of 2 Wb/s. Determine the induced voltage.

Since the turns are connected in series, the total induced voltage will be the sum of the voltages induced in the turns of the coil.

Total induced voltage = 100 × 2
= 200 V

THE LEFT-HAND GENERATOR RULE

It was shown earlier that, when the direction of the conductor's motion was reversed, the induced voltage was reversed. The same effect can be produced by reversing the magnetic lines of force. The polarity of the induced

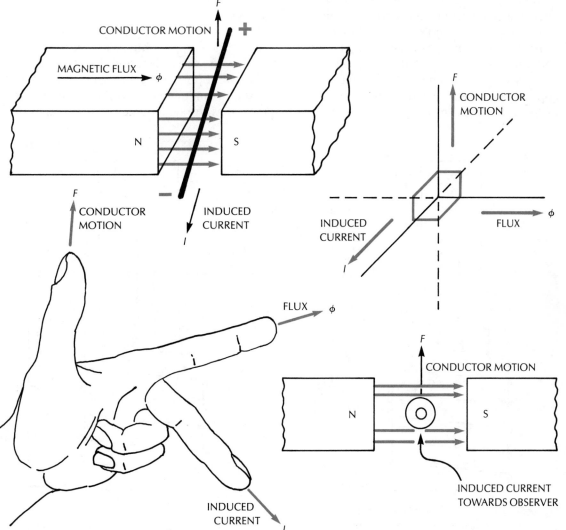

Figure 5-4 Left-hand generator rule.

voltage can be found experimentally by interpreting the direction of the voltmeter deflection caused by the induced voltage. These results are often illustrated by another of the left-hand rules. This one is called the left-hand rule for a generator and is illustrated in Fig. 5-4.

In the left-hand generator rule, the thumb, index, and middle fingers of the left hand are extended at right angles to one another. When the thumb is pointed in the direction of the conductor's motion, and the index finger pointed in the direction of the flux, the middle finger will point in the direction of the induced electron flow. The end of the conductor that gains electrons becomes negatively charged and the other end that loses electrons becomes positively charged. The rule is also applicable when the magnet is moved instead of the conductor. In this case, however, the

thumb must point in the direction of relative conductor motion.

LENZ'S LAW OF ELECTROMAGNETIC INDUCTION

Faraday's law states that an induced voltage is produced in a conductor whenever the magnetic field through the conductor is varied. Lenz's law shows that there is a definite relationship between the direction of this induced current and the variation of the field that causes it.

Lenz's law of electromagnetic induction states: "The current induced in a closed circuit will flow in such a direction that its magnetic field will oppose the applied flux which caused the induced current." For example, in Fig. 5-5, as the north pole of a bar magnet is

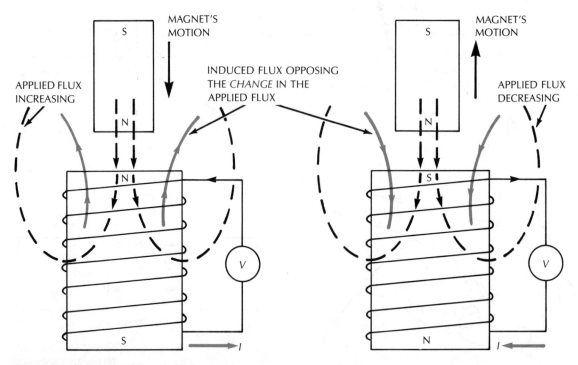

Figure 5-5 The current induced will flow in a direction such that the flux it produces would oppose the *change* in the applied flux which causes it.

inserted into the coil, the applied flux cuts through the coil circuit and induces a current flow through the coil circuit. Lenz's law states that this induced current will flow in such a direction that the magnetic field it sets up will oppose the applied flux, causing the induction. Conversely, as the north pole of the bar magnet is removed from the coil, its flux cuts the coil turns in the opposite direction. The electrons will now flow also in the opposite direction and, hence, the polarity of the coil's induced field is reversed. The coil's south pole now attracts the north pole of the bar magnet and opposes its removal.

The current induced in a coil can also be explained in terms of energy conservation. The electrical energy produced in the coil circuit must come from another part of the system. In this case, the energy transferred comes from the mechanical movement of the bar magnet. The transfer of energy involves doing work. This work is done, as the bar magnet must be moved against the opposing action of the induced field. For example, it was shown previously that when the north pole of the magnetic was withdrawn from the coil, the induced current in the coil sets up a south pole which opposed the removal of the magnetic.

MAGNETIC AND ELECTROMAGNETIC APPLICATIONS

It would be difficult to list all applications of magnetic and electromagnetic induction. However, we can examine the operation of a few common applications. Important devices such as the generator, the alternator, and the transformer are reserved for more detailed study in separate chapters.

THE MAGNETIC RELAY

The relay is one of the most useful and simplest of electromagnetic devices. Figure 5-6 shows how a relay operates. When the switch is closed, a current flows through the relay coil

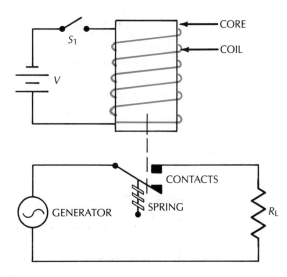

Figure 5-6 The electromagnetic relay.

and it sets up a magnetic field. This field will now attract the armature, pulling it up and thereby closing the contacts which control the generator circuit. When the switch is opened, the relay coil current is stopped and so its field collapses. The spring will now pull the contacts open, thus disconnecting the generator from the load.

This system shows how one circuit can be used to control another circuit, and at the same time be electrically isolated from the other. Because this relay coil can be energized from a relatively low voltage source and provides added safety by its isolation, relay circuits are often used to control high-voltage or high-current circuits. The property of the relay also makes it useful for remote control applications. For example, in some industrial and commercial buildings, relay circuits are used to control the main lighting circuits. In this type of system, the switch is located at one point and the other components are at some distance. The switch is used to operate a low-voltage relay, which in turn controls the higher voltage, higher current load circuit.

THE REED SWITCH

The reed switch is shown in Fig. 5-7. It consists

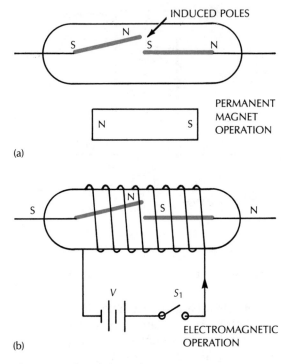

Figure 5-7 The reed switch.

of two contacts made of a ferromagnetic material, sealed in glass or plastic container. This switch can be activated either by a permanent magnet, or by an electromagnet, as shown in Fig. 5-7 (a) and (b). The electromagnet reed switch device is often called a reed relay.

The operation of this device is quite simple. When a magnet is placed near the switch, a magnetic field is induced in each contact by the flux lines from the magnet. Therefore, each contact will become a tiny magnet with magnetic poles, as shown. These poles will now attract, closing the contacts of the switch. The operation of the reed relay is similar to the reed switch, the only difference being that the magnetic force is now provided by an electromagnetic coil.

COMPUTER MEMORIES

A variety of electromagnetic devices are used to store information in computers. The magnetic cores illustrated in Fig. 5-8 are one such device. These cores are actually tiny pieces of a ferrous material which can be magnetized in either of two directions. A current flow through the wires, which are strung through the cores, can be made to impart the desired magnetic direction to the cores. Since each core may be magnetized in either of two directions, each core is therefore a binary device. One magnetic direction can be arbitrarily set to represent 0 and the other direction to represent 1. In binary computer logic, patterns of 0's and 1's are used to represent information such as letters and words. In Fig. 5-8, the pattern 1001 is illustrated.

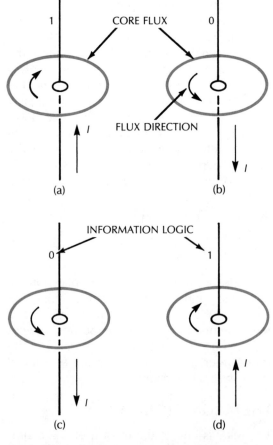

Figure 5-8 Magnetic cores in a computer memory.

SUMMARY OF IMPORTANT POINTS

1. Induction may be defined as the effect of one body on another without any physical contact between them.
2. Induction produced by an electrostatic field is called electrostatic induction.
3. Magnetism may be induced in a material such as soft iron by bringing a magnet close to it. This process is called magnetic induction.
4. When a conductor is placed in a varying magnetic field, a voltage will be induced between the ends of the conductor. This type of induction is called electromagnetic induction.
5. The voltage produced by electromagnetic induction depends on (a) the strength of the magnetic field, (b) the speed at which the flux cuts the conductor, (c) the length of the conductor, and (d) the angle at which the conductor cuts the flux.
6. Maximum induction occurs when the conductor cuts through the flux at right angles.
7. Faraday's law of electromagnetic induction states that the voltage induced in a conductor is directly proportional to the rate at which the conductor cuts the magnetic lines of force.
8. A voltage of one volt will be induced when a one-turn coil is cut by a magnetic flux at the rate of one weber per second.
9. In the left-hand generator rule, the thumb, index, and middle fingers of the left hand are extended at right angles to one another. When the thumb is pointed in the direction of the conductor's motion, and the index finger pointed in the direction of the flux, the middle finger will point in the direction of the induced electron flow in the conductor.
10. Lenz's law states that the induced current in a closed circuit will flow in such a direc-
tion that its magnetic field will oppose the applied flux which caused the induced current.
11. The magnetic relay uses an electromagnet which when energized is made to attract an armature. The armature's movement will either open or close switch contacts.
12. The magnetic relay can (a) provide circuit isolation and (b) use a low voltage to control a high voltage.
13. The reed switch consists of two contacts made of a ferromagnetic material and sealed in a glass or plastic container. The switch is activated when it passes close to a magnet.

REVIEW QUESTIONS

1. Briefly define each of the following:
   (a) electrostatic induction
   (b) magnetic induction and
   (c) electromagnetic induction
2. Give an application for each of the three induction methods in Q. 1.
3. Draw diagrams and briefly describe two methods whereby the external flux which cuts through a coil circuit may be varied.
4. What effect would each of the following circuit changes have on the circuit's induced current?
   (a) reversing the direction of a conductor's motion through a magnetic field
   (b) reversing the direction of the flux which cuts through a coil
   (c) allowing the magnetic field about a coil to collapse instead of to expand
5. Describe three methods of increasing the amount of induced voltage in a coil.
6. Briefly describe the left-hand generator rule.
7. Determine the voltage developed across a 200-turn coil when the flux which cuts the coil is 2 Wb/s.
8. Use Lenz's law and determine the direc-

tion of the induced current in each of the following circuits.

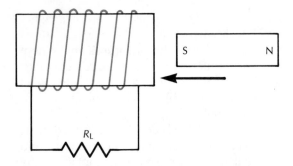

Figure 5-9(a) South pole of magnet entering coil.

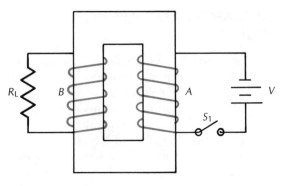

Figure 5-9(b) Induction at the instant the switch S_1 is closed.

9. You may need to do a little reading for this question. Briefly describe how the principles of magnetism and induction are used in each of the following:
   (a) tape recording and playback
   (b) inductive heating
   (c) the telephone receiver
   (d) the loudspeaker
   (e) transformer

10. Explain the advantages of using electromagnetic relays in some control systems.

CHAPTER SIX

DC ELECTRICAL MEASURING INSTRUMENTS

In this highly complex technological age, there is a need for various types of measurements such as pressure, temperature, humidity, current, voltage, resistance, etc. To understand electric circuitry, one will need to know about its voltages, currents, resistances, power, etc., and, consequently, the use and operation of electrical meters.

There are numerous types of electrical meters in use today; most measurements, however, are based on the properties of I, V, and R. In this chapter, the ammeter, voltmeter, and ohmmeter will be discussed.

These meters are similar in construction, since the meter movement used in all three instruments is basically a *galvanometer*. The galvanometer converts electrical energy to mechanical energy which provides an indication on some form of calibrated scale. Most measuring instruments operate on the basic principles of electromagnetism. These principles show how the magnetic fields of two magnetized objects can produce an attracting or repelling force between the two objects. The amount of force produced is proportional to the strength of the magnetic field, which in turn is dependent on the current flow that produces the magnetic force.

There are various types of electromagnetic current meter movements in use today of which the following are the most common: the *moving-coil meter*, the *moving iron-vane meter*, and the *electrodynamometer*. The latter two will be discussed in Chapter Twenty. The advent of microtechnology has also led to the development of digital meters, which also will be dealt with in Chapter Twenty.

THE GALVANOMETER

The galvanometer is a very sensitive device that is able to detect very small electric currents. This meter movement is also known as the d'Arsonval movement, named after Arsene d'Arsonval, a Frenchman who invented the device in 1882. It was quite delicate, very accurate, yet somewhat crude, and it could only measure very small currents. Dr. Edward Weston made improvements in the meter movement. He extended the range of

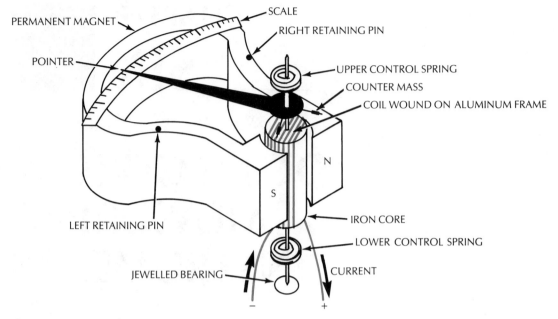

Figure 6-1 D'Arsonval meter movement.

the meter, made it rugged, inexpensive and easy to use. Figure 6-1 illustrates the basic construction features of the d'Arsonval meter movement of today.

The moving-coil meter operates on electromagnetic principles. It contains two magnets, one a stationary permanent magnet, and the other a moving coil electromagnet. The permanent magnet which produces a stationary magnetic field concentrates the field in the area of the moving coil.

The moving-coil (electromagnet) consists of several turns of fine wire wound on an aluminum frame. The aluminum frame is mounted on a shaft that is pivoted between two permanently mounted jewel bearings so that the coil can rotate freely between the poles of the permanent magnet. A pointer is attached to ·this assembly, and as it turns the pointer indicates the amount of current flow on a scale. Counterweights are attached to the pointer to achieve perfect balance, and to enable the meter to show the same reading in the vertical or horizontal position. To limit the movement of the pointer and coil, retaining pins are

located on either side of the meter. Spiral springs on each end of the shaft of the moving coil, oppositely wound, keep the meter pointer at zero when no current is flowing through the coil. Temperature changes can affect the springs, but by having two oppositely wound springs, the turning effects are cancelled out. When the moving coil turns, one spring tightens the tension to provide a braking force, while the other spring releases the tension. The spiral springs are also used to provide a path for current flow to the moving coil from the meter terminals.

To reduce the amount of flux leakage to a bare minimum and to allow the turning force (torque) to increase uniformly as the current increases, the permanent magnet poles have a semi-circular shape. This semi-circular shape allows a minimum amount of space between the permanent magnet and the moveable coil.

When current flows through the moving-coil, a magnetic field is set up with its own poles. These poles become attracted or repelled by the poles of the permanent magnet. The number of turns of wire in the moving coil and the current flowing through the moving coil will determine the strength of the magnetic field about the coil. A large current will produce a strong field, resulting in greater forces of attraction and repulsion (turning or twisting force) between the moving coil and the permanent magnet. As the moving coil rotates, the pointer rotates with it and swings across the scale that is calibrated to indicate the current flow. Thus, the greater the current flow through the moving coil, the further the coil will turn, and the pointer will indicate a larger reading on the scale.

As the pointer moves across the scale, it does not stop at a specified value, but oscillates about the correct value before coming to a full stop. To prevent this oscillation, the meter movement must be *damped*. Damping is similar to a breaking action and brings the pointer to a rest, thus eliminating the oscillatory action. The aluminum frame on which the coil is wound is a conductor, so that when the aluminum frame cuts across the lines of flux of the permanent magnet, a small current called an *eddy-current* is induced in the aluminum frame. These eddy-currents circulate around the closed circuit formed by the frame, producing their own magnetic flux. This magnetic field, which is set up around the aluminum frame, opposes the magnetic field of the moving coil and the motion that produces the eddy current. This action will reduce the meter's oscillation to a minimum value.

Another type of suspension system known as the *taut-band* suspension has become increasingly more popular. Figure 6-2 illustrates the construction of the taut-band meter movement.

In this assembly, the coil is suspended by two metal bands (torsion bands). The bands are connected between the moving coil and a tension spring that keeps the bands tight. When current flows through the moving coil, the bands become twisted, but when the current stops flowing, the bands untwist and the pointer returns to zero. This system does away with the bearings and spiral springs found in

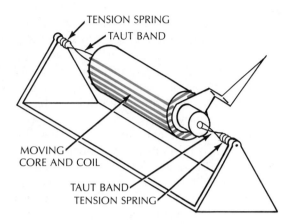

Figure 6-2 Taut-band meter movement.

the regular d'Arsonval meter movement; thus, it has less friction problems. As a result, the meter movement is more sensitive, is shock- and vibration-resistant, and is capable of retaining its accuracy for long periods. The bands also serve as conductors for the current flow to the moving coil.

METER ACCURACY AND SENSITIVITY

The *sensitivity* of a meter movement is the amount of current (I_M) necessary for full-scale deflection. The meter movement that requires the least amount of current for full-scale deflection (FSD) is the most sensitive. Of two meters, one rated at 1 mA and the other at 50 mA (FSD), the 1-mA meter is considered to be the more sensitive because it only requires 1 mA of current to produce full-scale deflection. Of course, any current greater than this value may cause the pointer to strike the right retaining pin and bend, or cause the coil to burn out, or both effects can occur.

The amount of current required for full-scale deflection depends upon the number of turns of wire on the moving coil. Smaller currents require more turns of wire in order to create a stronger magnetic field, while higher currents require heavier wire, making the coil too heavy. Every moving coil has a certain amount of resistance, which in turn is dependent upon the diameter of the wire and the number of turns on the coil. This resistance is called the *internal resistance* (R_m) of the moving coil. The internal resistance and sensitivity of any meter are inherent characteristics that cannot be changed without altering the design of the meter movement.

The reading on any meter should always be taken from a position at right angles to the meter face. Reading the pointer position from the right or left side of the meter may result in an incorrect reading. This incorrect reading is called *parallax*. Many meters have a mirror along the scale so that the meter is read at the point where the pointer and its mirror reflection are one. Thus, the optical error of parallax is eliminated.

THE BASIC AMMETER

The ammeter is an instrument used to measure current flow in a circuit. The moving coil through which the current flows has a very low resistance. Due to this low resistance, it must be used with a great deal of caution. If improperly connected, the meter movement can burn out. Certain rules for using the ammeter must be adhered to if damage to the meter is to be avoided.

1. The ammeter must be connected in series with the current being measured. The circuit under test must be broken so that the ammeter can be inserted, allowing the current to be measured to flow through the meter.
2. A dc meter must be connected observing correct polarity. This will allow the meter to deflect up-scale. A reversed polarity will cause the needle to deflect below zero, forcing the pointer against the left retaining pin.
3. The current ratings of the meter must never be exceeded.

The terminals of most meters are marked with a minus and plus sign, or red for positive and black for negative polarity. The current

flow must enter the negative terminal of the meter, through the moving coil, and leave from the positive terminal in order to read up-scale.

The negative terminal of the meter must be connected to a point in the circuit that can be tracked back to the negative side of the voltage source. Similarly, the positive terminal of the meter must be connected to the point in the circuit that returns to the positive terminal of the voltage source. Figure 1-16 in Chapter One shows the proper connections of an ammeter in a circuit.

INCREASING THE RANGE OF THE AMMETER

Each basic meter movement has a certain current rating (I_M). Very few movements can measure more than 10 mA. Thus, if a meter movement has a sensitivity of 1 mA (for full-scale deflection), the current through the movement must not exceed 1 mA. This meter movement by itself has a single usable range of 0 to 1 mA. It would be advantageous if this meter could measure currents greater than 1 mA. The simplest way to measure these large currents is to connect a low resistance in parallel (shunt) with the meter movement. This resistor is called a *shunt*. The low-resist-

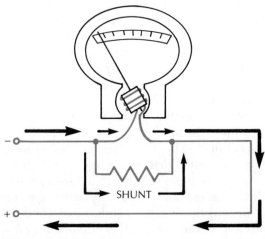

Figure 6-3 A dc ammeter with a shunt.

ance shunt will direct most of the current through the shunt while allowing a small current to flow through the meter movement. Thus, the coil and the shunt each carry a part of the total current. The circuit may be designed so that the shunt will carry a desired portion of the total current, and the meter scale is calibrated to indicate the total current, even though the meter movement carries only a small part of it. For example, if the moving coil conducts 0.1 of the total circuit current, the shunt passes the remaining 0.9, but the actual current marked on the scale is 10 times the current flowing through the moving coil. Figure 6-3 shows an ammeter with a shunt connected in parallel with the moving coil.

Shunts are made from two basic materials, manganin and nichrome. Both materials have a low temperature coefficient; that is, the resistance value of the shunt varies only slightly when exposed to a wide range of temperatures. For small currents, shunts are made of short lengths of manganin or nichrome, usually wound in the form of a coil with or without a plastic form. In applications where the amount of current flow is large, the shunts are constructed of a small plate of manganin or nichrome.

Shunts may be connected internally in meters that are designed to read up to 30 A, and externally in meters that read above 30 A.

Before calculating shunt resistances, a thorough understanding of the behaviour of currents and voltages in a parallel circuit is necessary. In Chapter Three, it was learned that the current will divide proportionally between the values of the two resistors connected in parallel. In Fig. 6-4, we see the current division with two resistors connected in parallel.

R_2 is twice as large as R_1, therefore the current flow through R_2 (5 A) is one-half the current through R_1 (10 A). Since R_1 is parallel to R_2, it follows that V_1 will be equal to V_2. Using Ohm's law $(V = IR)$, this relation may be expressed as:

$$V_1 = V_2$$

substituting $I_1R_1 = I_2R_2$

With a slight modification, the above equation can be used to develop a formula for calculating the value of a shunt for a moving coil for any application. The diagram in Fig. 6-4 is similar to a meter and shunt combination. It can now be redrawn with new labels, as shown in Fig. 6-5.

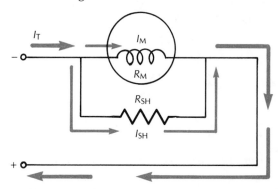

Figure 6-5 Schematic diagram showing the meter movement and the shunt.

Therefore, the equation $I_1R_1 = I_2R_2$ can now be rewritten as $I_{SH}R_{SH} = I_MR_M$. In this equation, as long as three out of four values are known, the fourth value can be calculated. Most of the time it may be necessary to find the value of the shunt resistance R_{SH}; the equation for R_{SH} as resolved from the equation $I_{SH}R_{SH} = I_MR_M$ becomes:

$$R_{SH} = \frac{I_M R_M}{I_{SH}} \qquad (6\text{-}1)$$

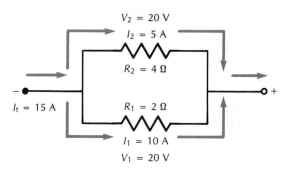

Figure 6-4 Current division in a parallel circuit.

This equation may be used to calculate the shunt when extending the range of any ammeter.

The meter sensitivity (I_M) is usually marked on the face of the meter or in the instruction manual. It should be obvious from Fig. 6-5 that the shunt current I_{SH} is equal to:

$$I_{SH} = I_T - I_M \qquad (6-2)$$

COMPUTING THE SHUNT RESISTANCE

Example 6-1: Assume that you want to extend the range of a 1 mA, 100-Ω meter movement to 10 mA (Fig. 6-6). Extending the range of the meter to 10 mA means that the total circuit current (I_T) is 10 mA (full-scale deflection).

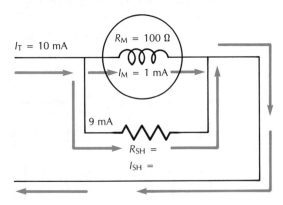

Figure 6-6 Typical problem for calculating R_{SH}.

Since the moving-coil current (I_M) is 1 mA at full-scale deflection, the shunt current (I_{SH}) must be 9 mA. The moving coil resistance (R_M) is given as 100 Ω.

The value of the shunt current may be calculated as:

$$I_T = I_M + I_{SH}$$

therefore $I_{SH} = I_T - I_M$
$$I_{SH} = 0.01 - 0.001$$
$$I_{SH} = 0.009 \text{ A}$$

Using the shunt equation for calculating R_{SH}:

$$R_{SH} = \frac{I_M R_M}{I_{SH}}$$
$$R_{SH} = \frac{0.001 \times 100}{0.009}$$
$$R_{SH} = 11.1 \ \Omega$$

Thus, a shunt resistance of 111.1 Ω is necessary to divert 9 mA of current around the meter movement when 10 mA is being measured.

The range of the 1 mA meter movement has been extended to 10 mA. With 10 mA of total current, the pointer deflects full scale to show 10 mA, even though only 1 mA flows through the meter movement. If this same meter were connected into a circuit carrying 5 mA, the current would still divide in the same ratio, that is, nine to one. One-tenth of 5 mA, which is 0.000 5 A, would flow through the meter movement and the other nine-tenths, which is 0.004 5 A, would flow through the shunt. In this case, the pointer would deflect one-half scale to indicate 5 mA.

Example 6-2: Calculate the highest current that can be read by a 0-10 mA meter movement (Fig. 6-7) when a 4-Ω shunt is placed across it. The internal resistance (R_M) is 100 Ω.

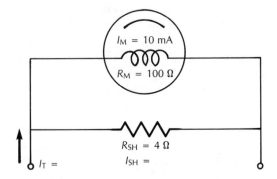

Figure 6-7 Calculating I_T.

(a) Apply Ohm's law to calculate the voltage across the moving coil.

$$V_M = I_M R_M$$
$$V_M = 0.01 \times 100$$
$$V_M = IV$$

Therefore, the voltage across the shunt resistor is

$$V_{SH} = IV$$

(b) Apply Ohm's law to calculate the current through R_{SH}.

$$I_{SH} = \frac{V_{SH}}{R_{SH}}$$
$$I_{SH} = \frac{1}{4}$$
$$I_{SH} = 0.25 \text{ A}$$

(c) Thus, the total current that can be read by the meter is the sum of the meter movement current and shunt resistor current.

$$I_T = I_M + I_{SH}$$
$$I_T = 0.01 + 0.25$$
$$I_T = 0.26 \text{ A}$$

Example 6-3: The ammeter shown in Fig. 6-8 is to measure a current of 100 mA. The meter movement sensitivity, $I_M = 0.01$ A, and the shunt resistance is 0.50 Ω. Calculate the value of the internal meter resistance (R_M).

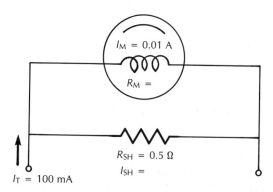

Figure 6-8 Calculating R_M.

(a) Determine the current flowing through the shunt.

$$\text{Given } I_T = I_M + I_{SH}$$

$$\text{therefore } I_{SH} = I_T - I_M$$
$$I_{SH} = 0.10 - 0.01$$
$$I_{SH} = 0.09 \text{ A}$$

(b) Find the voltage drop across the shunt resistance.

$$V_{SH} = I_{SH} \times R_{SH}$$
$$V_{SH} = 0.09 \times 0.50$$
$$V_{SH} = 0.045 \text{ V}$$

(c) Since R_M and R_{SH} are in parallel, therefore
$$V_M = 0.045 \text{ V}$$

(d) Applying the formula

$$R_M = \frac{V_M}{I_M}$$
$$R_M = \frac{0.045}{0.01}$$
$$R_M = 4.5 \text{ Ω}$$

MULTIRANGE AMMETERS

A multirange ammeter contains a basic meter movement and several shunts. A range switch selects the appropriate shunt for the desired current range. Thus, it is possible to change conveniently from one scale to another. To avoid possible meter damage when using the multirange ammeter, it is important to start at

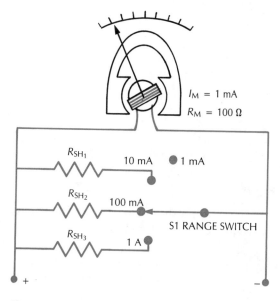

Figure 6-9 Basic multirange ammeter circuit.

the highest meter scale value and then switch to a lower scale value, one that will give a pointer deflection of about centre scale. Figure 6-9 shows a basic multirange ammeter circuit.

CALCULATING THE SHUNTS IN A MULTIRANGE AMMETER

The value of the shunt resistance for each range is calculated using the same method as the shunt for a single-range current meter.

Example 6-4: For this example, the values to be used are shown on Fig. 6-9. Assume that the range of this 0-1 mA, 100-Ω meter movement is to be extended to measure 0-10 mA, 0-100 mA, and 0-1 A by using shunts for each range.

(a)   No shunt resistance is required for the 0-1 mA range because the meter movement is rated at 1 mA, thus it can handle up to 1 mA of current.

(b)   For the 0-10 mA range, the total current entering the meter is 10 mA. The portion of current that will flow through the shunt is calculated as:

$$I_T = I_{SH_1} + I_M$$
$$I_{SH_1} = I_T - I_M$$
$$I_{SH_1} = 0.01 - 0.001$$
$$I_{SH_1} = 0.009 \text{ A}$$

Now R_{SH} is calculated.

$$R_{SH_1} = \frac{I_M R_M}{I_{SH_1}}$$
$$R_{SH_1} = \frac{0.001 \times 100}{0.009}$$
$$R_{SH_1} = 11.1 \ \Omega$$

Therefore, a shunt resistance of 11.1 Ω is required for the 10 mA range.

(c)   In the 100 mA range, the total current is 100 mA, thus the shunt current is calculated as:

$$I_{SH_2} = I_T - I_M$$
$$I_{SH_2} = 0.10 - 0.001$$
$$I_{SH_2} = 0.099 \text{ A}$$

For this range, 0.099 A must be diverted through the shunt resistance. Now the value of the shunt is computed as:

$$R_{SH_2} = \frac{I_M R_M}{I_{SH_2}}$$
$$R_{SH_2} = \frac{0.001 \times 100}{0.099}$$
$$R_{SH_2} = \frac{0.1}{0.099}$$
$$R_{SH_2} = 1.01 \ \Omega$$

(d)   Finally, in the 0-1 A range the shunt current is:

$$I_{SH_3} = I_T - I_M$$
$$I_{SH_3} = 1 - 0.001$$
$$I_{SH_3} = 0.999 \text{ A}$$

The value of the shunt for the 0-1 A range is:

$$R_{SH_3} = \frac{I_M R_M}{I_{SH_3}}$$
$$R_{SH_3} = \frac{0.001 \times 100}{0.999}$$
$$R_{SH_3} = \frac{0.1}{0.999}$$
$$R_{SH_3} = 0.10 \ \Omega$$

THE AYRTON (RING) SHUNT

The arrangement of shunts that were calculated in Fig. 6-9 is the simpler of two arrangements for calculating the value of shunt resistors. This method of connecting shunts has two disadvantages:

1. The shunt is momentarily disconnected from the circuit when the range switch is moved from one position to another, at which time the full-line current flows through the meter movement. This momentary surge of current could burn out the meter movement.

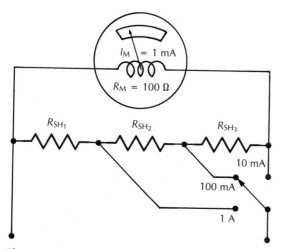

Figure 6-10 Ayrton (ring) shunt

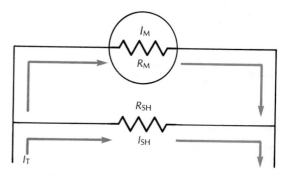

Figure 6-11 Developing a ring-shunt equation.

from Kirchhoff's current law

$$I_T = I_M + I_{SH} \qquad (6\text{-}4)$$

Example 6-5: For the Ayrton-ring shunt shown in Fig. 6-12, find the shunt resistances for the following ranges, when $I_M = 1$ mA and $R_M = 100 \ \Omega$. (a) 0-10 mA range, (b) 0-100 mA range, (c) 0-1 A range, (d) the "ring" resistances R_{SH_2} and R_{SH_3}.

(a) 0-10 mA range (Fig. 6-12(a) see page 96): The shunt resistance for this range consists of $R_{SH_1} + R_{SH_2} + R_{SH_3}$.

from Eq. 6-4 $I_{SH} = I_T - I_M$
$$I_{SH} = 10 \text{ mA} - 1 \text{ mA}$$
$$I_{SH} = 9 \text{ mA}$$

from Eq. 6-3 $R_{SH} = \dfrac{I_M R_M}{I_{SH}}$

$$R_{SH} = \dfrac{I_M R_M}{I_{SH}}$$

$$R_{SH} = \dfrac{(1 \text{ mA})(100)}{9 \text{ mA}}$$

$$R_{SH} = 11.11 \ \Omega$$

thus, $R_{SH_1} + R_{SH_2} + R_{SH_3} = 11.11 \ \Omega$

(b) 0-100 mA range (Fig. 6-12(b)): The shunt resistance for this range consists of $R_{SH_1} + R_{SH_2}$.

from Eq. 6-4 $I_{SH} = I_T - I_M$
$$I_{SH} = 100 \text{ mA} - 1 \text{ mA}$$
$$I_{SH} = 99 \text{ mA}$$

2. The contact resistance (the resistance of the range switch contacts) becomes significant. This resistance becomes an appreciable part of the total shunt resistance and could cause a significant error in a meter reading.

The ring shunt circuit disconnects the meter movement from the circuit under test when the range is switched from one position to another, thus isolating the meter movement from any surge currents that may be present. Figure 6-10 shows a typical Ayrton shunt connection.

DEVELOPING THE RING-SHUNT EQUATION

The current flowing in Fig. 6-11 will divide itself in a ratio inversely proportional to the ratio of R_M and R_{SH}. From the relationship:

$$V_M = V_{SH}$$
$$\text{and } I_M R_M = I_{SH} R_{SH}$$

The current and resistance may be expressed by the ratio

$$\frac{I_M}{I_{SH}} = \frac{R_{SH}}{R_M} \qquad (6\text{-}3)$$

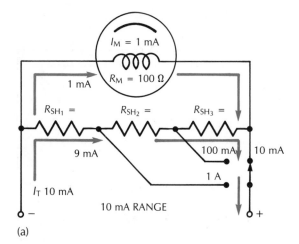

(a)

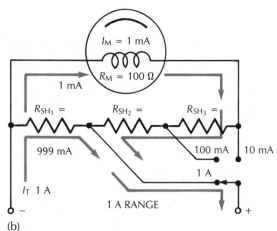

(b)

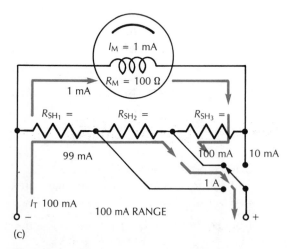

(c)

Figure 6-12 Calculating the shunt resistances of an Ayrton-ring shunt.

from Eq. 6-3 $R_{SH} = \dfrac{I_M R_M}{I_{SH}}$

$$R_{SH} = \dfrac{(1 \text{ mA})(100)}{99 \text{ mA}}$$

$$R_{SH} = 1.01 \ \Omega$$

thus $R_{SH_1} + R_{SH_2} = 1.01 \ \Omega$

(c) 0-1 A range (Fig. 6-12):
The shunt resistance for this range consists of only R_{SH_1}.

from Eq. 6-4 $I_{SH} = I_T - I_M$
$$I_{SH} = 1 \text{ A} - 1 \text{ mA}$$
$$I_{SH} = 1 \text{ A} - 0.001 \text{ A}$$
$$I_{SH} = 0.999 \text{ A (999 mA)}$$

from Eq. 6-3 $R_{SH} = \dfrac{I_M R_M}{I_{SH}}$

$$R_{SH} = \dfrac{(1 \text{ mA})(100)}{999 \text{ mA}}$$

$$R_{SH} = 0.10 \ \Omega$$

thus $R_{SH_1} = 0.10 \ \Omega$

(d) To calculate the "ring" resistances R_{SH_2} and R_{SH_3}:

$$R_{SH_2} = (R_{SH_1} + R_{SH_2}) - (R_{SH_1})$$

by substituting the results from part (b) and (c)

$$R_{SH_2} = 1.01 - 0.10$$
$$R_{SH_2} = 0.91 \ \Omega$$
$$R_{SH_3} = (R_{SH_1} + R_{SH_2} + R_{SH_3}) - (R_{SH_1} + R_{SH_2})$$

by substituting the results from part (a) and (b):

$$R_{SH_3} = 11.11 - 1.01$$
$$R_{SH_3} = 10.1 \ \Omega$$

therefore $R_{SH_1} = 0.10 \ \Omega$
$R_{SH_2} = 0.91 \ \Omega$
$R_{SH_3} = 10.1 \ \Omega$

VOLTMETERS

The basic current meter movement can also be used to measure voltage but, as in the ammeter, its range is limited. Every meter movement has a fixed resistance, and when current flows through the movement, a voltage drop is developed across it. In fact every meter movement has a certain voltage and current rating. The rated voltage drop across the moving coil will cause full-scale deflection. For example, a meter movement having a current sensitivity of 0–1 mA with a coil resistance of 500 Ω will have a rated voltage drop of:

$$V_M = I_M R_M$$
$$V_M = 0.001 \times 500$$
$$V_M = 0.5 \text{ V}$$

This meter will deflect full scale when connected across a voltage of 0.5 V. Since the deflection is proportional to the current, the meter may therefore be used to measure voltages from 0 to 0.5 V. The meter scale will be calibrated to measure units of voltage. If the above meter movement were connected across a higher voltage than 0.5 V, it would be damaged.

To measure voltages higher than the voltage range of the basic meter movement, a resistor called a *multiplier resistor* is connected in series with the meter movement. The purpose of the multiplier is to limit the current flow through the meter, so that it will not exceed the meter sensitivity rating. Figure 6-13 shows a multiplier connected in series with a meter movement to form a voltmeter.

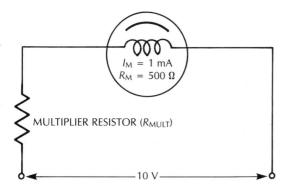

Figure 6-13 A voltmeter.

The multiplier resistor will cause the voltmeter to have a high resistance, unlike the

ammeter, which is a low-resistance device. This high-resistance property of a voltmeter can be reasoned from the method of its connections in a circuit. A voltmeter, unlike the ammeter, is connected in parallel, and this will tend to increase the circuit current, as shown in Fig. 6-14. In order to minimize any changes in the circuit operation, it should be obvious that this meter must be made to have a high resistance,

$$I_T = I_L \text{ (without meter)}$$
$$I_T = I_L + I_M \text{ (with meter connected)}$$

Therefore, R_M should be kept relatively large, making I_M very small.

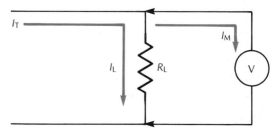

Figure 6-14 Connecting a high-resistance voltmeter across a resistance.

Example 6-6: Given the values shown in Fig. 6-14, the meter is to be connected across 10 V. Calculate the value of R_{mult} required in the circuit to limit the current flow to 1 mA.

$$R_T = \frac{V_{max}}{I_M}$$
$$R_T = \frac{10}{0.001}$$
$$R_T = 10\ 000 \text{ Ω}$$

The 10 000-Ω resistance is the total opposition required to limit the current flow to 1 mA. The multiplier resistance is:

$$R_{mult} = R_T - R_m$$
$$R_{mult} = 10\ 000 - 500$$
$$R_{mult} = 9500 \text{ Ω}$$

The above meter movement can now measure 0 – 10 V safely without being damaged. The scale is calibrated to read 0 – 10 V.

The correct polarity must be observed when connecting a dc voltmeter in the circuit.

The negative terminal of the voltmeter must be connected to the negative side of the voltage being measured and the positive terminal of the voltmeter to the positive side of the circuit.

MULTIRANGE VOLTMETERS

A voltmeter can be designed to have more than one scale by using several multiplier resistors. A range switch is used to select the proper multiplier for the required scale. Figure 6-15 illustrates a multirange voltmeter.

There are two methods used for calculating the values of multipliers (R_{mult}) for a multirange voltmeter. In the first method, as shown in Fig. 6-15, each multiplier is calculated as a separate voltmeter, just as in Ex. 6-6.

Example 6-7: Calculate the multiplier resistance required for each of the ranges shown in Fig. 6-15.

(a) 0-1 volt range

$$R_T = \frac{V_{max}}{I_M}$$

$$R_T = \frac{1}{0.001}$$

$$R_T = 1000 \ \Omega$$

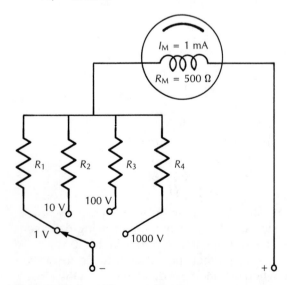

Figure 6-15 A multirange voltmeter.

Since R_M is 500 Ω, the multiplier resistance is calculated to be:

$$(R_1) \ R_{mult} = R_T - R_M$$
$$(R_1) \ R_{mult} = 1000 - 500$$
$$(R_1) \ R_{mult} = 500 \ \Omega$$

(b) 0-10 volt range

$$R_T = \frac{V_{max}}{I_M}$$

$$R_T = \frac{10}{0.001}$$

$$R_T = 10 \ 000 \ \Omega$$

therefore $(R_2) \ R_{mult} = R_T - R_M$
$$(R_2) \ R_{mult} = 10 \ 000 - 500$$
$$(R_2) \ R_{mult} = 9500 \ \Omega$$

(c) 0-100 volt range

$$R_T = \frac{V_{max}}{I_M}$$

$$R_T = \frac{100}{0.001}$$

$$R_T = 100 \ 000 \ \Omega$$

therefore $(R_3) \ R_{mult} = R_T - R_M$
$$(R_3) \ R_{mult} = 100 \ 000 - 500$$
$$(R_3) \ R_{mult} = 99 \ 500 \ \Omega$$

(d) 0-1000 volt range

$$R_T = \frac{V_{max}}{I_M}$$

$$R_T = \frac{1000}{0.001}$$

$$R_T = 1 \ 000 \ 000 \ \Omega$$

$$(R_4) \ R_{mult} = R_T - R_M$$
$$(R_4 \ R_{mult} = 1 \ 000 \ 000 - 500$$
$$(R_4) \ R_{mult} = 999 \ 500 \ \Omega$$

It can be seen from the results obtained for R_{mult}, that the higher the voltage range is, the higher the value of the multiplier resistance.

SERIES-MULTIPLIER ARRANGEMENTS

The second method of connecting multiplier resistors is the *series-multiplier arrangement*, shown in Fig. 6-16.

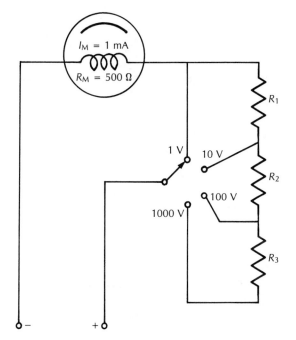

Figure 6-16 Multirange voltmeter with series-multiplier arrangements.

When a higher resistance is required for the higher voltage ranges, the switch merely adds the required series resistors. The multipliers are calculated as follows in Ex. 6-8.

Example 6-8: Calculate the multiplier resistance for each of the ranges shown in Fig. 6-16.

(a) 0-1 volt range

The meter movement indicates 1 V for a full-scale deflection, thus no multiplier is required for the 0-1 V range.

(b) 0-10 volt range

$$R_T = \frac{V_{max}}{I_M}$$

$$R_T = \frac{10}{0.001}$$

$$R_T = 10\ 000\ \Omega$$

The multiplier resistor R_1 for the 0-10 volt range is:

$(R_1)\ R_{mult} = R_T - R_M$
$(R_1)\ R_{mult} = 10\ 000 - 500$
$(R_1)\ R_{mult} = 9500\ \Omega$

(c) 0-100 volt range

$$R_T = \frac{V_{max}}{I_M}$$

$$R_T = \frac{100}{0.001}$$

$$R_T = 100\ 000\ \Omega$$

$(R_2)\ R_{mult} = R_T - R_M$
$(R_2)\ R_{mult} = 100\ 000 - 500$
$(R_2)\ R_{mult} = 99\ 500\ \Omega$

Now, the multiplier resistance for this range is actually made up of $R_1 + R_2$ in series. Since 99 500 Ω is needed for the multiplier resistance and R_1 was found to be 9500 Ω in part (a), therefore, R_2 must equal 90 000 Ω.

(d) 0-1000 volt range

$$R_T = \frac{V_{max}}{I_M}$$

$$R_T = \frac{1000}{0.001}$$

$$R_T = 1\ 000\ 000\ \Omega$$

$(R_3)\ R_{mult} = R_T - R_M$
$(R_3)\ R_{mult} = 1\ 000\ 000 - 500$
$(R_3)\ R_{mult} = 999\ 500\ \Omega$

But R_{mult} for this range is actually $R_1 + R_2 + R_3$, therefore:

$$R_1 + R_2 + R_3 = 999\ 500$$
$$9500 + 90\ 000 + R_3 = 999\ 500$$
$$99\ 500 + R_3 = 999\ 500$$
$$R_3 = 999\ 500 - 99\ 500$$
$$R_3 = 900\ 000\ \Omega$$

VOLTMETER SENSITIVITY (OHMS-PER-VOLT)

The ohms-per-volt rating is also called the sensitivity of the voltmeter, and the higher the rating, the more sensitive the meter. The ohms-per-volt rating is the amount of resistance $(R_M + R_{mult})$ needed for 1 V of deflection. Thus, the 1-mA, 500-Ω meter movement that was used in Fig. 6-16 has an ohms-per-volt rating of 500/1 or 500 ohms-per-volt. The ohms-per-volt rating is the same for all ranges because it is an inherent characteristic of the meter movement. It is determined by the full-scale current (I_M) of the meter movement.

The sensitivity in ohms-per-volt can be determined by dividing the full-scale current rating of the meter movement into 1 volt:

$$\text{ohms/volt} = \frac{1 \text{ volt}}{I_M} \qquad (6\text{-}5)$$

Thus, a 50-μA meter movement would have an ohms-per-volt rating of $\dfrac{1}{0.000\ 050} = 20\ 000$ ohms-per-volt, while a 1 mA meter would have an ohms-per-volt rating of $\dfrac{1}{0.001} = 1\ 000$ ohms-per-volt. The 50-μA meter movement is more sensitive than the 1-mA meter movement because it offers more resistance to obtain a deflection of 1 volt. It can be stated that the 50-μA meter requires less current from the circuit than the 1-mA meter to produce full-scale deflection. The ohms-per-volt rating is usually printed on the meter face.

LOADING EFFECT OF VOLTMETERS

To measure the voltage in a circuit, a voltmeter must be connected across the circuit under test. Some of the current flowing in the circuit must flow through the voltmeter, thus altering the behaviour of the circuit. Any change in circuit operation may alter the voltage being measured. For example, if a low ohms-per-volt meter is used in a high-resistance circuit, the measured voltage will not be the same as the actual voltage reading without the meter connected to the circuit. This effect is called the *loading effect* (Fig. 6-17). For example, in Fig. 6-17(a) two 100-kΩ resistors are connected in series across a 30-V source. Since the resistors are of equal value, a potential difference of 15 V is developed across each resistor.

Thus, it is expected that a voltmeter would read 15 V if it were connected across each of

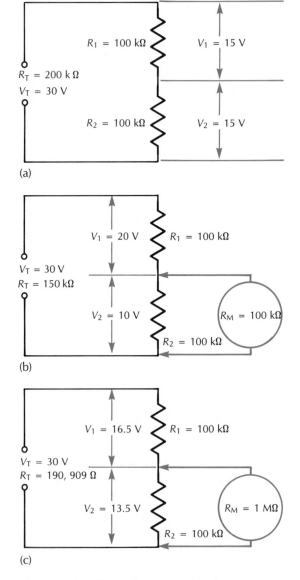

(a)

(b)

(c)

Figure 6-17 Loading effect of a voltmeter.

these resistors. However, if the voltmeter has an ohms-per-volt rating of 100 000 and it is connected across R_2, as shown in Fig. 6-17(b), the actual reading will not be correct. The reason for this inaccuracy is that when the meter with a resistance R_M is connected in parallel with R_2, the two resistors R_M and R_2 when added in parallel will effectively change the value of R_2 to 50 kΩ. As a result, the total circuit resistance is now 150 kΩ. Effectively, R_2 and R_M are now one-third of the total circuit resistance. The voltage distribution now changes, since R_1 is now larger than R_2 and R_M combined. Instead of reading 15 V, as shown in Fig. 6-17(a), the meter measures 0.33 of the source voltage, or 10 V. This represents an error of 33 percent. The loading effect of the meter movement (R_M) causes the voltage drop across R_2 to decrease. This loading effect may be minimized by using a voltmeter that has a resistance 10 times greater than the resistance across which the voltage is measured.

Figure 6-17(c) shows the same circuit with a 1-MΩ voltmeter connected across R_2. The effective resistance of R_2 and R_M added in parallel is 90 909 Ω. This is much closer to the value of R_2. The total resistance of the circuit now becomes 190 909 Ω. Therefore, the circuit operation is only slightly upset. The meter reading is 13.5 V across R_2, which represents an error of approximately 10 percent.

When using a multiple range voltmeter, the voltmeter resistance changes with the range selected. Higher ranges utilize greater values of multiplier resistance. Therefore, in high-resistance circuits, use the highest meter range that can be read accurately, for less loading effect.

THE OHMMETER

The ohmmeter is an instrument that measures resistance. In its basic form, an ohmmeter consists of an internal dry cell, a meter movement, and a current-limiting resistance. For measuring resistance in a circuit, the current in the circuit must be turned off, as the ohmmeter cell supplies current for deflecting the meter movement. Thus, if a known voltage is applied to a device and the current is measured by a meter movement, the amount of deflection is determined by the value of the resistance (Ohm's law). The scale can then be calibrated directly in ohms. Since the internal cell determines the direction of current flow through the meter movement, the test leads may be

connected to any polarity and still produce an up-scale reading.

SERIES OHMMETER

There are two types of ohmmeters, the *series ohmmeter* and the *shunt ohmmeter*. The series ohmmeter (Fig. 6-18) contains the same meter movement that was used in the ammeter and voltmeter, except that the scale is now calibrated to read in ohms.

When the meter leads are short-circuited, the meter movement deflects full scale to the right. Since there is zero resistance between the test leads, the full-scale deflection indicates zero resistance. When the ohmmeter leads are open, no current flows through the meter movement, and the meter movement does not deflect. Therefore, the ohmmeter will indicate infinite (∞) resistance, which means that there is infinite resistance (open circuit) between the test leads.

It should now be obvious that the meter scale will be marked zero ohms at the right end for full-scale deflection, and infinite ohms at the left end for very high resistance with no deflection.

Thus, the ohmmeter uses an inverse scale. The low-resistance is on the right of the meter scale and high resistance is indicated on the left.

This ohmmeter also has a nonlinear scale

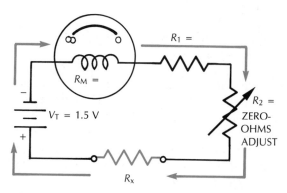

Figure 6-18 Series ohmmeter circuit.

because of the inverse relationship between resistance and current. This causes an expanded scale near the zero end of the meter and a crowded scale near the high-resistance end. Unlike the ohmmeter, the voltmeter and ammeter have linear scales because of the direct proportionality between current and voltage. The divisions in a linear scale are equal.

The fixed resistor R_1 (Fig. 6-18) acts as a current-limiting resistor to keep the current flow through the meter movement at a safe level. Variable resistor R_2 is called the *zero ohms* adjustment. The purpose for this zero adjustment is to compensate for aging cells. As the cells age, the output voltage drops, reducing the current flow through the meter movement, and the meter may no longer deflect full scale. By adjusting the value of R_2, the current flow can be increased to the proper level required for full-scale deflection.

For an ohmmeter to read higher values of resistance, either the existing cell can be replaced by another cell with a higher voltage output, or a meter movement that requires less current for full-scale deflection may be used.

MULTIPLE OHMMETER RANGES

Many ohmmeters have a range switch that can expand resistance measurements from less than 1 Ω up to many megohms. The range switch acts as the multiplying factor for an ohms scale. All that needs to be done to obtain the resistance value being measured is to multiply the scale reading by the R_x factor of the range switch.

Figure 6-19 shows how a higher resistance range can be implemented. The multiple-range ohmmeters replace the limited use of a single-range ohmmeter. A switch (S_1) is added to the switch between the two ranges. A higher voltage source (V_2) and a higher series resistor (R_3) increase the range of the ohmmeter by a factor of 10.

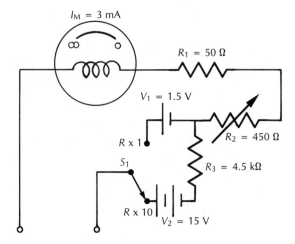

Figure 6-19 Creating higher resistance ranges.

For example, on the $R \times 1$ range, the source voltage is 1.5 V, thus the full-scale current

$$I_M = \frac{V_T}{R_T}$$
$$= \frac{1.5}{500}$$
$$= 0.003 \text{ A or 3 mA}$$

On the $R \times 10$ range, the source voltage is 15 V, thus the full-scale current is

$$I_M = \frac{V_T}{R_T}$$
$$I_M = \frac{15}{5000}$$
$$I_M = 0.003 \text{ A or 3 mA}$$

The resistance R_3 must be 10 times greater than R_2 when switching from the $R \times 1$ scale to the $R \times 10$ scale.

THE SHUNT OHMMETER

The basic ohmmeter can be modified to measure low values of resistance. This new circuit has the internal cell, meter movement, and external unknown resistance in three parallel paths. It is called the *shunt ohmmeter* and a circuit is shown in Fig. 6-20.

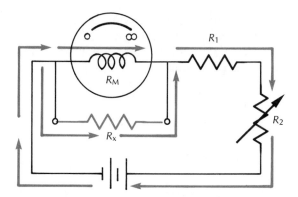

Figure 6-20 Shunt ohmmeter circuit.

The resistance to be measured is connected in parallel with the meter movement. The current flowing from the cell is divided into two paths, the meter movement (R_M) and the unknown resistance (R_x). This completely alters the characteristics of the ohmmeter. When no resistance is being measured between the test leads (infinite resistance), maximum current flows through the meter movement, deflecting the pointer to full scale. Full-scale deflection on this meter, therefore, indicates maximum (infinite) ohms. R_1 acts as a current limiter to protect the meter movement. R_2 is adjusted with test leads open to produce full-scale deflection or infinite-resistance reading.

When the test leads are shorted out (zero resistance), this causes the circuit current to bypass the meter movement. Since no current flows through the meter movement, then no deflection takes place and the meter reads 0 Ω. This is opposite to what happens in the series ohmmeter.

When measuring resistance, the current flow divides in a ratio inversely proportional to the meter resistance (R_M) and the unknown resistance (R_x). More current will flow through the smaller of the two resistances.

The scale of the shunt ohmmeter is non-linear, like that of the series ohmmeter.

One disadvantage of the shunt ohmmeter is that the cell discharges every time the ohm-meter is turned on (constant drain on the cell). The main advantage is that it can measure low values of resistance more accurately than the series ohmmeter.

THE WHEATSTONE BRIDGE

The Wheatstone bridge is used for precise measurement of resistance. Figure 6-21 shows a schematic diagram of a typical Wheatstone bridge, named after the English inventor Sir Charles Wheatstone (1802–1875).

The Wheatstone bridge consists of a voltage source (dry cell), a galvanometer, and four resistors connected in two parallel circuits called a *bridge circuit*. The source voltage is connected in series with a switch (S_1) and across the two junctions of the two parallel resistance branches (points A and B). The galvanometer has a switch (S_2) connected in series with it, and is connected across terminals C and D. The unknown resistance R_x is connected between points A and C. The unknown resistance R_x is balanced against an accurate variable resistor R_3. The resistance R_1 is equal to R_2.

Voltage is applied to the circuit by closing

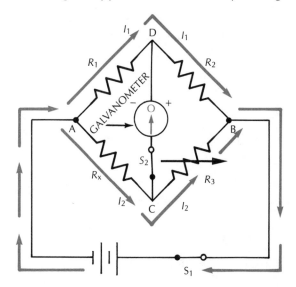

Figure 6-21 Schematic diagram of a Wheatstone bridge circuit.

switch S_1. Current flows from the voltage source to point A. From here the current divides into two paths, I_1 which flows through R_1 and R_2, and I_2 flowing through R_x and R_3. These currents will develop a voltage drop across each resistor in proportion to the value of each resistor. It is obvious that R_1 and R_2 form one voltage divider network, and R_3 and R_x form another voltage divider network. If resistors R_1 and R_2 are equal, and R_x is equal to R_3, then the current flow and voltage drop in both dividers will be equal. As a result, the potential difference between points C and D will be zero; thus, no current flows through the galvanometer, producing a zero reading. The Wheatstone bridge is said to be balanced when the current through the galvanometer is zero.

When R_3 is adjusted to produce a zero galvanometer reading, the value of resistance R_x is equal to the value of the known resistor R_3. The variable resistor R_3 has a calibrated scale which can be read to show its exact resistance value.

When the bridge is balanced, points C and D are at the same potential energy level; thus, the voltage drops. V_x will be equal to V_1 and V_3 will be equal to V_2. This relationship may be used to develop an equation for R_x for any balanced bridge circuit.

$$V_x = V_1 \text{ and } V_3 = V_2$$

therefore

$$\frac{V_x}{V_3} = \frac{V_1}{V_2}$$
$$\frac{I_2 R_x}{I_2 R_3} = \frac{I_1 R_1}{I_1 R_2}$$

I_1 and I_2 cancel, thus

$$\frac{R_x}{R_3} = \frac{R_1}{R_2}$$

Solve for R_x, the unknown resistance

$$R_x = \frac{R_1 R_3}{R_2} \tag{6-6}$$

Note that this equation applies only to a *balanced* bridge circuit.

Example 6-9: In a balanced Wheatstone bridge circuit similar to the one in Fig. 6-21, $R_1 = 20\ \Omega$, $R_2 = 1000\ \Omega$ and $R_3 = 100\ \Omega$. Find the value of the unknown resister R_x.

$$R_x = \frac{R_2 R_3}{R_1}$$
$$R_x = \frac{1000 \times 100}{20}$$
$$R_x = 5000\ \Omega$$

THE SLIDE WIRE BRIDGE

The operation of the slide wire bridge is similar to that of the Wheatstone bridge discussed earlier. Figure 6-22 shows the slide wire bridge circuit.

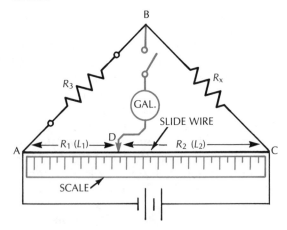

Figure 6-22 The slide wire bridge.

The wire is an alloy of uniform cross-section and resistivity. The most common alloy used is nichrome.

The operation of the slide wire bridge circuit depends on the ratio of two known resistances that are compared to the ratio of a known resistance and an unknown resistance. The point D is a sliding contact that is moved along the wire in order to balance the bridge. When the bridge is balanced, the galvanometer will indicate zero. Also, at this point, there is no difference in potential energy between points B and D.

R_1 represents the resistance of the straight wire from A to D (L_1) and R_2 represents the resistance from D to C (L_2).

The resistance of a straight conductor varies directly with the length. Thus wire marked as R_1 and R_2 will have a value of resistance that is proportional to their respective lengths.

$$\frac{R_1}{R_2} = \frac{\text{length of AD}}{\text{length of CD}} = \frac{L_1}{L_2}$$

As mentioned earlier, when the bridge is balanced, points B and D are at the same potential energy level and

therefore $V_3 = V_1$ and $V_x = V_2$

thus $\dfrac{V_x}{V_3} = \dfrac{V_2}{V_1}$

substituting IR for V

therefore $\dfrac{I_xR_x}{I_3R_3} = \dfrac{I_2R_2}{I_1R_1}$

and $R_x = \dfrac{R_2R_3}{R_1}$ or $\dfrac{L_2R_3}{L_1}$ (6-7)

A meter scale is mounted underneath the slide wire to provide a measuring scale.

Example 6-10: If $R_3 = 100\ \Omega$, L_1 is 25 cm, and $L_2 = 75$ cm, the unknown resistance is:

$$R_x = \frac{L_2R_3}{L_1}$$
$$R_x = \frac{75 \times 100}{25}$$
$$R_x = 300\ \Omega$$

The slide wire bridge has the advantage that a balance can be obtained very quickly, and any change in supply voltage will not change the setting of the bridge. However, it is not as accurate as the Wheatstone bridge because of the wearing of the wire.

THE MEGGER (MEGOHMMETER) INSULATION TESTER

The Megger (Fig. 6-23) is an instrument that is designed to measure very high values of resistance (megohm range) beyond the range of ordinary ohmmeters and the Wheatstone

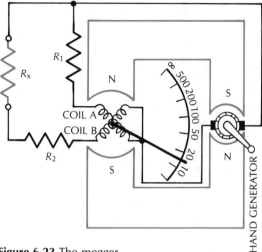

Figure 6-23 The megger.

bridge. These high values of resistance are found in conductor insulation, and between the windings in motors and transformers. To make this resistance test for insulation breakdown, a very high voltage is necessary, higher than the voltage delivered by an ohmmeter cell. This high voltage is applied across the conductor and the outside surface of the insulation.

The Megger consists of a hand-driven dc generator which supplies the necessary voltage for making the measurement and a meter movement to indicate the value of the resistance being measured. The hand generator can generate a voltage of 100 V, 500 V, 1000 V, 2500 V or 5000 V, depending on the model selected.

The meter has two windings, coils A and B, that oppose each other. As the current flows through these coils, a magnetic field is produced, which reacts with the permanent field, causing the coils to rotate. When a resistance is applied to the Megger, the current in the circuit flows through both coils A and B. Coil A causes the pointer to move in a clockwise direction, while coil B tends to move the pointer in a counter-clockwise direction. The pointer will come to rest at a point at which the two forces are balanced. If a very high resistance is connected across the input of the megger, most of the current will flow through coil A and very little through coil B, and the pointer will deflect towards infinity. However, if a very low resistance is connected across the input terminals, most of the current will flow through coil B and very little through coil A, and the pointer will deflect toward zero.

Resistor R_2 acts as a current limiter. It protects the meter movement when the test leads are short-circuited.

The speed at which the generator is driven does not affect the reading on the scale because the currents flowing through both coils change in the same proportion as the generated voltage is changed.

A Megger must be used very carefully. A 5000-V Megger should never be used to test insulation whose voltage rating is less. The high voltage produced may "rupture" the insulation and may result in serious damage.

SUMMARY OF IMPORTANT POINTS

1. Three types of electrical meters are: the permanent-magnet moving coil, moving-iron vane, and electrodynamometer.
2. The permanent-magnet movement is used in galvanometers to measure very small currents.
3. The D'Arsonval galvanometer contains a permanent magnet and a moving coil. As a direct current flows through the moving coil, an electromagnetic field is set up opposing that of the permanent magnet. The coil rotates. The amount of repulsion is proportional to the amount of current flowing in the moving coil.
4. The taut-band movement replaces the jewelled bearing and springs, making the meter movement more rugged.
5. An ammeter is a low-resistance meter placed in series in a circuit to measure current. The current range of an ammeter is expanded by using a resistance called a shunt, connected in parallel with the meter movement.

6. A voltmeter consists of a meter movement connected in series with a high resistance called a multiplier. The voltmeter is connected across the voltage source to measure the potential energy difference. The voltmeter is a high-resistance device to prevent any change in the circuit being tested. The sensitivity of a voltmeter is expressed in ohms per volt. It indicates the loading effect of the meter; the higher the sensitivity of the meter, the lower the loading effect.
7. Ohmmeters measure resistance. There are two types of ohmmeters used, the series and the shunt. In the series ohmmeter, the cell, meter movement, and unknown resistance are connected in series. Full-scale deflection indicates zero ohms, and infinity is indicated by no deflection. The shunt ohmmeter has the unknown resistance connected across the meter movement. It measures low values of resistances.
8. The Wheatstone bridge is one of the most sensitive instruments used for measuring resistance.
9. A slide wire bridge has the advantage of obtaining a balance very quickly; however, it is not as accurate as the Wheatstone bridge.
10. A Megger is an instrument that generates its own dc voltage. It is used for measuring very high values of resistance.

REVIEW QUESTIONS

1. Explain how a simple moving coil meter movement may be adapted to read a wide range of small or large currents.
2. State another name given to the moving-coil meter movement.
3. Describe how a galvanometer can be converted to a voltmeter.
4. Explain how a galvanometer may be converted to an ohmmeter.
5. Name the instrument that is used for measuring low values of resistance.

6. State how the sensitivity of an ammeter is indicated.

7. Define volt meter sensitivity.

8. Indicate how an ammeter is connected in a circuit.

9. Explain how a voltmeter is connected in a circuit.

10. State the advantage a taut-band suspension has over the jewelled bearing suspension system.

11. Explain what would happen if an ammeter were connected as a voltmeter.

12. Describe what would happen if a meter were connected with reverse polarity.

13. Draw a diagram of a permanent-magnet moving coil.

14. Describe, what is meant by the word "damping" and why is damping necessary in a meter movement?

15. Define the term "linear scale."

16. Write in full the term "FSD."

17. Explain the differences between a shunt and a multiplier resistor.

18. Describe how an electric current causes the coil of a galvanometer to move.

19. Explain the force that causes the needle to return to zero when the circuit is broken.

20. Draw a diagram of the Wheatstone bridge and explain the operation of the bridge to show a balanced and unbalanced condition.

21. Name the basic parts of an ohmmeter.

22. Explain why the insertion of a voltmeter in a circuit tends to alter the conditions in the circuit.

23. Describe how the "loading effect" of a voltmeter may be minimized.

24. Explain (a) why a low shunt resistance increases the range of an ammeter, (b) why a high multiplier resistance increases the range of a voltmeter.

25. Describe a slide wire bridge and explain why it is very similar to a Wheatstone bridge.

26. Define the term "Megger".

27. State the advantage of the Megger over an ohmmeter for measuring resistances.

28. Draw the circuits of a series and shunt ohmmeter and explain the basic difference between the two circuits.

29. Explain the relationship between the ohms per volt rating and the current that is necessary to produce FSD in a voltmeter.

30. State the purpose of the zero-ohms adjust control in the ohmmeter.

PROBLEMS

1. Find the value of a shunt resistance required to increase the range of a 1 mA, 100-Ω meter movement to read 100 mA.

2. If a current of 10 mA is applied to an ammeter and 50 μA of current flow through the meter movement, find the current flowing through the shunt resistance.

3. The sensitivity of a meter movement is 1 mA and its resistance is 50 Ω. Calculate the value of the shunt resistance required to extend the range of the meter movement to read a current of 10 mA.

4. A 3-Ω shunt extends the range of a 0.5 mA meter movement to 5 mA. What is the internal resistance of the meter movement (R_M)?

5. What value of shunt resistance is required to extend the range of a 1 mA meter movement to 10 mA if the meter movement has a resistance of 75 Ω?

6. Calculate the value of shunt resistance required to extend the range of a meter movement that has a sensitivity of 10 mA and a resistance of 100 Ω, to read (a) 100 mA, (b) 1 A, (c) 10 A.

7. Given a meter movement with a sensitivity of 10 mA and a resistance of 75 Ω, what is the maximum current that can be measured by the meter, if the shunt resistance is 0.5 Ω?

8. Find the maximum current that a 100-mA meter movement with a resistance of 100 Ω and a 1-Ω shunt resistance can measure.

9. What value of multiplier resistance is required by a 1-mA, 100-Ω meter movement to measure 250 V?

10. What current would be required for a 20 000 ohms-per-volt meter to obtain FSD?

11. A meter movement with a sensitivity of 1 mA and a resistance of 100 Ω is used in conjuction with a 6-V battery to measure resistance. Find the total internal resistance of the meter.

12. How much loading resistance would a 200 ohms-per-volt voltmeter produce on a circuit when set to a 5-V range?

13. Calculate the voltage required to produce a full-scale deflection when a 10 000-ohm resistor is connected in series with a meter movement rated at 1 milliampere with an internal resistance of 100 ohms.

14. In a Wheatstone bridge, as shown in Fig. 6-21, the value of R_1 is 100 Ω, R_2 is 25 Ω, and R_3 is 500 Ω. What is the value of R_x?

15. In the slide wire bridge circuit shown in Fig. 6-22, R_1 is 80 Ω, L_1 is 20 cm, and L_2 is 80 cm. Find the unknown resistance R_x.

CHAPTER
SEVEN

DC GENERATORS

INTRODUCTION

Electrical energy is one of the most convenient forms of energy. It does not appear naturally, but can be converted from other forms of energy. Of the various electrical energy converters in use today, the generator is used to produce the bulk of the world's electrical energy requirements. Batteries, discussed in Chapter Two, convert chemical energy into electrical energy. They are portable and some types can be recharged, making them suitable for certain applications. However, their supply of energy is very limited. Thermocouples convert heat energy into electrical energy, solar cells convert light energy into electrical energy, and piezoelectric crystals will produce electrical energy when a varying mechanical pressure is applied to them. The electrical energy from thermocouples, solar cells, and piezoelectric devices is relatively limited and they are not used for supplying large loads. However, because of their small physical size, these devices are used in many control and instrumentation systems.

Generators, on the other hand, convert mechanical energy into electrical energy. They can supply much energy and may be connected to any prime mover capable of producing rotary motion. This rotary motion can be supplied by steam turbines, gasoline or diesel engines, electric motors, flowing water, wind energy, geothermal wells, or atomic reactors. Basically, a generator produces a voltage by the rotation of a group of conductors in a magnetic field. It uses the principle of electromagnetic induction to convert the input mechanical energy into electrical energy. This principle was discussed in the previous chapter on electromagnetic induction.

GENERATING AN AC SINE WAVE

Figure 7-1 illustrates a simple ac generator. It consists of a single loop of wire which is rotated within a fixed magnetic field. Each end of the single loop coil is connected to a copper ring. These rings are called slip rings, and they are mounted on the shaft, but are insulated from it, and from each other. This rotating assembly, (coil, shaft, and slip rings) in a generator is called the *armature*. The armature coil is connected to an external circuit by means of two brushes which are positioned to rub against the slip rings as the armature is rotated.

It was shown previously that when a conductor is moved to cut through magnetic flux, a voltage will be induced in the conductor. It was also pointed out that the maximum induc-

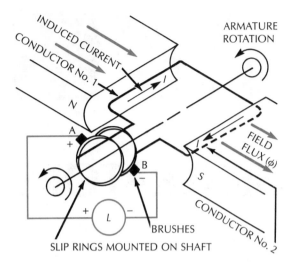

Figure 7-1 A simple ac generator.

tion will occur when the conductor cuts through the flux at right angles, and no induced voltage will be produced when the conductor moves in a direction parallel to the flux flow. The polarity of this induced voltage may be obtained from applying the left-hand generator rule, as described in Chapter Five. In this rule, the thumb, index, and middle finger of the left hand are positioned at right angles to one another. When the thumb is pointed in the direction of the conductor's motion, and the index finger pointed in the direction of the field flux, the middle finger will point in the direction of the induced electron flow from the conductor.

In Fig. 7-1, the two conductors of the single loop armature coil are numbered 1 and 2, and they are connected to slip rings to make contact with brushes A and B, respectively. When conductor #1 moves downward through the flux, conductor #2 will move upward through the flux. From the left-hand generator rule it follows that the currents induced at this instant will be flowing into conductor #1 and out of conductor #2. These directions are illustrated in Fig. 7-1. The two conductors are connected in series and the induced current direction indicates that the voltage between terminals A and B is the sum of the voltages induced in

conductors #1 and #2. The voltage between terminals A and B, *at any instant* during the rotation of the loop is equal to the sum of the voltage induced in conductor #1 and the voltage induced in conductor #2.

The rotating armature coil may therefore be considered a voltage source whose output voltage is connected to an external circuit by means of the slip rings and brushes. Since the current at the instant of armature rotation shown in Fig. 7-1 is flowing out of brush B and into brush A, it follows that the armature output voltage at this instant would be negative at terminal B and positive at terminal A. The magnitude and polarity of this armature voltage will now be investigated for a complete rotation of the armature coil.

Figure 7-2 shows the position of the two conductors in a single loop armature at one-quarter intervals for one complete rotation of the armature. The diagrams have been simplified by showing only the cross-sectional view of the two conductors and by omitting the brushes and slip rings. It must be understood that conductor #1 is always in contact with brush A by means of the slip ring connected to this conductor. Similarly, conductor #2 is always in contact with brush B.

At startup, the conductors are moving parallel to the field flux; therefore, no flux lines are being cut by the conductors, and the induced voltage is zero. As the loop rotates toward the one-quarter turn position, the rate and angles at which the field flux is cut by the conductors increases and reaches a maximum at the instant the loop is at one-quarter revolution. Therefore, the induced voltage would increase from zero to a maximum during this first one-quarter armature revolution. As the loop proceeds to the one-half turn position, the rate at which the field flux is cut by the conductors decreases and the angle of cutting decreases from 90° to 0° At the instant of one-half armature revolution, the conductors are again moving parallel to the flux. Therefore, the induced voltage will decrease from a maxi-

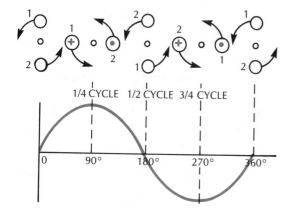

1/4 CYCLE 1/2 CYCLE 3/4 CYCLE

0   90°   180°   270°   360°

Figure 7-2 For each rotation of the single loop armature, one cycle of sine wave voltage is induced in the armature.

mum value to zero as the armature rotates from its one-quarter turn position to the one-half turn position.

It should be apparent now that during the first one-half revolution of the armature the induced voltage increases from zero to a maximum and then decreases back to zero. This voltage pulse is shown in Fig. 7-2 as a positive pulse by virtue of the way the voltmeter is connected to the brushes. If the voltmeter connections were reversed, then, obviously, the meter deflection would indicate the opposite or negative voltage pulse.

During the second half of armature rotation, the armature conductors experience a rate and angle of cutting of the field flux similar to that during the first half revolution. That is, the rate and angle of the flux cut by the armature first increases and reaches a maximum at the instant the armature passes the three-quarter turn position, and then the rate and angle of the flux cut decreases and is again zero at the instant the armature completes the revolution. Therefore, like the first half of armature rotation, a voltage pulse is also induced in the conductors during the second half of armature rotation. However, the voltage pulse induced during the second half of armature rotation is opposite in polarity to the voltage induced during the first half of armature rotation.

This reversal of polarity in the induced voltage is due to the interchange of the conductors' position for the second-half of armature rotation compared to the first-half of rotation. As shown in Fig. 7-2, the motion of conductor #1 is downward through the flux during the first half of armature rotation, but in the second half of rotation, its action is upward through the flux. Therefore, the output polarity of conductor #1 during the second half of armature rotation is reverse to that of the first half of rotation. The terminal polarity of conductor #2 is reversed for the same reason. Thus, the induced voltage pulse during the second half of armature rotation is opposite to that of the first half of armature rotation. It should be apparent now that for each complete revolution of the armature one cycle of sine-wave voltage is induced. This is an alternating voltage, and this type of generator is commonly called an *alternator*. The alternator will be discussed in more detail in a later chapter.

THE SIMPLE DC GENERATOR

A direct current generator is simply an ac generator provided with a *commutator* instead of slip rings. The commutator is a device for reversing the brush contact with the armature coil terminals every time the induced current in the coil reverses, so that the output current taken by the brushes is always in the same direction. The simplest commutator is a split ring, or a ring cut into two segments. This is illustrated in Fig. 7-3 (a) and (b) on page 112.

The simple generator shown in Fig. 7-3 contains a single-loop armature which is rotated within a magnetic field. Therefore, as was described for the simple ac generator, a cycle of sine wave voltage is induced in the loop for each rotation of the armature. The terminal ends of the armature loop are connected one to each segment of the split ring commutator. The brushes for this generator are positioned in an axis where the induced voltage in the armature coil is reduced to zero and is about

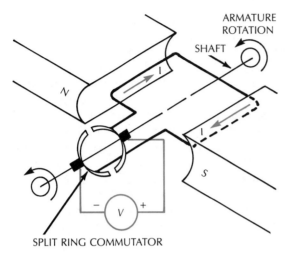

Figure 7-3(a) The simple dc generator with split ring commutator.

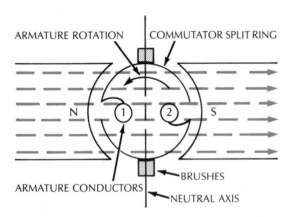

Figure 7-3(b) Brushes positioned in the neutral axis with the armature conductor terminals connected to the commutator segments.

to reverse in polarity. This axis is called the neutral axis, and it is in this plane that commutation occurs. That is, the segments exchange brush contact so as to keep the polarity of the brushes the same. It should also be observed that the neutral axis lies midway between the field poles and is perpendicular to the field flux.

A complete revolution of the armature shown in Fig. 7-3 will now be investigated. The simplified drawings in Fig. 7-4 show the

positions of the armature conductors together with their commutator segment at intervals of one-quarter rotation for one complete revolution. The brushes are also included to emphasize commutation or the switching action of the segments with the brushes.

Since this generator contains a single-loop armature which is rotated within a magnetic field, a cycle of sine wave voltage is induced in the armature conductors for each rotation of the armature in the same way as was previously described for the simple alternator. However, the actual voltage output at the brushes is now forced to maintain a constant direction by the process of commutation.

When the armature conductors are in position (a) shown in Fig. 7-4, they are not cutting through any field flux, and so the induced voltage in the conductors is zero. This is therefore described as zero on the induced voltage curve. Also, observe that at this instant, the conductors are in the neutral plane and both commutator segments are in contact with the brushes. Therefore, at this instant both armature conductors are being shorted.

As the armature is rotated from position (a), through position (b), and on to position (c), the voltage induced in the conductors increases, and reaches a maximum at position (b), and then the voltage decreases and reaches zero again at position (c). From the left-hand generator rule, the terminal end of conductor #1 will be positive and the terminal end of conductor #2 will be negative during this first half of armature rotation. Therefore, since segment #1 is in contact with brush B, brush B will be positive, and similarly, brush A will be negative because it is in contact with segment #2.

As the armature revolves to complete the second half of its cycle, a voltage pulse is again induced in the conductors. This voltage reaches a maximum at position (d) and then reduces, and reaches zero at position (e). From the left-hand generator rule, it is observed that segment #2 is now positive and segment #1 is negative. This is opposite to what took place

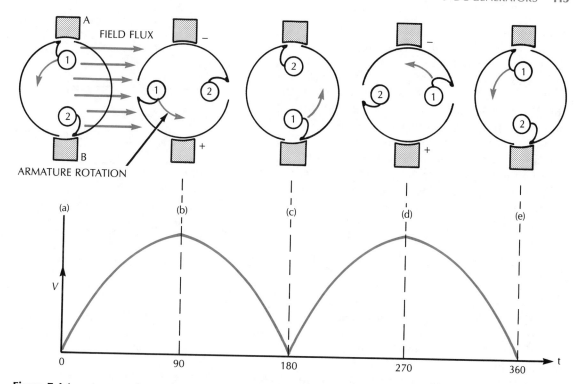

Figure 7-4 Armature rotation and output voltage from a single-loop dc generator.

during the first half of armature rotation. However, as shown in Fig. 7-4, segment #2, which is now positive, is at this time in contact with brush B, and so brush B remains positive during the second half of the armature rotation. Similarly, brush A remains negative.

It should be obvious now that even though the actual voltage pulses induced in the armature conductors reversed for each half cycle of armature rotation, the same as for the simple ac generator, the switching action of the commutator segments is such that the polarity of both brushes is kept constant during the complete rotation of the armature. This switching action of the commutator segments is called *commutation* or commutator action. During commutation, the segments reverse contact with the brushes just as the induced current is about to reverse in direction, maintaining the output current in a constant direction. Therefore, the output voltage pulses at the brushes

are varying value dc. This is shown in Fig. 7-4 by drawing the voltage wave pulses all in the same direction.

It should be observed that during commutation the two segments, and therefore the two conductors, are momentarily shorted by the brushes. If there is any induced voltage in the conductors at this instant of commutation, then an exceedingly high and dangerous current would flow through the shorted conductors. This high current can damage the winding and is responsible for sparking at the brushes. However, as shown in Fig. 7-4, the segments and brushes are arranged so that during commutation the two conductors are also passing through the neutral plane, and in this axis the induction in the conductors is zero.

The simple generator described in Fig. 7-4 is not suitable for most applications because its output voltage is pulsating, as well as very low.

Also, the current capacity of a single-loop armature is very limited. It was pointed out in Chapter Five, that the magnitude of the induced voltage in a conductor is dependent on the rate of flux cut by the conductor and the angle of cutting. Therefore, the output voltage from a generator may be increased by either increasing the revolutions per minute of the armature or by increasing the field flux itself in the generator circuit. However, since most generators are designed to be operated at some rated speed, increases in speed beyond their rated value are seldom used for increasing their output voltage. On the other hand, variation in field flux provides the major method for regulating the output voltage of a generator. The methods of producing field flux and the effects it has on the operation of a generator will be discussed later in this chapter.

Simple changes can be made to the elementary dc generator, to illustrate how generators are designed for supplying a higher and more constant dc voltage. For one thing, the single-loop armature can be replaced with a coil having many turns. Since the turns are connected in series, the output voltage at any instant will be equal to the sum of the voltages induced in each of the turns. The voltage fluctuations, often called the ripple voltage, can also be reduced by using more than one multiloop coil in the armature.

Figure 7-5 shows an armature with two rotating coils positioned at right angles to each other. Each end of the two coils is connected to a separate commutator segment, so the commutator will now have a total of four segments. There are still only two brushes, and the action of the segments, as the armature is rotated, is such that one brush is again always made positive and the other brush negative. Therefore, the output of this generator is dc, but with twice the number of voltage pulses per revolution of the armature, compared with the single coil generator. The voltage wave

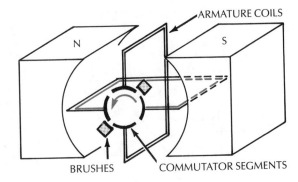

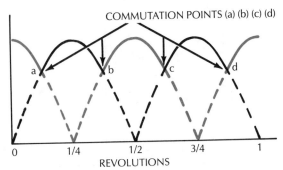

Figure 7-5 A dc generator with two armature coils at 90° to each other.

diagram of Fig. 7-5 shows that the actual induced voltage pulses in the two coils are 90° out of phase. Hence, when one voltage pulse is increasing, the other pulse is decreasing. From the points of commutator action shown on the wave diagram, it is obvious that the net output voltage is more steady and has less voltage ripple than the output voltage of the single coil generator.

PRACTICAL DC GENERATORS

GENERATOR ELECTRIC CIRCUIT

Figure 7-6 shows the component parts of a typical modern generator. The basic principle of operating, though, remains very much the same as was described for the simple generator. Field flux produced by the field windings is established in the frame, pole cores, air gap,

Figure 7-6 Component parts of a dc generator.   Courtesy: John Dubiel

and armature core, all of which form what is known as the *magnetic circuit* of a generator. The armature consists of a core and a commutator mounted on a shaft. Coil windings are fitted into the core slots and their ends are connected to the commutator segments. This assembly is called the *armature* and it is rotated within the field flux. The voltage induced in the armature windings is connected to an external circuit by brushes which are positioned to make contact with the commutator segments. The brushes and the armature bearings are mounted on the end shields, which, in turn, are fastened to the cylindrical frame. The armature windings, commutator, brushes, and field windings form the *electric circuit* of the generator.

THE BRUSHES

The brushes used in generators are usually made from a compound of carbon and graphite. They are made in various sizes, and their graphite content provides a self-lubricating action as they rub against the commutator. The main function of the brushes is to transfer the energy present at the commutator to an external circuit. To ensure good electrical contact, each brush is placed in some type of spring-loaded holder and a braided copper wire, called a pigtail, is used to connect the brush to the holder. Connections to an external circuit are then made from the brush holders. The spring mechanism in the holder helps to maintain the proper tension on the brushes as they rub against the commutator, and also to feed the brushes toward the commutator as they wear out. The brush holders are bolted to one of the generator end shields, but they are insulated from it. The holders are positioned so the brushes will be in the generator's neutral axis.

THE COMMUTATOR

The commutator is part of the armature assembly. It is made up of a number of copper segments mounted on a cylindrical form, and insulated from each other with thin wedges of mica. The commutator is then fitted to, but insulated from, the armature shaft, as shown in Fig. 7-6. The function of the commutator, it may be recalled, is to provide a reversing action to the induced voltage just as this voltage itself is about to reverse. This causes the generator output at the brushes to be maintained in a constant direction.

THE ARMATURE CORE

The armature core is cylindrical, or drum-shaped, and it is made of soft iron. The core is actually made from thin strip laminations of soft iron. These strips are insulated with a varnish and then pressed together to form the core. The core strips are designed, and stacked together, so that grooves or slots are formed on the longitudinal surface of the core. The armature coils are fitted into these core slots.

The reason for the use of a soft-iron armature core is that the soft iron helps to reduce an energy loss in a generator called hysteresis loss. As the armature is rotated, the core's electron dipoles must constantly move in order to maintain their alignment with the field flux. The core's opposition to this constant realignment of its magnetic dipoles is responsible for this loss. Since soft iron has a low reluctance, its opposition to magnetic alignment with the field flux is at a minimum, and so hysteresis loss is kept to a low value.

The advantage of *laminating* the iron core is

to minimize another generator core loss called the eddy-current loss. This loss is due to the currents induced in the core material itself, as the armature is rotated within the flux. By laminating the iron core, currents can only be induced in the thin laminations, and these very small currents can only circulate in the laminations. This restriction on the flow of these currents greatly reduces the eddy-current power loss in the generator.

THE ARMATURE WINDING

The armature winding which is fitted to the core is very involved, and a complete analysis is beyond the scope of this text. However, an understanding of some of the important properties of armature windings will help you to appreciate the operation of all rotating electrical machines.

One of the first types of armature windings used in generators was the ring-wound type. This ring-wound armature is no longer in use because the winding is both inefficient and difficult to assemble. As shown in Fig. 7-7, only the outer half of each turn is cut by the field flux, and so induction occurs in only 50 percent of the armature winding. The inner half of the winding is inactive and so adds mostly weight to the armature. Secondly, the ring-winding itself is difficult and time-consuming to produce since the turns must be threaded through the ring loop. In place of the ring-wound armature, the cylindrical or drum-type armature is found in all modern generators. In the drum-type armature, the windings are placed on the core's surface, and this allows both halves of each armature turn to cut through the flux as the armature is rotated. The windings may also be externally pre-formed and then fitted into the core slots.

The actual winding of an armature is made up of several coils which are positioned in the core's slots. The arrangement of these coils, and the connection of their ends to the commutator segments, may be made in several different ways. However, because of certain

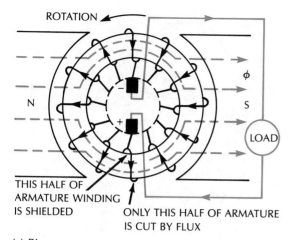

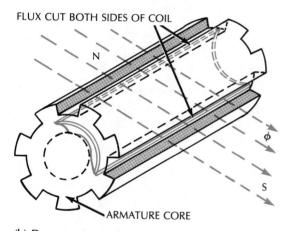

(a) Ring-type armature

(b) Drum-type armature

Figure 7-7 Ring- and drum-type armatures.

common properties, these different armature-coil arrangements may be divided into three groups: the wave-type winding, the lap-type winding, and a combination of the wave and lap windings, sometimes called the frog-leg winding. The arrangement of the armature coils and their connection to the commutator segments determines to a large extent the voltage and current characteristics of the generator.

In all armature drum-type windings, the coils are always positioned to maintain armature symmetry. This, of course, ensures a smooth motion of the armature. A second property of all drum-type coil arrangements is that each coil is positioned so that when one side of the coil passes over one field pole, the other side of the same coil will simultaneously pass over the opposite field pole. This could be described by saying that the sides of a coil are arranged so that they will span the same distance as the distance between opposite field poles. A third common property, as will be shown, is that the number of armature coils is equal to the number of commutator segments.

THE WAVE-TYPE ARMATURE WINDING

The wave-type armature winding gets its name from the appearance of this winding on the drum. Although there are many variations to the wave winding, they all share certain common properties. In a wave-wound armature, one end of a coil is connected to a commutator segment, which is separated by a distance equal to two field poles from the segment to which the other end of the coil is connected.

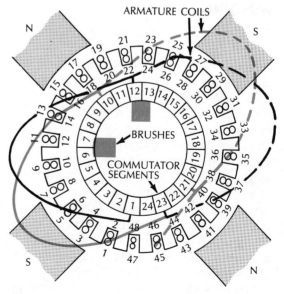

Figure 7-8 Four armature coils in a wave-wound armature.

This is illustrated in the simplified wave diagram of Fig. 7-8. Also, observe that each segment is connected to the ends of two different coils, and that these coils are positioned on opposite sides of the armature. This arrangement of the armature coils places all the coils which are under the influence of similar pole pairs in series; thus, there will be two parallel branches formed in this type of winding.

It should be obvious that the generator brushes would be positioned at the junctions of these two parallel voltage branches, and that these two junctions would also lie in the neutral axis of the generator. Regardless of the number of field poles, the wave-wound armature has only two parallel voltage branches and, also, only one pair of brushes is usually used. Since each parallel branch consists of a number of coils connected in series, their voltages are cumulative and the output voltage from a wave-wound armature is relatively high. However, the current capacity of this armature is low because there are only two current paths.

THE LAP-TYPE ARMATURE WINDING

The lap-type armature windings get their name from the manner in which the coils are arranged on the armature core, and the manner in which the coil ends are connected to the commutator segments. As shown in Fig. 7-9, each coil of a lap winding "overlaps" the previous coil. The arrangement of the coils in the armature slots for the lap-type winding is very similar to that for the wave winding. You will recall that for both types of windings the sides of a coil must be positioned to span the same distance as between opposite poles. However, the coil ends of these two types of windings are connected to the commutator segments differently. The resulting voltage and current characteristics are also very different.

In the lap winding, unlike the wave winding, the two ends of any coil are always connected

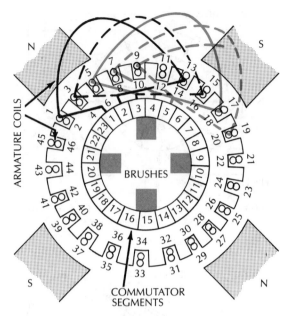

Figure 7-9 The lap-wound armature.

to adjacent commutator segments and each segment is connected to the ends of two adjacent coils. Therefore, as was the case for the wave winding, the number of armature coils is again equal to the number of commutator segments. However, in the lap-winding, coils under similar pole pairs are placed in parallel. Therefore, in the lap winding, there are as many parallel branches in the armature as there are field poles. This is unlike the wave winding, in which the coils under similar pole pairs were in series and, thus, there were always only two parallel branches regardless of the number of field poles.

It should be obvious that there is one situation in which both the lap- and wave-wound armatures would share the same characteristics. This occurs when the generator has only two poles. For either type of armature winding, a two-pole generator would have only two parallel armature paths. In a four-pole generator, however, the lap-wound armature would have four parallel paths, and the wave-wound armature would have only two parallel paths. The number of brushes required by a lap-

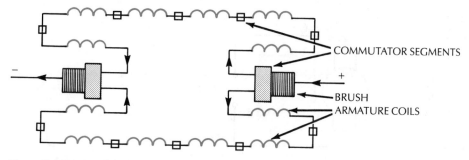

Figure 7-10(a) Simplified diagram showing the brushes and parallel paths in a 4-hole lap-wound armature.

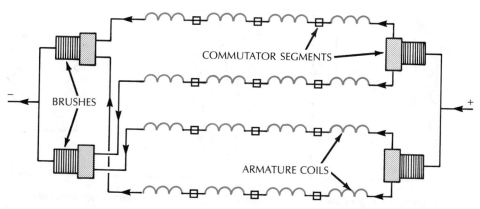

Figure 7-10(b) Simplified diagram showing the brushes and parallel paths in a multi-pole wave-wound armature.

wound generator is also different than that required for the wave-wound generator. Whereas the wave-wound generator generally uses only two brushes, the lap-wound generator requires as many brushes as there are poles. The brushes in a lap-wound generator are connected within the generator, so that there is still only one pair of output voltage terminals.

Since the lap-wound armature has as many parallel paths as field poles, this means that in general there will be a greater number of current paths in this type of armature winding. However, each parallel branch will now have a smaller number of coils in series compared to a wave-wound armature. Therefore, it can be concluded that, whereas the wave-wound

armature has a high voltage and low current rating characteristic, the lap-wound armature, on the other hand, has a lower voltage but a higher current rating characteristic.

GENERATOR FIELD EXCITATION

The generator cores and the frame are constructed with a low reluctance material, such as soft iron, in order to concentrate the field flux and to produce a stronger magnetic field. Since the field flux of a generator must be kept in a constant direction, permanent field magnets can be used. A generator that uses permanent field magnets is called a magneto. However, permanent magnets are bulky and, besides, it is not possible to vary the flux from

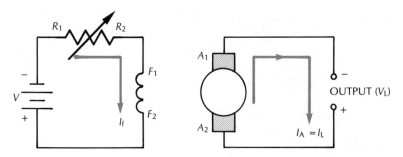

Figure 7-11(a) The separately excited dc generator.

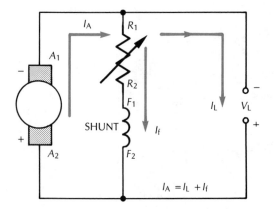

Figure 7-11(b) The self-excited shunt generator.

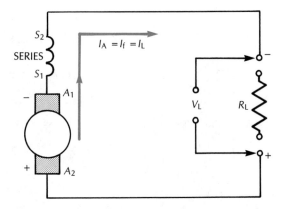

Figure 7-11(c) The self-excited series generator.

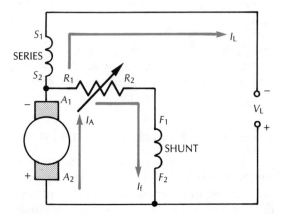

Figure 7-11(d) The short shunt dc compound generator.

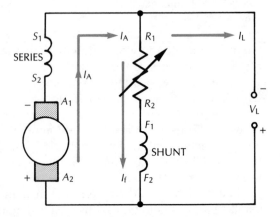

Figure 7-11(e) The long shunt dc compound generator.

a permanent magnet. Therefore, most generators are constructed, instead, with electromagnetic field coils.

Since the field flux in a generator must be

kept in a constant direction, it follows that the field coils must be energized from a dc source. Generators take their names from the type of dc field excitation that is used. When the field is excited from a separate source, such as a battery, as shown in Fig. 7-11(a), the generator is called a separately excited generator. When a generator provides its own field excitation, it is called a self-excited generator. There are three types of self-excited generators: the shunt, series, and compound generator. The self-excited shunt generator has its field winding connected in parallel across the armature. When the field is in series with the armature, the generator is a self-excited series generator, and when the shunt and series connections are combined, the generator is called a compound generator. Compound generators may be connected in short shunt with the shunt field in parallel with only the armature, or in long shunt with the shunt field in parallel with both the armature and series field. These self-excited generators are shown in Figs. 7-11(b), (c), (d) and (e).

The field coils of a generator are usually externally pre-formed, and then fitted over the pole cores. These coils are connected in series so that they will produce alternate field poles, as shown in Fig. 7-12. A generator may be designed to have many field poles, but the number of field poles is always an even number. For each north pole in a generator, there is

a corresponding south pole. By increasing the number of field poles in a generator, the effect of core saturation can be reduced. This is so because an increase in field poles will cause more flux paths to be created in the core and frame of the generator. Therefore, by increasing the number of poles, the core and frame of a generator can be made thinner, and this means that the generator can be made more compact. Another advantage of increasing the number of field poles is that the generator's voltage itself can be increased.

GENERATOR-INDUCED VOLTAGE

The average induced voltage of a generator may be determined from the equation:

$$V_g = T\phi n \text{ volts} \tag{7-1}$$

where ϕ = flux per pole

n = the r/min of the armature or $2\pi n$ radians per minute

and T = generator constant. This constant will depend on factors such as the number of generator field poles and the design and number of armature windings

It should be apparent that Eq. (7-1) is simply a restatement of Faraday's law of induction: the induced voltage is directly proportional to the rate at which flux is cut by the armature circuit. Equation (7-1) shows that the induced voltage of a generator would increase in direct proportion to the increase in the generator's flux per pole and to the increase in the generator's r/min. Thus, doubling the flux per pole would double the induced voltage, and similarly doubling the r/min would also double the induced voltage.

THE MAGNETIZATION CURVE

Since a generator is usually operated at some constant speed, the method most often used for regulating a generator's voltage is through its field flux. From Eq. (7-1), you saw that the generator-induced voltage is directly propor-

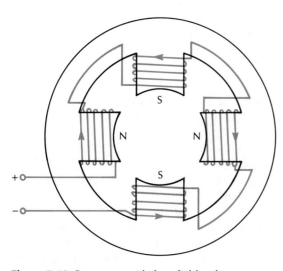

Figure 7-12 Generator with four field poles.

tional to the flux per generator pole. Since this flux is produced when a current flows in the pole winding, the generator's induced voltage, therefore, depends directly on the amount of field current. However, due to the saturation effect of the iron core, the field flux produced by the field windings is not directly proportional to the field current. Therefore, it should be obvious that a generator's voltage is not exactly proportional to the current in the field windings.

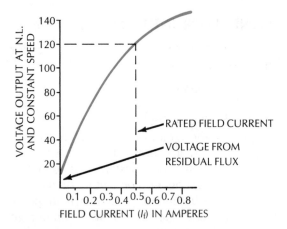

Figure 7-13 Generator magnetization curve.

The graph in Fig. 7-13 illustrates the effect of core saturation on the output voltage of a generator. It is often called the generator magnetization curve, or no-load characteristic curve. To obtain this curve, the generator is operated at constant speed and no-load, and the field current is varied from zero to 125 percent of rated value. The field current in a generator is controlled by means of a field rheostat. It should be varied carefully so as to avoid any overshoot from a required value. Exceeding a required value, and then returning to it, may cause some change in the residual flux pattern in the core. This, in turn, would introduce some error in the results.

The reason for core saturation, and hence the nonproportionality of the flux produced versus the applied current, was mentioned in

an earlier chapter on magnetism. It will be recalled, according to the Domain theory of magnetism, that when a current flows in the field windings of a generator, the core becomes magnetized, as many of the iron's magnetic dipoles are made to align themselves in the direction of the field flux. This electron dipole alignment increases the concentration of flux lines and so increases the magnetic field strength. The relatively straight line portion of the magnetization curve of Fig. 7-13 indicates that the flux produced is almost proportional to the initial increases in the field current. However, beyond the upper limit, the number of unaligned electrons in the core atoms is now quite small. It has also become increasingly more difficult to align these remaining electrons. This point on the magnetization curve is called the *saturation point*. Above the saturation point, much larger increases in current flow are required for corresponding increases in flux in the core.

It should also be observed that the curve in Fig. 7-13 was not started at the point of zero-induced voltage. Even with the field current reduced to zero, there is still some flux remaining in the core of a generator from a previous operation. This residual flux is responsible for the small output voltage of a generator when the field current is at zero. As will be shown later, the residual flux in a generator plays an important role in the startup of self-excited generators.

GENERATOR INTERNAL VOLTAGE LOSSES

The actual output voltage available at the terminals of a generator is less than the voltage induced in the armature itself. This difference in voltage is due to the internal voltage losses of the generator. The losses are caused by the resistance in the armature circuit, by a net field flux distortion called armature reaction, and from a reduction in field flux resulting from a combination of the first two losses.

RESISTANCE VOLTAGE DROP

The voltage loss due to the armature branch circuit resistance is also called the *IR* generator voltage loss. This internal armature circuit resistance consists of the resistance of the armature windings, the series field and interpole windings, if used, and the brushes. These resistances are in series and so may be combined.

$$R_A = R_a + R_b + R_s \qquad (7\text{-}2)$$

It should be obvious that the internal resistance voltage drop of a generator would increase as the load current is increased. At no-load, for the separately excited generator shown in Fig. 7-14, the *IR* voltage loss would be zero since the current through the armature is zero. Therefore, at no-load, the terminal voltage of this generator is equal to the generated voltage. However, as the load current is increased, this *IR* voltage drop would increase. This is illustrated in the graph shown in Fig. 7-14.

From the circuit of Fig. 7-14(a),

$$V_L = V_g - I_A R_A \qquad (7\text{-}3)$$

Combining Eq. 7-3 and Eq. 7-1 provides what is often called the fundamental generator equation, because it contains all the main factors that govern the terminal voltage of the generator.

$$V_L = K\phi n - I_A R_A \qquad (7\text{-}4)$$

Example 7-1: For the generator shown in Fig. 7-14, determine the voltage loss and the generated voltage if the generator supplies a load of 5 A at 120 V.

where voltage loss = $I_A R_A$
　　　　voltage loss = 5 A × 12 Ω
　　　　voltage loss = 60 V

Since $V_L = V_g - I_A R_A$

　　　$V_g = 120$ V + 60 V
　　　$V_g = 180$ V

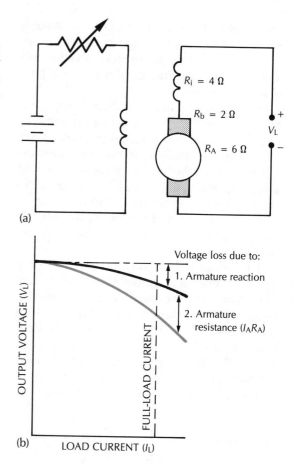

(a)

(b) LOAD CURRENT (I_L)

Voltage loss due to:
1. Armature reaction
2. Armature resistance ($I_A R_A$)

Figure 7-14 Voltage drop due to the internal resistance in the armature branch of a generator.

ARMATURE REACTION

Armature reaction causes both a net loss in generator voltage and also sparking between the brushes and the commutator segments. This sparking, if not controlled, can cause quite severe damage to the brushes and commutator segments.

Armature reaction is caused by the interaction between armature flux and field flux. The field flux is produced when the field current is applied to the field windings and this flux is required for inducing the generator's armature voltage. The armature flux, however, is produced when the induced armature voltage is connected to supply a load current. Consequently, the effects of armature reaction will increase with corresponding increases in generator load current.

In armature reaction, as shown in Fig. 7-15, the armature flux and field flux combine to produce a new and distorted magnetic field. As shown, the direction of this new generator flux is such that the commutating axis is shifted away from the neutral axis. In a generator, the shift in this axis is always in the direction of armature rotation, and the amount of this axis displacement is dependent on the load current.

It is obvious that commutation will now occur at this new commutating load axis. Therefore, armature reaction will necessitate a repositioning of the generator brushes each time the load current is changed. This procedure, however, is not practical unless, of course, the generator current is kept constant.

When armature reaction is not corrected, it causes commutation to occur outside the commutating axis. This means that an armature coil undergoing commutation will not be quite at zero voltage. Since at commutation this coil is momentarily shorted, a current will flow through this shorted coil. The armature coils of a generator have relatively low resistance and this short circuit current could be dangerously high. This current could damage the winding. It is also responsible for sparking at the brushes. Secondly, since this shorted coil's voltage forms part of the generator's output, armature reaction therefore would cause a net reduction in the armature voltage.

SELF-INDUCTION IN THE ARMATURE

Self-induction in the armature coils is another generator property which causes very similar effects to that of armature reaction. You will recall from the chapter on induction that when the magnetic flux about a coil collapses, a voltage is induced in the coil. This is exactly what happens when an armature coil passes

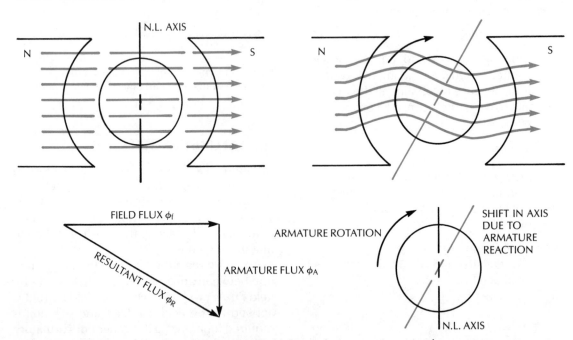

Figure 7-15 The effect of armature reaction on the field flux and commutating axis of a generator.

through the neutral axis. The voltage thus set up is called a self-induced voltage. Since this coil is momentarily shorted by the brushes, a self-induced current will flow through this winding. This self-induced voltage should not be confused with the generator's induced output voltage as they are two separate voltages.

Since an armature coil is surrounded by iron, it will have a high inductance value. This property of inductance is the topic of a later chapter in this text. However, the higher inductance causes the self-induced voltage to be out of phase, and so the flux set up by this voltage will therefore be out of phase. This is similar to the flux produced by the armature-induced voltage. In fact, the armature-induced flux is out of phase for the same reason. Therefore, this self-induced voltage will cause a further shift in the neutral axis of the generator, as shown in Fig. 7-16. This, of course, will add to both the sparking at the generator brushes and to the reduction in the armature output voltage.

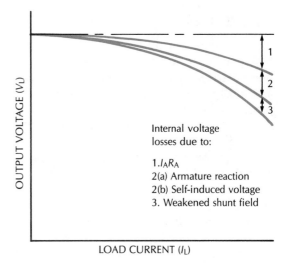

Figure 7-16(b) Voltage losses in a self-excited generator.

the field of such a generator depends on the output voltage for its excitation, any reduction in the output voltage will consequently reduce the field current. Therefore, the losses due to the generator's internal armature resistance, to armature reaction, and to armature self-induction will, in effect, cause a further reduction in the generator's output voltage. This voltage loss is shown in Fig. 7-16(b).

CORRECTING ARMATURE REACTION

It was seen that the internal voltage losses of a generator increase with the increases in load current. To reduce the IR voltage drop, armatures are designed to have as low a resistance as possible. Armature reaction and armature self-induction, on the other hand, set up magnetic fields that produce a net shift in the neutral axis. Therefore, any generator additions, which are used to reduce the voltage losses and sparking caused by this shift in the neutral axis, will involve provisions for cancelling these magnetic fields. Also, since the amount of shift in the neutral axis changes with changes in the generator load current, whatever method is used should be self-regulating.

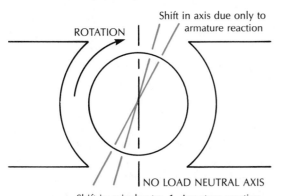

Figure 7-16(a) The effects of (1) armature reaction and (2) self-induced voltage in the armature.

VOLTAGE LOSS DUE TO REDUCTION IN GENERATOR OUTPUT

This additional form of generator voltage loss occurs only in self-excited generators. Since

INTERPOLES

One method of reducing the deviation between the commutating axis and the actual neutral axis is by the use of *interpoles*. These are small windings that are wound on pole pieces in the generator housing, as shown in Fig. 7-17. These windings are located in the no-load neutral axis. They are connected in series-opposing with the armature. That is, even though they have the same current as the armature, the interpole flux will oppose any armature flux in the vicinity of the neutral axis. Since self-induction occurs in this axis, it will now be cancelled by this interpole flux. Hence, the use of interpoles in a generator will reduce the shift in the neutral axis caused by self-induction in the armature. The effect of the interpoles is also self-regulating because they are connected in series with the armature. That is, when the armature flux changes due to a change in load current, similar changes will occur in the interpoles.

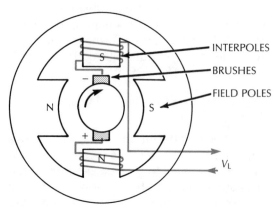

Interpoles reduces mainly the field distortion caused by armature self-induction.

Figure 7-17 The location and connection of interpoles in a generator.

THE USE OF COMPENSATING WINDINGS

The use of interpoles cancels armature self-induction but produces little or no change in

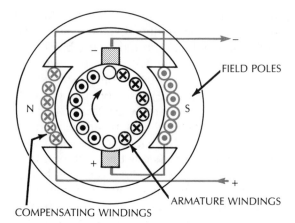

Compensating windings reduces mainly the field distortion caused by armature reaction. It only partially cancels armature self-induction.

Figure 7-18 The use of compensating windings to reduce armature reaction.

the problem of armature reaction. Armature reaction, unlike self-induction, is caused by armature flux which exists all around the armature. A second type of windings, called *compensating windings*, is used to help cancel this armature flux. These are small windings which are set in the generator's main pole pieces. Like the interpoles, they too are connected in series-opposing with the armature. Thus, their effect will both oppose the armature flux and also be self-regulating.

The use of both interpoles and compensating windings will eliminate most of the shift in the neutral axis caused by changes in the load current. In some small generators, only interpoles are used, and in some others, only compensating windings are used. In most large generators, however, both forms of winding are often included in their construction.

VOLTAGE REGULATION— LOAD CHARACTERISTICS

The percent voltage regulation of a generator is an important property to know in order to determine the suitability of the generator for a particular application. It is defined as:

percent voltage regulation   (7-5)

$$= \frac{\text{no-load voltage} - \text{rated full-load voltage}}{\text{rated full-load voltage}} \times 100$$

This quantity indicates the extent of change in the generator's output voltage and its load is increased from no-load to full-load. For example, for a lighting circuit, which requires a relatively constant voltage over a wide load range, a generator having a small percent voltage regulation should be selected. A small percent voltage regulation means that as the generator's load current is increased from no-load to full-load, the change in output voltage will be minimal.

You have now seen that the addition of a load to a generator will cause its output voltage to change, unless some provision is made to keep this voltage constant. Changes in either the speed or in the field current may be used to regulate the output voltage of a generator under varying load conditions. A graph which describes the output voltage of a generator for different values of load current is called the load-voltage characteristic curve or the voltage-regulation characteristic. These results for a generator can be obtained by measuring the output voltage for different values of load current. For this test, the generator must be operated at constant rated speed and with the field current so adjusted that rated voltage is obtained at full-load. The load-voltage characteristics for different generators are discussed in the following sections.

THE SEPARATELY EXCITED GENERATOR

The field excitation for the separately excited generator is supplied from an independent dc source, such as a storage battery or a separate dc generator. The diagram and a typical load-voltage characteristic curve for this generator are shown in Fig. 7-19, on page 128.

As shown by the characteristic curve, the output voltage of this generator decreases as

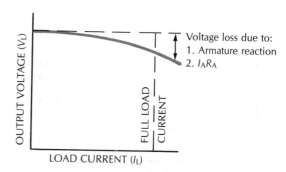

Figure 7-19 Load-voltage characteristic curve. (separately excited generator)

the load current is increased. The decrease in output voltage is caused by the internal voltage losses which increase with corresponding increase in load. It was shown that for the separately excited generator, these voltage losses are caused by the armature circuit resistance and by armature reaction. (Note: the term "armature reaction" is used to reflect the combined effect of both armature reaction and armature self-induction. This is commonly done because both properties produce similar results.)

The separately excited generator, because of its independent field current, will provide quick output response to changes in its field current. It will also respond to a wide range of speeds and of field currents. However, the separate dc source required for this generator usually makes it more expensive than a comparable self-excited generator. Hence, its use is limited mainly to experimental and precise control circuit applications.

THE SELF-EXCITED SHUNT GENERATOR

This generator provides its own field excitation. The armature is connected directly across the field coils and a field rheostat, as shown in Fig. 7-11(b). The field rheostat can be used to vary the field current. The characteristic curve of a typical shunt generator is shown in Fig. 7-20.

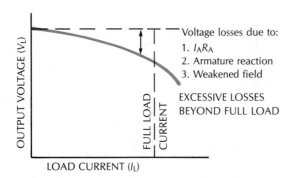

Figure 7-20 Load voltage characteristic curve.

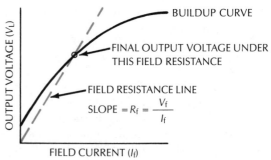

Figure 7-21 Voltage buildup in a self-excited shunt generator.

All self-excited generators depend on the presence of some residual field flux for their startup operation. This residual flux is the magnetic flux that remains in the field core from a previous operation of the generator. If there is no residual flux present in the field as in a new generator or one that was not in use for a long time, it will be necessary to first "flash" the field before the generator will operate satisfactorily. To flash the field, it should be disconnected from the armature and then connected across a dc source for a few minutes. The current flowing in the field coil will set up a magnetic field, and when this current is removed, the field core will remain slightly magnetized.

VOLTAGE BUILDUP ACTION

At startup of this generator, the residual flux will produce a small output voltage. Since the output voltage is also connected across the field coils, it will cause a small current to flow through the field coils. The flux produced by this current will strengthen the field. Hence, the output voltage will increase, which, in turn, will further strengthen the field and cause another increase in the output voltage. You can see that the initial small voltage set up by the residual flux at startup of this generator causes a chain buildup in the output voltage. This generator property is sometimes referred to as voltage buildup action.

The voltage buildup in a shunt generator does not continue forever for a given genera-

tor speed. The final output voltage depends on the field-circuit resistance and on core saturation. Too high a field resistance will limit the field current and so reduce the buildup voltage. This is illustrated in Fig. 7-21. For rated voltage, the field current of this generator is usually set between 0.5 and 5.0 percent of the full-load current.

Under certain conditions, the shunt generator can fail to experience buildup action. You have seen that residual flux is necessary at startup. If, when the armature is rotated, the output voltage is zero, this indicates that there is probably no residual flux, and that the field needs to be flashed. On the other hand, if there is a small output voltage but still no buildup action, then the problem is other than that of no residual flux. One probable cause of buildup failure may be an open field circuit or too high a field-circuit resistance. Even with residual flux, if the field is open, no field current can flow to cause voltage buildup. Another cause could be that the field is connected in reverse with the armature. When this happens, instead of the field current strengthening the field, it actually weakens it. This is because the two fields are now opposing each other. To correct this problem will involve the simple reversal of the field connections. Other conditions that may prevent buildup are a dirty commutator, dirty or loose brush connections, or too low an armature speed.

LOAD CHARACTERISTICS

The load characteristic curve for this generator, shown in Fig. 7-20, is quite similar to that for the separately excited generator. However, the voltage losses, which increase with load, are slightly greater. For the shunt generator, these losses are caused by the armature circuit resistance, by armature reaction, and by the net decrease in field strength when the output is decreased from the first two losses.

The voltage regulation of this generator, especially those with interpoles, remains fairly good up to its rated full-load. However, beyond rated load, the output decreases rapidly. This property provides the generator with its own self-protection against overloads. For example, if the output is accidently shorted, this will cause an extreme overload on the generator. However, since the generator's output voltage is reduced to zero, the field current will also decrease to zero and this in turn will simply disable the generator.

Shunt generators are used for charging storage batteries, for supplying the excitation current for large alternators, and for supplying other loads that are relatively small and close to the source.

THE SERIES GENERATOR

In a series generator, the armature, field, and load are connected in series. Therefore, the load current will flow through both the armature and the field coils. Since the series-field coils must carry the full armature current, the series-field coil, unlike the shunt-field coil, is constructed of a few turns of heavy wire. The shunt-field resistance, therefore, would be much greater than the series-field resistance. The shunt-field coil is constructed of many turns of fine wire. The series generator load characteristic curve is shown in Fig. 7-22.

As shown by the load characteristic curve, the series generator has extremely poor voltage regulation. At no-load, the field current is zero, and the output voltage is very low. The small output at no-load is due to the residual

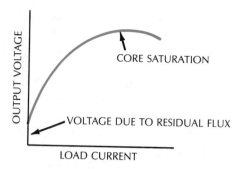

Figure 7-22 Load voltage characteristic curve.

flux in the generator. As the load is increased, the field current also increases, and so the output voltage increases. The voltage continues to increase with corresponding increases in load until the core reaches saturation. Beyond this point, the generated voltage should stay relatively constant with further increases in load. However, the generator voltage losses continue to increase with load. Therefore, the actual output of the generator will begin to decrease for load currents beyond the saturation limit.

Due to its poor voltage regulation, the series generator has very few direct applications. However, as you will see in the next section, the characteristic of the series generator is often combined with the shunt generator to produce a generator with excellent voltage regulation characteristics.

THE COMPOUND DC GENERATOR

It was seen that as the load increases, the output voltage of a shunt generator decreases and the output voltage of a series generator increases. If these two characteristics are combined, then the loss in voltage encountered by one will be compensated for by the increase in voltage of the other. The result is a generator having excellent regulation characteristics. Such a generator is called a *compound generator*. When the shunt field is connected in parallel with the series combination of the armature and series field, the compound generator is said to be connected in long shunt.

When the shunt field is connected in parallel with only the armature, the compound generator is said to be connected in short shunt. The output characteristics of the long- and short-shunt generator are almost identical. These two compound generator connections were shown in Figs. 7-11(d) and (e).

Even though the connections of the shunt and series windings in a compound generator are different, these two windings are nonetheless located on the same pole pieces of the generator. It is also important that these two windings be so arranged that their flux will *aid* each other. That is, when the load is increased, the increase in series-field flux must compensate for the loss in flux that normally would occur in the shunt field. When the two fields aid each other, the compounding is called *cumulative*. If these two windings are connected in reverse, so that their magnetic fields oppose each other, the generator would have extremely poor voltage regulation. This is called *differential compounding* and is not used in generators.

DEGREES OF COMPOUNDING

The level of compounding in a generator can be altered by changing the amount of current through the series field. In order to vary only the series-field current, a bypass rheostat is connected, as shown in Fig. 7-23. When the amount of series-field current is such that it causes the full-load output voltage to be about the same as the no-load voltage, the generator is said to be "flat-compounded." When the full-load voltage is greater than the no-load voltage, the generator is "overcompounded" and when the full-load voltage is less than the no-load voltage, the generator is "undercompounded." The load characteristic curves for these three levels of compounding are illustrated in Fig. 7-23(b).

Since compound generators may be designed to have a wide variety of characteristics, they are more extensively used than other types of generators. For example, if the generator must supply a distant load, the compound generator can be over-compounded. In this

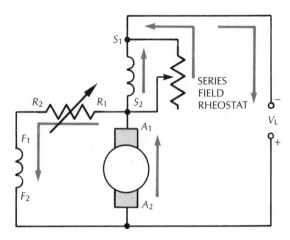

Figure 7-23(a) Varying the series field current.

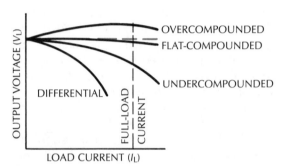

Figure 7-23(b) Load characteristics for compound generators.

way, the increase in generator terminal voltage will compensate for the voltage drop in the feeder lines. Another advantage of the compound generator is that it is self-regulating. When the load is changed, the series-field current also changes so that the correct amount of generator flux is maintained. This property of the generator helps to keep the terminal voltage more constant under variable load operation.

GENERATOR POWER LOSSES —EFFICIENCY

EFFICIENCY

In any machine, the output power is always not quite equal to the input power. This difference is due to the power losses that occur whenever energy is transmitted or transformed from one form into another. Power loss in a machine is energy wasted and it shows up as heat in the machine. The ratio of the output power of a machine to its input power is defined by its efficiency.

$$\text{percent efficiency} = \frac{\text{output power}}{\text{input power}} \times 100 \qquad (7\text{-}6)$$

COPPER LOSSES

It should be obvious that the operating efficiency of a machine is an important consideration. For a large machine, in particular, it is usually more convenient to determine the machine's efficiency by first finding its power losses. The input and output of a machine is related by:

$$\text{input power} = \text{output power} + \text{total power losses} \qquad (7\text{-}7)$$

The total power losses of a generator consist of the generator's copper and mechanical losses. Copper losses are caused by the resistance in the armature, shunt, series and interpole windings. These losses vary with current and are also called I^2R losses.

Total copper losses
$$= I_A^2 R_A + I_f^2 R_f + I_s^2 R_s + I_i^2 R_i \qquad (7\text{-}8)$$

MECHANICAL LOSSES

Mechanical losses are caused by rotation. They are not affected by current, but rather are affected by speed. Since a generator is usually operated at some rated speed, these losses would remain relatively constant. As a result, the mechanical losses are sometimes referred to as the generator "fixed losses."

CORE LOSSES

The mechanical power losses of a generator consist of hysteresis losses, eddy-current losses, and friction losses. The combined losses from hysteresis and eddy currents are sometimes called the generator's "core losses." It was described earlier in the chapter that hysteresis losses are caused by the resistance of the core's electrons to magnetic alignment. This opposition to magnetic alignment produces heat in the core and is therefore an energy loss. Hysteresis losses are minimized by constructing the generator's magnetic core with a high permeability steel.

In an earlier section, it was seen that eddy currents were the small currents induced in the iron core, as the core cuts through the flux. These currents produce heat in the core, and therefore they constitute a power loss. Eddy currents in a generator are minimized by laminating the core. Thin steel strips are insulated, usually with a varnish, and then stacked together to form the core. Lamination reduces the eddy-current paths in the core, and thus reduces the power losses due to eddy currents.

FRICTIONAL LOSSES

Frictional losses are caused by bearing friction, brush friction, and windage or air friction. Friction produces heat and thus it constitutes a power loss in a generator.

$$\begin{aligned} \text{total mechanical losses} &= \text{core losses} \qquad (7\text{-}9) \\ &\quad + \text{friction losses} \\ &= \text{hysteresis} \\ &\quad + \text{eddy current} \\ &\quad + \text{friction} \end{aligned}$$

The fixed losses of a generator can be determined by first measuring the input power to the prime mover, such as an electric motor, while operating uncoupled from the generator and at rated speed. The input power to the motor is then measured again, this time while coupled to the generator at no-load and turning at its rated speed. The difference between these two input powers to the motor, less the small generator copper losses, is equal to the energy required to overcome the mechanical losses of the generator. This loss will remain relatively constant as the generator's load is increased to full-load. Only the copper losses will increase. The following examples illustrate the calculation of power losses in a generator.

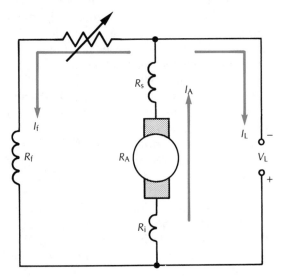

Figure 7-24 A compound generator with interpoles.

Example 7-2: A compound generator, as shown in Fig. 7-24, was tested for losses and efficiency and the following readings were taken:

FIXED LOSSES TEST
1. Input to motor uncoupled and at rated speed: $V_L = 115$ V, $I_L = 0.6$ A
2. Input to motor coupled with generator: $V_L = 115$ V, $I_L = 1.5$ A
3. Generator field current and voltage at rated voltage: $V_f = 120$ V, $I_f = 0.4$ A

WINDING RESISTANCE: VOLT-AMPERE METHOD:

$$R_A = 6\ \Omega, \quad R_f = 300\ \Omega, \quad R_S = 2\ \Omega,$$
$$R_i = 1.5\ \Omega$$

Determine the losses and efficiency of the generator when it is operated at its rated full load of 2.5 A, and 120 V.

1. Total no-load losses
   = motor input (coupled) − motor input (uncoupled)

= $(115 \times 1.5) - (115 \times 0.6)$
= $172.5 - 69 = 103.5$ W

2. Total no-load copper losses
   (Note: $I_A = I_f$ at no-load)
   = $I_f^2 R_f + I_A^2 (R_A + R_S + R_i)$
   = $(0.4)^2 300 + (0.4)^2 (9.5)$
   = $48 + 1.5 = 49.5$ W

3. Total mechanical losses = $103.5 - 49.5$
                                     = 54 W

4. Total copper losses at full-load:
   (Note: $I_A = I_L + I_f$)
   = $(I_f^2 R_f) + [I_A^2 (R_A + R_S + R_i)]$
   = $(0.4)^2 300 + [(2.9)^2 (9.5)]$
   = $48 + 79.9 = 127.9$ W

5. Total losses at full-load = $127.9 + 54$
                                     = 181.9 W

6. % Efficiency
$$= \frac{\text{full-load output}}{\text{full-load output} + \text{total losses}} \times 100$$
$$= \frac{120 \times 2.5}{(120 \times 2.5) + (181.9)} \times 100$$
$$= 62.3\%$$

SUMMARY OF IMPORTANT POINTS

1. The generator converts mechanical energy into electrical energy by the principle of electromagnetic induction.
2. Maximum induction occurs when the conductor cuts the flux at 90°.
3. The left-hand generator rule may be used to determine the direction of the induced current.
4. The output voltage from the ends of an armature loop is sinusoidal. The polarity reverses as the loop passes through the neutral plane. The neutral plane is perpendicular to the field flux.
5. The generated ac voltage is converted to dc voltage by commutator action. This is a switching action performed between the commutator segments and the brushes.
6. The commutator is mounted on the generator shaft and its segments are con-

nected to the armature windings. The brushes are mounted on the frame and are positioned in the neutral plane.

7. The generator magnetic circuit consists of: the field windings, the frame, pole cores, the air gap, and the armature core.

8. The generator electric circuit consists of: the armature windings, the commutator, brushes, and the field windings.

9. The low reluctance of the iron cores help to reduce the hysteresis energy loss.

10. Eddy-current losses are reduced by laminating the cores.

11. Three types of armature drum windings are the wave, lap, and "frog leg." All armature windings are designed to span the same distance as between two unlike poles.

12. The wave winding is a series-type winding. It forms only two parallel branches. It has a high voltage capacity.

13. The lap winding is a parallel-type winding. It forms as many parallel branches as there are poles. It has a high current capacity.

14. The field windings may be separately or self-excited. For self-excitation, the field may be connected to produce series, shunt, or compound generators.

15. For self-excitation, the field must contain some residual flux for voltage build-up action.

16. The series and differential compound generators have poor voltage regulation.

17. There are always an even number of field poles. Increasing the field poles helps to reduce core saturation and increases the output voltage.

18. The generator voltage will increase by increasing the field flux or increasing the armature speed. Voltage control is achieved mainly through flux variation.

19. Armature reaction is caused when the field flux combines with the armature flux. The net field is distorted and the neutral axis is shifted in the direction of armature rotation. Self-induction in the armature coils adds to the effect of armature reaction.

20. Armature reaction increases with load. It causes sparking at the brushes and reduces the output voltage.

21. Interpoles and compensating windings are used to correct armature reaction. These windings are connected in series with the armature.

22. Generator power losses may be divided into (a) mechanical and (b) resistance losses.

23. Mechanical losses increase with speed. They include hysteresis, eddy current, friction, and windage.

24. Resistance losses increase with load. They are caused by the resistance in the windings. They are also called copper or I^2R losses.

25. Percent voltage regulation

$$= \frac{\text{N.L. volt} - \text{F.L. volt}}{\text{F.L. voltage}} \times 100$$

26. Percent efficiency $= \dfrac{\text{Output power}}{\text{Input power}} \times 100$

REVIEW QUESTIONS

1. Name five energy converter devices and give an application for each device.

2. Name four prime movers which may be used to supply the mechanical input to a generator.

3. Name the parts of (a) the electric circuit, and (b) the magnetic circuit of a generator.

4. Describe (a) the left-hand generator rule, and (b) Faraday's law of induction.

5. Explain why the two conductors in a single-loop armature are considered to be connected in series.

6. What is the basic difference between the simple alternator and simple generator?

7. Explain why, when a coil is rotated within a magnetic field, an alternating voltage is induced in the coil.

8. Describe (a) commutator action, and (b) the neutral axis of a generator.

9. Draw diagrams to show the one-quarter turn positions for a single-loop, two-pole generator, and draw its output waveform for one cycle.

10. Draw the output waveforms for one cycle of *reverse* rotation for the armatures shown in (a) Fig. 7-3 and (b) Fig. 7-4.

11. Why is a generator's armature constructed with more than one coil?

12. Describe the construction assembly of the commutator, and state what governs the number of commutator segments that are required by a generator.

13. State three methods for increasing the output voltage of a generator.

14. Draw a diagram to show two sine waveforms that are 90° out of phase.

15. Describe the construction, assembly, and location of the brushes in a generator.

16. Why is the armature core constructed of laminated soft iron?

17. Why is the ring-wound armature no longer used?

18. What property do the lap- and wave-type windings have in common? Which of these two windings has the end of an armature coil connected to adjacent segments?

19. Compare the voltage and current ratings of the lap- and wave-wound armatures for a generator with (a) two poles and (b) four poles.

20. Name three methods of providing field excitation for a generator.

21. Draw diagrams to show three types of self-excited generators.

22. What are the advantages of increasing the number of field poles in a generator?

23. What is the fundamental generator equation? Which of the factors in this equation are variable in a generator?

24. Draw a graph of a typical magnetization curve. Why is this graph not a straight line?

25. State three causes for internal voltage losses in a generator.

26. Describe the cause and effects of (a) armature reaction, and (b) armature self-induction.

27. Describe how the effects of armature reaction and armature self-induction are reduced in a generator.

28. Draw the load-voltage characteristics curves for (a) a separately excited generator, (b) a shunt generator, and (c) a series generator. Indicate the different internal voltage losses for each of these generators.

29. Describe "voltage buildup" in (a) self-excited shunt generator and (b) series generator.

30. Name four factors which can cause voltage buildup failure. State how each of these can be corrected.

31. Describe how a shunt generator provides its own self-protection against an extreme overload.

32. Compare the number of turns and the resistance between the shunt and series-field windings of a generator.

33. Why does the series generator have extremely poor voltage regulation?

34. How can the degree of compounding be varied in a compound generator? Draw graphs to illustrate the load-voltage characteristics for (a) overcompounding (b) flat-compounding, and (c) undercompounding.

35. List the different power losses which occur in a generator. Which of these are classified as (a) core losses, (b) rotational losses, and (c) copper losses? Which of these losses are affected by (a) speed and (b) load current?

36. Describe what causes (a) eddy-current losses and (b) hysteresis losses in a generator. How are these losses minimized?

PROBLEMS

1. When a generator with 2000 armature conductors is rotated at 900 r/min, the generated voltage is 100 V. What would be the generated voltage if the armature conductors were increased to 4000 and the r/min increased to 1800? Assume all other generator properties to remain constant.

2. A shunt generator operating at a speed of 1200 r/min has a generated voltage of 200 V. If both the r/min and field flux to this generator are doubled, what would be the new generated voltage?

3. A generator is rated at 2.5 kW, 220 V. What is its rated full-load current?

4. A generator supplies a load of 8 A at 210 V. What is its kilowatt output?

5. A shunt generator supplies a load of 12 A at 220 V. If the field resistance is 200 Ω, what is the armature current?

6. A separately excited generator supplies a load of 6 A at 120 V. If the field resistance is 160 Ω, what is the armature current?

7. A separately excited generator supplies a load of 4 A at 120 V. Assuming that the internal voltage losses are due only to its internal resistances, find the armature-generated voltage if the armature and brush resistances are 6 Ω and 2 Ω, respectively.

8. A shunt generator supplies a load of 5 A at 120 V. Find the generated voltage if the armature circuit and field resistances are 6 Ω and 150 Ω, respectively. Assume voltage loss due to armature reaction equals 12 V.

9. A shunt generator with compensating windings supplies a load of 8 A at 220 V. Find the generated voltage if the armature circuit, compensating windings, and field resistances are 6 Ω, 3 Ω, and 200 Ω, respectively. (Note: armature reaction is now negligible.)

10. A separately excited generator supplies a load of 6 A at 220 V. If the armature circuit and field resistances are 4 Ω and 180 Ω, respectively, find (a) the no-load voltage and (b) the percent voltage regulation.

11. At no-load and rated speed, the generated voltage of a shunt generator is 143 V. The internal resistances of the generator are $R_A = 6$ Ω, and $R_f = 200$ Ω, and the loss due to armature reaction is 8 V. If the rated load of the generator is 2 A, and the field current is kept constant at 0.6 A, determine (a) the no-load output voltage, (b) the full-load output voltage, and (c) the percent voltage regulation.

12. The internal resistances of a compound generator with interpoles are $R_a = 6$ Ω, $R_s = 3$ Ω, $R_i = 2$ Ω, and $R_f = 200$ Ω. The rated full-load of the generator is 4 A at 120 V. Determine the generated voltage at (a) no-load, and (b) at full-load. Assume that the series field is adjusted for flat compounding.

13. Determine the total copper losses for the generator in Q. 12 at (a) no-load and (b) at full-load.

14. A shunt generator with interpoles is rated at 3 A, 115 V. The internal resistances of the generator are $R_a = 5$ Ω, $R_i = 2$ Ω, and $R_f = 150$ Ω. If the generator's fixed losses are 54 W, determine the generator's efficiency at full-load.

15. The following results were obtained in testing a shunt generator for efficiency.
    (a) Motor input uncoupled...
       $V_L = 120$ V, $I_L = 0.8$ A.
    (b) Motor input coupled....
       $V_L = 120$ V, $I_L = 1.9$ A.
    (c) Generator internal resistances...
       $R_a = 5$ Ω, $R_f = 150$ Ω.
    (d) Generator-rated load is 3 A at 115 V, and I_f is maintained constant. Determine (a) the total rotational losses, (b) the full-load copper losses, and (c) the percent efficiency at full-load.

16. A short-shunt compound generator is rated at 220 V, 5 A. Its internal resistances are $R_a = 8$ Ω, $R_s = 3$ Ω, and $R_f = 170$ Ω. Determine (a) the series-field voltage drop, (b) the shunt-field current, (c) the armature current.

17. If the rotational losses for the generator in Q. 16 are 62 W, determine its percent efficiency at rated load.

DC MOTORS

INTRODUCTION

The electric motor is a machine that converts electrical energy into mechanical energy. It transforms electricity into a rotary motion that is used to perform work. It uses magnetism produced by electric current to turn its armature, resulting in mechanical rotation.

The dc motor and dc generator are very similar, in fact a dc generator can be utilized as a dc motor. When generators are utilized, they are usually set up so that the rotation of the armature is in one direction, whereas a motor may be required to turn in either direction. Dc generators require electric current for the field coils, either from an external source

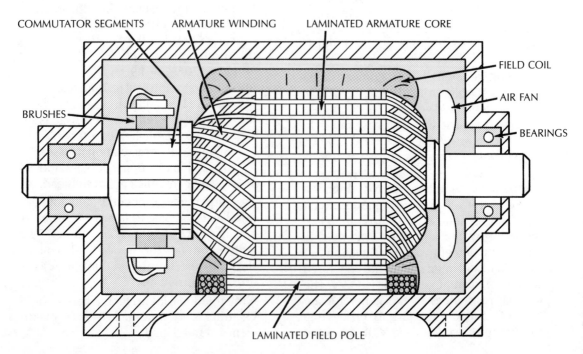

COMMUTATOR SEGMENTS   ARMATURE WINDING   LAMINATED ARMATURE CORE

FIELD COIL

AIR FAN

BRUSHES

BEARINGS

LAMINATED FIELD POLE

Figure 8-1 Cross-sectional view of a dc motor.

(separately excited) or from within (self-excited). The dc motor, on the other hand, must have electric current supplied to both field and armature circuits from an external source. Fig. 8-1 shows a cross-section of a dc motor.

Direct current motors are used in applications where exacting control characteristics are very important, such as cranes, elevators, numerical control, tape drives, and data processing. They are also used in the field of transportation, in vehicles such as electric buses and streetcars.

BASIC MOTOR PRINCIPLE

In Chapter Six, it was shown that the basic meter movement has a coil suspended between two poles of a permanent magnet. When current flows through the coil, a magnetic field is set up around the moving coil. This magnetic field reacts with the permanent magnetic field, producing a force which causes the moving coil to turn. This is the basic principle for all dc motors. (Fig. 8-2)

The elementary dc motor is made up of a turn of wire mounted on a turning shaft located between the poles of a permanent

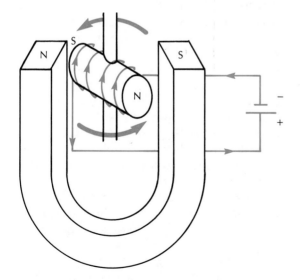

Figure 8-2 Magnetic poles producing a twisting force.

magnet. To understand the action between the turn of wire and the permanent magnetic field, let's examine the effect of a current-carrying conductor that is placed in a magnetic field between two poles.

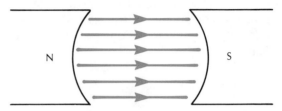

Figure 8-3 Field flux.

When a current is passed through a conductor, a circular magnetic field is produced around the conductor. Using the left-hand rule, we can determine in what direction the flux lines will flow around the conductor. In Fig. 8-3, a magnetic field is shown between two poles, the lines of force travel from north to south. In Fig. 8-4, a cross section of a current-carrying conductor shows the direction of current flow and the direction that the magnetic field is acting around the conductor. If the current-carrying conductor is placed between the poles of the magnet as in Fig. 8-5(a), both magnetic fields interact and will be distorted. Above the conductor, the magnetic lines of force of the conductor repel the lines of force from the N-pole, causing them to bend below the conductor. Thus, the denser magnetic field exerts a force (upward) on the conductor in the direction of the area with the lower magnetic field density. The force exerted upon the conductor depends on the strength of the magnetic field between the poles and on the amount of current flowing through the conductor.

If the current flowing through the conductor

Figure 8-4 Flux around conductors.

is reversed, the direction of the magnetic field around the conductor is reversed. The density of the magnetic field below the conductor is now weakened, and thus the conductor will move downward (Fig. 8-5(b)). If the current in the conductor is left alone and the polar field is reversed, the conductor will also move in the opposite direction to the one shown in Fig. 8-5(a).

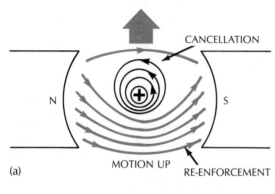

(a)

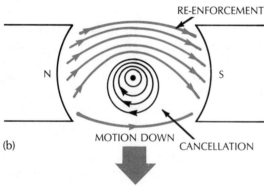

(b)

Figure 8-5 Force acting on a current-carrying conductor.

Reversing both the polar and conductor fields will cause the conductor to continue moving in the same direction.

The direction in which the conductor moves may be determined by Fleming's right-hand rule for dc motors. The rule is stated as follows: *Place the thumb, index finger, and the middle finger at right angles to each other. Place the hand so that the index finger points in the direction of the magnetic field flux lines, the*

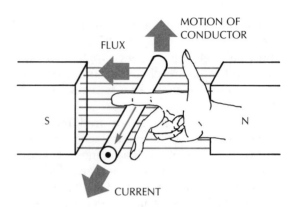

Figure 8-6 Right-hand rule for motors.

middle finger in the direction of the current in the wire, and the thumb will point in the direction of motion of the conductor. (See Fig. 8-6.)

TORQUE AND ROTARY MOTION

The resultant turning action on the single (turn) armature as it cuts the magnetic field at right angles causes rotation and is called *torque*. Figure 8-7 shows a magnetic field produced around the two conductors acting in the direction indicated.

The polar magnetic field interacts with the magnetic field of the conductor, causing the twisting force (torque). Because the armature turn is mounted on a shaft that pivots on its axis, the armature rotates away from the field poles, in a counter-clockwise direction, Fig. 8-7(b). If the direction of the current flowing through the turn, or the polarity of the magnetic field is reversed, the direction of the torque reverses. However, if both the polar field and the direction of the current are reversed, rotation will be in the same direction.

When the armature has rotated to a position where it is moving parallel to the lines of force of the polar field, then there is no interaction between the two magnetic fields. This position is called the *neutral plane*, Fig. 8-8.

Since there is no interaction between the two magnetic fields in the neutral plane, no

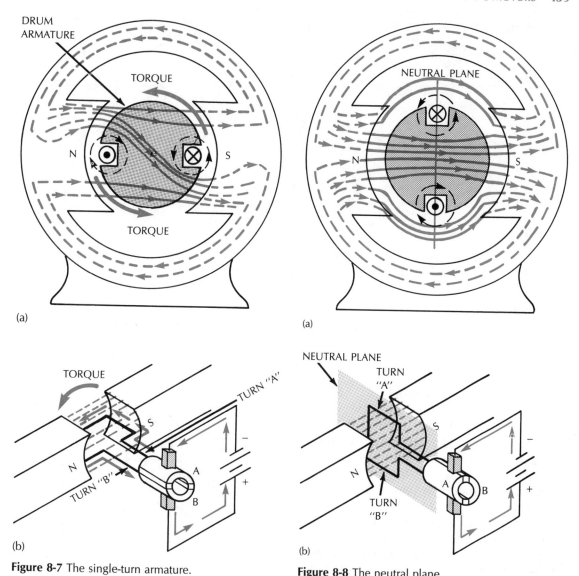

Figure 8-7 The single-turn armature.

Figure 8-8 The neutral plane.

torque is produced and the armature will stop rotating. Conversely, to get the elementary dc motor started, it is necessary to move the single turn armature out of the neutral plane. In order to achieve a continuous rotation, the direction of current flow in the turn must reverse when the turn reaches the neutral plane This reversal is accomplished by a switching device called a *commutator* (discussed in Chapter Seven on dc generators).

COMMUTATION

In Fig. 8-7(b), assume current flows into side A of the armature turn and out of side B. Applying Fleming's right-hand rule, the torque produced will cause the single turn to rotate as shown. When the turn reaches the neutral plane shown in Fig. 8-8(a), the torque on the single turn will now be zero, but inertia will carry it past this point. At the same time, commutator segments A and B will break contact

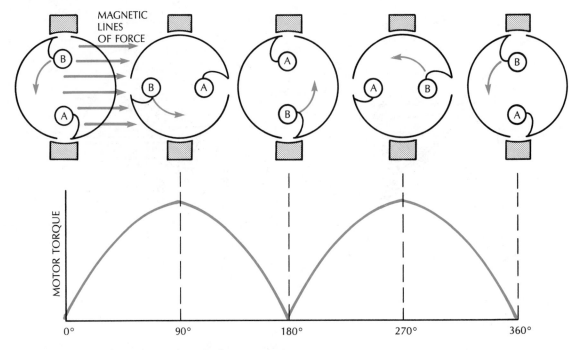

Figure 8-9 Torque characteristics for a single-turn armature

with the brushes and then re-establish contact, but with the opposite brush. This switching action is called *commutation*. Since the two armature conductors have exchanged sides and the two segments have switched contact with the brushes, the single turn will continue to produce a torque in the same direction.

The single-turn armature has the disadvantage of producing an irregular torque, as shown in Fig. 8-9. When the turn is horizontal, maximum torque is produced; when the turn is in the neutral plane, no torque is produced. To eliminate this effect and keep the level of torque more constant, we can make an armature with two or more turns.

In Fig. 8-10(a), the two-turn armature has the two turns placed at right angles to each other. Thus, when one turn is in the neutral plane position and disconnected from the power source, the other turn is in a plane where it is connected to the power source, and maximum torque is developed. The commutator in a two-turn armature is divided into four seg-

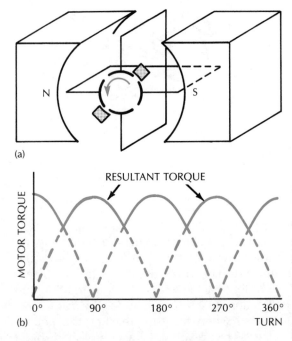

Figure 8-10 Torque characteristics for a two-turn armature.

ments, and switches the current to one turn while disconnecting the current from the other turn. Thus, with two turns the torque developed is steadier than that of a single-turn armature, but still somewhat erratic, as can be seen in Fig. 8-10(b). The extra weight of the turn that does not develop any torque offsets the advantage of additional torque being developed. The two-turn armature will start by itself when current is applied because at least one of the two turns will be out of the neutral plane at any time to produce torque.

If a third turn is added to the elementary motor that has one set of brushes, two turns are disconnected from the power source and become dead weight, while one turn is developing torque to drive the motor. One would quickly ascertain that to overcome this disadvantage all that needs to be done is to provide a set of brushes for each armature turn. But this solution is not a very practical one, due to the extra cost and maintenance problems. This problem can be overcome by using one pair of brushes, and connecting all the armature turns in series, allowing the current to flow through all the turns at the same time, except the ones disconnected from the main power source by the brushes. All the other windings not disconnected will produce torque at the same time. Most dc motors use one set of brushes and multi-turn armatures wound in series-parallel arrangements. This produces a greater as well as a more constant torque.

In Fig. 8-11, a practical four-turn armature is shown. In part (a), it can be seen that a parallel circuit exists. The current flows through all four turns at the same time, thus all four coils produce torque. In Fig. 8-11(b), when turns A and C are in the neutral plane, they are shorted out by the brushes, thus no current flows in turns A and B. The absence of current in these two turns means that they are not producing torque. At the same instant, turns B and D are still connected in the circuit and conduct current; thus, they are producing the torque which keeps the armature turning.

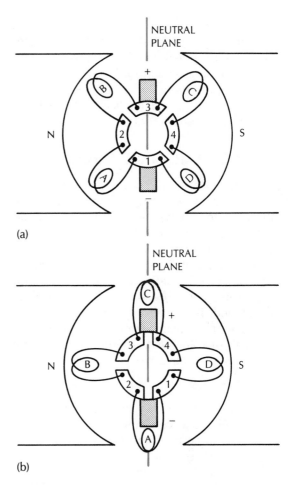

(a)

(b)

Figure 8-11 A practical four-turn armature.

Because of the symmetrical arrangement, two of the armature windings are always shorted out at one time when they arrive at the neutral plane.

In an armature that has many windings, for example 16 windings, 14 windings would conduct current and produce torque during the short time that 2 of the windings were inactive.

LAP AND WAVE WINDINGS

The armature winding for a dc motor is very much the same as that of a dc generator. In fact, a dc motor may be used as a dc generator with a few minor changes. The drum-wound armature can have the coils wound in two different configurations: the lap or parallel winding, and the wave or series-winding. Both are discussed in greater detail in Chapter Seven on dc generators. Essentially, the parallel lap winding is used for low-voltage, high-current applications, while the series wave-winding is used in circuits that have high-voltage, low-current requirements.

CALCULATION OF TORQUE

The amount of torque or force developed on the armature is due to the combined action of the polar (main) magnetic field (ϕ), the magnetic field around the rotating armature (this is expressed in terms of the armature current I_A), and the physical dimensions (electromechanical energy conversion constant) of the motor (K_T). Thus, it can be stated that the torque developed is directly proportional to the strength of the main stator field flux and the strength of rotor field. Therefore the torque equation is written as:

$$T = K_T \times \phi \times I_A \qquad (8\text{-}1)$$

where T = torque in newton metre (N · m)
    K_T = design constant
    ϕ = total number of lines of force per pole (Wb)
    I_A = armature current, (A)

This equation is the same one used to calculate the counter torque of a generator.

Example 8-1: Given a dc motor with a manufacturer's constant rating of $K_T = 16.4 \times 10^{-8}$. The polar magnetic field is 6×10^6 Wb, and the full-load armature current rating is 5 A. Find how much torque can be developed.

$$T = K_T \times \phi \times I_A$$
$$T = (16.4 \times 10^{-8}) \times (6 \times 10^6) \times (5)$$
$$T = 492 \times 10^{-2}$$
$$T = 4.92 \text{ N} \cdot \text{m}$$

Example 8-2: A 220-V, 3730-W motor develops a torque of 12 N · m when the current is 10 A. The motor constant $K_T = 15.2 \times 10^{-8}$. Find the total lines of flux developed in the magnetic field.

$$T = K_T \times \phi \times I_A$$
$$12 = (15.2 \times 10^{-8}) \times (\phi) \times (10)$$
$$\phi = \frac{12}{15.2 \times 10^{-8} \times 10}$$
$$\phi = \frac{12 \times 10^8}{15.2 \times 10}$$
$$\phi = 0.789 \text{ Wb}$$

TORQUE MEASUREMENT

The performance of a motor under varying load conditions may be determined experimentally. The torque developed by a motor may be measured by a device called a *prony brake*, Fig. 8-12.

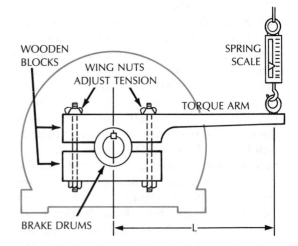

WOODEN BLOCKS

WING NUTS
ADJUST TENSION

SPRING SCALE

TORQUE ARM

BRAKE DRUMS |←————— L —————→|

Figure 8-12 Prony brake used for measuring torque developed by a motor.

There are various prony brake designs available. In Fig. 8-12, the prony brake drum encased in wooden blocks is attached to the motor pulley. By tightening the wing nuts, the brake can be adjusted on the pulley for the desired load on the motor. The brake drum has an extension torque arm that is fastened to a spring scale which measures in newtons. The torque output of the motor can be determined from the applied load. The torque is the product of the net force on the scale (in newtons) times the effective length (ℓ) of the torque arm in metres.

Torque = force × length   (8-2)

Example 8-3: A prony brake arm is 0.4 m in length. The wing nuts on the brake are tightened on the motor pulley, creating a force of 50 newtons. What torque is being developed by the motor?

Torque = force × length
     = 50 × 0.4
     = 20.0 N · m

ARMATURE REACTION

In Chapter Seven, it was shown that the generator developed a problem of arcing at the brushes, because of *armature reaction* when under load. In a similar fashion, when a dc motor is loaded the armature current flux and the field flux interact so as to distort the main field. This causes the perpendicular neutral plane to shift opposite to the direction of rotation of the armature. Since the brushes may now be out of alignment with the neutral plane, this results in sparking at the brushes and a net reduction in torque. Recall that in dc generators, armature reaction caused the neutral plane to shift in the same direction as armature rotation. This shift in the neutral plane is opposite to that of the dc motor. The amount of armature reaction varies with the

load applied to the motor, the larger the load, the larger the armature current, and the greater the shift of the neutral plane, Fig. 8-13.

With the motor turning in a clockwise direction, the brushes must be moved opposite the direction of rotation to minimize sparking. When a dc motor is being operated in only one direction and at constant speed, the brushes can be placed at the new full-load position of the neutral plane, Fig. 8-13(c), to minimize sparking. However, if the motor is being operated with a varying load and speed, the armature current will vary. This will cause the armature reaction to vary and, along with it, the neutral plane. To produce perfect commutation, the brushes would have to be rotated to a new position every time the neutral plane shifts. This constant realignment of brushes is not a practical solution. To overcome armature reaction, special windings, called *interpoles*, and *compensating windings* are utilized to reduce the field distortion. These two types of windings were discussed in Chapter Seven, dc generators (Figs. 7-17 and 7-18).

INTERPOLES

Interpole windings perform the same function in a motor and generator, that is, to overcome the effects of armature reaction.

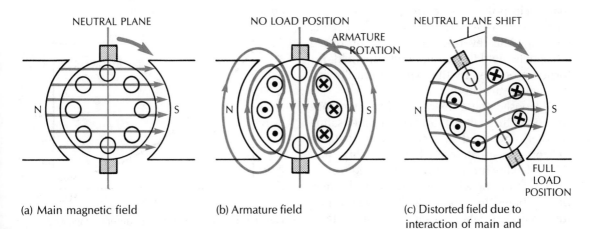

(a) Main magnetic field   (b) Armature field   (c) Distorted field due to interaction of main and armature magnetic field

Figure 8-13 Armature reaction.

Interpole windings are connected in series with the armature winding but placed between the main poles. They produce a magnetic field that cancels the armature magnetic field in the vicinity of the interpoles. Thus, the interpole field opposes the tendency of armature reaction to shift the neutral plane. The action of the interpole winding is also self-regulating because it is connected in series with the armature. Therefore, the cancelling effect will self-adjust with any changes in the armature load current.

The only difference between a generator and a motor, as far as interpoles are concerned, is that in the generator the interpole has the same polarity as the main field pole *in front of it* in the direction of rotation; in a motor, the polarity of the interpole is the same as the main field pole *that it has just rotated past*.

COMPENSATING WINDINGS

As discussed in Chapter Seven, interpoles tend to correct mainly armature self-induction. Armature reaction, unlike self-induction, is caused by armature flux that exists all around the armature. A second winding, called *compensating winding*, is used to cancel the armature flux.

These windings are also in series opposing with the armature but are placed within the main field poles. The current flow from the armature is reversed in the compensating windings. This reversed current flow produces a magnetic field that counteracts the field distortion produced by the load current flowing through the armature.

Obviously, these extra windings increase the cost of the motor. However, compensating windings are used in large motors that are utilized in situations where stability and dependability must be strictly adhered to.

Compensating windings and interpoles may be combined in a motor to overcome armature reaction where exacting commutator action is an important criterion.

COUNTER VOLTAGE (V_C) IN A MOTOR

As the armature of a dc motor rotates in the polar magnetic field, the windings cut the flux lines, inducing a voltage (electromotive force) in the windings. This induced electromotive force opposes the applied voltage and is known as the counter voltage. The counter voltage induced by the *generator action* is proportional to the speed of rotation, and the field strength. The greater the flux and the faster the rotating speed, the greater will be the counter voltage. This relationship may be expressed by the equation:

$$V_T = V_C + I_A R_A \qquad (8\text{-}3)$$

where V_T = motor terminal voltage
V_C = counter voltage
I_A = armature current
R_A = armature resistance
and is derived from Kirchhoff's Second Law, which states that the total voltage applied to a series circuit must equal the sum of the individual voltage drops around the circuit.

When a voltage is applied to a stationary armature, there is no countervoltage generated. The counter voltage will always be less than the applied voltage. The only opposition in the circuit is the armature resistance which is comparatively small; thus the armature current will be very large. This excessive current could burn out the armature.

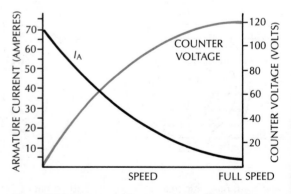

Figure 8-14 Motor speed vs. counter voltage and armature current.

As the armature rotates, a counter voltage is generated that opposes the applied voltage, which in turn limits the armature current to a level that will still produce a torque in the motor, but not necessarily large enough to overheat the armature windings. Figure 8-14 shows the relationships between motor speed, the counter voltage, and armature current. This can be illustrated by the following example:

Example 8-4: A dc motor has an armature circuit resistance of 2 Ω. The motor is operated on 220 V and its rated armature load current is 20 A. Determine (a) the counter voltage at the instant it is started, (b) the armature current at this instant, (c) the counter voltage at rated speed and load.

(a) At the instant of starting, the counter voltage is zero.

(b) $\therefore V_T = I_A R_A$

$$220 = I_A \times 2$$
$$I_A = \frac{220}{2}$$
$$I_A = 110 \text{ A}$$

(c) At rated speed and load
counter voltage $V_C = V_T - I_A R_A$
$$V_C = 220 - (20 \times 2)$$
$$V_C = 220 - 40$$
$$V_C = 180 \text{ V}$$

FACTORS AFFECTING MOTOR SPEED

LOAD EFFECT

When a load is applied to a motor, the motor tends to slow down. The amount of counter voltage produced in the armature is reduced. The applied voltage then causes an increase in armature current, which will produce a greater armature flux. The torque increases, causing a corresponding increase in speed. As the motor speed increases, the counter voltage increases, and the motor tends to regulate its speed.

Conversely, as the applied load is reduced, the motor speeds up, causing the induced counter voltage to increase, reducing the armature current. The armature flux decreases, resulting in a reduction of torque and motor speed. The slower motor speed reduces the counter voltage, and the motor stabilizes itself.

FIELD EFFECT

In the preceding paragraph, it was shown that when the armature current was increased, the result was an increase in torque and speed. Assuming a constant load, an increase in field strength will cause a greater voltage to be induced in the armature. The increase in the counter voltage will cause the armature current to decrease and reduce the torque and speed of the motor.

If the field strength is reduced, the value of the counter voltage will decrease. This will cause the armature current, and hence the motor speed, to increase. Although the speed of a motor may be regulated by changing the applied voltage, this method will also drastically affect the motor torque. Instead, variation of the field current is the most common method of dc motor speed control (Fig. 8-15).

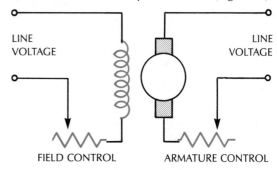

LINE VOLTAGE   LINE VOLTAGE

FIELD CONTROL   ARMATURE CONTROL

Figure 8-15 Methods of speed control.

SPEED REGULATION

A motor that is able to maintain its speed at a constant level when a variable load is applied to it is said to have good *speed regulation*. It is a built-in characteristic of a motor and remains the same as long as the applied voltage does not vary.

Speed regulation of a motor is a comparison of its no-load speed to its full-load speed, and is usually expressed as a percent as follows:

$$\text{percent speed regulation} \qquad (8\text{-}4)$$
$$= \frac{\text{no-load speed} - \text{full-load speed}}{\text{full-load speed}} \times 100$$

Example 8-5: If the no-load speed of a dc shunt motor is 157.05 rad/s (1500 r/min) and when the motor carries its rated load, the speed drops to 146.58 rad/s (1400 rpm). Calculate the percent speed regulation. (One revolution per minute is equal to $2\pi/60$ rad/s or 0.1047 rad/s.)

percent speed regulation
$$= \frac{\text{no-load speed} - \text{full-load speed}}{\text{full-load speed}} \times 100$$
$$= \frac{157.05 - 146.58}{146.58} \times 100$$
$$= \frac{10.47}{146.58} \times 100$$
$$= 0.0714 \times 100$$
$$= 7.14\%$$

The lower the speed-regulation percentage figure of a motor, the better is the speed regulation. If the speed-regulation percentage figure is high, then its speed regulation is poor.

CLASSIFICATION OF DC MOTORS

Direct current motors are classified into three groups, drawing their name from the way in which their field windings are electrically connected to the voltage source. Each type has its own specific load-speed characteristics, as well as certain preferred applications. The three categories are the *shunt*, *series*, and *compound* dc motors. (See Fig. 8-16.)

MOTOR RATINGS

Direct current motors are rated according to their voltage, current, speed, and power output.

MOTOR LOSSES

The losses in a dc motor are similar to the losses that occur in a dc generator. These losses were described in Chapter Seven, dc generators. As in the generator, motor losses

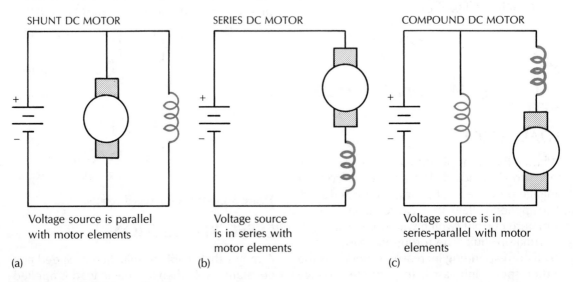

SHUNT DC MOTOR

Voltage source is parallel with motor elements

(a)

SERIES DC MOTOR

Voltage source is in series with motor elements

(b)

COMPOUND DC MOTOR

Voltage source is in series-parallel with motor elements

(c)

Figure 8-16 Classification for dc motors.

may be divided into mechanical and copper losses.

1. Copper losses, or I^2R losses, are present in the armature windings, the shunt field, the series field, and interpole field windings. The I^2R losses depend on the effective resistance of these windings. Power is used whenever a current flows through these resistances. These resistances may be found by the ammeter-voltmeter method; they are then multiplied by the currents squared to obtain the copper losses.

2. Mechanical or rotational losses are divided into two groups: the (a) iron or core losses, (b) friction losses.

    Iron or core losses (a) consists of the (i) eddy current and (ii) hysteresis losses.

The cause and correction of eddy current, hysteresis, and frictional losses were described in detail in Chapter Seven, dc generators. These are exactly the same for dc motors.

EFFICIENCY

Efficiency relates to how well a motor will convert electrical energy into mechanical energy.

The power output of a motor is less than its power input due to inherent losses that are built into the motor. Some of the power is lost through eddy currents, hysteresis, armature reaction, resistance of the windings, and mechanical losses due to friction and air resistance. Power that is not transferred to the load is called a *power loss*. Thus, the output power can be computed by subtracting the power losses from the input power.

The efficiency of a motor may be computed by the following equation:

$$\text{percent efficiency} = \frac{\text{output power}}{\text{input power}} \times 100 \qquad (8\text{-}5)$$

Example 8-6: Calculate the efficiency of a shunt dc motor at full-load which produced the following test results: (a) for the armature and brush resistance (volt-amp) test, the armature current was 15 A with an applied voltage of 6 V; (b) for the no-load test the motor produced an armature current of 4 A, and a shunt field current of 0.5 A with an applied voltage of 115 V; (c) with full-load applied to the motor the armature current was 15 A at 115 V.

(a) Winding resistance:

   (i) shunt field resistance $R_{SH} = \dfrac{V_L}{I_A}$

$$R_{SH} = \frac{115}{0.5}$$
$$R_{SH} = 230 \ \Omega$$

   (ii) armature resistance (volt-ampere test)

$$R_A = \frac{V_A}{I_f}$$
$$R_A = \frac{6}{15}$$
$$R_A = 0.4 \ \Omega$$

   (iii) I_L (at no-load) $= I_A + I_f$
$$= 4 + 0.5$$
$$= 4.5 \ A$$

(b) Core losses
$$= \begin{bmatrix} \text{input power at} \\ \text{no-load} \end{bmatrix} - \begin{bmatrix} \text{resistance losses at} \\ \text{no-load} \end{bmatrix}$$
$$= (V_L I_L) - (I_f^2 R_{SH} + I_A^2 R_A)$$
$$= (115 \times 4.5) - (0.5^2 \times 230 + 4^2 \times 0.4)$$
$$= 517.5 - (57.5 + 6.4)$$
$$= 517.5 - 63.9$$
$$= 453.6 \ W$$

(c) (i) I_L (at full-load) $= I_A + I_f$
$$= 15 + 0.5$$
$$= 15.5 \ A$$

   (ii) Total losses at full-load
$$= \text{total copper losses} + \text{core losses}$$
$$= (I_f^2 R_f + I_A^2 R_A) + 453.6$$
$$= (0.5^2 \times 230 + 15^2 \times 0.4) + 453.6$$
$$= (57.5 + 90) + 453.6$$
$$= 601.1 \ W$$

(d) Percent efficiency
$$= \frac{\text{output power}}{\text{input power}} \times 100$$

$$= \frac{\text{input power} - \text{total losses}}{\text{input power}} \times 100$$

$$= \frac{V_L I_L \ (\text{full-load}) - \text{total losses}}{V_L I_L} \times 100$$

$$= \frac{(115 \times 15) - 601.1}{115 \times 15.5} \times 100$$

$$= \frac{1782.5 - 601.1}{1782.5} \times 100$$

$$= \frac{1181.4}{1782.5} \times 100$$

$$= 66.28\%$$

Generally, the higher the efficiency of a motor, the lower the losses.

THE SHUNT DC MOTOR

In the shunt dc motor, since the field windings and armature windings are connected in parallel across the voltage source, this means that incoming current will divide itself, some flowing through the field winding and the bulk of it flowing through the armature winding because of its low resistance.

LOAD EFFECT

When a load is applied to the shunt motor, the motor slows down. The slight decrease in speed causes a corresponding reduction in the counter voltage, which allows the effective armature voltage to increase, thus increasing the armature current. With a higher armature current and a constant polar field strength, torque is increased to bring the armature speed back up to drive the increased load. The speed of the motor will then remain relatively constant at the new value of the load.

Conversely, if the load on the shunt motor is decreased, the motor speeds up causing a corresponding increase in the counter voltage, thereby decreasing the armature current and torque. The speed of the motor will change to match the new value of the load, where the motor is again in electrical balance.

A shunt dc motor develops high torque at any speed; it can develop 150 percent of its rated torque with the proper starting circuits. A starting resistance connected in series with the armature will limit the armature current to a low value until the speed of the armature increases up to the level where an effective counter voltage can be built up.

Since a shunt dc motor can maintain a constant speed within a specific load range, it is regarded as a constant-speed motor. Thus, a shunt motor is said to have good speed regulation, (5 to 10 percent from no-load state to full-load state), which makes it desirable for driving machine tools that require constant-speed operation under varying load conditions.

FIELD EFFECT

By inserting a resistance in the field winding, the shunt motor can be operated at various speeds. More resistance reduces the field current, thereby weakening the field flux, and the motor speeds up. (Fewer flux lines are cut by the rotating armature, producing a lower counter voltage, resulting in an increase in armature current and torque causing the motor to speed up.)

A lower value of field resistance will result in a stronger magnetic field, a greater counter voltage, and a decrease in motor speed. Figure 8-17 shows the speed-load and torque-load characteristics of a shunt dc motor.

Inserting resistance in the field circuit provides a smooth means of varying the speed of a shunt dc motor from low to high speed.

It was shown that a shunt dc motor speeds up as the current through the field winding is decreased (due to a lowering of the counter voltage and an increase in armature current and torque). If the field winding is opened, this will reduce the field strength to the value of residual magnetism in the field-winding core. The loss of the field flux will cause the torque and speed to increase to very high values, causing the motor to *run away*, and destroy itself through centrifugal forces.

The speed of a shunt dc motor may also be changed by placing a variable resistance in the armature circuit. This method is not desirable

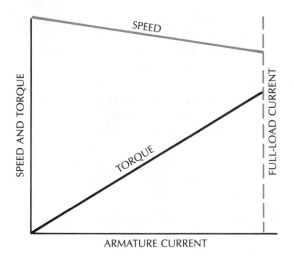

Figure 8-17 Speed-load and torque-load characteristics of a shunt dc motor.

because the armature current is larger than the field current, resulting in a greater power loss across the armature rheostat. As a result, less current is made available for actual motor operation. In addition, an armature rheostat causes the motor to have poor speed regulation.

ROTATION

The direction of rotation of a shunt dc motor may be changed by changing the direction of current flow through the field or armature windings. If interpole or compensating windings are found in the motor, it may be easier to reverse the direction of current flow through the armature windings.

When a shunt dc motor is started, a starting resistance must be connected in series with the armature to limit the armature current until the proper motor speed is attained to build up the necessary counter voltage. This will be discussed in greater detail in Chapter Nine on dc motor control.

Shunt dc motors are used in applications where the speed must be kept fairly constant as the applied load varies. Some of these applications are driving machine tools used in a machine shop (lathes, grinders, shapers, etc.), fans and blowers, and the prime mover in motor-generator sets.

SERIES DC MOTORS

The series motor is sometimes referred to as the workhorse in the dc motor industry, as it is usually used to perform work where extremely high torque is required, under heavy overload conditions.

The series dc motor has its field winding connected in series with the armature and load, and a common current flows through both windings. Therefore, an increase in load current is automatically felt by the armature and field winding.

The field winding is wound with few turns of heavy wire because it must carry all the current flowing through the motor.

If the load on a series motor is increased, the motor slows down, thus the counter voltage decreases. The armature current, field strength, and torque are increased, thus building up the counter voltage again. The interconnected series of events stops the motor speed from building up.

On the other hand, if the load is decreased, the motor will speed up, increasing the induced counter voltage, which in turn will cause the armature current, torque, and speed to decrease.

If the load is completely removed, the armature speeds up and a higher counter voltage is induced in the armature. The current through the armature and field windings decreases, causing a weaker magnetic field. The weakened magnetic field in turn causes the armature to rotate faster, increasing the induced counter voltage. This increased counter voltage causes the armature to rotate still faster. The motor cannot rotate fast enough to generate the proper amount of counter voltage that is required to restore the balance; thus, the armature speeds up till the windings break apart due to centrifugal forces. For this reason, the load should never be disconnected from the series motor. Series dc motors are seldom used with belt drives where the load can be removed if the belt breaks. Usually, the load is connected directly or through gears to the series motor.

The speed-load and torque-load characteristics of a series dc motor are shown in Fig. 8-18.

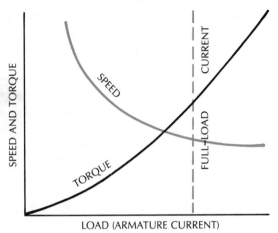

Figure 8-18 Speed-load and torque-load characteristics of a series dc motor.

As shown in Fig. 8-18, in a series dc motor, torque increases as the load and armature current increases, while the speed decreases. The shunt motor, however, cannot slow down with an increase in load; thus, it is more susceptible to overloading conditions. For example, if a crane driven by a dc motor were lifting a heavy load and the motor was a shunt motor, it would try to maintain a constant speed and become overloaded. The series motor, on the other hand, would decrease the speed and counter voltage, causing the armature current and torque to increase.

The series dc motor is used in applications where torque and speed requirements vary substantially. Jobs that require a heavy starting torque and a high rate of acceleration as in traction equipment such as hoists, cranes, electric locomotives and heavy construction trucks (Euclids) make use of this motor.

COMPOUND DC MOTORS

A compound dc motor is a series and shunt dc motor combined. Like a compound generator, it has a series and shunt field. The shunt-field winding has many turns of fine wire and it is connected across the armature. The series field winding has few turns of heavy wire and is connected in series with the armature.

As with dc generators, the shunt field may be connected in parallel with the armature in one of two different ways:

1. If the shunt-field winding is connected in parallel with the series combination of the armature and series field, then it is called a *long shunt* (Chapter Seven, dc generators).
2. If the shunt field winding is connected in parallel with the armature alone, then it is called a *short-shunt*.

The operation of the long-shunt and short-shunt compound motors is almost the same.

The compound dc motor has similar characteristics as the series and shunt dc motors combined. If the windings are connected so that the series field aids the shunt field, the motor is called a *cumulative compound* dc motor. However, if the series winding is connected so that its field opposes that of the shunt field, the motor is said to be *differentially compounded*.

When a load is added to a cumulative compound dc motor, the motor speed decreases at a greater rate than that of the shunt dc motor but less than in a series dc motor. The field flux increases as in the series dc motor and at the same time the counter voltage decreases. This combination of events produces a stronger torque. The speed-load and torque-load characteristics are shown in Fig. 8-19.

The cumulative-compound dc motor has variable load characteristics similar to the series dc motor. However, if the load is removed, it has a maximum speed that keeps the motor from running away. It has a fairly constant speed, and develops a strong torque

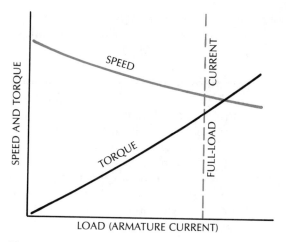

Figure 8-19 Speed-load and torque-load characteristic curves of a cumulative-compound dc motor.

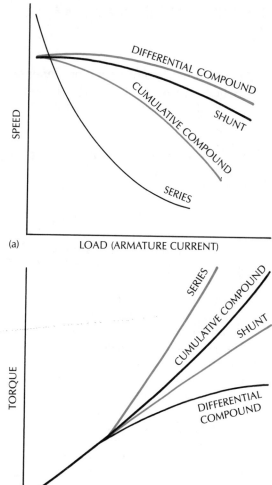

(a)

(b)

Figure 8-20 Speed-load, torque-load characteristics of shunt, series, and compound dc motors.

on heavy loads, as required by large presses and reciprocating shearing machines.

In a differential-compound dc motor, the series field opposes the shunt field, thus weakening the total magnetic field.

If a load is applied to a differential-compound dc motor, the motor slows down resulting in a lower value of counter voltage to be developed. The load and armature current flow will increase and cause a stronger reaction to take place between the two fields, weakening the overall magnetic field still further. The result will be an increase in motor speed, up to a safe operating speed.

The differential-compound dc motor characteristics are similar to those of the shunt motor, only somewhat exaggerated. Thus, it has a nearly constant speed under variable loads and poor torque characteristics at excessive loads. Since the shunt dc motor has a sufficiently constant speed and much better torque characteristics, it is used in many motor applications in lieu of differential compound motors. Figure 8-20 shows the speed-load and torque-load characteristics of a shunt, series, and compound dc motors.

SUMMARY OF IMPORTANT POINTS

1. Direct current motor operation depends on the interaction of two magnetic fields: one produced around the armature conductors and the other a fixed magnetic field supplied by the stator.
2. The right-hand rule for motors is used to indicate the direction in which a current-carrying conductor will move in a magnetic field.
3. A dc motor produces a turning force called torque. The amount of torque developed is directly proportional to the strength of the main magnetic field and the armature current.
4. The four principal parts of a dc motor are a magnetic field, an armature, a commutator, and brushes.
5. The commutator reverses the armature current flow at a point when the poles of the armature are in the neutral plane.
6. Field windings may be connected in series, parallel, or series-parallel with the armature windings.
7. When a motor is running, the armature generates a counterelectromotive force in the armature which opposes the applied voltage.
8. The counter voltage limits the flow of armature current.
9. The interaction of the armature field and the main field in a dc motor causes the neutral plane to shift (armature reaction) opposite to direction of rotation.
10. With constant-speed motors, the brushes can be moved to the new neutral plane, but with varying-speed motors, interpoles and compensating windings are used to minimize the effects of armature reaction.
11. Armature windings are connected to the commutator in lap (parallel low-voltage, high-current) or wave (series high-voltage, low-current) arrangements.
12. The speed of a dc motor is usually varied by a variable resistance connected in series with the field winding.
13. Series dc motors have the field windings (few turns of heavy wire) and armature windings connected in series. When a load is applied, the speed is reduced causing the counter voltage to decrease and armature current to increase, as well as the field current.
14. The series dc motor develops a high starting torque, but poor speed regulation. Without a load the series motor will run away and self-destruct due to centrifugal forces.
15. Shunt dc motors have the field windings (many turns of fine wire) connected in parallel with the armature windings. The field strength is independent of the armature current. The shunt dc motor speed varies slightly with variable loads but has less starting torque than a series dc motor.
16. Compound dc motors have the series winding connected in series with the armature, and the shunt winding connected in parallel with the armature. The speed and load characteristics can be changed by connecting the series and shunt-field windings so that the fields either aid or oppose each other. Compound dc motors combine the high starting torque of series dc motors with the good speed regulation of shunt dc motors.
17. Dc motor losses are due to: armature reaction, resistance of windings, friction, eddy currents, hysteresis, brush resistance, flux leakage.

REVIEW QUESTIONS

1. State briefly the function of a motor.
2. Describe briefly, the fundamental principle of operation of a dc motor.
3. Show how one can determine the direction of rotation of the armature in a dc motor.

4. Define torque and state its unit of measurement.
5. Describe the development of torque and state why it is important in dc motors.
6. What two factors determine the amount of torque developed in a dc motor?
7. Explain how torque in a dc motor may be measured.
8. Describe armature reaction and its effects in a dc motor.
9. Why is it necessary to have more than one winding in an armature?
10. In what direction is the neutral plane shifted by armature reaction?
11. Describe the function of a commutator.
12. How can armature reaction be minimized?
13. Why does a motor produce a counterelectromotive force in the armature?
14. Describe how the counter voltage counteracts load changes.
15. Explain how the speed of a dc motor may be varied.
16. Define (a) eddy-current losses, (b) hysteresis losses, and state how these losses may be reduced.
17. How are lap and wave windings connected in a circuit?
18. When are lap or wave windings used in dc motors?
19. Define speed regulation.
20. Describe the factors that affect the speed of a shunt dc motor as a load is applied.
21. Explain how a shunt dc motor differs from a series dc motor in the construction and connection of the field and armature windings.
22. How are interpoles connected in a dc motor circuit?
23. What motor would you select if a constant speed was necessary when a variable load was applied—series or shunt dc motor?
24. Does a shunt or series dc motor produce a high starting torque?
25. Describe what would happen to a series dc motor if the load were removed.

26. What does the load-speed characteristic curve of a dc motor illustrate?
27. Describe what results would occur if a shunt dc motor were overloaded.
28. What major advantage does a shunt dc motor have over the series dc motor?
29. Describe the construction of a compound dc motor.
30. Explain what is meant by (a) cumulative compound, (b) differential compound.
31. What are the advantages of a compound dc motor over the series and shunt dc motors?
32. Explain how the direction of a motor may be reversed.
33. Give examples of where a series dc motor, shunt dc motor, and compound dc motor are used.
34. By what units are dc motors rated?
35. Draw the speed-load characteristic curves for a cumulative-compound and differential-compound dc motor.
36. Explain in detail the change in the speed curves as shown in the graph of Q. 35.
37. Explain why the speed regulation of a series dc motor is inferior to that of a shunt dc motor.
38. Describe why a series dc motor has a higher starting torque than a shunt dc motor.
39. Compare the speed regulation of the compound dc motor with that of a (a) series dc motor, (b) shunt dc motor.
40. Draw the torque-load characteristic curves for a cumulative-compound and differential-compound dc motor and explain why the two curves differ.

PROBLEMS

1. A shunt dc motor having an armature resistance of 4 Ω, with 120 V applied draws a line current of 3 A and a field current of 0.2 A. Calculate (a) the power loss in the armature winding, (b) the value of the counter voltage.
2. If the power loss in the field winding in the

motor of Problem 1 is 25 W, and losses due to friction are 10 W, find (a) the power input, (b) power output, (c) efficiency of the motor.

3. Given the relationship $P = 2\pi nT$, where P = power output in watts, n = rad/s, and T = torque in N·m, find the torque produced in Q.1 and Q.2 if the motor rotates at 178.02 rad/s.

4. Find the power delivered by a motor when the armature current is 25 A, the applied voltage is 115 V, and the counter voltage is 110 V.

5. A motor rated at 7.46 kW has a shunt-field resistance of 115 Ω and a field current of 3 A. Calculate the value of the applied voltage.

6. A dc motor has a constant rating $K_T = 15.2 \times 8^{-10}$ N·m/A. The magnetic strength of the field windings is 7×10^6 per Wb. At full load the armature current is 6 A. Calculate the amount of torque that is developed.

7. A 1500-W motor, connected to a 110-V line, draws a current of 5 A and develops a torque of 12 N·m. The motor constant $K_T = 17.1 \times 10^{-8}$ N·m/A. Find the total number of flux lines developed in the magnetic field.

8. A prony brake attached to a motor pulley has a brake arm length of 1.3 m. When the speed of the motor is 120.43 rad/s (1150 r/min) the scale indicates a net force of 55 newtons. Calculate the torque being developed.

9. The applied voltage across a motor armature is 125 V; the counter voltage is 122.5 V and the current is 18 A. Find the resistance of the armature.

10. A motor has an armature resistance of 0.3 Ω, and the applied voltage is 115 V and draws a current of 24 A. Calculate the value of the counter voltage.

11. If the speed in Problem 10 were reduced by one-half, and the field remained the same, find the current that would flow in the armature.

12. If the no-load speed of a dc motor is 104.72 rad/s (1000 r/min) and the speed drops to 100.53 rad/s (960 r/min) when a load is applied, calculate the percent speed regulation.

13. A dc motor is rated at 6.6 kW. Its losses are 1035 W. Find the efficiency of the motor.

14. A motor rated at 11 000 W draws a current of 50 A from a 240-V line. Calculate the motor efficiency.

15. The efficiency at rated load of a 3-kW, 600-V shunt motor is 85 percent. Calculate the input current flow.

DC MOTOR CONTROL

In studying the dc motor (Chapter Eight), it was stated that the armature resistance is very low —approximately one ohm. It was also pointed out that if this resistance were the only opposition to current flow, the resulting armature current would be very high. As soon as the motor starts rotating, a counter voltage is generated in the armature, limiting the armature current to a reasonable value. However, at the point when the motor is just starting to rotate, the counter voltage is very low (almost zero), and the starting current is excessively high. This high current produces a great stress on the armature windings, brushes, and commutators, to the point that it can cause burnouts in any of these three components.

DC MOTOR STARTERS

To prevent this excessive starting current, a special device called a *dc motor starter* is used. The dc motor starter contains a variable resistance that is connected in series with the armature at startup.

The dc motor starter is exactly what the name implies; it is used to start the motor and performs no other function once the motor reaches its operating speed. There are other devices available that combine the functions of starting and variable speed control; they are called *dc motor controllers*. These will be discussed later in this chapter.

The dc motor starter, in addition to limiting the value of the starting current, also contains a mechanism for automatically disconnecting the motor from the voltage source should the field circuits become open or the line voltage fails. The starting resistance is automatically disconnected every time the motor stops, but is available for the next time the motor is to be started.

Direct current motor starters and controllers are classified into the following groups: 1. type of operation—*manual or automatic*, 2. type of construction—*faceplate and drum controllers*, 3. types of enclosures—*open type or watertight*, 4. number of terminal connections—*three-point and four-point starters*.

MOTOR-STARTING REQUIREMENTS

Small dc motors (375 W or less) can usually be started by a simple across-the-line switch, while larger dc motors require a reduced voltage starter.

When starting dc motors, it is necessary to protect the motor from the inrush of heavy current flow. The starting current usually is limited to 150 percent of the normal full-load operating current of the motor. Also, the starting torque should be increased to a high value so that motor is brought to full speed as soon as possible.

As described earlier, there is no counter voltage when the motor is at a standstill, because the only opposition to current flow is the armature resistance. The amount of current flowing through an armature may be determined by the following equation.

$$I_A = \frac{V_T - V_C}{R_A} \qquad (9\text{-}1)$$

where I_A = armature current
V_T = applied voltage
R_A = armature resistance
V_C = counter voltage

Example 9-1: Find the armature current in a 3.73-kW motor if the armature resistance is 0.5 Ω, and the applied voltage is 230 V. (The value of V_C is zero at standstill at the instant the motor is connected to the line.)

$$I_A = \frac{V_T - V_C}{R_A}$$
$$I_A = \frac{230 - 0}{0.5}$$
$$I_A = 460 \text{ A (starting current)}$$

If the full-load current of a motor of this size were 20 A, the ratio of starting current to full-load current would be 460/20 or 23 to 1. This large inrush current will produce excessive torque and heat which may damage the motor and its attached load and burn the armature winding, brushes, and commutators.

A variable resistance connected in series with the armature can limit the starting current to approximately 150 percent of the full-load current.

The value of the series resistance required to limit the armature current to 150 percent of full-load current may be computed using the following equation:

$$R_S = \frac{V_T}{I_A} - R_A \qquad (9\text{-}2)$$

where R_S = starting resistance
V_T = applied voltage
I_A = required armature starting current
R_A = armature resistance

In Ex. 9-1, the starting current would have to be limited to 150 percent of 20-A full-load current or 30 A. Thus, the value of starting resistance must be:

$$R_S = \frac{V_T}{I} - R_A$$
$$R_S = \frac{230}{30} - 0.5$$
$$R_S = 7.66 - 0.5$$
$$R_S = 7.16 \ \Omega$$

This resistance must be connected in series with the armature circuit at the instant the motor is connected to the applied voltage. As the motor increases in speed, the counter voltage builds up, and this resistance must now be gradually reduced as the motor reaches its rated speed.

Maximum torque development is controlled by a field rheostat, which provides a variable value of field flux. Thus, a motor that has maximum field strength and armature current will develop maximum torque. This will cause the motor to reach rated speed in the shortest time possible under existing load conditions.

MANUAL STARTERS

Starting and control are important features in the operation of dc motors, and these properties have been designed to match the needs of the various dc motors. Face-plate starters are manually operated. They are used in applications where frequent starting and stopping is not mandatory. There are two basic types of face-plate starters: the three-point and four-point starters.

THREE-POINT STARTER

The three-point starter gets its name from the three terminal connections made between it and the shunt motor (Fig. 9-1).

The starting rheostat has a tapped resistor. A wiper arm makes contact with these taps as the handle is turned. An iron bar attached to

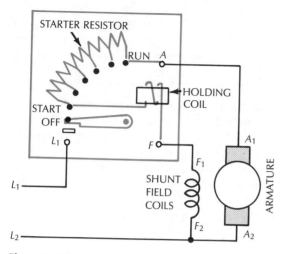

Figure 9-1 Three-point starter for a shunt dc motor wired for no-voltage and field loss protection.

the wiper arm moves over the taps' contact buttons to cut in or out portions of the tapped resistor. When the arm is moved to the first contact (**start** position), the shunt field and no-voltage holding coil are connected in series and directly across the applied voltage, while a starting resistance is placed in series with the armature. As the wiper arm is advanced successively from one contact button to another, the starting resistance is gradually cut out and the motor speed increases. The handle should be moved slowly, about two seconds per contact button. If it is moved too quickly, a current surge will be produced; moving it too slowly may result in a burned out starting resistance. When all the resistance has been cut out, the armature is connected directly across the line and the wiper arm is now in the **run** position. The starter remains in the full **run** or **on** position as long as the holding coil is energized. If the line voltage falls below a predetermined value or fails completely, the holding coil becomes de-energized. The wiper arm is then pulled back to the **off** position by a spring. The holding coil provides several distinct advantages:

1. It holds the wiper arm in the **run** position.
2. If the applied voltage drops to a low value or fails, the holding coil will release the

wiper arm, sending it to **off** position.

3. If an open circuit should occur in the shunt field coil while the motor is running, the holding coil will release the wiper arm shutting the motor **off**.

The proper way of turning a dc motor off is to open the main switch. The wiper arm will then return to the **off** position by itself. The wiper arm should never be forced to break contact with the holding coil as severe arcing between the wiper arm and contact buttons will occur.

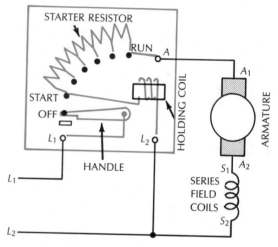

Figure 9-2 Three-point starter connected to a series dc motor, wired for no-voltage protection.

The circuit of Fig. 9-2 shows a three-point starter connected to a series dc motor. The starter provides *no-voltage protection*. If the line voltage should drop to a low level, the starter will function to turn the motor off.

However, if the series motor loses its load, it will race to self-destruction since it is not provided with overspeed protection.

FOUR-POINT STARTER

The speed of a dc motor may be controlled by a rheostat connected in series with the field coil. When a rheostat is connected in series with the field coil, it may cause the field current to drop to a low enough value so that the holding coil would not be able to hold the wiper arm in the **run** position. To overcome this disadvantage, a four-point starter is used.

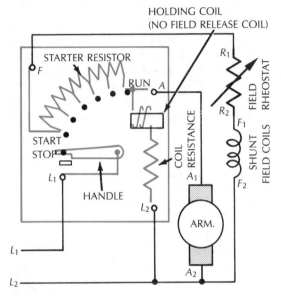

Figure 9-3 Four-point starter with variable speed control.

Figure 9-3 shows a four-point starter which permits speed control by a field rheostat in the field circuit.

The holding coil is not connected in series with the shunt field as in the case of a three-point starter. The holding coil and a series resistor form one branch of a parallel circuit and are connected directly across the line voltage. Thus the holding coil current provides *low-voltage release* and is made independent of the field current. The field coil, with its rheostat which varies the speed of the motor, forms the second branch of a parallel circuit, while the armature with its starting resistor form a third branch of the parallel circuit.

This circuit permits speed control by varying the field current, but does not provide protection against motor racing if the field circuit should become open.

The four-point starter uses the same procedure to start the motor as described for the three-point starter. Once the wiper arm is in

the **run** position, the field rheostat is used to adjust the motor speed to the desired value. To stop the motor, the field rheostat must be reset to cut out all the resistance which will reduce the motor speed to its normal operating speed. Then the supply voltage can be turned off to stop the motor completely. The rheostat provides only above-normal speed control. This ensures that a strong field will be present when the motor is started again.

MANUAL STARTING RHEOSTATS FOR DC SERIES MOTORS

Special dc starting rheostats and controllers are used for series dc motors. These starting rheostats perform the same function as the three- and four-terminal starting rheostats used in shunt and compound motors but differ in the internal and external connections.

The series dc motor starters are broken down into two categories: no-voltage protection, and no-load protection.

SERIES DC MOTOR WITH NO-VOLTAGE PROTECTION

A series dc motor with no-voltage protection is utilized to accelerate the motor to rated speed. The holding coil is connected across the applied voltage. If the applied voltage should drop to a predetermined level, the electromagnet of the holding coil is de-energized. The spring quickly returns the wiper arm to the **off** position, protecting the motor from damage due to low-voltage conditions. Thus, to turn the motor **off** the main line switch should be opened.

SERIES DC MOTOR WITH NO-LOAD PROTECTION

In another type of series dc motor starting rheostat with no-load protection, the holding coil is wound with few turns of heavy wire and is connected in series with the armature, field winding, and applied voltage source. If the load connected to the motor drops, it will result in a corresponding drop in armature current which is sensed by the holding coil. The holding coil electromagnet will weaken, releasing the wiper arm back to the **off** position. This feature prevents the series motor from reaching excessive speeds and damaging the motor under light or no-load conditions.

COMBINATION MANUAL STARTERS AND SPEED CONTROLLERS

In many industrial applications, for example cranes, elevators, conveyors, transportation, etc., controllers are used for regulating the operation of electrical equipment. A controller must perform the functions of starting and stopping, accelerating and decelerating, reversing, regulating the speed, and dynamic braking or plugging. In some of these applications (elevator), the motor must be frequently started, stopped, reversed, and have its speed varied continually.

FACE-PLATE CONTROLLER

Controllers may be used with either a shunt or compound motor. There are two basic types of manual speed controllers in use: (a) above-normal speed controllers, and (b) above-and-below normal speed controllers. The above-and-below normal speed controller will be dealt with in this chapter. Figure 9-4 shows a manual starter with above-and-below normal speed controller connected to a compound DC motor operating in the below-normal speed mode. This controller is able to offer a wide range of motor speeds.

The moveable wiper arm is connected to two rows of button contacts. The lower row of contacts is connected to a tapped resistor used with the armature circuit. The upper row of contacts is connected to a tapped resistor used in the shunt field circuit. The contacts are mounted on the front of the panel and the

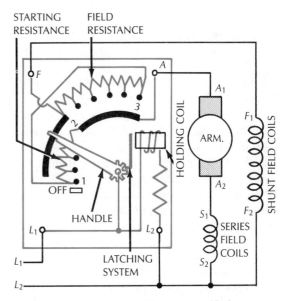

Figure 9-4 Manual starter with above-and-below normal speed controller connected to a compound dc motor (set for below-normal speed operation).

tapped resistors are housed in a ventilated box behind the panel.

When the wiper arm is moved in a clockwise direction, with main voltage applied, the motor starts and gradually accelerates because sections of starting resistance are cut out in steps from the armature circuit while the shunt field has full-line voltage applied. In Fig. 9-4, the wiper arm is located between points 1 and 2, causing the motor to operate at below-normal speed. When the wiper arm reaches point 2, the motor has attained normal speed of operation. There is then no resistance in either armature or field circuit. Thus full-line voltage is applied to both armature and shunt field circuits.

The latching system which operates in conjunction with the holding coil will lock the moveable arm at any contact point desired, until moved to some other point.

For below-normal speed, the field is at full voltage; however, resistance is added to the armature branch.

When a motor is operating at below-normal speed with a load, there is a large current flowing in the armature circuit. As a result, the armature or starting resistors must be large enough to radiate the heat produced by the large current. The physical size of this type of controller, therefore, is larger for a given wattage rating than a normal manual starter.

As the wiper arm moves above point 2, it makes contact with the radial conductor on the inner or starting resistance, thus applying full-line voltage to the armature. At the same time, the wiper arm also makes contact with the outer field-resistance buttons, adding sections of resistance in series with the shunt field winding. This causes the magnetic field to become weaker and the motor speeds up. When the wiper arm reaches point 3, maximum speed is attained and the motor is now operating at above-normal speed.

When the line switch is opened, the holding coil releases the latching system and the spring pulls the wiper arm to the **off** position. The controller is now in a position where it can be used to restart the motor once the voltage is applied.

DRUM CONTROLLERS

Although drum controllers are being replaced by solid-state controls, there are still a number of them in use; thus, they will be dealt with briefly in this chapter.

Drum controllers are used to start, stop, reverse, and vary the speed of a motor. They are found in elevators, cranes, electric trolleys, machine-tool industry, etc. Whereas face-type controllers are designed to be turned **on** and left alone, drum controllers are built ruggedly to withstand the wear and tear that comes with continued motion, as in the case of a crane that is constantly changing speed and direction.

The construction of a drum controller is shown in Fig. 9-5(a), and the schematic diagram in Fig. 9-5(b) shows a drum controller connected to a compound motor.

Inside the switch is a series of stationary

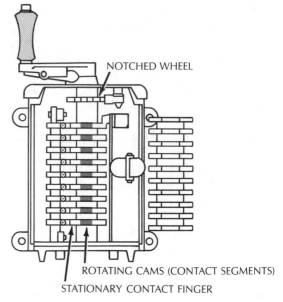

(a) Construction

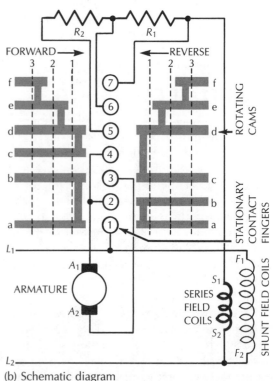

(b) Schematic diagram

Figure 9-5 Drum controller.

finger contacts to which the motor, resistors, and applied voltage are connected. These fingers are insulated from each other. A series of contact segments are mounted on the rotating drum. They may vary in length, and are arranged to come in to contact with the stationary fingers at any desired position as the drum is turned. A spring arrangement ensures that a good electrical connection is made between the stationary fingers and the rotating contact segments. The drum assembly has a notched wheel that is keyed to the central shaft. The cylinder always stops in a position where a roller is forced into one of the notches by a spring, ensuring a proper electrical connection, and at the same time indicating to the operator the proper position of the handle.

In Fig. 9-5(b), the lettered cams are the rotating contact segments mounted on the drum shaft. These are connected by jumpers, as shown. The numbered contacts are the stationary fingers. As the cams are rotated, they make an electrical connection with the stationary fingers and form the required circuits between the applied voltage, resistance, and the motor.

If the motor is to move forward, the handle is moved forward in the direction indicated by the arrow. The fingers are now in position 1 and the current flows from L_1 through all the resistance, motor, and back to L_2. The motor starts rotating, and as the handle is moved further to position 2, the resistance is gradually being reduced in the armature circuit and inserted into the field circuit. When the handle is turned to notch 3, all the resistance has been cut out of the armature circuit and transferred to the field circuit for across-the-line full-speed operation.

For reverse operation, the cams are moved in the direction shown by the reverse arrow.

Drum controllers are usually connected with dynamic braking resistors, magnetic contactors, protective circuits, and blowout coils or arc barriers. These will be subsequently covered in this chapter.

INTRODUCTION TO AUTOMATIC MOTOR CONTROL

At the beginning of the Industrial Revolution, machines were operated chiefly by hand, wind, steam, and water energy. With the advent of electricity and the electric motor, motor control was relatively simple; an array of machines was deriving energy from a common line shaft which was started and stopped a few times daily. With the demand for increased production, the line shaft gave way to individual drive where the motor could be started and stopped more frequently with greater flexibility.

A motor controller could be designed to fit the needs of that particular machine to which the motor was connected. Operations were added that automatically started, stopped, reversed, varied the speed of the motor, sensed a number of conditions such as changes in temperature, liquid levels, over-current, under-voltage, over-voltage, protection, etc. These operations are referred to as an *automatic motor control* system in that they control some operation of an electric motor.

An automatic control system consists of two basic devices:
1. Pilot control devices
2. Primary control devices

The pilot control devices may be manually or automatically operated, such as push buttons, thermostats, limit switches, float switches, timers, photo cells, etc. They will turn the primary control devices **on** or **off**.

The primary control devices connect the load to the applied voltage. These consist of switches, relays, contactors, starters, etc., and can be manually or automatically operated as well.

ELECTRICAL DIAGRAMS AND SYMBOLS

Wiring diagrams serve a number of specific purposes. Diagrams are used by the factory wiremen who build the equipment in the manufacturing operations. Then, a technician must test the equipment before shipping. The electrician who installs the equipment must follow the diagrams. Finally, the maintenance personnel use diagrams when they have to locate the faulty components after the apparatus breaks down. Symbols are used on diagrams to represent components in electric circuits. Numbers and letters are used to identify each specific component. These drawings will also serve as the official record of the particular equipment.

There are usually three basic diagrams that are prepared when a design of a particular machine is undertaken:

SCHEMATIC DIAGRAM

A schematic or elementary diagram shows all circuits and components of the machine. It shows the sequence of operation of the various components in the electric circuit.

WIRING DIAGRAM OR CONNECTION DIAGRAM

The wiring diagram shows the exact location of each component. It also identifies these components and terminals.

INTERCONNECTION DIAGRAM

The interconnection diagram shows the external connections from a controller to associated equipment.

SYMBOLS

Graphic symbols illustrate the physical details and function of an electric circuit. Thus, it may be necessary to be able to read and write the symbols which represent the various components before any work can be accomplished in motor control.

Electrical symbols used on schematic and wiring diagrams bear a significant relationship to each other and yet are distinct.

On the schematic diagram, symbols show the basic components of the circuit such as switch, relay, coil, etc., while on the wiring diagrams these same symbols are arranged into groups to show a pictorial representation of the component such as the contacts and coils drawn in their proper physical location. Figure 9-6 shows many of the most commonly used symbols in schematic or elementary diagrams.

DEVICE DESIGNATIONS

Since a schematic circuit or wiring diagram shows the electrical and physical position of a component, that component must be identified by a number or letter and its contacts should be identified in the same manner. The letters are usually the first letters of the words that describe the function of that component. For example, a push button would be designated PB. Table 9-1 shows a partial list of device designations of some of the most common components used. (See page 194.)

SCHEMATIC (ELEMENTARY) DIAGRAMS

The schematic diagrams also referred to as line diagrams are used to represent the correct electrical sequence of operation. Each component is shown where it is located in the electric circuit and no attempt is made to show the components in their relative physical locations. The schematic diagram is ideal for understanding the operation of a particular control circuit. All control components are located between vertical lines that represent connections to the applied voltage. The main circuit lines are usually drawn with heavy lines and the control wires with light lines.

When reading or drawing a schematic diagram, always start from the left-hand line L_1 and work to the right-hand line L_2. When you have a diagram with multi-sequence operation the control circuit may have several lines to it. Thus you read or draw one line first and when that line is finished move on to the next

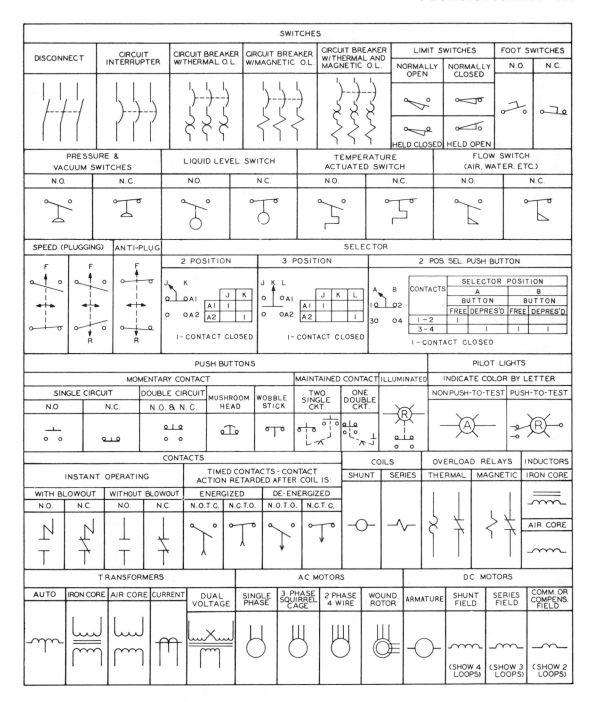

Figure 9-6 Standard diagram symbols.

Courtesy: Square D Company Canada Limited.

Table 9-1 Component Designations

Component	Designation
Accelerating	A
Ammeter	AM
Braking	B
Brush on commutator	A_1, A_2
Capacitor	C
Circuit breaker	CB
Control relay	CR
Diode	D
Disconnect switch	DS
Dynamic braking	DB
Field accelerating	FA
Field decelerating	FD
Field protection	FP
Forward	F
Hoist	H
Holding coil	HC
Jog	J
Limit switch	LS
Limit switch forward	LSF
Main contactor	M
Overcurrent	OC
Overload	OL
Overvoltage	OV
Plugging	P
Push button	PB
Rheostat	RH
Reverse	R
Shunt field	F_1, F_2
Starting contactor	S
Series field	S_1, S_2
Time-delay relay	TR
Undervoltage	UV

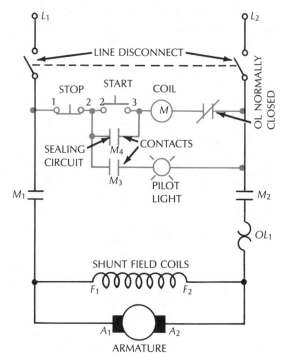

Figure 9-7 Schematic diagram of an across-the-line starter.

one, starting at the left until you have completed the circuit. Figure 9-7 shows a typical schematic diagram of an across-the-line starter control circuit.

TYPICAL CONTROL CIRCUIT

Figure 9-7 shows an elementary control circuit with start and stop push buttons and a sealing circuit M_4. The following sequence describes the operation of this circuit.

1. When the start button is pressed, a complete circuit is made between contacts 2–3. Current flows from L_1 through the normally closed stop button, through contacts 2–3 (start button), through the coil M, and through the normally closed contact of overload relay (OL) to L_2.
2. The current flowing through coil M energizes the coil causing contacts M_1, M_2, M_3, M_4 to close. The sealing circuit around contacts 2–3 of the start button closes. Releasing the start button will not open the circuit, as coil M remains energized and holds contact M_4 closed or seals the circuit.
3. Contacts M_1 and M_2 connect the motor to the applied voltage, while contact M_3 turns the pilot light **ON**.
4. The overload relay (OL) has normally closed contacts and they will only open at a predetermined level of current when the motor is overloaded. Once the overload

relay contacts opens, then coil *M* is de-energized, all *M* contacts open, and the motor is disconnected from the voltage source.

When the main contacts M_1 and M_2 are closed and the stop button is pressed, contacts 1–2 are opened, and coil *M* is de-energized, opening contacts M_1, M_2, M_3, M_4, and the motor stops. Coil *M* cannot be energized until the start button completes the circuit between contacts 2–3.

On a schematic diagram all switches and contacts are usually shown in their normal or inoperative position.

WIRING DIAGRAMS

In a wiring diagram the connection points are usually drawn in the same physical location as they actually appear on the component itself. The draftsman then draws the location of the wire between the proper connection points, producing a circuit that resembles the actual wiring. A wiring diagram, of an across-the-line starter is shown in Fig. 9-8. This circuit is the same one shown in the schematic diagram of Fig. 9-7.

MAGNETIC CONTACTORS

The *magnetic contactor* is a primary control device that is the workhorse of the industry. It is used in automatic controls for switching circuits or relatively high current loads in ac and dc circuits. A contactor does not provide overload protection. A *magnetic starter* contains the same magnet assembly with coil and contacts and in addition provides overload protection. A *control relay* is also an electromagnetic device, similar in operating characteristics to a contactor, but operates at lower current levels. Relays are used in control circuits of magnetic starters, contactors, solenoids, etc. These devices contain a low-

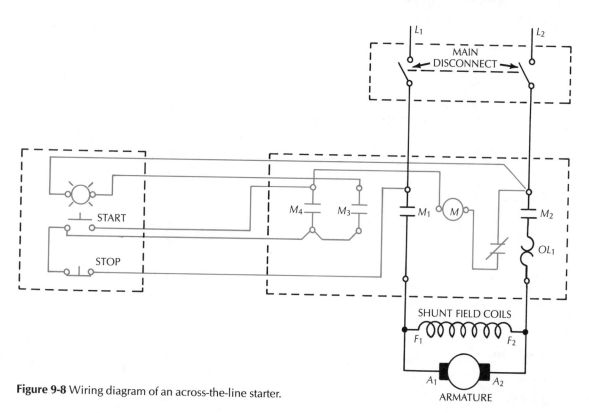

Figure 9-8 Wiring diagram of an across-the-line starter.

voltage control circuit that is used to operate other devices in a high-voltage circuit.

The contactor is activated by a push button that is usually located at a position remote from the motor and controller. A contactor can also be made to operate automatically and can be actuated by various sensing devices such as a thermostat, limit switch, pressure switches, float switches, flow switches, time clocks, etc.

The magnetic contactor has a coil placed on an iron core. When current flows through the coil, the iron core becomes magnetized, which then attracts the moveable armature. Figure 9-9 shows a diagram of clapper and solenoid-type contactors with normally open (N.O.) contacts.

When a coil is de-energized and the contacts are open, they are called Normally Open (N.O.) contacts, and are indicated by two short parallel lines. However, when a coil is de-energized and the contacts are closed, they are called Normally Closed (N.C.) contacts, and this is shown as a line drawn across the parallel lines.

Magnetic switches have several advantages:

1. It is possible to have a motor controlled from more than one position, thus reducing the costs involved.
2. If the distance from the push button to the motor is great, the magnetic switch may be mounted near the motor. The control wires from the push button to the contactor may be small diameter, reducing the cost. With manual controls, heavy conductors have to be run to the point of control.
3. The operator is protected by operating a low current circuit instead of manipulating a heavy switch which carries large currents.
4. Magnetic contactors may be activated automatically by sensing devices (e.g., float switch), thus eliminating the need for an attendant. Manual switches cannot be operated automatically.
5. Automatic safe acceleration is easily attained by means of contactors, thus reducing damage to the motor.

Magnetic contactors are built in various sizes, with each size rated at a certain power-handling capability.

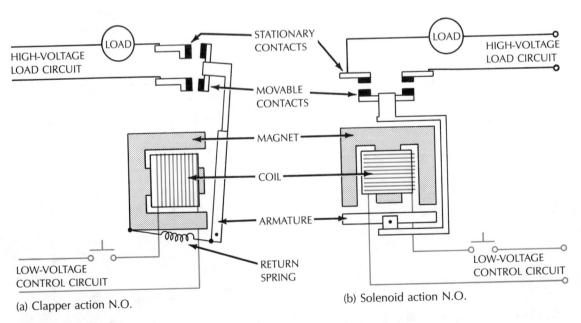

(a) Clapper action N.O.

(b) Solenoid action N.O.

Figure 9-9 Contactors.

As the armature is moved toward the core, the moving contacts make a connection with the stationary contacts, completing a secondary circuit which will supply voltage to an electrical load. The armature may have a number of contacts to operate several circuits. The coil may have a few turns of heavy wire so that it can be connected in series with the circuit and be operated by the current flow, or it may have many turns of fine wire, in which case it is connected in parallel with the circuit and is voltage-operated. When the operating coil is de-energized, the armature is returned to the open position by a return spring.

The control circuit, usually a low-voltage circuit, uses fine control wires, while the load circuit uses heavy wire.

BLOWOUT COIL

Contacts quite often make and break heavy currents which produce large sparks that could damage the contact points. To minimize heavy arcing and prolong the life of the points, the contactor is equipped with a blowout coil and an arc chute to protect surrounding equipment. Figure 9-10 shows how the blow-

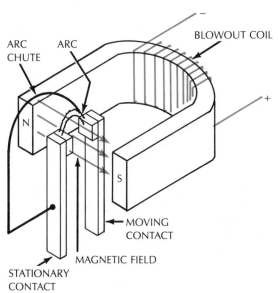

Figure 9-10 Blowout coil with arc chute.

out coil helps extinguish the arc across the contacts with the aid of an arc chute. The arc sets up a magnetic field that reacts with the magnetic flux produced by the coil. The two magnetic fields repel each other, forcing the arc up, thus lengthening the spark until it reaches the arc chute, where it is quickly extinguished. By reducing the time or persistence of the arc, less damage will be done to the contacts.

OVERLOAD PROTECTIVE DEVICES

Motors can be damaged and their effective life reduced when subjected to excessive currents higher than their full-load currents rating. Motors can withstand the *inrush* starting current for short periods of time without damage to insulation and windings. Any current that is greater than the full-load current rating of the motor can be classified as an overcurrent. An overload occurs when a motor is running and the load current exceeds the current required to develop rated torque. The motor then draws a higher current than full-load current.

There are several devices that can be used to prevent a motor from being damaged by excessive current, such as fuses, circuit breakers, and overload relays.

1. Fuses and circuit breakers are designed to provide short-circuit protection. They cannot distinguish between temporary inrush starting current and an overload current. If a fuse or circuit breaker were chosen to protect a motor on the basis of the full-load current rating, it would open the circuit every time the motor started. But if they were chosen to withstand the large inrush starting current, they would not provide protection against small overloads.
2. Overload relays are the heart of motor protection. They have built-in characteristics, which permit a relay to remain closed during startup periods when the inrush current rises to a high value, yet provide

protection on small overloads above the full-load current rating when the motor is running. Unlike the fuse, the overload relay can withstand repeated tripping and can be reset; whereas the fuse is nonrenewable.

Overload relays may be classified as either thermal or magnetic. Magnetic overload relays respond only to excessive current and are not affected by temperature. Thermal overload relays react to high temperatures caused by the overload current. There are basically three types of thermal relays: (a) bimetallic, (b) melting alloy, and (c) thermistor.

MAGNETIC OVERLOAD RELAY

Magnetic overload relays are current-sensitive devices consisting of a coil and a normally closed set of contacts (Fig. 9-11). There are two basic types of magnetic overload relays, (a) instantaneous-trip type, and (b) time-delay trip type.

Both relays have a moveable magnetic core inside a coil which carries the motor current. When the current through the coil becomes abnormally high, a strong magnetic field is set up around the coil and pulls the core (iron

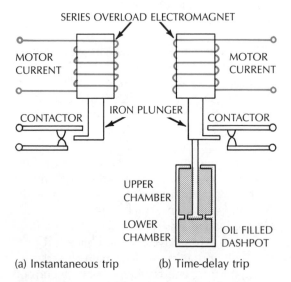

(a) Instantaneous trip   (b) Time-delay trip

Figure 9-11 Magnetic overload relay.

plunger) upward. When the core rises, it trips a set of contacts, thus opening the circuit.

The time-delay relay trip is similar to the instantaneous trip, except that the movement of the plunger core is slowed by a piston working in an oil-filled dashpot, which provides the time-delay feature (Fig. 9-11(b)). As the plunger moves up, it encounters opposition produced by the oil in the upper chamber. The oil from the upper chamber slowly flows to the lower chamber through the holes in the piston, allowing the plunger to rise slowly and open the contactor. The tripping time may be varied by uncovering the oil bypass holes in the plunger. Because of the time delay, magnetic overload relays are used to protect motors having long accelerating times.

THERMAL OVERLOAD RELAYS

Bimetallic Relay: A bimetallic relay is shown in Fig. 9-12(a). It is made up of two strips of dissimilar metals, aluminum and steel, welded together. Both metals have different coefficients of expansion. One end of the bimetal is fastened securely, while the other end is free to bend as the temperature rises. When a heavy overload persists, causing the temperature of the motor to rise, the aluminum expands at a faster rate than the steel strip, causing the bimetallic strip to bend sufficiently to open the contacts. When the contacts open, the holding coil is de-energized, thus opening the applied voltage circuit to the motor.

One advantage of the thermal relay is that it will not trip instantaneously when subjected to momentary high starting currents or short overloads.

Some bimetallic relays are reset manually and others are automatically reset. Automatic reset relays should not be installed where automatic start could be dangerous. The unexpected restarting of a motor could damage the device connected to the motor, or hurt the operator of the machine.

Molten Alloy Relay: The molten alloy relay also

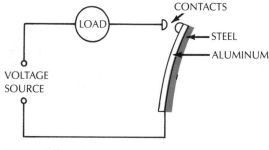

(a) Bimetallic relay

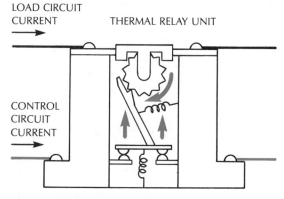

(b) Molten alloy relay

Figure 9-12 Thermal overload relays.

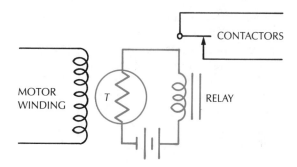

Figure 9-13 Thermistor overload relay.

referred to as the solder pot relay, is illustrated in Fig. 9-12(b). A spring is held under tension by a pawl engaged with a stationary ratchet wheel which is soldered with an alloy to the shaft. Excessive motor overload current flows through a small heater, producing sufficient heat to melt the solder alloy, allowing the ratchet wheel to spin free. This allows the spring to open the contacts to the holding coil, de-energizing it and stopping the motor. After the alloy cools and solidifies, the relay can be reset by manually pushing a reset button.

Overload relays are usually rated at 125 percent of the motor full-load current.

THERMISTOR OVERLOAD RELAY

A thermistor is a solid-state device whose resistance varies as the temperature changes.

It is made up of various ceramic materials. Some thermistors have a negative temperature coefficient (NTC), which means that as the temperature of the thermistor increases, the resistance decreases. A positive temperature coefficient (PTC) indicates that the thermistor resistance increases as the temperature rises.

Thermistors have many applications. They are used for temperature measurement, liquid level measurements, altimeters, power measurement, and voltage control where any change in temperature would produce a corresponding change in resistance and current flow.

In motor control, thermistors are usually placed in direct contact with the motor winding. A cold thermistor will conduct enough current to energize a relay. Figure 9-13 shows a thermistor overload relay. If the motor windings get hot, the thermistor resistance increases, reducing the current flow. The relay opens and so does the main contactor which shuts the motor **off**.

AUTOMATIC REVERSING

Direct current motors may be reversed by reversing the current flow through the armature circuit, thereby reversing the polarity of the armature with respect to the field. To accomplish this, the control circuit must have two contactors, one for the forward connection and one for the reverse connection. However, to prevent both contactors from being energized simultaneously and causing a short circuit, two safeguard methods of "interlocking" are incorporated in the circuit. These are the mechanical interlock and electrical interlock.

The *mechanical interlock* is the most widely used, and it consists of (a) an interlock between the contactors, and (b) a push button interlock.

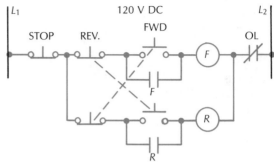

Figure 9-14(b) Double circuit push buttons are used for push button interlocking.

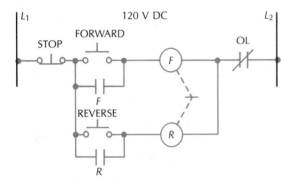

Figure 9-14(a) Mechanical interlock between the two contactors, preventing the two contactors from closing simultaneously.

The interlock between the contractors in Fig. 9-14(a) is shown as a broken line between the coils. It consists of a positive mechanical interlock preventing the coils F and R from being energized simultaneously. When the forward contactor F is energized and closed, then the reverse contact R is prevented from closing by means of the mechanical interlock. Contactor F is prevented from closing in the same manner.

Pressing the forward push button will result in coil F being energized and the motor will run. To reverse the motor, it must be brought to a full stop, then the reverse push button may be pressed.

Another type of reversing interlock is actually an electrical method, and is referred to as a *push button interlock*, Fig. 9-14(b). It incorporates double-circuit push buttons that have N.O. and N.C. contacts. They are wired so that if both push buttons are depressed simultaneously, coils F and R will not energize.

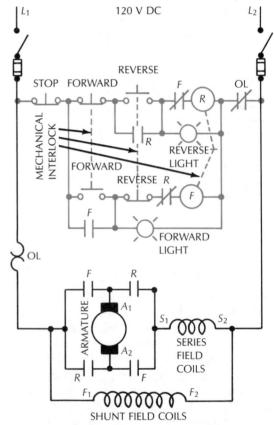

Figure 9-14(c) Reversing a compound dc motor with mechanical interlocked control circuit.

If the forward push button is pressed, coil F is energized, closing the sealing contacts F. If the reverse push button is now pressed, it will first open the circuit to coil F, stopping the motor, then energizes coil *R*. Thus, to reverse the motor it is not necessary to press the stop push button.

It should be pointed out that repeated reversals could cause the motor to overheat, and may result in damage to the driven machinery.

The final type of interlock, Fig. 9-14(c), makes use of N.C. auxiliary contacts on both the forward and reverse contactors. When the forward push button is pressed, coil *F* is energized closing the N.O. sealing contact *F* and opening the N.C. contact *F*, which prevents the reverse coil *R* from being energized. The same operation will occur for the reverse condition. This method of contact interlock uses electrical contacts, thus it is called electrical interlock.

The various types of interlocks may be used in combination with each other. Usually, mechanical interlock is used in combination with electrical interlock to provide extra safeguards.

It is important that a motor be brought to a full stop before the reverse button is pressed. To prevent damaging a motor, magnetic starters also come equipped with a timing relay which prevents a motor from being reversed before it comes to a full stop.

ANTI-PLUGGING RELAY

Plugging is the reversal of a motor while it is running. This method, to be discussed later in this chapter, is used to provide braking action for a motor and bringing it to a quick stop. In a dc motor circuit having large loads, the motor must be brought to a full stop before reverse voltage is applied. If a reverse voltage is applied to the motor while it is running forward, a large counter voltage will build up in the armature circuit adding to the voltage being applied. This will result in a current flow that will be double the normal starting current, thus damaging the armature windings. Also, a severe strain on the shaft will develop from the strong magnetic forces. To prevent reverse

voltage from being applied while the motor is still running forward, an anti-plugging relay (AP) is utilized. The anti-plugging relay (AP) is connected in parallel with the armature, as shown in Fig. 9-15.

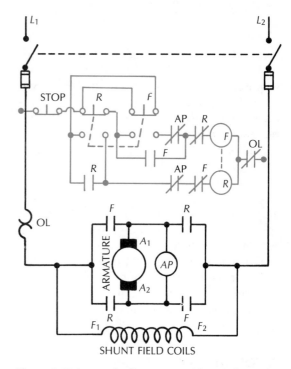

Figure 9-15 Across-the-line starter with anti-plugging relay to prevent reverse voltage from being applied.

When the forward push button is pressed, current flows from L_1 through the stop push button, the reverse push button, normally closed AP and *R* contacts, coil *F* and the OL to L_2. Coil *F* is energized, closing the *F* contacts which apply voltage to the armature and the AP relay. The sealing circuit around the forward push button and the AP relay is closed. The AP relay opens the normally closed AP contacts in series with the reverse contactor. If the reverse push button is pressed, relay *R* will not be energized because the AP contacts are open. When the motor is turned off, the AP relay does not drop out immediately because it is still receiving current from the counter voltage being induced in the armature until it stops rotating. When the armature speed drops, the counter voltage is reduced to a low value and the AP relay loses its energy. At this particular point, the motor can be reversed.

REDUCED VOLTAGE DC STARTERS

It was indicated at the beginning of this chapter that small dc motors of 375 W or less do not require starting resistors. For motors larger than 375 W, some means of limiting the starting current must be utilized. A starting limiter resistance must be connected in series with a dc motor armature to prevent damage due to excessive torque and heat.

Reduced voltage starters are designed to cut out sections of the starting resistance in the armature circuit, as the motor comes up to speed. This gives the motor smooth torque without producing current surges.

Contactors are activated by one of several common methods: the counter voltage starter, the definite-time acceleration, the current-limit method, lockout contactors, and voltage-drop acceleration.

1. COUNTER VOLTAGE CONTROLLER

The counter voltage controller in Fig. 9-16 has several relays (A_1, A_2, A_3) connected in parallel with the armature. At the instant of start, the counter voltage across the armature is very small, and not large enough to energize the relays. As the motor accelerates, the counter voltage builds up in the armature, and at a certain level will be strong enough to energize the relays. Each coil is energized in succession at a different voltage as the armature accelerates, and the relay contacts cut out the starting resistance in several steps to provide smoother acceleration.

The counter voltage controller automatically adjusts itself to different starting periods, depending upon the size of the load. A large load reduces the motor speed; thus it takes a longer time to build up the counter voltage,

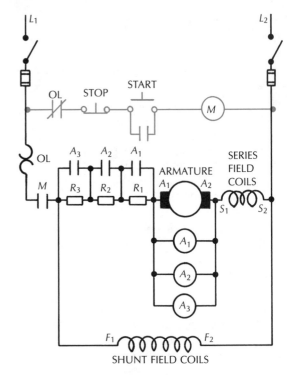

Figure 9-16 Counter voltage controller providing three-step control of a motor.

lengthening the starting period.

The disadvantage of the counter voltage starter is that if the line voltage fluctuates, the rate of acceleration will vary. For example, if the line voltage rises, the starting resistance may be removed too soon, and if the line voltage drops, the contactor may not get activated at all. The contactors will close at a one voltage level, and will not function properly if the line voltage fluctuates.

2. DEFINITE-TIME ACCELERATION

In definite-time acceleration, or time-limit acceleration, the starting resistor is cut out in stages by contactors that operate in succession at definite time intervals, allowing the motor to accelerate smoothly.

The coil of the contactor has an iron core that is surrounded by a heavy copper sleeve. When the coil is de-energized, the magnetic field of the coil collapses and induces a large current in the copper sleeve, which produces a magnetic field opposite to the original field which is trying to collapse. Thus, the decay of the original field is slowed down, extending the time during which the coil retains control of its contactor until the motor has had time to accelerate. The contactor has normally closed contacts. Thus, when the coil is energized, the contacts open; when the coil is de-energized, the contacts close after a definite time. Figure 9-17 shows a schematic wiring diagram of a definite-time acceleration controller.

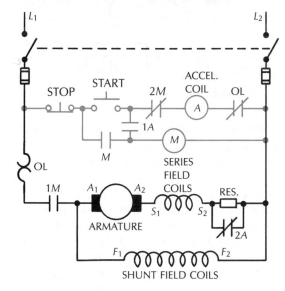

Figure 9-17 Schematic diagram of a definite-time accelerator.

When the start push button is pressed, current flows from L_1, through the N.C. stop button, start button, normally closed contacts $2M$, and accelerating coil A, causing the N.C. accelerating contacts $2A$ to open, and the N.O. contacts $1A$ to close. When contacts $2A$ open, resistance is added into the circuit. Coil M is now energized and closes the line contacts $1M$ and the sealing contacts M. Closing the line contactor $1M$ causes the line current to flow through the armature, series-field coils, and resistance. N.C. contacts $2M$ are opened and cause the accelerating coil (A) to become

de-energized. The accelerating contacts ($2A$) drop back and close after a time-delay period, shorting the resistance out of the circuit, thus connecting the motor across the applied voltage.

The advantage of this type of controller is that the acceleration is independent of speed or current flow, making definite-time acceleration the most widely used. Its disadvantage is that the resistance is removed at the same rate whether the load is heavy or light.

The timing devices used vary from a mechanical mechanism, dashpot mechanism, magnetic time-delay, capacitor discharge, relays, and motor-driven timers.

3. CURRENT-LIMIT ACCELERATION
Current-limit acceleration of dc motors (Fig. 9-18) utilizes contactors with current-sensitive relays. The contactor remains open as long as the current remains at a certain level; once the current falls below that predetermined level, the contactor closes. The operating coils used here are designed to be connected in series with the armature, and are very fast in operation.

The time required to activate the contactor

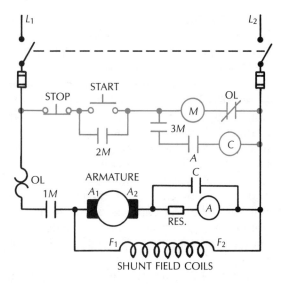

Figure 9-18 Current-limit acceleration starter.

and cut out the starting resistance depends upon the speed of the motor.

When the start push button is pressed, current flows from L_1, through the stop, start push buttons, and through coil M. Coil M is energized, thus contacts $1M$, $2M$, $3M$ close. Contact $2M$ provides a sealing circuit around the start push button. The closing of contact $1M$ connects the armature across the voltage source and current flows from L_1, through $1M$, the armature, resistor, and coil A. At this point, current is at maximum value as no counter voltage has been developed. When coil A is energized, contact A closes, completing the circuit to coil C. The motor accelerates and the counter voltage builds up. The current flow is now diverted as the energized coil C closes contact C which cuts out the resistance and the series relay A.

Several current levels can be built into the starter if motor acceleration is to be achieved by changes in current. Each level or step must have its own current relay that drops out at a specific value of current.

One drawback of this system is that it is necessary to have quite a few electrical interlocks if the circuit is to function properly. This creates difficulty for the maintenance people.

4. LOCKOUT CONTACTORS

Another type of relay that operates on current flowing through the armature circuit is the *lockout relay*. It has two coils that are connected in series with the armature. This type of a controller is designed for use as an acceleration means for dc motors. The contactor is normally open and remains open until the inrush current has dropped to a predetermined level. The closing coil holds the contacts closed while the motor is accelerating and the starting current is reduced. The lockout coil holds the contacts open during the period of high starting current. Figure 9-19 shows a control circuit of a single-coil lockout starter.

In Fig. 9-20, when the start button is pressed, coil M is energized, closing the main

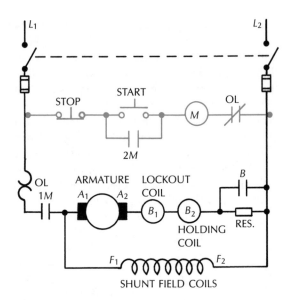

Figure 9-19 Control circuit of a single-coil lockout starter.

$1M$ contacts and the sealing contacts $2M$. Current flows from L_1 through the closed contact $1M$, armature, lockout coil, holding coil, and the resistance to L_2. The high starting current energizes the lockout coil which keeps the contacts open. As the motor accelerates, the counter voltage increases, reducing the current flow. At a predetermined current level, the magnetic field of the holding coil overrides that of the lockout coil, closing the contactor (B). This will short out the resistance, and place the armature across the full-line voltage.

Lockout acceleration does not require interlocking features as in the case of the series relay. However, since the lockout and holding coils must carry the armature current, they have to be wound with heavy wire. Thus, a different winding will be required for individual motor sizes.

5. VOLTAGE-DROP ACCELERATION

A voltage-drop accelerator controller utilizes a two-coil lockout contactor, one for closing and the other for holding the contactor open. The closing coils $1LA$, $1LB$, and $1LC$ are connected to the line voltage through the accelerating contacts A, B, and C. The holding coils A, B,

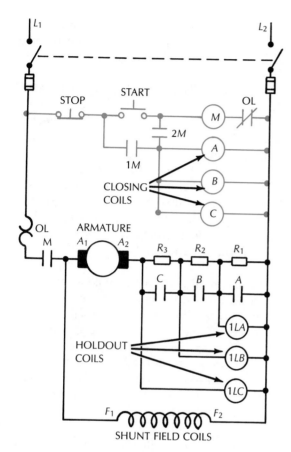

Figure 9-20 Voltage-drop acceleration.

and C have their contacts connected across the starting resistors R_1, R_2, and R_3, respectively, as shown in Fig. 9-20.

On starting, the current flow in the armature is high, producing a large voltage drop across the starting resistors. At this instant the counter voltage drop across the armature is relatively low. As the motor accelerates and the counter voltage builds up, the armature current decreases, causing the voltage drop across the starting resistors to decrease. The voltage drops across the resistors closes coils 1LA, 1LB, 1LC, initiating the time delay.

When the start push button is pressed, current flows through the control circuit, energizing coils M, A, B, C. The contact M applies voltage to the motor while contacts 1M and

2M seal the starting push button, and coils A, B, C will want to close accelerating contacts A, B, C. However, the heavy starting current flowing through the armature and starting resistors produces a large voltage drop across the resistors. This, in turn, causes a large current to flow through the holdout coils 1LA, 1LB, 1LC, holding the accelerating contacts A, B, C open because the magnetic fields of the holdout coils overrides the magnetic fields of closing coils A, B, C. As the motor accelerates, the counter voltage builds up, reducing the current flow through R_1. The magnetic field of holdout coil 1LA is now weaker than the field of closing coil A, thus the accelerating contact A closes, shorting out resistor R_1.

As resistor R_1 is shorted out, the current in the armature circuit increases, causing the voltage drop and current across R_2 to increase. The current is large enough to cause the holdout coil 1LB to override the closing coil B, and keep the contact B from closing.

As the motor accelerates further, the counter voltage builds up to a greater value, causing the current to drop to a new value. The decrease in current in the armature circuit reduces the voltage drop across resistor R_2, weakening the holdout coil 1LB. Since it is now weaker than closing coil B, contact B now closes, shorting resistor R_2, causing the armature current and the voltage drop across R_3 to increase. The increase in voltage drop across R_3 causes the holdout coil 1LC to be stronger than the closing coil C. This keeps the contact C open until the motor accelerates further, increasing the counter voltage which reduces the armature current. The holdout coil 1LC is now weaker than closing coil C. Thus the contact C now closes and shorts out resistor R_3. The motor has now attained rated speed and is connected across the line voltage.

DYNAMIC BRAKING

Braking is a form of action in which a motor is slowed down or brought to a full stop. The braking action of an electric motor may be accomplished through mechanical friction or by electrical means. Mechanical brakes (Fig. 9-21) require maintenance due to heat and wear on braking surfaces. In electrical braking, the motor produces a magnetic field which opposes the rotation of the armature.

Figure 9-21 Bubenzer double shoe brake operated by "Eldro" thrust actuator.

Courtesy: AEG - Telefunken Corporation.

There are three basic methods of achieving electrical braking: regenerative, dynamic, plugging.

Regenerative braking takes place automatically whenever a mechanical load is applied to a motor. Consider a dc motor driving a conveyor in a steel mill at a high rate of speed. If the operator reduces the speed of the motor suddenly, the torque produced by the motor is quite small compared to the torque produced by the load turning due to inertia. Now, the load drives the motor, which behaves like a generator. A counter voltage is generated that is greater than the applied voltage. The counter voltage causes the current to flow in the opposite direction through the armature, producing a new torque that opposes the torque produced by the motor itself, reducing the

speed of the motor. When the motor has slowed down to its new speed, the braking action disappears and the motor continues to drive the conveyor.

Dynamic braking is similar to regenerative braking and yet has several minor differences. In dynamic braking, a resistance is connected across the armature when it is disconnected from the line and the shunt field is left connected across the line. The motor is converted to a generator with the resistor connected across the armature serving as a load on the generator, braking it to a stop. Figure 9-22 shows a schematic diagram of a full-voltage starter with dynamic braking: (a) running position, (b) braking condition.

To accomplish dynamic braking, the stop button is pushed, de-energizing contractor M, shutting down the motor. At the same instant, contact $3M$ closes, connecting the resistor across the armature. The energy produced by the generator action establishes a current in the armature. This current and field flux develop a torque which opposes rotation, and slows the motor to a quick stop.

The degree of braking is controlled by the value of the resistor. Increasing the resistance decreases the braking effect and the motor will take longer to stop. Decreasing the resistance increases the braking effect, stopping the motor quickly.

Dynamic braking provides a simple and safe means for stopping machinery such as trolleys, cranes, elevators, etc., driven by dc motors. Dynamic braking allows braking to take place more frequently with less maintenance and shock to the driven machinery than with mechanical brakes. The braking action is faster and provides a smoother stop with the shaft free to rotate after the motor has completely stopped. On high-speed machinery, it is important to make use of dynamic braking to reduce the speed of the machine before applying mechanical brakes.

Plugging as a means of braking involves the reversing of rotation of a motor by reversing

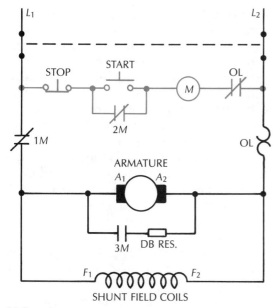

(a) Running

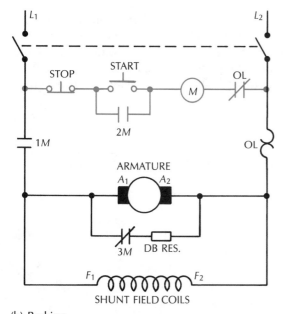

(b) Braking

Figure 9-22 Voltage starter with dynamic braking.

the voltage applied while the field winding remains connected. The motor develops a counter torque, exerting a retarding force.

Plugging may be used to bring a motor to a quick stop or a quick reversal of direction of rotation. The motor is usually stopped in less than one reverse revolution.

Plugging may be used on small motors or motors attached to light loads. If plugging is used on motors with heavy loads, the motor shaft is subjected to severe mechanical strain and the armature current doubles, thereby subjecting the windings to large amounts of heat. Figure 9-23 shows a starter with a plugging circuit.

In this circuit the anti-plugging relay disconnects the motor when the motor reaches zero speed and does not allow the motor to go into reverse. When the stop button is pressed, the current flow is interrupted to the forward(F) contactor and the motor is now disconnected from the voltage source. As the rotating armature slows down, it continues to produce a

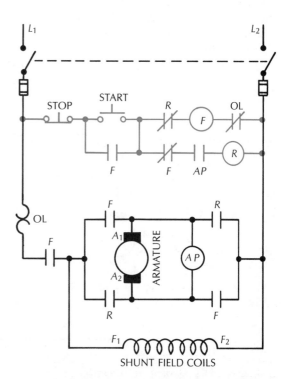

Figure 9-23 Plugging circuit for dc motor.

counter voltage which is applied to the armature with reverse polarity. The counter voltage continues to energize the anti-plugging relay. The reverse current produces a reverse torque on the armature, causing its speed to decrease quickly. As the armature slows down, the counter voltage decreases and the anti-plugging relay is de-energized, disconnecting the motor completely.

A resistance may be added in series with the armature to limit the heavy surge of current when the reverse power is applied.

WARD-LEONARD SYSTEM OF SPEED CONTROL

The Ward-Leonard or adjustable-voltage system of speed control is used in applications that require large motors with rapid changing speeds. This system is used in steel mill and paper mill drives, hoisting equipment, elevators, etc., where a precise control of speed over a wide range is required. Figure 9-24 shows a simplified diagram of the Ward-Leonard System.

A separate motor generator set is used to supply a variable voltage to the armature of the controlled dc motor. A dc generator is coupled to and driven by a constant-speed ac 3-phase induction motor. The fields of the generator and the dc motor are separately excited and controlled by the field rheostats R_1 and R_2. Field rheostat R_1 controls the output voltage of the dc generator, which is applied to the armature of the dc motor. This voltage will vary the speed of the dc motor at below-normal motor speeds. Field rheostat R_2 varies the armature voltage of the dc motor, allowing a wide range of above-normal speeds. With this dual-control system, it is possible to obtain precise speed control over a wide range as required in steel mill drives, where speed changes occur frequently. If the generator voltage drops to zero, the dc motor will come to a standstill.

A reversing switch in the dc generator, may be used to reverse the direction of the dc motor.

The disadvantage of the Ward-Leonard System is the high initial cost and the low overall efficiency of the ac motor, dc generator and the controlled dc motor. However, it has good stability and flexibility as it responds to changes in speed.

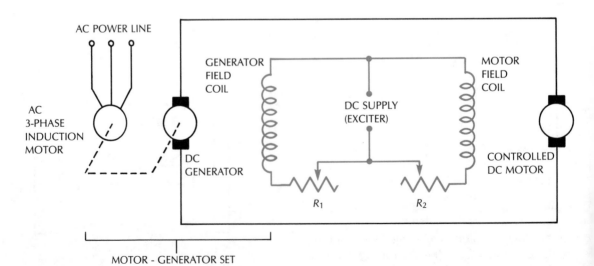

Figure 9-24 Ward-Leonard method of speed control.

SOLID-STATE MOTOR CONTROL

As we have seen in this chapter, automated machinery requires a means of switching and controlling a motor. With modern equipment such as computers, numerical machines, tape drives, etc., the control apparatus requires a means of switching that is very fast and can operate at a high repetitive rate. Mechanical relays cannot operate at these high speeds without breaking down quickly. Solid-state devices have no moving parts, they are small, low cost, and use small amounts of electric energy.

Solid-state motor control will be dealt with again in Chapter Twenty-Two.

SUMMARY OF IMPORTANT POINTS

1. Direct current motors up to 1500 watts can be started with full-line voltage. Motors rated over 1500 W must employ reduced voltage starting to protect the motor.
2. Resistance must be connected in series with the armature of a dc motor to reduce current surge during starting.
3. As the dc motor accelerates, it produces counter voltage, which reduces the starting surge currents.
4. Starting resistance can also be shorted out of the circuit automatically by counter voltage, timer, or current limiting controllers.
5. The speed of a dc motor operating at below-rated speed may be controlled by a resistance in the armature circuit.
6. The speed of a dc motor operating at above-rated speed may be controlled by a resistance in the field circuit.
7. The direction of rotation of a dc motor may be changed by reversing the direction of current flow through the armature while maintaining the same polarity of the field windings.

8. Wiring diagrams show the physical location of the wiring and devices, while schematic diagrams show sequence of operation.
9. Large magnetic contactors contain a blowout and arc chute to extinguish the arc to minimize damage to the contacts.
10. Overload protection is supplied by thermal and magnetic overload relays.
11. A three-point starter provides no-field, no-voltage, and low-voltage release protection to dc motors.
12. A four-point starter can provide speed control above and below rated speed.
13. A counter voltage starter has a coil connected in parallel with the armature, while a current-lockout accelerator has a coil connected in series with the armature.
14. A definite-time accelerator operates on the principle of delayed decay of the magnetic field in a coil.
15. Dashpot starters (timers) use oil in a dashpot, which retards the movement of a plunger when a coil is energized.
16. There are three types of electrical braking: regenerative, dynamic, and plugging.
17. The Ward-Leonard (adjustable voltage) system of speed control is used in applications that require large motors with rapid changing speeds.

REVIEW QUESTIONS

1. Describe the two functions of a motor starter.
2. Explain why a small dc motor may be started by connecting it across the full-line voltage, while large dc motors must be started with reduced voltage.
3. What is the difference between an elementary diagram and a wiring diagram?
4. Give four uses of diagrams.
5. How are the main voltage lines (load circuit) usually drawn in a schematic (line) diagram?
6. Define the following: (a) mechanical inter-

lock, (b) electrical interlock, (c) normally open, (d) normally closed.

7. What is the purpose of a magnetic blow-out and arc chute on a contactor? Explain how the arc is extinguished.

8. How does a magnetic contactor provide safety for the operator of electrical equipment?

9. What is the function of a copper sleeve mounted around the core inside a relay coil?

10. Draw a complete schematic (line) diagram of a circuit with a running indicator light and low voltage protection.

11. Explain the operation of a three-point face plate starter.

12. What is the advantage of a three-point starter?

13. Give one disadvantage of a three-point starting rheostat.

14. Draw a diagram of a four-point starting rheostat.

15. Give one advantage of a four-point starting rheostat.

16. When is the arm of a four-point starter released?

17. Name two types of manual speed controllers commonly used with shunt and compound dc motors.

18. How is overload protection provided in a motor?

19. Identify the devices described by the following designations: (a) CR, (b) B, (c) A, (d) DB, (e) HC, (f) OL, (g) M, (h) PB, (i) TR.

20. What is the difference in construction of a voltage relay and a current relay?

21. State where a drum controller is used.

22. Draw a diagram of a magnetic circuit breaker and explain how it works.

23. Name and explain the difference between the three thermal relays which provide overload protection.

24. Explain how a counter voltage controller operates.

25. What is a lockout controller? How does it work? Where is it used?

26. Explain with a diagram the principle of a definite-time magnetic accelerator.

27. What advantage does a definite-time accelerator have over the lockout controller?

28. Explain with the aid of a diagram the principle of dynamic braking.

29. In what circuit are the coil and contacts of an overload relay connected?

30. What is the purpose of a mechanical interlock on a forward-reverse contactor?

31. What is meant by plugging?

32. What type of relay prevents the reversing of a dc motor while it is running?

33. Draw a straight line diagram of a reversing voltage starter.

34. Explain the basic principle of regenerative braking.

35. Explain in detail why a dc motor draws more current when standing still than it does when running.

36. A shunt dc motor connected to a 120-V circuit draws an armature current of 6 A when the armature is accelerating. The armature resistance is 1.5 Ω. Calculate the value of the counter voltage developed by the motor.

37. How are thermal overload elements and their contacts connected in a motor control system?

38. A shunt dc motor draws a current of 3.5 A from a 120-V source. The armature circuit resistance is 0.5 Ω and the field circuit resistance is 245 Ω. Find the counter voltage when the armature is at standstill and when it is rotating.

39. Find the armature current of a shunt dc motor, connected to a 125-V source. The counter voltage is 120 V and the armature resistance is 0.5 Ω.

40. Calculate the counter voltage of a motor that has an armature circuit resistance of 0.75 Ω. When the motor is connected to a 220-V source, the armature draws a current of 50 A.

41. A shunt dc motor with an armature resist-

ance of 0.05 Ω is connected to a 120-V source and develops a counter voltage of 110 V. Find the armature current.

42. Design a circuit that will meet the following requirements:
    (a) The motor must be protected from overloads and weak magnetic fields.
    (b) The motor must be equipped with dynamic braking.
    (c) The motor must have a reduced voltage starting in the forward and reverse mode.
    (d) The starting time must be adjustable for various motor loads.

CHAPTER
TEN

ALTERNATING CURRENT FUNDAMENTALS

ALTERNATING CURRENT VERSUS DIRECT CURRENT

In this chapter the basic properties of alternating voltage and current will be discussed. You will study the use of ac versus dc, properties of the sine wave, angular measurement, instantaneous and effective values, ac effective resistance, ac power and phasors.

More than 90 percent of the electrical energy used in the world is generated and distributed as alternating current. This extensive use of alternating current, compared to direct current, is due to many reasons. Some of these reasons are:

1. Alternating voltage generators can be built with much larger power and voltage ratings than direct current generators. The ac generator, commonly called an *alternator* may be constructed with its armature or output winding mounted stationary. Since the armature is stationary, it can be connected directly to the external distribution circuit. This type of connection makes it possible for alternators to be designed with extremely high power ratings. Armature voltages as high as 13 800 V are common

and currents of considerable magnitude may be obtained. The magnetic field winding is made the rotating part of this type of alternator. The field coils, which require a relatively low dc voltage, are connected through slip rings to an external dc source.

In a dc generator, the armature must be rotated. The output, therefore, must be taken through a brush and commutator assembly. These connections limit the maximum voltage and current ratings to relatively low values. Direct current generators with a top rating of about 750 V are common.

2. Alternating voltage can be easily and efficiently stepped up or stepped down by means of transformers. With direct current, transformers cannot be used. A series resistor network may be used to change a dc voltage, but this system experiences power losses. Motor-generator sets can also be used to either reduce or increase dc voltages, but their efficiencies are also relatively low.

3. In general, ac motors and controls are

Figure 10-1 Power transformer.
Courtesy: Canadian General Electric Co. Ltd.

simpler, more reliable, and lighter in weight than dc equipment of similar ratings. For example, the ac induction motor has no commutator or brushes. It is of simpler construction, more rugged, and cheaper than a dc motor of comparable power and speed ratings.

4. The ease and efficiency with which ac voltages may be either stepped up or stepped down make it possible to transmit ac energy economically over long distances.

When ac energy has to be conducted to distant load centres, its voltage is first stepped up to a high value. Increasing the voltage results in a lowering of the line current with a corresponding reduction in the I^2R losses in the line. At the load centre, the voltage is then stepped down to the required value. This means that huge alternators can be located close to their prime moving source, such as a waterfall, where ac energy can be cheaply generated, and then conducted economically to distant consumers.

5. Alternating current energy is also well suited for most consumer loads, such as constant speed motors, lighting and heating equipment. These devices can all be designed to operate efficiently with ac.

Although alternating current is used extensively, there are many applications for which direct current must be used or where it does the job better than ac. The following are a few specific examples of where the use of dc has a definite advantage:

1. For the field excitation in alternators and synchronous motors.

2. For various electrochemical processes such as electroplating, charging of batteries, electrolysis, refining of some metals, and production of industrial gases.

3. For high-intensity searchlights and projectors.

4. For motors requiring precise speed control such as motors used in papermaking machines, metal rolling mills, high-speed printing presses, and high-speed gearless elevators.

5. For traction motors on such systems as subway cars, locomotives, and trolley buses. The use of dc motors on these systems can eliminate the need for transmission gearing, clutches, and drive shafts.

It was noted above that it is not easy to raise or lower dc voltages. For this reason, dc voltages are usually produced on location. For example, in a large locomotive, a diesel engine is used to drive dc generators which, in turn, supply the energy to the dc traction motors. For many applications, the required dc is simply obtained by converting the easily available ac into dc by means of rectifiers.

ALTERNATING VOLTAGE

It was shown in Chapter Seven that when a loop conductor cr coil is rotated within a magnetic field, an alternating voltage is induced in the coil. This voltage is then brought to the outside by a pair of slip rings and brushes. When this voltage is connected to a load, it sets up an alternating current flow.

The ac generator is more frequently called an alternator. Most alternators are constructed with their armature coil windings stationary and with a rotating field. This design makes it possible for alternators to have much higher voltage and current ratings.

As the alternator is rotated, the field flux cutting through the coil increases from a minimum at position 1–1A shown in Fig. 10-

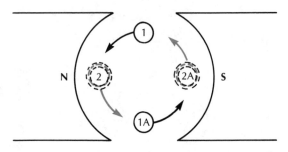

Figure 10-2(a) The elementary alternator.

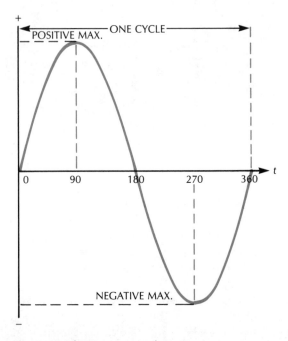

Figure 10-2(b) Sine-wave voltage.

2(a) to a maximum at position 2–2A and then decreases back to a minimum at position 1A–1. During the second half of rotation, the flux cut varies in the same way. It increases from a minimum to a maximum and then decreases back to a minimum. However, observe that during this interval of rotation, the coil conductors are in the opposite position to that during the first half of rotation. Therefore, the output from an alternator is continuously changing with respect to time as shown in Fig. 10-2(b). During one interval, the voltage increases from zero to a positive maximum and then decreases back to zero. During the next half rotation, the voltage again increases from zero, but this time to a negative maximum and then once again decreases back to zero. This completes the rotation of the coil with respect to the two field poles. The voltage wave will now repeat itself for equal intervals of time. The voltage wave pattern as shown in Fig. 10-2(b) is that of a mathematical function called the trigonometric sine wave function. Alternators are designed so that their outputs closely follow this function.

CYCLE, FREQUENCY, AND PERIOD

A complete wave pattern as shown in Figs. 10-3(a) and 10-3(b) is often called a cycle as it represents a complete electrical rotation of

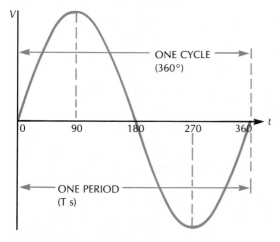

Figure 10-3(a) A wave, cycle, and period.

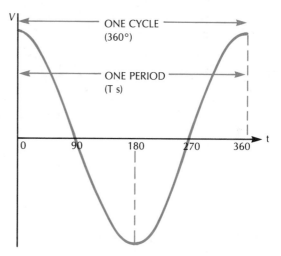

Figure 10-3(b) A wave, cycle, and period.

360°, and the time which elapses during this cycle is called the period. The number of cycles that the wave completes in one second is called the frequency and it is measured in hertz (Hz). The frequency (f) and period (T) of a wave are related by the equation;

$$f = \frac{1}{T} \qquad (10\text{-}1)$$

In North America the most commonly used frequency is 60 Hz and in Europe it is 50 Hz.

FREQUENCY, R/MIN, AND FIELD POLES

The output voltage frequency of an alternator is dependent on both the alternator's rotational speed and the number of field poles in the alternator. It should be obvious that if the rotational speed of an alternator is doubled, then the number of voltage cycles induced in one second would also double. This is because the armature will be cutting through the flux at twice the rate. Hence, an alternator's frequency is directly proportional to its (r/min) revolutions per minute.

In the two-pole alternator shown in Fig. 10-2(a), you saw that one mechanical rotation of the alternator induced one cycle of voltage. If

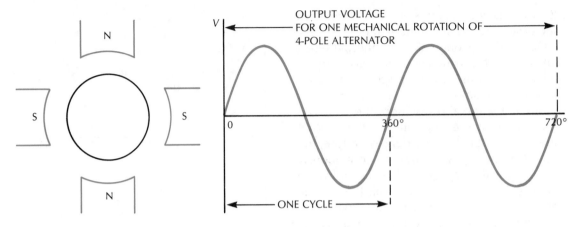

Figure 10-4a   A 4-pole alternator.   **Figure 10-4(b)**

the number of field poles are doubled, as shown in Fig. 10-4(a), then the armature will cut through two pairs of field poles during one mechanical rotation. This causes the alternator to induce two cycles of ac voltage for each rotation of the armature. Therefore, if the field poles of an alternator are doubled, then the output voltage frequency of the alternator also doubles. Hence, the output frequency of an alternator is directly proportional to both its r/min and to its pole pairs. The relationship between frequency, r/min, and field poles may be expressed mathematically as;

$$f = \frac{PS}{60} \qquad (10\text{-}2)$$

where f = frequency in Hz,
   P = number of pole pairs
   S = revolutions per minute

and the division by 60 is to convert r/min to revolutions per second.

Example 10-1: Determine the voltage frequency of a four-pole alternator which is rotating at 180 r/min.

Solution:

$$f = \frac{PS}{60} \qquad P = 2 \text{ pole pairs}$$

$$f = \frac{2 \times 1800}{60}$$

$$f = 60 \text{ Hz}$$

MECHANICAL AND ELECTRICAL DEGREES

In the two-pole alternator shown in Fig. 10-2(a), one mechanical rotation of the alternator induced one ac voltage cycle. However, in the four-pole alternator shown in Fig. 10-4(a), one mechanical rotation of the armature produces two ac voltage cycles. Hence there is a distinction between mechanical and electrical degrees. Mechanical degrees refers to the angular rotation of the armature, whereas electrical degrees refers to the movement of an electrical ac wave with respect to time.

BASIC TRIGONOMETRIC FUNCTIONS

A knowledge of the basic trigonometric functions, sine, cosine, and tangent, usually abbreviated sin, cos and tan, is essential in order to analyze and solve many ac circuits. For example, you have already seen that the voltage output from an alternator is a mathematical sine wave function. Therefore a brief review of these functions is in order.

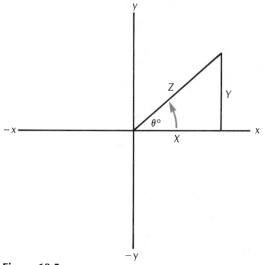

Figure 10-5

All trigonometric functions are based on the ratios of certain specific sides of a right triangle. By definition and in reference to Fig. 10-5, these ratios are:

$$\sin \theta = \frac{\text{opposite}}{\text{hypotenuse}} = \frac{Y}{R} \qquad (10\text{-}3)$$

$$\cos \theta = \frac{\text{adjacent}}{\text{hypotenuse}} = \frac{X}{R} \qquad (10\text{-}4)$$

$$\tan \theta = \frac{\sin \theta}{\cos \theta} = \frac{\text{opposite}}{\text{adjacent}} = \frac{Y}{X} \qquad (10\text{-}5)$$

The ratios for different angles, θ, have all been computed and are readily available from trigonometric tables. One such table may be found at the back of this text. This trigonometric table lends itself to the solutions of many right triangle problems.

Example 10-2: If in Fig. 10-5, $Z = 100 \ \Omega$ and $\theta = 30°$, determine (a) Y, (b) $\cos \theta°$ (c) X, and (d) $\tan \theta°$:

    from trig. tables sin 30° = 0.5000
    from Eq. 10-3, $Y = Z \sin \theta°$
                   $= 100 \times 0.5000$
                   $= 50 \ \Omega$

    from trig. tables cos 30° = 0.8660
    from Eq. 10-4, $X = Z \cos \theta°$
                   $= 100 \times +0.8660$
                   $= 86.60 \ \Omega$

    from trig. tables tan 30° = 0.5770 or
    from Eq. 10-5, $\tan \theta° = \dfrac{Y}{X}$

                        $= \dfrac{0.5000}{0.8660}$

                        $= 0.5770$

VOLTAGE AND CURRENT VALUES

There are five important values for voltage and current in alternating current circuits. They are: (a) the maximum or peak value, (b) the peak to peak value, (c) the instantaneous value, (d) the effective or rms value, and (e) the average value.

PEAK VALUE

The maximum value of voltage or current is the value from the reference level to the peak. The maximum value is also called the peak value. The peak to peak value is equal to twice the maximum value. These two ac values are illustrated in Fig. 10-6.

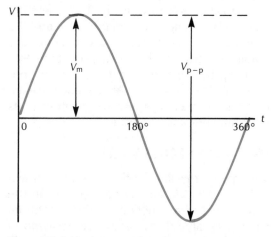

Figure 10-6 Maximum and peak to peak values.

Points	Angle θ	Sin θ	Points	Angle θ	Sin θ
0	0°	0	4	120	0.866
1	30	0.5	5	150	0.5
2	60	0.866	6	180	0
3	90	1.0			

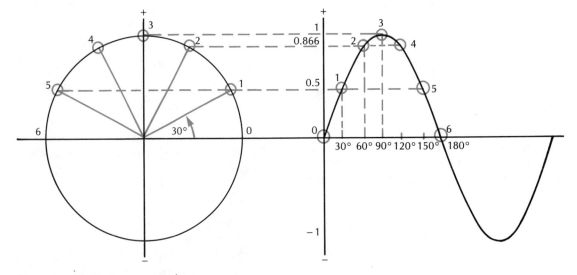

Figure 10-7 Instantaneous ac values.

One complete rotation of a circle contains 360° and one complete ac sine wave is also equal to 360°. As shown in Fig. 10-7, there is a direct relationship between the circle and the sine wave. The circle can be divided into four equal segments, each of which contains 90°. Similarly, the sine wave can be broken into four equal 90° sections. At the end of each one-quarter section, the wave is either at its maximum or minimum value.

INSTANTANEOUS VALUE

The maximum value for the sine wave shown in Fig. 10-7 is 1.0, and all other values in the waveform are a function of this maximum value. The value at any instant in the waveform is called the *instantaneous value*. For example

at point 1, the instantaneous value can be written as (1)(sin 30°) = 0.5, and at point 8, the value of this instant can be written as (1)(sin 240°) = (1) · (−0.8660) = −0.8660. Similarly, if the maximum value of the waveform was 100, then the instantaneous value at point 1 will be (100) (sin 30°) = (100)(0.5) = 50, and the instantaneous value at point 8 will be (100)(sin 240°) = (100)(−0.8660) = −86.60.

We can now generalize that if the maximum value of a sinusoidal voltage is known, then the voltage at any instant in the waveform is given by the equation:

$$V_{instant} = V_m \sin \theta° \tag{10-6}$$

where $\theta°$ = any angle, $V_{instant}$ = voltage at angle $\theta°$, and V_m = maximum or peak voltage. The corresponding instantaneous current equation is:

$$I_{instant} = I_m \sin \theta° \qquad (10\text{-}7)$$

Example 10-3: What is the instantaneous value of a sinusoidal voltage at 300° if the maximum voltage is 500 volts?

$$
\begin{aligned}
V_{instant} &= V_m \sin \theta° \\
&= 500 \sin 300° \\
&= (500)(0.8660) \\
&= 433.01 \text{ V}
\end{aligned}
$$

Example 10-4: What is the instantaneous value of an ac current at 150° if the maximum current is 40 A?

$$
\begin{aligned}
I_{instant} &= I_m \sin \theta° \\
&= 40 \sin 150° \\
&= (40)(0.5) \\
&= 20 \text{ A}
\end{aligned}
$$

THE EFFECTIVE OR RMS VALUE

The effective value, also called the *rms value*, is the most frequently used ac wave value. The effective value of an ac current is based on its heating effect. It is equivalent to a dc current of the same value. For example, 2 A of effective current in an ac circuit will produce the same amount of heat in a 6-ohm resistor as 2 A of direct current. The voltage value that causes the effective current is an effective voltage. This is equivalent to a dc voltage with the same value. Note that the effective value, as shown in Fig. 10-8, is always at a constant positive level even though the waveform itself changes direction.

The effective value of an ac wave is also called the rms value because of the method used in its computation. Rms is actually the abbreviation for root-mean-square. This means finding the square root of the mean of the square of the instantaneous values. This is the same as the square root of the mean of the square of the maximum value. It can be shown that this computation simplifies to the following effective current and voltage equations:

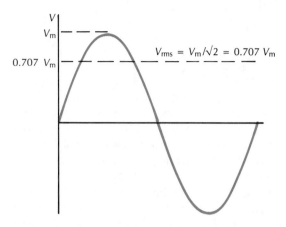

Figure 10-8 The effective or rms value of an ac wave.

$$
\begin{aligned}
I_{eff} &= I_{rms} \\
&= \frac{I_m}{\sqrt{2}} \\
&= 0.707 \, I_m \qquad (10\text{-}8)
\end{aligned}
$$

and

$$
\begin{aligned}
V_{eff} &= V_{rms} \\
&= \frac{V_m}{\sqrt{2}} \\
&= 0.707 \, V_m \qquad (10\text{-}9)
\end{aligned}
$$

Example 10-5: What is the effective value of an ac current with a peak value of 8 A?

$$
\begin{aligned}
I_{eff} &= \frac{I_m}{\sqrt{2}} \\
&= \frac{1}{\sqrt{2}} \times I_m \\
&= 0.707 \times 8 \\
&= 5.66 \text{ A}
\end{aligned}
$$

Example 10-6: A peak ac voltage of 160 V produces a peak current of 8 A through a resistive circuit. Determine the effective ac voltage and current.

$$
\begin{aligned}
V_{rms} &= \frac{V_m}{\sqrt{2}} \\
&= (0.707)(160) \\
&= 113.12 \text{ V}
\end{aligned}
$$

$$I_{rms} = \frac{I_m}{\sqrt{2}}$$
$$= (0.707)(8)$$
$$= 5.66 \text{ A}$$

Most ac instruments are designed to indicate effective values instead of the maximum or peak to peak values. Therefore, the scales on ac voltmeters and ammeters will be calibrated to read in effective values. It is common practice to write the symbols for effective voltage and current without the use of the subscript. To represent the effective values for current and voltage, we will write simply I and V, respectively, instead of I_{eff} and V_{eff}. The symbols I_m and V_m will continue to indicate maximum values and V_{p-p} will represent peak to peak values.

THE AVERAGE VALUE

The average value of an ac current or voltage is sometimes required in some instruments constructed to measure both ac and dc quantities. These types of meters are designed so that when an ac quantity is being measured, the ac

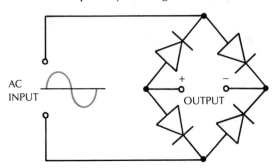

Figure 10-9(a) Full-wave bridge rectifier.

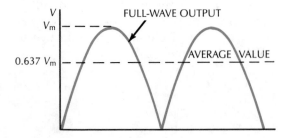

Figure 10-9(b) Average value.

quantity is first rectified. For this, the meter circuit must include, in part, a rectifier network. A full-wave bridge rectifier circuit is shown in Fig. 10-9. The rectifiers permit electron flow in only one direction.

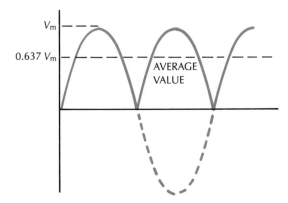

Figure 10-10(a) Full-wave rectifier output.

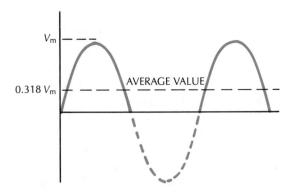

Figure 10-10(b) Half-wave rectifier output.

A meter, which uses a rectifier circuit, will respond to the average of the ac pulses taken from the rectifier's output. The average value, when the full-wave rectifier circuit is used, has been found to be 0.637 of the maximum instantaneous value of the sine wave. When the half-wave rectifier is used, the average reduces to half of 0.637 or 0.318 of the instantaneous maximum.

$$V_{av} = 0.637 \, V_m \qquad (10\text{-}10)$$

$$I_{av} = 0.637 \, I_m \qquad (10\text{-}11)$$

Meters which are designed to measure ac

voltage by responding to the average voltage have the added advantage of having linear voltage scales. The same is also true for the average current meters. On the other hand, ac voltmeters and ac ammeters, which respond directly to ac, lack this property. Their scales are nonlinear and, as a result, it is sometimes difficult to obtain accurate readings, particularly near the lower end of these scales.

Since effective voltage is equal to 0.707 V_m and average voltage is equal to 0.637 V_m, it follows that effective voltage will be 1.11 times greater than the average voltage. The same constant would also apply to a sinusoidal current. Therefore, the voltage and current scales on this type of rectifier meter may be recalibrated to read directly in rms values by multiplying the average scale readings by the constant, 1.11.

Example 10-7: Determine the average voltage for an ac voltage having a peak value of 140 volts.

Note: For finding average voltage, we will always assume full-wave rectification unless specified otherwise.

$V_{av} = 0.637 \ V_m$
$\quad = (0.637)(140)$
$\quad = 89.18 \ V$

Example 10-8: The scale readings on an average current are to be recalibrated to read in rms values. What is the new reading for an average current value of 4 A?

$I = (1.11)I_{av}$ (10-12)
$\quad = (1.11)(4)$
$\quad = 4.44 \ A$

VOLTAGE AND CURRENT PHASE RELATIONSHIPS

When an alternating voltage is applied to a circuit, it causes an alternating current of the same frequency to flow through the circuit. The waveforms for the voltage and current in

the circuit may show the same relationship to each other, in which case they are said to be *in-phase*, or they may not have the same relationship, in which case the two waveforms are said to be *out-of-phase*. It is also possible for two or more currents or voltages in the same circuit to be out-of-phase with each other. Phase difference is important in many ac circuit analyses. For example, it must be considered when out-of-phase quantities are added or subtracted.

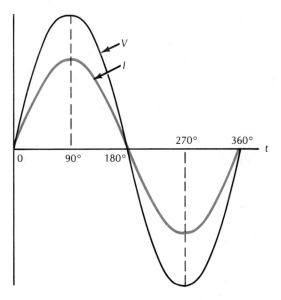

Figure 10-11 Voltage and current in-phase.

In Fig. 10-11, the current and voltage are in-phase because the two waveforms pass through their zero values and increase in the same direction to their maximum values at the same time. Hence, the phase difference between waves that are in-phase is zero.

It will be shown in later chapters that in some types of ac circuits the current and voltage can be out-of phase with each other. There are two possible out-of-phase relations between current and voltage in a circuit:

1. *Current may lag voltage* by some phase angle, $\theta°$, as shown in Fig. 10-12. In this relationship, the current waveform passes through zero in a positive direction at a

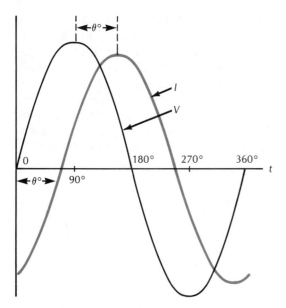

Figure 10-12 Current lags voltage by $\theta°$.

period of time later than the voltage waveform. Another way of looking at the lagging current is in terms of when the current waveform passes through its maximum value in a positive direction.

2. *Current may lead voltage* by some phase

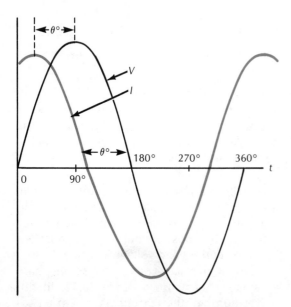

Figure 10-13 Current leads voltage by $\theta°$.

angle, $\theta°$, as shown in Fig. 10-13. In this relationship, the current waveform passes through zero and reaches its maximum value in a positive direction before the voltage waveform.

The student should note that it is customary to express phase difference in time between two waveforms in terms of electrical degrees. This of course is possible because, as was shown in an earlier section, the time axis of a waveform may also be described in terms of electrical degrees. Phase difference measured in degrees is also called the *phase angle*, and its common symbol is the Greek letter theta (θ). It will be shown, in later discussions, that the phase angle by which a current lags or leads a voltage may vary in different circuits from zero degree (in-phase) to 90°.

PHASORS (VECTORS) AND ADDITION OF PHASORS

The drawing of waveforms to illustrate the phase relationship between a current and a voltage is very cumbersome. Instead, the use of phasors is universally employed to illustrate and solve ac circuit properties. A phasor is a line segment used to represent the magnitude and direction of a current or voltage. The direction of the line represents the direction of the phasor and the length of the line represents its magnitude. A combination of phasor lines is called a phasor diagram. Phasors are nearly always drawn to represent effective values of current and voltage, but they can also be used to illustrate maximum values. The following are a few simple rules and conventions which are essential for the consistency in results when drawing phasor diagrams:

1. When two waveforms are in-phase, they have the same direction and so their phasors are drawn on the same line. Therefore, the in-phase current and voltage waveforms shown in Fig. 10-11 may be represented by the phasor diagram of Fig. 10-14(a).

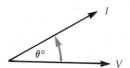

Figure 10-14(a) Current and voltage are in-phase.

Figure 10-14(b) Current leads voltage by $\theta°$.

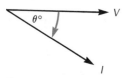

Figure 10-14(c) Current lags voltage by $\theta°$.

2. The counter-clockwise direction is considered as the positive direction of rotation. A phasor which is rotated in a counter-clockwise direction from a given phasor is said to lead the given phasor. Similarly, the clockwise direction is considered as the negative direction of rotation. A phasor which is rotated in a clockwise direction from a given phasor is said to lag the given phasor. Therefore, the waveforms in Fig. 10-13, where the current leads the voltage by $\theta°$, may be represented by the phasor diagram of Fig. 10-14(b). Likewise, the waveforms in Fig. 10-12, where the current lags the voltage by $\theta°$, may be represented by the phasor diagram of Fig. 10-14(c).

3. The magnitude of a phasor is given by a scaled length of the phasor line. It is not necessary to use the same scale for both the current and voltage phasors. However, if there are more than one current phasor, then a common scale must be used for these current phasors. Similarly, if more than one voltage phasor exists in a circuit, then a common scale must be used for these voltage phasors.

4. In a series circuit, the current is constant through all parts of the circuit. Hence, it is convenient to draw the current phasor on a horizontal line and use it as the reference

phasor for other phasors in the same diagram.

5. Similarly, in a parallel circuit the voltage is the same across parallel branches. Hence, the voltage phasor is drawn on the horizontal line and used as the reference for other phasors.

The addition of phasors is different from the addition of ordinary (scalar) quantities. When adding phasors, both the magnitude and direction of the phasors must be taken into account, whereas in scalar addition only the magnitude of the numbers is important. There are various mathematical concepts which, at times, can be applied in the addition of phasors. The following examples will review just enough material which will enable the reader to solve, by one method or another, the different phasor-related problems in this text.

TRIANGULAR METHOD

Phasor addition can be done by a graphical method often called the *triangular method*. For example, the phasor sum or resultant of two phasors may be obtained by drawing each of the phasors to length with respect to a convenient scale. The first phasor is started from the zero or origin point and the second phasor is then drawn starting out at the end of the first phasor. The resultant or phasor sum is the line which closes the triangle, hence the name triangular method. This method may be expanded to add more than two phasors. Succeeding phasor is simply drawn starting out from the end of the last phasor. The resul-

SCALE:
$1 \text{ cm} = 4 \text{ A}$
$I_1 = 20 \text{ A} = 5 \text{ cm}$

$I_2 = 16 \text{ A} = 4 \text{ cm}$
$I = 8.7 \text{ cm} = 34.8 \text{ A}$
$\theta° = 13.5°$

$\therefore I = 34.8 \text{ A}$ and I leads I_1 by $13.5°$

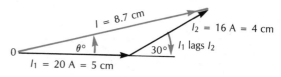

Figure 10-15 Phasor addition — triangular method.

tant is again the line which closes the figure.

Example 10-9: The two branch currents, I_1 and I_2, in a parallel circuit are 20 A and 16 A, respectively, with I_1 lagging I_2 by 30°. Determine the resultant or total circuit current.

THE PYTHAGOREAN RELATIONSHIP

In some circuits, two phasors may have a phase difference of exactly 90°. The sum of two such phasors may be obtained by using *Pythagorean relationship*. This relationship states that in a right triangle the hypotenuse squared is equal to the sum of the squares of the other two sides.

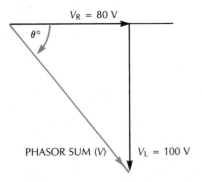

Figure 10-16 Phasor sum—Pythagorean relationship.

Example 10-10: In a series R-L circuit, the resistive voltage (V_R) and inductive voltage (V_L) are respectively 80 and 100 volts. In this circuit the inductive voltage lags the resistive voltage by 90°. Find the phasor sum of these two vol-

$$V = \sqrt{(V_R)^2 + (V_L)^2} \qquad (10\text{-}13)$$
$$= \sqrt{(80)^2 + (100)^2}$$
$$= 128.06 \text{ V}$$
$$\theta° = \tan^{-1}\frac{V_L}{V_R} \qquad (10\text{-}14)$$
$$= \tan^{-1}\frac{100}{80}$$
$$= 51.3°$$

Therefore $V = 128.06$ V and lags V_R by 51.3°

PHASORS ON THE SAME LINE OF ACTION

In some circuits, two or more phasors may be *exactly in-phase or exactly 180° out-of-phase*. In such special cases, these phasors will all lie on the same "line of action," and the phasor sum is simply the ordinary addition (in-phase) or subtraction (180° out-of-phase) of these phasors.

Example 10-11: A series circuit contains three voltage drops,

$$V_1 = 60 \text{ V}, V_2 = 80 \text{ V and } V_3 = 30 \text{ V}$$

V_1 is in-phase with V_2 and 180° out-of-phase with V_3. Find the phasor sum of these three voltages.

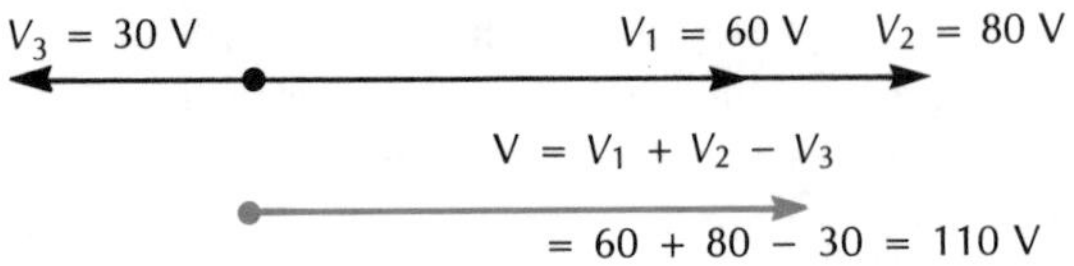

$V_3 = 30$ V $\qquad\qquad V_1 = 60$ V $\quad V_2 = 80$ V

$$V = V_1 + V_2 - V_3$$
$$= 60 + 80 - 30 = 110 \text{ V}$$

Figure 10-17 Phasor sum for phasors on the same "line of action."

From Fig. 10-17, it is obvious that:

$$V = V_1 + V_2 - V_3$$
$$= 60 \text{ V} + 80 \text{ V} - 30 \text{ V}$$
$$= 110 \text{ V} \qquad (10\text{-}15)$$

PHASOR ADDITION BY THE COMPONENT METHOD

You have now seen that phasors which are exactly 90° out-of-phase or exactly on the same "line of action" may be added relatively easy. However, in many ac circuits, the phasors are neither exactly on the same "line of action" nor exactly 90° out-of-phase. For such cases, phasor addition by the component method will be found to be useful. This method combines the use of basic trigonometric functions with the Pythagorean relationship and the addition of phasors on the same "line of action".

Example 10-12: The currents in each of the two branches of a parallel circuit are $I_1 = 12$ A and $I_2 = 10$ A. If I_1 leads the source voltage V by 30° and I_2 lags the source voltage by 50°, find the net source current, I_T. The phasor diagram for the source voltage and branch currents are shown in Fig. 10-18(a).

Using trigonometric functions find the x and y components for the phasors I_1 and I_2. These components are illustrated in Fig. 10-18(b), and may be substituted for I_1 and I_2.

$$I_{1_x} = I_1 \cos 30° \qquad (10\text{-}16)$$

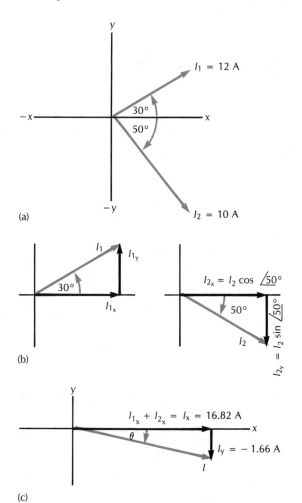

(a)

(b)

(c)

Figure 10-18 Phasor addition by the component method.

$$= 12 \text{ A} \times 0.866$$
$$= 10.39 \text{ A}$$

$$I_{2_x} = I_2 \cos 50°$$
$$= 10 \text{ A} \times 0.643$$
$$= 6.43 \text{ A}$$

$$I_{1_y} = I_1 \sin 30° \qquad (10\text{-}17)$$
$$= 12 \text{ A} \times 0.5$$
$$= 6 \text{ A}$$

$$I_{2_y} = I_2 \sin 50°$$
$$= 10 \text{ A} \times 0.766$$
$$= 7.66 \text{ A}$$

The x-components of the two phasors have the same "line of action." Similarly, the y-components have the same "line of action." However, note that the y-component for I_2 is negative. Find the sum of each of these two sets of components:

$$\text{Sum of the x-components} = I_x$$
$$= I_{1_x} + I_{2_x}$$
$$= 10.39 \text{ A}$$
$$+ 6.43 \text{ A}$$
$$= 16.82 \text{ A}$$
$$\text{Sum of the y-components} = I_y$$
$$= I_{1_y} + I_{2_y}$$
$$= 6 \text{ A} - 7.66 \text{ A}$$
$$= -1.66 \text{ A}$$

The component sums, I_x and I_y, are illustrated in Fig. 10-18(c). Note that the phasor sum of I_1 and I_2 has now been reduced to the phasor sum of I_x and I_y and that these two new phasors are at right angle. The solution should now be obvious.

From the Pythagorean relationship

$$I = \sqrt{(I_x)^2 + (I_y)^2}$$
$$= \sqrt{(16.82)^2 + (-1.66)^2}$$
$$= 16.9 A$$

$$\theta = \tan^{-1} \frac{I_y}{I_x}$$
$$= \tan^{-1} \frac{1.66}{16.82}$$
$$= \tan^{-1} 0.099$$
$$= 5.64°$$

Therefore $I = 16.9$ A and lags V by 5.64°.

MATHEMATICAL EXPRESSION OF A PHASOR

It is often convenient to simplify the way in which a *phasor may be expressed mathematically*. The following convention is commonly used:

$$\text{Voltage phasor} = V\underline{/\pm\theta°} \qquad (10\text{-}18)$$

$$\text{Current phasor} = \underline{/\pm\theta°} \qquad (10\text{-}19)$$

The V and I represent the respective rms values for the voltage and current phasor. $\theta°$ indicates a lead phase angle or a counter-clockwise rotation of $\theta°$, and $-\theta°$ represents a lag phase angle or clockwise rotation of $\theta°$. The positive x-axis of the x-y coordinate system is used as the reference direction. This method of expressing a phasor is sometimes referred to as the polar form. For example, the answer to Ex. 10-12, which was stated as:

$$I = 16.9 \text{ A and lags } V \text{ by } 5.64°$$

may be expressed mathematically as:

$$I = 16.9 \text{ A} \underline{/-5.64°}$$

AC CIRCUITS WITH ONLY RESISTIVE LOADS

Alternating current circuits with only resistive loads are perhaps the simplest of ac circuits. The analysis of such an ac circuit is very much the same as the analysis of a similar dc circuit, the reason being that in a purely resistive circuit all of the circuit's phasors are in phase. That is, the phase angle between the phasors of such a circuit is zero.

POWER IN AN IN-PHASE CIRCUIT

The power curve in Fig. 10-19 shows the instantaneous power values for a circuit in which the current and voltage are in phase. The power at any instant is found by multiplying the value of the current and voltage at that instant. Since, for an in-phase (purely resistive) circuit, when the current is positive, the volt-

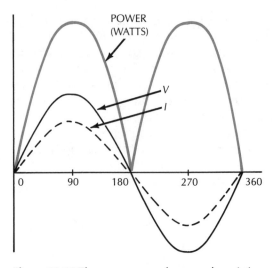

Figure 10-19 The power curve for a purely resistive circuit.

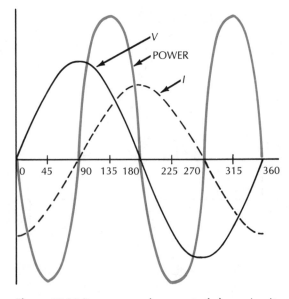

Figure 10-20 Power curve for an out-of-phase circuit.

age is also positive, and when the current is negative, the voltage is also negative, the instantaneous power values will always be in a positive direction as shown by the power curve. For such a circuit, the average power is measured in watts, and it is found in the same way as for a dc circuit.

Watts $= V \times I$ (10-20)

$\quad\quad = I^2R$ (V and I are in phase)

$\quad\quad = \dfrac{V^2}{R}$

In later chapters, you will study circuits in which the current and voltage are not in phase. The resulting power curve for such an out-of-phase circuit will not always be in a positive direction. This is shown in Fig. 10-20. Hence, the equation for finding the average power in watts will have to be modified to take into account the phase angle between the current and voltage in the circuit.

Example 10-13: For the circuit shown in Fig. 10-21, determine (a) the circuit's resistance, R_T, (b) the currents, I, I_2, and I_3, and (c) draw the phasor diagram to show I, I_2, I_3, V, and R_T.

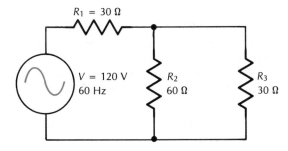

$R_1 = 30\ \Omega$

$V = 120\ V$
$60\ Hz$

R_2
$60\ \Omega$

R_3
$30\ \Omega$

Figure 10-21 A purely resistive ac circuit.

y

$I_2 \quad I_3 \quad R_T \quad I \quad V \quad W$

x

Figure 10-22 Phasor diagram for the circuit in Fig. 10-21.

$R_T = R_1 + \dfrac{R_2 \times R_3}{R_2 + R_3}$

$\quad\ = 30 + \dfrac{60 \times 30}{90}$

$\quad\ = 50\ \Omega$

$I = \dfrac{V}{R_T}$

$\quad = \dfrac{120\ V}{50\ \Omega}$

$\quad = 2.4\ A$

$I_2 = I \times \dfrac{R_3}{R_2 + R_3}$

$\quad\ = 2.4 \times \dfrac{30}{60 + 30}$

$\quad\ = 0.8\ A$

$I_3 = I \times \dfrac{R_2}{R_2 + R_3}$

$\quad\ = 2.4 \times \dfrac{60}{60 + 30}$

$\quad\ = 1.6\ A$

Power in watts $= V \times I = 120 \times 2.4$
$\quad\quad\quad\quad\quad\quad\quad\quad = 288\ W$

or

Power in watts $= P_1 + P_2 + P_3$
$\quad\quad\quad\quad = I^2R_1 + I_2^2R_2 + I_3^2R_3$
$\quad\quad\quad\quad = (2.4)^2(30)$
$\quad\quad\quad\quad\quad + (0.8)^2(60) +$
$\quad\quad\quad\quad\quad (1.6)^2(30)$
$\quad\quad\quad\quad = 172.8 + 38.4 + 76.8$
$\quad\quad\quad\quad = 288\ W$

AC EFFECTIVE RESISTANCE

All resistances, whether they are in a dc or in an ac circuit, oppose the flow of current and produce heat. That is, the resistance in any circuit consumes power, and this power is measured in watts.

In later chapters, you will read about other circuit devices which not only oppose the flow of an ac current but, unlike resistance, also cause the current and voltage in the circuit to move out-of-phase. The power used by these devices is different from the power (in watts) used by resistance.

The total resistance of an ac circuit, sometimes called the effective resistance or ac resistance, may be several times larger than the total pure dc resistance itself in the same circuit. This increase in resistance is caused by the eddy-current losses, hysteresis loss, skin effect, and dielectric loss which may be present in an ac circuit. These effects will vary with the frequency, current, and voltage of the circuit. They all produce heat, that is, they use power in watts, and so their presence in a circuit will increase the total effective resistance in that circuit.

EDDY-CURRENT LOSSES

Eddy-current losses are caused by the small currents (eddy currents) which are induced in the iron core of a generator or transformer. These currents are induced by the alternating flux which is produced when an ac current flows through the circuit. Laminated cores are often used to limit this type of power loss in a circuit.

HYSTERESIS LOSS

Hysteresis loss is the power used to overcome the friction caused by the constant reversal of the millions of electron dipoles in the nearby iron of a circuit. These electron dipoles of the iron atoms in the core of a device like the transformer must continuously reverse their magnetic directions as the ac current through the circuit changes its direction. Special steels with relatively low friction such as silicon steel are used in the construction of the cores found in most ac equipment in order to minimize this power loss.

SKIN EFFECT

Skin effect is the term used to describe the tendency of an ac current to flow along the surface of a conductor, compared to a dc current which uses the entire cross-sectional area of the conductor. Skin effect, therefore, reduces the effective conducting area of a circuit's conductors and, hence, causes an increase in the circuit's resistance.

DIELECTRIC LOSS

Dielectric loss is a very small power loss compared to the other circuit losses. It is caused by

a voltage stress which is placed on the insulation of a conductor, as the ac voltage continuously increases and decreases to and from its peak values.

The following practical circuit problem will illustrate the difference between the dc resistance of a circuit and its effective or ac resistance.

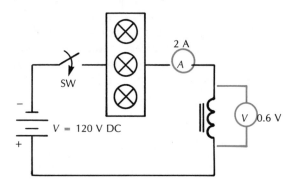

Figure 10-23 The dc resistance of a coil.

Example 10-14: A coil with a laminated steel core is connected to a dc source, as shown in Fig. 10-23. For dc, a coil is virtually a short-circuit, and so a limiting resistor or a load such as a bank of lamps is used to reduce the circuit current to a safe value.

Determine the dc resistance of the coil using the volt-ampere method.

$$R_{dc}\text{ (coil)} = \frac{V_c}{I}$$
$$= \frac{0.6 \text{ V}}{2 \text{ A}}$$
$$= 0.3 \text{ }\Omega$$

Next, the coil is now connected to a 120-V, 60-Hz supply, as shown in Fig. 10-24. Since a coil, also called an inductor or a reactor, opposes the flow of an ac current, the limiting resistor or lamp bank is no longer necessary. The inductor and its effects in a circuit will be discussed in the next chapter. The problem is investigated for (a) the laminated core removed from the coil, and (b) the core re-inserted into the coil. A sensitive wattmeter is used to measure the power used by the coil

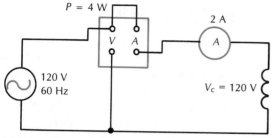

(a) Iron core removed from the coil

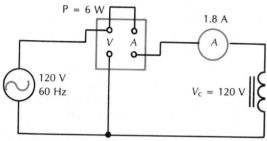

(b) Core inserted into the coil

Figure 10-24 The ac resistance of a coil.

for each of the two cases. Note: with the laminated core in the coil, the circuit current decreases but the actual power consumed increases. Determine the effective resistance (R_e) of the coil in each of the two circuits.

R_e (with core removed)	R_e (with core inserted)
$= \dfrac{W}{I^2}$	$= \dfrac{W}{I^2}$
$= \dfrac{4}{(2.0)^2}$	$= \dfrac{6}{(1.8)^2}$
$= 1\,\Omega$	$= 1.85\,\Omega$

Note, that the dc resistance of the coil in this example, with or without the core, is 0.3 Ω, whereas the ac resistance without the core is 1.0 Ω and the ac resistance with the core inserted is 1.85 Ω. The increase in the resistance of the coil without the core in Fig. 10-24 over its dc resistance in Fig. 10-24 is due mainly to skin effect and dielectric losses in the circuit. A further increase in the coil's resistance when the core was inserted is due to the additional power losses caused by the

effects of hysteresis and eddy currents. These are the core or iron losses.

In all future discussions, the resistance of an ac circuit will be assumed to be the effective resistance, unless stated otherwise.

SUMMARY OF IMPORTANT POINTS

1. The alternator is designed to produce an ac sinusoidal waveform.
2. There are numerous advantages of ac over dc:
   (a) easy to step-up or step-down ac voltages
   (b) fewer losses during transmission
   (c) ac machines are less complex and more compact

Properties of ac voltage

3. One cycle of ac equals 360 electrical degrees, and the time for one cycle is the period $T = \dfrac{1}{f}$. Mechanical degrees refer to the number of mechanical rotations.
4. The frequency is the number of cycles per second: the common frequency in N.A. is 60 Hz. In an alternator the output frequency is $f = \dfrac{PS}{60}$.
5. The instantaneous voltage $V_i = V_m \sin \underline{/\theta}$
   (a) the rms or effective voltage $V_{rms} = V\frac{m}{\sqrt{2}}$ $= 0.707\,V_m$
   (b) the rms value is based on its heating effect compared to that of dc
   (c) the average value based on full-wave rectification is $V_{av} = 0.637\,V_m$
6. Three basic trigonometric functions are:

   (a) $\sin \underline{/\theta} = \dfrac{\text{opp.}}{\text{hyp.}}$

   (b) $\cos \underline{/\theta} = \dfrac{\text{adj.}}{\text{hyp.}}$

   (c) $\tan \underline{/\theta} = \dfrac{\text{opp.}}{\text{hyp.}}$

Phasors

7. When quantities are in-phase, their phasors are drawn in the same direction.

8. A leading phase angle is measured counter-clockwise $\underline{/+\theta}$ and a lagging phase angle is measured clockwise $\underline{/-\theta}$.

9. In a series circuit, the current phasor is used as the reference phasor, and in a parallel circuit the voltage phasor is used as the reference.

10. Phasors may be added using the graphical method or the phasor component method.

11. In a purely resistive circuit, V, I, R and W are in-phase.

12. In an out-of-phase circuit, V and I are out of phase. I, R, and W are still in phase.

13. In an ac circuit, true power $= I^2R = VI \cos \underline{/\theta}$

Ac effective resistance = pure de resistance + ac effects

14. Ac resistance may be several times its pure dc resistance due to the added effects of hysteresis, eddy currents, skin effect, and dielectric loss. These effects generally increase with frequency.

REVIEW QUESTIONS

1. List some of the reasons why the generation and use of ac voltage is so much more common than the use of dc voltage.

2. Briefly explain why an ac voltage is first stepped up to a very high value before it is transmitted over a long distance.

3. List six applications for dc where its use has a definite advantage over ac.

4. Describe two examples of how a dc voltage may be varied.

5. Describe with the aid of diagrams the output of a simple four-pole alternator.

6. Compare the output from a two-pole alternator with that of a four-pole alternator for one complete rotation of the armature.

7. Find the period for the waveforms with the following frequencies:
   (a) 90 Hz and (b) 220 Hz

8. Determine the voltage frequency of a six-pole alternator which is rotated at 1200 r/min.

9. In order to parallel the outputs from two alternators, they must have the same voltage frequencies. If one alternator has four-poles and turns at 2400 r/min, what should be the r/min of the second alternator which has six poles?

10. Explain the difference between mechanical and electrical degrees for a four-pole alternator.

11. Draw and label a right-triangle in standard position with a phase angle, $\theta°$, and then define the three basic trigonometric functions with respect to this angle.

12. Draw the phasor diagram and find the x- and y-components for each of the following phasors (a) 12 A $\underline{/20°}$ (b) 120 V $\underline{/80°}$ (c) 20 A $\underline{/-30°}$ (d) 60 V $\underline{/-60°}$ (e) 16 A $\underline{/0°}$

13. Find the rms value for each of the following peak ac values: (a) 100 V (b) 300 V (c) 1200 V

14. Find the peak value for each of the following effective ac values: (a) 10 A (b) 12 A (c) 16 A

15. An ac voltage has a peak value of 400 V. Find the instantaneous value at (a) 40° (b) 60° (c) 90° (d) 45° and (e) 200°

16. Explain how the term "root-mean-square voltage" was derived.

17. Draw a diagram to illustrate the output from (a) a half-wave rectifier circuit, and (b) a full-wave rectifier circuit.

18. The scale on an average current meter is to be recalibrated to read in rms values. What is the rms value for the average scale readings of (a) 5 a, (b) 12 A, and (c) 10.81 A?

19. Draw the sinusoidal waveforms to show an ac current lagging an ac voltage by 60 electrical degrees.

20. Draw the phasor diagram for the waveforms in Q. 19.

21. Draw the sinusoidal waveforms to show an ac current leading an ac voltage by 30 electrical degrees.
22. Draw the phasor diagram for the waveforms in Q. 21.
23. Find the phasor sum of two phasors, $V_1 = 60$ V and $V_2 = 90$ V, which are 180° out-of-phase. Assume V_1 to be in-phase with the positive x-axis.
24. Find the phasor sum of two voltages, $V_1 = 60$ V and $V_2 = 90$ V, of which V_1 leads V_2 by 90 electrical degrees, and V_2 lies on the horizontal x-axis.
25. The two-load voltages in a series circuit are $V_1 = 60$ V and $V_2 = 90$ V. If V_1 leads V_2 by 60°, find the magnitude of the source voltage using the triangular method. Assume V_2 to be in-phase with the positive x-axis.
26. Find the magnitude of the source voltage in Q. 25, using the "phasor component method."
27. A parallel circuit has branch currents, $I_1 = 8$ A and $I_2 = 20$ A. If I_1 leads the voltage by 60° and I_2 lags the voltage by 40°, find the phasor sum of the two currents. Draw a phasor diagram to show all three currents.
28. The load voltages in a series circuit are $V_1 = 60$ V $\underline{/20°}$, $V_2 = 100$ V $\underline{/60°}$, and $V_3 = 120$ V $\underline{/-30°}$. Find the phasor sum of these three voltages and draw the phasor diagram to show all four voltages.
29. For the circuit given in Fig. 10-25, determine:
    (a) total resistance (R_T) (b) source current, I_T (c) branch currents, I_1, I_2, I_3, and I_4 and (d) the voltage drops V_1, V_2, V_3 and V_4.
30. For the circuit in Q. 29, determine (a) the power used by each of the load resistors and (b) the total circuit power.
31. Draw a phasor diagram to show all the voltages and currents in Q. 29.
32. Five 100-W lamps are connected in parallel and across an ac voltage with a peak value of 200 V. Determine:
    (a) the total circuit power
    (b) the effective circuit voltage
    (c) the effective circuit current
    (d) the circuit's resistance
    (e) the peak circuit current
33. (a) What ac circuit properties cause its effective resistance to be greater than its pure dc resistance? (b) what effect does frequency have on these properties?
34. Briefly describe the cause and effects of each of the following ac circuit properties: (a) skin effect (b) hysteresis losses (c) eddy-current losses and (d) dielectric loss.
35. An electromagnetic coil with a laminated core takes 5 A, when connected to 110 V dc. When connected to a 110 V, 60 Hz ac source, the current is 2 A, and a wattmeter reads 106 W. Determine:
    (a) the dc resistance of the coil
    (b) the ac effective resistance of the oil
    (c) Explain the reasons for the difference between these two resistance values.
    (d) Explain why the coil's current decreased from 5 A when connected to the 110-V dc to 2 A when connected to the 110 V, 60 Hz ac.

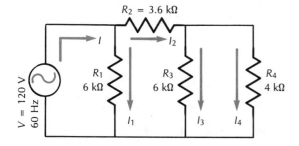

Figure 10-25 A purely resistive coil.

INDUCTANCE AND INDUCTIVE REACTANCE

Two important laws on electromagnetic induction were discussed in Chapter Five. They are: (a) Faraday's law of electromagnetic induction, which states when the flux through a coil or conductor is varied, a voltage will be induced in the coil or conductor, and (b) Lenz's law, which states that the induced voltage which is set up by a changing magnetic field always acts in a direction such that it will oppose that which is causing it. In this chapter, the effects of these two properties in a circuit will be further investigated.

INDUCTANCE

In Fig. 11-1, the changing source voltage will cause a changing source current. This, in turn, will set up a changing magnetic field through the coil. Therefore, from Faraday's and Lenz's laws, a voltage will be induced in the coil, and this voltage will oppose the source voltage. That is, when the source voltage is *increasing positively*, the induced voltage will be *increasing negatively*, as shown by the graphs in Fig. 11-1. For this reason, the induced voltage, V_L, may be called a counter voltage.

This property of a circuit or component having an induced voltage that opposes any

change in the applied voltage or current is called *inductance*. The inductance of a coil may be described as a form of electrical inertia. It is the property of a circuit or an inductor to store energy in a magnetic field. The symbol, L, is used to represent inductance, and it is

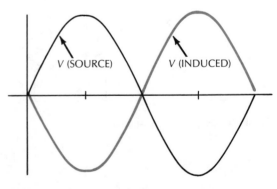

Figure 11-1 The induced voltage in a coil.

measured in henrys (H). Other physical factors which determine the inductance of a coil are: the length of the coil, the type of core material, and the cross-sectional area of the coil.

An inductance of one henry is defined as the inductance of a component of a circuit which will cause a current change of one ampere per second to induce a voltage of one volt.

INDUCTANCE IN DC CIRCUITS

In a circuit having a steady dc supply, as shown in Fig. 11-2, there are only two instances when the effects of any circuit inductance will be noticeable. These are the moment the switch is closed, or opened. In both cases, the current will be changing and as a result induced voltages can be set up in the circuit. When the circuit current is steady, the flux about the coil will be constant and, hence, there will be no induced voltage.

At the instant the switch, in the circuit shown in Fig. 11-2, is closed, the lamps will light up but will be dim at first, and then will slowly increase to their maximum brilliancy. This effect is due to the counter voltage set up in the circuit, because at the instant the switch is closed, the circuit current will begin to increase (change), from zero to its steady-state value. During this time, energy is taken from the source and stored in the expanding mag-

netic field about the inductor. This expanding field will induce a voltage in the coil, and this voltage will act to oppose the change (increase) in the source current.

When the switch is closed, the current will initially increase very rapidly, and hence the induced voltage (V_L) will be very high. However, as the current increases toward its steady-state value, its rate of change decreases and the induced voltage will also decrease. This explains why the lamps will first be dim and then gradually increase in brilliancy. When the current is at steady-state, its rate of change will be zero, and hence the induced voltage will be zero and the lamps will be at maximum brightness. It should be noted that at steady-state current, the inductor will be virtually a short-circuit. For this reason, care must be exercised when a dc voltage is to be connected directly across a coil. The graphs in Fig. 11-3(a) illustrate the general current and induced voltage characteristics for the instant the switch is closed.

Figure 11-3(b) shows the general current and induced voltage characteristics for the moment the switch in Fig. 11-2 is opened. At this point the circuit current must now decrease from its steady-state value to zero.

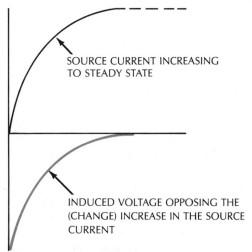

SOURCE CURRENT INCREASING TO STEADY STATE

INDUCED VOLTAGE OPPOSING THE (CHANGE) INCREASE IN THE SOURCE CURRENT

Figure 11-3(a) Induced voltage for an increasing source current (switch closing).

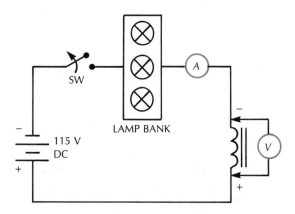

Figure 11-2 An inductor connected in a dc circuit.

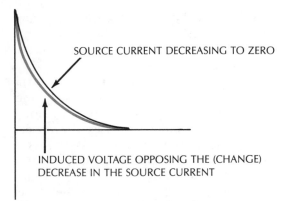

Figure 11-3(b) Induced voltage for a decreasing source current (switch opening).

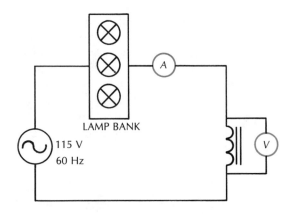

Figure 11-4 An inductor connected in an ac circuit.

Therefore, the field about the inductor will begin to collapse, and the energy stored in the field will now be released back into the circuit. Since the field is collapsing (changing), a voltage will be induced once again in the coil. This induced voltage will oppose the circuit current from decreasing (changing). That is, the induced voltage will now act in such a way as to aid or maintain the circuit current. This property is responsible for the spark which may be seen at the switch contacts at the instant the switch is opened.

INDUCTANCE IN AC CIRCUITS

In an ac circuit the source voltage is continuously changing, and hence it will set up a magnetic field around the conductor that will expand and collapse. As a result, the effects of inductance will be more pronounced in ac circuits than in steady-state dc circuits.

The effects of inductance in ac circuits can be demonstrated by a circuit such as that shown in Fig. 11-4. The source voltage is of the same effective value as that used in the dc circuit of Fig. 11-2. The inductor and lamps are also the same. In this circuit, the induced counter voltage (V_L), will always be present once the circuit is turned on. Since this induced voltage opposes the source voltage, the effective voltage (V_R) and the effective

current in the circuit will be greatly reduced. As a result, the lamps in this circuit will be dim.

If the core is removed from the coil, the circuit current and the brilliancy of the lamps will increase. Removal of the core will increase the coil's reluctance, reducing the amount of flux which cuts the coil, and the induced counter voltage (V_L). Therefore, removing the core has the effect of reducing the circuit's inductance.

Reducing the source frequency will produce a similar effect. It will cause a decrease in the rate of flux variation about the inductor. Therefore, the induced counter voltage (V_L) will decrease, and as a result, the circuit current and hence the brilliancy of the lamps will increase. Frequency does not affect the inductance of a coil, but rather, as it will be shown later, frequency is directly related to the coil's reactance.

From the previous discussion we can see that an inductor will allow free passage of a steady dc current in a circuit, but will oppose or choke the flow of an ac current. For this reason, inductors are widely used in filter networks. Inductors are constructed in various shapes, as shown in Fig. 11-5. They have numerous applications in both electrical and electronic systems. For example, you have seen the use of induction coils in generators, dc motors, and magnetic relays and switches.

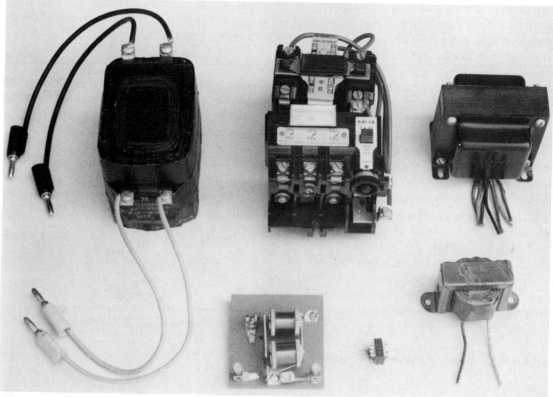

Figure 11-5 Types of inductors. *Courtesy: John Dubiel*

In later chapters you will also study their use in transformers and in other devices.

The inductance of a coil or circuit, however, is not always a useful circuit property. For example, in later sections you will see that inductance in a circuit will cause the circuit current to lag the voltage. In some circuits, this property may cause an unnecessary increase in the source current with a resultant increase in the system losses. Inductance may also cause unwarranted interferences in some circuits. For example, it is possible for a magnetic field from one coil to link with another coil and induce a voltage in this coil. This is called *mutual induction*. Mutual induction is the basis for the operation of transformers, but in some circuits the effect of this property can cause undesirable interference in the operation of these circuits. The positioning of any coils and even the wiring will therefore be critical in such a circuit.

Almost all ac circuits will have at least some inductance. Of course, for a purely resistive circuit, the assumption used is that any circuit inductance is extremely small and therefore negligible. In many other circuits, however, the property of inductance must be taken into account. For example, most industrial loads tend to be highly inductive. Even a long, straight wire also has a certain amount of inductance, and calculations of transmission line performance must take into account the inductance of the line.

INDUCTIVE REACTANCE (X_L)

It was seen that for a steady dc current the inductor is basically a short-circuit. In an ac circuit, however, the current is changing continuously, and therefore is continuously inducing a counter voltage in any circuit inductance. Because this induced voltage will continuously oppose the flow of the ac current in the circuit, it has a similar effect as resistance. This opposition by an inductor to the flow of an ac current is called *inductive reactance* and, like resistance, it too is measured in ohms. The symbol for inductive reactance is X_L.

It will be seen, however, that inductive reactance differs in many ways from resistance. Since the induced counter voltage of an inductor varies with the coil's inductance and with the frequency of the varying current, it should be obvious that inductive reactance will depend on both the inductance and the frequency. An increase in either or both of these two properties will increase the inductive reactance. Inductive reactance in ohms may be calculated from the equation.

$$X_L = 2\pi f L \qquad (11\text{-}1)$$

where X_L = inductive reactance in ohms (Ω)
$\quad \pi = 3.14$
$\quad f$ = frequency in hertz (Hz)
$\quad L$ = inductance in henry (H)

INDUCTANCES IN SERIES

In a purely inductive circuit, the induced counter voltage in the reactance will be equal and opposite to the effective voltage applied to the inductor. This voltage may be calculated using the Ohm's law equation.

$$V_L = I X_L \qquad (11\text{-}2)$$

where V_L = effective voltage across the reactance
$\quad I$ = effective current flowing through the reactance and
$\quad X_L$ = inductive reactance

Applying Kirchhoff's voltage law to the circuit shown in Fig. 11-6, we obtain the equation

$$V = V_{L_1} + V_{L_2} + V_{L_3} \qquad (11\text{-}3)$$

therefore $I X_{L_T} = I X_{L_1} + I X_{L_2} + I X_{L_3}$

From which it is obvious that the total reactance in the series circuit may be obtained from the equation

$$X_{L_T} = X_{L_1} + X_{L_2} + X_{L_3} \qquad (11\text{-}4)$$

This equation is similar to that for calculating the total resistance for resistances connected in series. Since $X_L = 2\pi f L$, Eq. 11.4 can be expressed as

$$2\pi f L_T = 2\pi f L_1 + 2\pi f L_2 + 2\pi f L_3$$

Therefore, when inductances are connected in series, the total inductance may be found by simple addition.

$$L_T = L_1 + L_2 + L_3 \qquad (11\text{-}5)$$

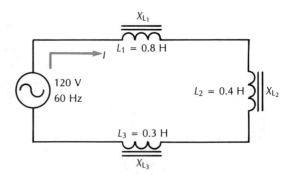

Figure 11-6 Inductances in series.

Example 11-1: Determine the following for the circuit shown in Fig. 11-6:

(a) L_T (b) X_{L_1} (c) X_{L_2} (d) X_{L_3} (e) X_{L_T} (f) I (g) V_{L_1} (h) V_{L_2} and (i) V_{L_3}.

$$\begin{aligned}
L_T &= L_1 + L_2 + L_3 \\
&= 0.8H + 0.4H + 0.3H \\
&= 1.5H
\end{aligned}$$

$$\begin{aligned}
X_{L_1} &= 2\pi f L_1 \\
&= 2 \times 3.14 \times 60 \times 0.8 \\
&= 301.59\ \Omega
\end{aligned}$$

$$\begin{aligned}
X_{L_2} &= 2\pi f L_2 \\
&= 2 \times 3.14 \times 60 \times 0.4 \\
&= 150.80\ \Omega
\end{aligned}$$

$$X_{L_3} = 2\pi f L_3$$
$$= 2 \times 3.14 \times 60 \times 0.3$$
$$= 113.10 \ \Omega$$

$$X_{L_T} = X_{L_1} + X_{L_2} + X_{L_3} = 2\pi f L_T$$
$$= 301.59 + 150.80 + 113.10$$
$$= 565.49 \ \Omega$$

$$I = \frac{V}{X_{L_T}}$$
$$= \frac{120}{565.49}$$
$$= 0.212 \ A$$

$$V_{L_1} = I X_{L_1}$$
$$= 0.212 \times 301.59$$
$$= 63.94 \ V$$

$$V_{L_2} = I X_{L_2}$$
$$= 0.212 \times 150.80$$
$$= 31.97 \ V$$

$$V_{L_3} = I X_{L_3}$$
$$= 0.212 \times 113.10$$
$$= 23.98 \ V$$

INDUCTANCES IN PARALLEL

In a parallel circuit, the voltage is the same across parallel branches and the net circuit current is equal to the sum of the branch currents. Therefore, in the circuit shown in Fig. 11-7

$$V = V_1 = V_2 = V_3 \tag{11-6}$$

and

$$I = I_1 + I_2 + I_3 \tag{11-7}$$

Using Ohm's law, we can write Eq. 11-7 as

$$\frac{V}{X_{L_T}} = \frac{V}{X_{L_1}} + \frac{V}{X_{L_2}} + \frac{V}{X_{L_3}}$$

It should now be obvious that the total inductive reactance for inductances connected in parallel may be calculated from the reciprocal equation.

$$\frac{1}{X_{L_T}} = \frac{1}{X_{L_1}} + \frac{1}{X_{L_2}} + \frac{1}{X_{L_3}} \tag{11-8}$$

This equation is similar to that used for calculating the total resistance when the resistances are in parallel. For two inductances in parallel, Eq. 11-8 simplifies to

$$X_{L_T} = \frac{X_{L_1} \times X_{L_2}}{X_{L_1} + X_{L_2}} \tag{11-9}$$

From Eq. 11-8, the total inductance for inductances in parallel will be

$$\frac{1}{L_T} = \frac{1}{L_1} + \frac{1}{L_2} + \frac{1}{L_3} \tag{11-10}$$

It should be apparent that the total circuit inductance and, hence, the total inductive reactance will reduce when the inductances are connected in parallel.

Example 11-2: Determine the following for the circuit shown in Fig. 11-7:

(a) L_T (b) X_{L_T} (c) I (d) I_1 (e) I_2 and (f) I_3

$$\frac{1}{L_T} = \frac{1}{L_1} + \frac{1}{L_2} + \frac{1}{L_3}$$
$$= \frac{1}{0.8} + \frac{1}{0.4} + \frac{1}{0.3} = \frac{17}{2.4}$$
$$L_T = \frac{2.4}{17}$$
$$= 0.14 \ H$$

The reactances for each of the three inductors were determined in Ex. 11-1.

$$\frac{1}{X_{L_T}} = \frac{1}{X_{L_1}} + \frac{1}{X_{L_2}} + \frac{1}{X_{L_3}}$$
$$= \frac{1}{301.59} + \frac{1}{150.80} + \frac{1}{113.10}$$
$$= 0.0033 + 0.0066 + 0.0088$$
$$= 0.0187$$

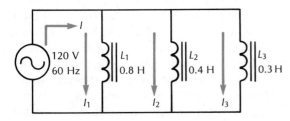

Figure 11-7 Inductances in parallel.

$$X_{L_T} = \frac{1}{0.0187}$$
$$= 53 \ \Omega$$

or

$$X_{L_T} = 2\pi f L_T$$
$$= 2 \times 3.14 \times 60 \times 0.14$$
$$= 53 \ \Omega$$

$$I_1 = \frac{V}{X_{L_1}}$$
$$= \frac{120}{301.59}$$
$$= 0.40 \ A$$

$$I_2 = \frac{V}{X_{L_2}}$$
$$= \frac{120}{150.80}$$
$$= 0.80 \ A$$

$$I_3 = \frac{V}{X_{L_3}}$$
$$= \frac{120}{113.10}$$
$$= 1.06 \ A$$

$$I = I_1 + I_2 + I_3$$
$$= 0.40 + 0.80 + 1.06$$
$$= 2.26 \ A$$

or

$$I = \frac{V}{X_{L_T}}$$
$$= \frac{120}{53}$$
$$= 2.26 \ A$$

Example 11-3: Determine the following for the circuit shown in Fig. 11-8:

(a) X_{L_T} (b) I (c) V_1 (d) V_2 (e) I_2 and I_3

$$X_{L_T} = X_{L_1} + \frac{X_{L_2} \times X_{L_3}}{X_{L_2} + X_{L_3}}$$
$$= 301.59 + \frac{150.80 \times 113.10}{150.80 + 113.10}$$
$$= 366.22 \ \Omega$$

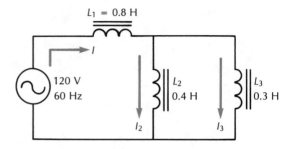

Figure 11-8 Inductances in series-parallel.

$$I = \frac{V}{X_{L_T}} = \frac{120}{366.22}$$
$$= 0.33 \ A$$

$$V_1 = I X_{L_1}$$
$$= 0.33 \times 301.59$$
$$= 99.52 \ V$$

$$V_2 = V - V_1$$
$$= 120 - 99.52$$
$$= 20.48 \ V$$

$$I_2 = \frac{V_2}{X_{L_2}}$$
$$= \frac{20.48}{150.80}$$
$$= 0.14 \ A$$

$$I_3 = I - I_2$$
$$= 0.33 - 0.14$$
$$= 0.19 \ A$$

PHASE RELATIONSHIP IN A PURELY INDUCTIVE CIRCUIT

In an inductive ac circuit, the current cannot increase immediately with the applied source voltage because it is opposed by the induced counter voltage in the inductor. In a purely inductive circuit, as shown in Fig. 11-9, when the applied current is passing through zero, such as at points a, c, and e, the instantaneous rate of change of the current is at its maximum. Hence, the induced counter voltage at these instances will be at its maximum. When the current passes through its peak, the rate of

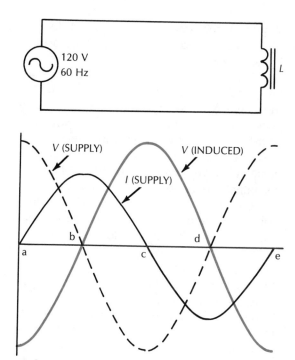

Figure 11-9 Voltage and current phase relation for a purely inductive circuit.

complete cycle, it will be found that the induced voltage is a sine wave which lags 90° behind the applied sine wave current.

In order for an applied current to flow through the circuit, there must be an applied voltage. This voltage, however, will be equal to but directly opposite the induced voltage, as shown in the graphs of Fig. 11-9. Therefore, the applied voltage and the induced voltage are 180° out-of-phase with each other. When the waves for the applied voltage and applied current are now compared, it is seen that the applied current lags the applied voltage by 90°. Therefore, in a purely inductive circuit, the circuit current lags the applied voltage by 90°. This relationship may be illustrated by a phasor diagram as shown in Fig. 11-10.

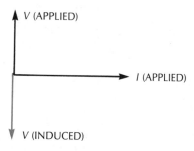

Figure 11-10 In a purely inductive circuit, the current lags the voltage by 90°.

POWER TAKEN BY A PURE INDUCTANCE

In previous chapters, it was shown that the power in watts taken by a resistive circuit is equal to the product of the voltage and the current. This is possible because in a purely resistive circuit the voltage and current are in-phase. As a result, the product of the current and voltage will always be positive, as shown in Fig. 11-11(a). This kind of positive power is often called *in-phase* or *true power* and it is measured in watts.

In a purely inductive ac circuit, the current lags the voltage by 90°. As a result, the power taken by an inductor is not always positive, as

change of the current is instantaneously zero. As a result, the induced counter voltage at points such as b and d will be instantaneously zero.

At point a the current is increasing from zero in a positive direction and at its maximum rate. At this instant, according to the laws of electromagnetic induction, the induced voltage will be at its negative maximum, as shown in the graph of Fig. 11-9. However, as the current increases towards its peak value, its rate of change decreases and the induced voltage will also decrease. At maximum current, point b, the rate of change of the current is now zero and so the resulting induced voltage will also reduce to zero. At point c the current is again increasing from zero at its maximum rate, but in the negative direction. Therefore, the induced voltage will now be at its positive maximum, as shown in the graph. If the induced wave pattern is developed for the

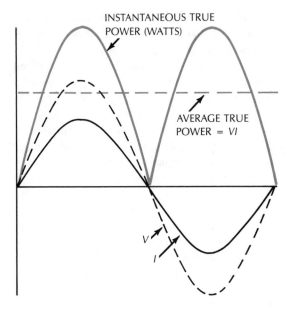

Figure 11-11(a) Power in a purely resistive circuit (watts).

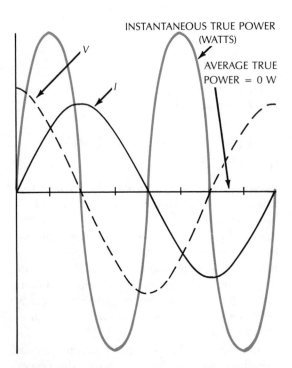

Figure 11-11(b) Power in a purely inductive circuit (VARs).

shown by the graphs in Fig. 11-11(b). In fact, the average power in watts taken by a pure inductor is zero.

In Fig. 11-11(b), the current and voltage are both positive between 0° and 90°. As a result, the power will be positive, as shown, at any instant between these two points. However, from 90° to 180°, the current and voltage are of opposite polarities, and this will result in a negative power pulse, as shown. When the instantaneous power is plotted for the entire cycle, a power wave, as shown in Fig. 11-11(b), will be obtained. The power wave shows that each voltage cycle will produce two positive and two negative power pulses. Since the positive power pulses are equal and opposite to the negative power pulses, the net power in watts taken by a pure inductor will be zero.

The student may recall from an earlier section that an inductor has the ability to store energy in its magnetic field during the period when the current is increasing, and then to release this energy back into the circuit at another period when the current is decreasing. Observe in Fig. 11-11(b) that when the current is increasing from zero to its peak value in either a positive or negative direction, the power is positive. Hence, the positive power pulses represent energy taken from the source by the inductance and stored in the inductor's magnetic field. Similarly, when the current is decreasing from its peak value to zero, the power pulses are negative. Therefore, the negative power pulses will represent the stored energy being returned back to the source. It is this release of stored energy which maintains the current in a direction that is opposite to that of the voltage.

INDUCTANCE REACTIVE POWER (VARS)

For a purely resistive circuit, the power in watts taken by the circuit is equal to the product of the voltage and current and, also, the power in watts taken by a pure inductance is equal to zero. However, since the product of the voltage and current for a pure inductance is not mathematically equal to zero, this product for a pure inductance must represent some different kind of power. The power associated with an inductance is a kind of "magnetic power," and it is called *reactive power*. Reactive power is measured in *VARs*, which is the abbreviation for volt-ampere-reactive.

Inductance *reactive power*
$$= V_L \times I_L \text{ VARs} \qquad (11\text{-}11)$$
$$= I_L^2 X_L \text{ VARs}$$

where V_L = voltage applied across the inductance

I_L = current flow through the inductance

and X_L = inductive reactance

The phasor diagrams for a purely resistive circuit and for a purely inductive circuit are shown in Fig. 11-12.

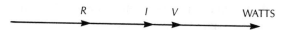

Figure 11-12(a) Phasor diagram for a purely resistive circuit.

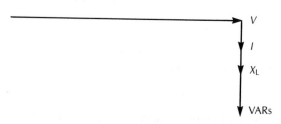

Figure 11-12(b) Phasor diagram for a purely inductive circuit.

THE EFFECTIVE RESISTANCE OF A COIL

In the same sense as it is true to say that all ac resistive circuits have at least some inductance, similarly all inductive circuits will have at least some resistance. The effective resistance in an inductor is due to the resistance of the coil's wire, and to the resistance caused by ac effects such as hysteresis and eddy currents.

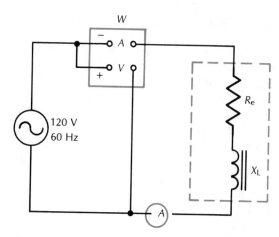

Figure 11-13 The effective resistance of a coil.

(Resistance from ac effects was discussed in the previous chapter.)

Since ac resistive effects increase with the increase in frequency, the effective resistance of an inductor will also increase when the frequency is increased. The effective resistance of a coil is considered to be in series with the inductance, as shown in Fig. 11-13. Since any true power (watts) in the circuit will be due only to circuit resistance, the effective resistance of the coil may be obtained from the equation.

$$R_e = \frac{W}{I^2} \qquad (11\text{-}12)$$

where R_e = effective resistance of the coil

W = watts used by the coil's resistance

I = current through the coil

THE Q-FACTOR OF A COIL

In some ac circuits, the resistance of a coil can be an undesirable property. For such a circuit, the coil may have to be specially designed in order to minimize its resistance. The ratio of a coil's reactance to its resistance is sometimes called the coil's Q-factor. Therefore, a coil with a high Q-factor will have a high reactance compared to its resistance.

$$Q\text{-factor of coil} = \frac{X_L}{R} \qquad (11\text{-}13)$$

R-L CIRCUIT IMPEDANCE

In a purely resistive circuit, any inductive effects are considered negligible. Similarly, in a purely inductive circuit any resistive effects are considered extremely small, and as a result they are omitted from any calculations. In most industrial circuits, however, the load is actually a combination of both resistance and inductance. That is, the circuit can no longer be treated as either purely resistive or as purely inductive. Hence, circuit equations will have to.be modified in order to account for both the resistive and inductive components of the circuit.

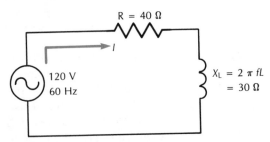

Figure 11-14 Resistance and inductance in series.

The total opposition to current flow in an ac circuit is called *impedance*. As shown by Fig. 11-14, this total opposition can be due to a combination of both resistance (R) and inductive reactance (X_L). The symbol for impedance is Z, and like resistance and reactance, it too is measured in ohms. From Ohm's law, the impedance of a circuit will be equal to the effective source voltage divided by the effective circuit current.

$$Z = \frac{V}{I} \tag{11-14}$$

It was previously shown that the current flowing through a pure resistance was in phase with the voltage across the resistance, and that the current through a pure inductance lagged the voltage across the inductance by 90°. Therefore, in the series circuit shown in Fig. 11-14, the circuit current will be in phase with the resistance voltage drop, V_R, and it will lag the inductance voltage drop, V_L, by 90°. This phase relationship is illustrated by the phasor diagram of Fig. 11-15. From Ohm's law, these two voltage drops may be calculated from the equations:

$$V_R = IR \tag{11-15}$$

and

$$V_L = IX_L \tag{11-16}$$

It is obvious from Fig. 11-15, that the voltage, V_R lags the voltage V_L by 90°. This phase difference must be taken into account when two such voltages are being added. Therefore, phasor addition must be used in adding the voltage drops in the loop equation for the circuit shown in Fig. 11-14.

$$V = V_R + V_L \text{ (phasor addition)} \tag{11-17}$$

Since V_R and V_L are exactly 90° out-of-phase, as shown in Fig. 11-15, this phasor addition is that of a right triangle. Hence,

$$V = \sqrt{(V_R)^2 + (V_L)^2} \tag{11-18}$$

This equation may be expanded to the form:

$$IZ = \sqrt{(IR)^2 + (IX_L)^2} \tag{11-19}$$

and cancelling I from each term gives the impedance equation:

$$Z = \sqrt{(R)^2 + (X_L)^2} \tag{11-20}$$

The impedance phasor triangle will be similar to the voltage phasor triangle. These two triangles are illustrated in Fig. 11-15. From the impedance equation (Eq. 11-19), it is obvious that the impedance, Z, of a series R-L circuit is equal to the phasor sum of the circuit's resistance, R, and its reactance, X_L. These two values, R and X_L will always be 90° out-of-phase, and hence their phasor addition is sometimes expressed as:

$$Z = R + jX_L \tag{11-21}$$

This is another method of expressing impedance. The "j" operator in X_L implies that X_L is

90° out-of-phase with R. This operator may be used to simplify many electrical calculations. However, it will not be used or discussed any further in this chapter or in any of the following chapters. A more thorough discussion of the "j" operator is given in the Appendix at the back of the text.

Example 11-4 If in the series R-L circuit of Fig. 11-14, $V = 120$ V, $R = 40$ Ω, and $X_L = 30$ Ω, find the following: (a) Z, (b) I, (c) V_R, (d) V_L, and (e) draw the phasor diagram.

The phasor diagram for this circuit is shown in Fig. 11-15.

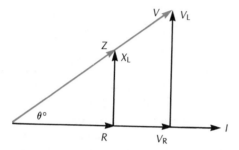

Figure 11-15 The impedance and voltage phasors for an R-L series circuit.

(a) $Z = R + X_L$ (phasor addition, R and X_L at right angle)

$$= \sqrt{(R)^2 + (X_L)^2}$$
$$= \sqrt{(40)^2 + (30)^2}$$
$$= 50 \text{ Ω}$$

(b) $I = \dfrac{V}{Z}$

$$= \dfrac{120}{50}$$
$$= 2.4 \text{ A}$$

(c) $V_R = I\,R$

$$= 2.4 \times 40$$
$$= 96 \text{ V}$$

(d) $V_L = I X_L$

$$= 2.4 \times 30$$
$$= 72 \text{ V}$$

Check: $V = V_R + V_L$ (phasor addition, V_R and V_L at right angle)

$$= \sqrt{(V_R)^2 + (V_L)^2}$$
$$= \sqrt{(96)^2 + (72)^2}$$
$$= 120 \text{ V (agrees with the given value)}$$

(e) The phasor diagram is given in Fig. 11-15.

APPARENT POWER (VOLT-AMPERE)

It is obvious that the total power for a R-L series circuit, as shown in Fig. 11-14, will contain both a true power (watts) component and a reactive power (VARs) component. This power combination in a circuit is called *apparent power* and it is measured in *volt-amperes* (VA). The apparent power for a circuit may be determined from the equation:

$$\text{apparent power (VA)} = VI = I^2Z \qquad (11\text{-}22)$$

Since the true power and reactive power components are 90° out-of-phase, as shown in Fig. 11-16, the circuit apparent power can also be obtained from the phasor addition of these two power components.

$$\begin{aligned}&\text{apparent power (VA)}\\ &= \sqrt{(\text{watts})^2 + (\text{VARs})^2}\end{aligned} \qquad (11\text{-}23)$$

The phasor triangle formed by the power phasors will also be similar to the voltage and impedance triangles for the circuit as shown in Fig. 11-16.

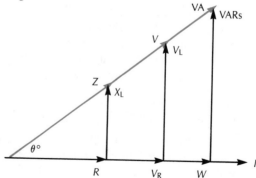

Figure 11-16 Power, voltage and impedance phasor triangles for the circuit shown in Fig. 11-14.

Example 11-5: Determine (a) the true power, (b) inductive reactive power, and (c) the apparent power for the circuit shown in Fig. 11-14.

(a) true power $= I^2 R$
$$= (2.4)^2 \times 40$$
$$= 230.4 \text{ W}$$

(b) inductive reactive power
$$= I^2 X_L = (2.4)^2 \times 30$$
$$= 172.8 \text{ VARs}$$

(c) apparent power $= VI$
$$= 120 \times 2.4$$
$$= 288 \text{ VA}$$

or apparent power $= I^2 Z$
$$= (2.4)^2 \times 50$$
$$= 288 \text{ VA}$$

or apparent power $= \sqrt{(\text{watts})^2 + (\text{VARs})^2}$
$$= \sqrt{(230.4)^2 + (172.8)^2}$$
$$= 288 \text{ VA}$$

POWER FACTOR AND PHASE ANGLE

You have seen that the circuit of Fig. 11-14 is neither purely resistive nor purely inductive. This is because it has both a resistive and an inductive component. Therefore, the circuit current will lag the source voltage by an angle, $\theta°$, which lies somewhere between 0° (purely resistive) and 90° (purely inductive), as shown in Fig. 11-16. This angle, $\theta°$, is commonly called the *phase or power factor angle* of the circuit, and the cosine function of this phase angle is termed the *power factor* of the circuit. Since the current is lagging the voltage, the power factor is more accurately described as a *lag power factor*.

$$\text{power factor} = \cos \theta° \qquad (11\text{-}24)$$

From Fig. 11-16, it is obvious that the circuit's power factor may be found from any of the relationships:

$$\text{P.F.} = \cos \theta°$$
$$= \frac{R}{Z}$$

or

$$= \frac{\text{true power (watts)}}{\text{apparent power VA}}$$

or

$$= \frac{V_R}{V} \qquad (11\text{-}25)$$

Of course, the phase angle, $\theta°$, can be found first by using one of the other trigonometric functions, and then the power factor can be determined by taking the cosine of this phase angle. For example:

$$\theta° = \tan^{-1} \frac{X_L}{R} \qquad (11\text{-}26)$$

The student may have noticed that most of the equations used in this chapter are actually extensions to the basic equations developed in previous chapters. Because of the phase difference between the circuit current and voltage, these basic equations are now modified to include the use of phasor addition of a right triangle, and the use of basic trigonometry. Consequently, the reader will find that the practice of first sketching a phasor diagram of a problem will invariably help in establishing many of the equations required for the solution of the question. For example, in the phasor diagram shown in Fig. 11-16, if the power factor, voltage, and circuit current are known, we can write the equations:

true power
$$= VI \cos \theta° \qquad (11\text{-}27)$$
$$= \text{apparent power} \times \text{power factor}$$

$$\text{reactive power} = VI \sin \theta° \qquad (11\text{-}28)$$

and others.

Example 11-6: An ac induction motor draws 1.2 A when it is operated on a 120-V supply. If the motor's power factor is 0.8 lag, determine: (a) the true power, (b) the reactive power, (c) the motor's impedance, (d) the effective resistance, and (e) the motor's inductive reactance. Since the motor has a lag power factor of 0.8, it will consume both watts and VARs. That is, the system is inductive with a phase angle:

$$\theta° = \cos^{-1} 0.8 = 36.9°$$

The phasor triangle can be easily sketched, as shown in Fig. 11-17. Since the current is constant throughout the circuit, the current phasor is used as the reference phasor.

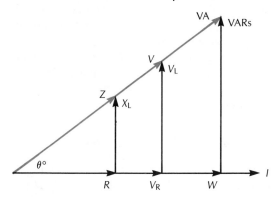

Figure 11-17 Sketch of a phasor diagram for Ex. 11-6.

(a) true power = $VI \cos \theta$.
$$= 120 \times 1.2 \times .8$$
$$= 115.2 \text{ W}$$

(b) Reactive power $(R) = VI \sin \theta$
$$= 120 \times 1.2 \times 0.6$$
$$= 86.4 \text{ VARs}$$

(c) impedance $(Z) = \dfrac{V}{I} = \dfrac{120}{1.2}$
$$= 100 \ \Omega$$

(d) effective resistance $(R) = Z \cos \theta$
$$= 100 \times 0.8$$
$$= 80 \ \Omega$$

(e) inductive reactance $= \sqrt{(Z)^2 - (R)^2}$
   or $\qquad\qquad\qquad = Z \sin \theta$
$$= 100 \times 0.6$$
$$= 60 \ \Omega$$

SUMMARY OF IMPORTANT POINTS

Principles of electromagnetism:
1. When the applied current to a coil is varied, a counter voltage is induced in the coil. This induced voltage always acts in such a direction as to oppose the change which is causing it (Lenz's law).

Inductance of an inductor or a circuit:
1. It is a form of electrical inertia.
2. It is the ability to store energy in a magnetic field.
3. The unit is the henry (H), which is the inductance of a component that will cause a current change of one ampere per second to induce a voltage of one volt.
4. Inductance in a dc circuit is noticeable only at the instances the circuit is turned **on** and turned **off**.

Inductive reactance in ac circuits:
1. Caused by the induced counter voltage which always acts such as to oppose the circuit current.
2. Inductive reactance $(X_L) = 2\pi f L$, and is measured in ohms.
3. In a purely inductive circuit the current lags the voltage by 90°. The applied voltage and induced voltage are 180° out-of-phase.
4. Pure inductive reactance does not use watts, but instead uses VARs.
5. Reactive power $= I^2 X_L$, and is measured in VARs.

The impedance of an R-L series circuit:
1. All circuits contain at least some resistance and some inductance.
2. Resistance and inductive reactance are 90° out-of-phase.
3. Total opposition is equal to the phasor sum of R and X_L, and it is called Impedance (Z)

$$Z = \sqrt{(R)^2 + (X_L)^2}$$
$$Z = \frac{V}{I}$$

4. The phasors of R, X_L, and Z form a right triangle.

The Q-factor of a coil:

1. Q-factor $= \dfrac{X_L}{R}$

2. A low Q coil will use more watts than a high Q coil.

3. The Q of a coil increases with increases in frequency.

Phase angle and power factor:

1. In a R-L circuit, the current lags the voltage by some angle, $\theta°$, which is somewhere between $0°$ (purely resistive) and $90°$ (purely inductive). This is the circuit's phase angle.

2. The circuit's power factor is the cosine of this phase angle.

$$P.F. = \cos \theta$$

3. An R-L circuit will use both watts and VARs. These two power components are $90°$ out-of-phase.

4. The total circuit power is the phasor sum of the watts and VARs' power components. This total power is called Apparent Power and it is measured in volt-amperes (VA).

$$\begin{aligned} &\text{apparent power}\\ &= VI\\ &= \sqrt{(\text{watts})^2 + (\text{VARs})^2} \end{aligned}$$

Phasor triangles of R-L series circuit:

1. The impedance, voltage, and power phasors form similar right triangles.

2. Since the circuit current is constant, the current phasor is used as the reference phasor.

3. Numerous circuit equations may be summarized from the phasor diagram. Some of these are:

(a) $P.F. = \cos \theta° = \dfrac{R}{Z} = \dfrac{\text{watts}}{\text{volt-amperes}}$

(b)   true power $= VI \cos \theta°$

(c)   reactive power $= VI \sin \theta°$

(d) $Z = \sqrt{(R)^2 + (Z)^2}$

(e) $V = \sqrt{(V_R)^2 + (V_L)^2}$

(f)   apparent power $= \sqrt{(\text{watts})^2 + (\text{VARs})^2}$

REVIEW QUESTIONS AND PROBLEMS

1. Define:
   (a) Faraday's law of electromagnetic induction and
   (b) Lenz's law

2. Describe the term "inductance of a coil" and state the factors which determine this property of a coil.

3. Draw graphs to show (a) when the effects of inductance may be noticeable in a dc circuit, and (b) the effects of this inductance.

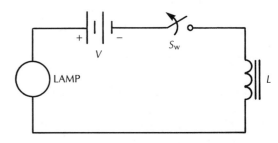

Figure 11-18 Inductance in a dc circuit.

4. Explain why the lamp in a circuit as shown in Fig. 11-18 may be observed to light up dimly at the instant the switch is closed and then to increase gradually to its maximum brightness.

5. Compare and comment on the brightness of the lamp in Fig. 11-18 with its brightness if the source were to be replaced with an ac voltage of the same effective value.

6. Assume the source in Fig. 11-18 is ac. What effect will (a) an increase in source frequency and (b) removal of the coil's iron core have on the circuit's current?

7. Name six devices which use the property of inductance.

8. Describe (a) mutual induction and (b) name a device which uses this circuit property.

9. Define (a) inductive reactance and (b) state two factors on which this property depend.

10. A circuit contains two inductors, $L_1 = 0.4$ H and $L_2 = 0.8$ H, connected in

series and across a 120-V, 60-Hz source. Determine (a) the total inductance and (b) the total inductive reactance.

11. If the two coils in Q. 10 were connected in parallel, find (a) the total inductance and (b) the total inductive reactance.

12. A circuit contains three inductors, $L_1 = 0.2$ H, $L_2 = 0.6$ H and $L_3 = 1.8$ H, connected in series across a 120-V, 60-Hz supply. Determine (a) the total inductance and (b) the total inductive reactance.

13. If the three coils in Q. 12 were connected in parallel, find (a) the total inductance, and (b) the total inductive reactance.

14. A circuit contains two inductive reactances $X_{L_1} = 40$ Ω, and $X_{L_2} = 30$ Ω, connected in series across a 120-V, 60-Hz supply. Determine (a) the total circuit reactance, (b) the circuit current, and (c) the voltage drop across each reactor.

15. If the two reactances in Q. 14 were connected in parallel, find (a) the total reactance, (b) the current through each reactor, and (c) the total circuit current.

16. Draw the ac waveforms to show the relationship between (a) the applied voltage, (b) the induced voltage, and (c) the current flow through a pure inductor.

17. Draw an ac current waveform, and indicate on it where its instantaneous rate of change is at (a) a maximum (b) a minimum.

18. If a current with a waveform as in Q. 17 is applied to a pure inductor, indicate on the graph where the induced counter voltage will be instantaneously at (a) a positive maximum, (b) a negative maximum, and (c) zero.

19. Draw the power waveforms and describe the power properties for (a) a purely resistive ac circuit and (b) a purely inductive ac circuit.

20. A series R-L circuit has a resistance of 25 Ω and an inductive reactance of 20 Ω. Find (a) the circuit impedance and (b) draw and label the phasor impedance triangle.

21. If an inductor which contains a resistance of 12 Ω and an inductance of 1.6 H is con-

nected across a 120-V dc supply, determine
(a) the current flow through the inductor and
(b) the circuit's true power

22. If the coil in Q. 21 was connected across a 120-V, 60-Hz ac supply, find:
(a) the circuit's impedance, (b) the circuit's current, and (c) the circuit's true power.

23. A series R-L circuit has a resistance of 30 Ω and an inductive reactance of 40 Ω. If the circuit is supplied from a 240-V, 60-Hz source, determine (a) Z, (b) I (c) V_R, and (d) V_L.

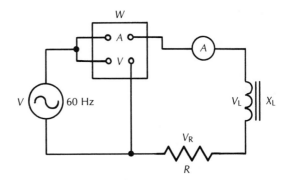

Figure 11-19 R-L series circuit.

24. In the circuit shown in Fig. 11-19, $V = 120$ V, 60 Hz, $I = 1.2$ A and the true power $= 80$ W. Determine: (a) Z, (b) R, (c) X_L and L.

25. For the circuit shown in Fig. 11-19, if $V_R = 80$ V, $V_L = 60$ V, and the true power $= 320$ W, determine: (a) V, (b) I, (c) R, (d) X_L, and (e) Z.

26. For the circuit given in Q. 25, determine (a) reactive power, (b) apparent power, and (c) the circuit's power factor.

27. Draw the phasor diagram for Q. 26, to show (a) the impedance triangle, (b) the voltage triangle, and (c) the power triangle.

28. An ac motor draws 4 A when it is operated at 240 V, 60 Hz and with a lag power factor of 80 percent. Determine: (a) the true power, (b) the reactive power, (c) the

apparent power, and (d) the circuit's phase angle.

29. A system has a resistance, $R = 50\ \Omega$, and an inductive reactance, $X_L = 30\ \Omega$, connected in series across a supply of 240 V, 60 Hz. Determine: (a) Z, (b) I, (c) P.F., (d) V_R, (e) V_L, and (f) the phase angle.

30. In the circuit of Q. 29, determine (a) the true power, (b) the reactive power, (c) the apparent power, and (d) draw the phasor diagram to show the circuit current, voltages, power, and impedance.

31. If a motor is rated at 5 kVA, 200 V, and 0.7 lag power factor, determine (a) the rated line current (b) the true power, (c) the reactive power, (d) the effective resistance, and (e) the inductive reactance of the motor.

CAPACITORS AND CAPACITIVE REACTANCE

THE CAPACITOR

Capacitors, also called condensers, may be found in as many arrangements in electric circuits as resistors. They are used in numerous electric and electronic circuits. There are different types of capicitors and they are made in various sizes and shapes. Capacitors are basically ac circuit devices. However, there are some which are specially designed for dc cir-

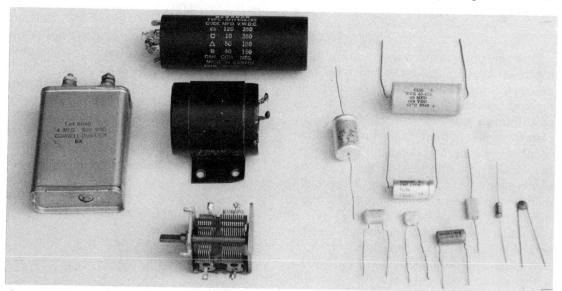

Figure 12-1 Types of capacitors. *Courtesy: John Dubiel.*

cuit applications, and these are called electrolytic capacitors.

A capacitor is formed whenever two conductors are separated by an insulating material. The conductors are called the *plates* and the insulating material is called the *dielectric*. Some commonly used dielectric materials in capacitors are paper, mica, bakelite, glass, and air.

A capacitor has the ability to store an electric charge. It will store energy in an electrostatic field. This property of a capacitor is called its capicitance, and the unit for capacitance is the *farad*, named in honour of the scientist Michael Faraday. In practical applications, a capacitance of one farad is physically very large. Therefore, capacitors in common use have values expressed in microfarads (μF), micro-micro farads ($\mu\mu$F) or picofarads pF).

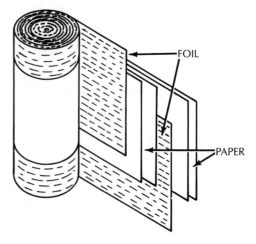

Figure 12-3 Paper and foil capacitor.

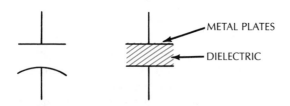

Figure 12-2 A simple capacitor.

One type of capacitor is shown in Fig. 12-2. It consists of two metal plates separated by a thin dielectric such as mica or a ceramic material and housed in a plastic case. Two leads pass through the case and connect to the plates. In some of these capacitors, a number of plates are used. The plates are placed between sheets of dielectric with alternate plates connected together. This arrangement increases the plate area, and there is a resultant increase in capacitance.

Another type of capacitor is the tabular type shown in Fig. 12-3. This capacitor is made of two strips of tin or aluminum foil separated by a strip of waxed paper or plastic. The assembly is then rolled up and encased in a metal or cardboard cylinder. This arrangement results

in a capacitor of small physical size but comparatively large capacitance.

The variable air capacitor is commonly used in tuning radio and TV receivers, where it is necessary to vary the capacitance in extremely low values. The assembly consists of one set of plates which can be rotated between a set of fixed plates. Rotating the moveable plates will vary the effective capacitor plate area and, hence, produce a change in capacitance. Note that the dielectric, which separates the plates, is air.

Another variable type of capacitor is the *trimmer capacitor*. The assembly contains two metal plates and an insulated screw which is used to vary the distance between the plates. When the thickness of the dielectric is increased, the capacitance of the capacitor decreases, and vice versa. The symbol for a variable capacitor is shown in Fig. 12-4.

Capacitors are basically ac devices; however *the electrolytic type of capacitor* is designed solely for use in dc applications. This type of capacitor consists of an anode or posi-

Figure 12-4 Symbol for a variable capacitor.

tive plate and a cathode or negative plate. The anode is an aluminum foil which is immersed in an electrolyte solution or surrounded by a gauze soaked with the electrolyte. A second aluminum foil, which is separated from the anode plate by the electrolyte, forms the cathode or negative plate. When this capacitor is energized from a dc source, an oxide film develops on the positive foil. This thin oxide film serves as the dielectric. Because this dielectric is very thin, the capacitance value of such a capacitor can be relatively high compared to its physical size.

It is important to always connect the negative plate of an electrolytic capacitor to the negative of the dc circuit. An incorrect connection may permanently damage the capacitor and an excessive reverse voltage may even cause some electrolytics to explode dangerously into pieces. The aluminum electrolytic capacitor is used in applications such as in power supply filter circuits, coupling and decoupling circuits in audio and television systems, and in timing and delay circuits.

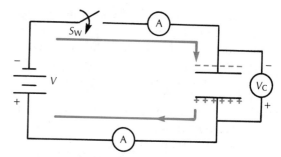

Figure 12-5(a) Charging a capacitor.

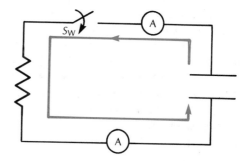

Figure 12-5(b) Discharging a capacitor.

CAPACITANCE AND ENERGY STORAGE

When a capacitor is connected to a dc source, as shown in Fig. 12-5, a current will be observed to flow momentarily and then decrease to zero. During this interval, there is a movement of electrons from the negative of the source to one of the capacitor's plates as shown and, at the same time, electrons are removed from the other plate and flow back to the positive of the supply voltage. This movement of electrons constitutes a current flow through the circuit, even though there is actually no electron flow through the dielectric of the capacitor. This current flow is called the *charging current*. The buildup of charges across the plates sets up a voltage (V_C) which opposes the source voltage. Therefore, as the capicitor voltage builds up, the charging current will decrease simultaneously. When the capacitor's voltage equals the source voltage,

the charging current will have been reduced to zero.

The energy stored by a capacitor in a circuit, as shown in Fig. 12-5(a), can be illustrated by carefully removing the capacitor from the circuit and connecting it to a low resistance, as shown in Fig. 12-5(b). The instant the switch is closed, the ammeter will indicate a momentary current flow through the circuit but in a direction opposite to the charging current. In this circuit, the electrons stored on the negative capacitor plate will now move to distribute themselves on the positive capacitor plate. This movement of electrons results from the energy stored by the capacitor and is called the *discharging current*.

It should be noted that a capacitor will store a charge for a relatively long period of time. As a result, a voltage can be present across a capacitor in a circuit even though the circuit itself is switched off. Hence, to reduce the possibility of a shock, it is good practice to discharge a capacitor before it is handled. To discharge a capacitor, its two terminals are simply shorted together with a piece of wire.

It was mentioned earlier that the property of a capacitor to store a charge is called its capacitance, and the unit of measurement is the farad (F). A capacitor has a capacitance of one farad when an applied voltage of one volt across its plates results in a charge buildup of one coulomb. The farad is an extremely large electrical unit, and the more practical units for the measurement of capicitance in a circuit is the microfarad and picofarad.

The charge that a capacitor will store is directly proportional to both its capacitance and the applied voltage. Therefore, the charge buildup at a capacitor plate will double if either the capacitance or the charging voltage is doubled. The charge stored by a capacitor is given by the equation:

$$Q = C \times V \qquad (12\text{-}1)$$

where Q = the charge in coulombs
　　　C = the capacitance in farads
　　　V = the charging voltage in volts

The capacitance of a capacitor can be increased by: (a) increasing the plate area, (b) moving the two plates closer together with a resultant decrease in the thickness of the dielectric, and (c) using a dielectric with a higher dielectric constant.

THE DIELECTRIC OF A CAPACITOR

The *dielectric constant* of a material measures its effectiveness when used as the dielectric of a capacitor. Air is assumed to have a dielectric constant of 1, and all other dielectrics are compared to this standard. For example, if the air dielectric of an air capacitor is replaced with a mica dielectric and the capacitance increases seven times, then the dielectric constant for mica will be 7. Below is a table of constants for a few commonly used dielectrics.

A capacitor ratings include: (a) its capacitance and (b) the maximum voltage which may

MATERIAL	DIELECTRIC CONSTANT (K)
air	1.0
bakelite	4.0 to 10.0
lucite	2.0 to 3.0
mica	6.0 to 7.0
paper	2.0 to 3.0
insulating oils	2.0 to 5.0
paraffin	1.5 to 2.0

be safely applied across the plates. If the voltage across a capacitor's plates becomes too high, the dielectric may rupture. This will result in permanent damage to the capacitor. The voltage rating of a capacitor, therefore, depends on the insulating strength of its dielectric. The insulating strength of a material is called its *dielectric strength*. Dielectric strength is not to be confused with dielectric constant, as these terms are entirely different. The dielectric strength of an insulator is described by the voltage which is required to break down a piece of this material having a thickness of 1 cm.

DIRECT-CURRENT *R-C* CIRCUITS

Figure 12-6(a) shows a resistance and a capacitance connected in series to a battery. It was shown that the capacitor will charge and acquire the voltage, V_C. However, the capacitor will not charge instantly. The rate at which the capacitor charges depends on the circuit resistance, R. A high resistance will slow down the movement of electrons to the capacitor and, hence, the capacitor voltage, V_C, will build up gradually.

From Kirchhoff's voltage law:

$$V = V_R + V_C \qquad (12\text{-}2)$$

At the instant the switch is closed in Fig. 12-6, the voltage across the capacitor V_C is zero, because at this instant there will be no charge build-up at the capacitor. Therefore, at this instant;

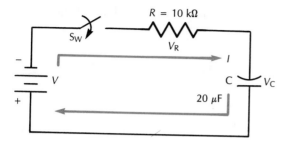

Figure 12-6(a) Charging a capacitor.

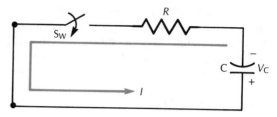

Figure 12-7(a) Capacitor discharging.

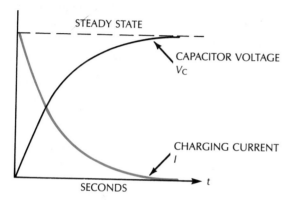

Figure 12-6(b) Voltage and current curves.

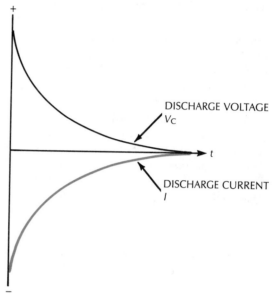

$$V_C = 0, \; V = V_R, \; \text{and} \; I = \frac{V}{R} \qquad (12\text{-}3)$$

However, as the capacitor charges, V_C will increase, and since V_C opposes the source voltage, the circuit current will decrease. That is, since Kirchhoff's law applies at any instant, as the capacitor voltage V_C increases, the resistance voltage drop, V_R, will decrease and the resultant electron flow in the circuit will also decrease. Figure 12-6(b) shows the typical graphs for the charging current and voltage of a capacitor when it is connected to a dc supply. It is obvious that when the capacitor voltage equals the supply voltage, the circuit current will have decreased to zero. Hence, with the exception of the temporary charging current, a capacitor can be said to block the flow of dc.

Figure 12-7(a) shows a charged capacitor being discharged through a resistance R. It was shown that the discharging current will

Figure 12-7(b) Discharging current and voltage curves.

flow in the opposite direction to the charging current. When the switch in Fig. 12-7(a) is closed, the stored charge in the capacitor does not decrease instantaneously to zero but, rather, the charge is distributed from the negative to the positive plate at a rate dependent on the circuit's resistance. A capacitor will discharge more slowly through a circuit with a high resistance then through one with a low resistance.

From Kirchhoff's voltage law, V_C will be equal to the V_R at any instant during discharge. At the instant the switch is closed, V_C will be at its maximum, and hence V_R and the discharge current will also be at their maximum. However, as the capacitor loses its charge, V_C will decrease and, therefore, the resultant circuit current and resistance voltage drop will also decrease. Figure 12-7(b) shows the typical voltage and current graphs for a discharging capacitor.

TIME CONSTANT

It is easy to demonstrate that the speed with which a capacitor charges in an R-C circuit is a function of the electrical size of both R and C. For example, in Fig. 12-6, if the time is measured for the capacitor to charge to approximately equal the supply voltage for various values of R and C, it will be noticed that the charging time increases when either R or C or both R and C are increased.

The time constant of an R-C circuit is defined as the time, in seconds, required for the capacitor to charge to a value of 63.2 percent of the source voltage. The value of this time constant is simply the product of R and C.

$$\text{time constant (seconds)} = R \times C \quad (12\text{-}4)$$

where R = circuit resistance in ohms
$\quad\quad C$ = capacitance in farads

Example 12-1: (a) Find the time constant for the values of R and C given in Fig. 12-6.

$$\begin{aligned}
\text{time constant} &= RC \\
&= 10\ 000 \times (20 \times 10^{-6})\text{F} \\
&= 0.2\ \text{s}
\end{aligned}$$

This means that at the end of 0.2 s (one time constant) the capacitor will have charged to 0.632 of the source voltage.

(b) Find the capacitor voltage at the end of the first time constant.

$$\begin{aligned}
V_C \text{ (at 1 time constant)} &= 0.632 \times V \\
&= 0.632 \times 120 \\
&= 75.84\ \text{V}
\end{aligned}$$

As the capacitor's voltage increases from 0 to 75.84 V in one time constant, the resistance voltage will decrease from 120 V to 44.16 V.

(c) Find the resistance voltage at the end of the first time constant.

$$\begin{aligned}
V &= V_R + V_C \\
V_R \text{ (at 1 time constant)} \\
&= 120\ \text{V} - V_C \text{ (at 1 time constant)} \\
&= 120\ \text{V} - 75.84\ \text{V} \\
&= 44.16\ \text{V}
\end{aligned}$$

The voltage of 44.16 is 36.8 percent of the source voltage. Therefore, during the first time constant, the capacitor's voltage increases to 0.632 of the source voltage, and the resistance voltage decreases to 0.368 of the source voltage. These values are shown in the charging curves of Fig. 12-8.

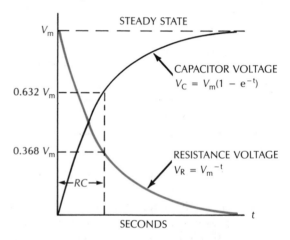

Figure 12-8 Time constant (charging circuit).

The time constant for a discharging R-C circuit is the same as for a charging circuit for the same values of R and C. That is, the time constant for a discharging circuit is equal to the time required for the capacitor to decrease its voltage by 63.2 percent of its charged value. This time constant is again simply equal to the product of R and C.

Example 12-2: In the discharging circuit of Fig. 12-7 (a), assume R = 10 000 Ω, C = 20 μF, and that the capacitor is charged to 120 V. Find (a) the circuit's time constant, and (b) the capacitor's voltage at the end of the first time constant,

where time constant

$$= R \times C$$
$$= 10\ 000\ \Omega \times (20 \times 10^{-6})F$$
$$= 0.2\ s$$

The capacitor's voltage decreases by 63.2 percent of its charged value during the first time constant. Therefore, the capacitor will be left with 36.8 percent of its charged value by the end of the first time constant.

$$V_C \text{ (at end of first time constant)}$$
$$= 0.368 \times 120$$
$$= 44.16\ V$$

This means that the capacitor will lose 75.84 V of its 120 V during the first time constant. These values are illustrated in the discharge curve of Fig. 12-9.

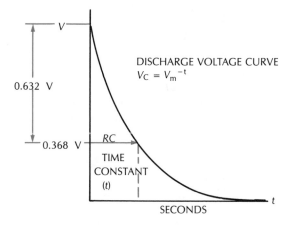

Figure 12-9 Time constant (discharging circuit).

The student must realize that because the rate of charging of a capacitor decreases as the capacitor charges, the capacitor's voltage will never charge up to equal exactly the value of the supply voltage. Similarly, the discharge voltage will never decrease to exactly zero. However, experiments have shown that after five time constants, the capacitor is essentially fully charged. Similarly, for practical purposes,

the capacitor is considered to be fully discharged after five time constants.

EXPONENTIAL EQUATIONS

Because of the resistance in the R-C circuit, the rate of increase in the capacitor voltage decreases as the capacitor charges. Similarly, the rate of decrease in the capacitor voltage decreases as the capacitor discharges. This decrease in rate is the reason for the curved type of graphs obtained during charging and discharging of the capacitor. Curved graphs such as these are called *exponential curves*, and they can be described by mathematical equations in a similar way as it is possible to describe straight line graphs. The equation for an exponential curve is called an exponential equation, and it can be easily recognized because it will always contain, in part, the constant, e. The constant, e, is an important constant like the constant, π.

$$e = 2.718 \text{ (constant)} \tag{12-5}$$

The exponential equations for these charging curves are shown in Fig. 12-8. The capacitor's voltage at any instant during charging is given by the equation:

$$V_C = V\ (1 - e^{-t}) \tag{12-6}$$

where V_C = capacitor voltage at (t) time constants
  V = source voltage
  $e = 2.718$
  and t = number of time constants

This equation describes an exponential curve which begins at the value $V_C = 0$, at time $t = 0$, and increases to the value $V_C = V$ at time $t = \infty$. Recall that for most practical applications, V_C is assumed equal to V after five time constants.

As the capacitor's voltage increases exponentially, the resistance voltage and, hence, the charging current, decreases exponentially. The charging current at any instant is given by the equation:

$$I_C = \frac{V_R}{R} = \frac{V}{R} (e^{-t}) \qquad (12\text{-}7)$$

where I_C = charging current at (t) time constants

V_R = resistance voltage at (t) time constants

V = source voltage

R = circuit's resistance

e = 2.718

and t = number of time constants

This equation describes an exponential curve which begins at its maximum value ($I_C = \frac{V}{R}$) at time $t = 0$, and decreases to zero ($I_C = 0$) at time $t = \alpha$. For most practical applications, I_C is assumed equal to zero after five time constants.

Example 12-3: A series R–C circuit has a resistance of 80 kΩ, a capacitance of 40 μF, and a dc supply voltage of 240 V. Determine (a) the capacitor's voltage and (b) the charging current, 6.4 seconds after the circuit is energized.

time constant = $R \times C$
= 80 000 $\Omega \times$ (40 x 10^{-6})F
= 3.2 s

number of time constants (t) = $\frac{6.4}{3.2}$ = 2

V_C (at $t = 2$) = $V (1 - e^{-t})$
= 240 $(1 - e^{-2})$
= 240 (0.865)
= 207.52 V

I_C (at $t = 2$) = $\frac{V}{R}(e^{-t})$

= $\frac{240}{80\ K}(e^{-2})$

= $\frac{240}{80\ K} (0.135)$

= 0.41 mA

Example 12-4: A 15-μF capacitor, which is charged to 150 V, is connected for discharge through a 200-kΩ resistance. Determine (a)

the capacitor voltage and (b) the discharge current, 4.5 seconds after the discharging switch is closed.

time constant = $R \times C$
= 200 000 $\Omega \times$ (15 $\times$ 10^{-6})F
= 3 s

number of time constants (t) = $\frac{4.5}{3}$ = 1.5

V_C will decrease as the capacitor discharges. Therefore, the exponential equation will be:

$V_C = V (e^{-t})$
= 150 $(e^{-1.5})$
= 150 (0.223)
= 33.45 V

The discharge current will also decrease as the capacitor discharges and $V_C = V_R$ at any instant. Therefore, the current equation will be:

$I_C = \frac{V}{R} (e^{-t})$

= $\frac{150\ V}{200\ 000\ \Omega} (e^{-1.5})$

= $\frac{150\ V}{200\ 000\ \Omega\ (0.223)}$

= 0.17 mA

or

$I_C = \frac{V_C}{R}$

= $\frac{33.45\ V}{200\ 000\ \Omega}$

= 0.17 mA

R-L TIME CONSTANT

This is perhaps an appropriate time to look also at the time constant of an R-L series circuit. The effects of inductance in a circuit were discussed in the previous chapter. It was shown that an inductor was basically a short circuit when connected to a supply as in Fig. 12-10 (a). This is opposite to the effect of a capacitor under the same conditions, because the capacitor will act basically as an open

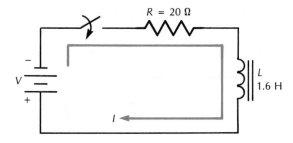

Figure 12-10(a) *R-L* circuit.

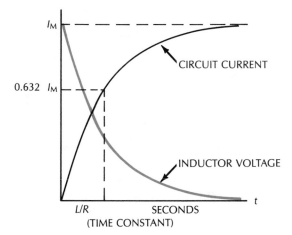

Figure 12-10(b) *R-L* time constant.

circuit. However, the current in the *R-L* circuit does not rise instantaneously to the maximum circuit value, but instead it rises gradually, as shown in Fig. 12-10 (b). This effect is due to the induced voltage set up across the inductor while the circuit current is increasing. The time required for the circuit current to reach 63.2 percent of the steady state circuit value is called the *circuit's time constant*. This time in seconds for an *R-L* series circuit is simply equal to L/R.

$$\text{time constant} = L/R \text{ s} \tag{12-8}$$

where L = inductance in H
and R = resistance in Ω

The graphs for the initial circuit current and the initial inductor voltage are exponential type curves as shown in Fig. 12-10 (b). They are described by the equations:

$$I = \frac{V}{R}(1 - e^{-t}) \tag{12-9}$$

$$V_L = V(e^{-t}) \tag{12-10}$$

where I = circuit current at (t) time constants
V = source voltage
V_L = induced inductor voltage
R = circuit's resistance
e = 2.718
and t = number of time constants

CAPACITORS IN AC CIRCUITS

In the previous sections, it was shown that when a dc voltage is applied across a capacitance, there is a momentary current flow through the circuit. The current decreases rapidly to zero as the capacitor charges and its voltage builds up to equal the applied voltage. The current will now remain at zero when the capacitor voltage equals the applied voltage and the applied voltage is constant.

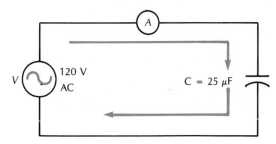

Figure 12-11 An ac capacitive circuit.

When a capacitance is connected across an ac supply as shown in Fig. 12-11, the capacitor will continually charge and discharge as the ac voltage changes direction. Therefore, the meter will indicate a constant ac current flow through the circuit. Note that there is no actual electron flow through the dielectric of the capacitor; however, there will be electron flow to and from the plates as the capacitor charges and discharges during each cycle.

It can now be concluded that whereas a capacitor will basically block the flow of dc, it will allow the flow of ac.

CAPACITIVE REACTANCE

As the charges build up across a capacitor's plates, the capacitor's voltage increases, and this voltage opposes the current flow in the circuit. Therefore, the capacitor voltage may be considered as a kind of counter voltage just like the induced counter voltage across an inductance. The current-limiting effect of an inductor in an ac circuit is called *inductive reactance*, and likewise, the current-limiting effect of a capacitance in an ac circuit is called *capacitive reactance*. The symbol for capacitive reactance is X_C, and the unit is the ohm, which is the same for resistance and inductive reactance.

The reactance of a capacitor can be found from the equation:

$$X_C = \frac{1}{2\pi fC} \qquad (12\text{-}11)$$

where f = supply frequency in hertz (Hz)
$\quad C$ = capacitance in farads (F)
$\quad X_C$ = capacitive reactance in ohms (Ω)

Equation 12-11 indicates that the capacitive reactance of a capacitor is inversely proportional to its capacitance and to the applied source frequency. That is, X_C will decrease if either the capacitance and or the frequency is increased.

When the capacitance is increased, the capacitor will take more ac charge ($Q = CV$) from the source in the same time. Therefore, the rate of ac charge flow in the circuit and, hence, the circuit current flow will increase. Since an increase in capacitance causes an increase in current for the same voltage and frequency, this means that the reactance must decrease when the capacitance is increased.

If the supply frequency is now increased, this will cause the same ac charge to flow, but in a shorter time. This means that the rate of charge flow is increased. Hence, an increase in frequency will also cause an increase in current for the same voltage and capacitance. Therefore, an increase in the frequency will

result in a decrease in the reactance of the capacitor, which is the same effect as when the capacitance itself is increased.

CAPACITANCE IN SERIES

When capacitors are connected in series, the charge which moves to each capacitor will be the same because there is only one current path. Therefore, for capacitors connected in series, as in Fig. 12-12,

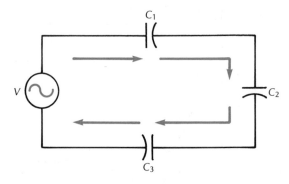

Figure 12-12 Capacitors connected in series.

$$Q_T = Q_1 = Q_2 = Q_3 \qquad (12\text{-}12)$$

where Q is the charge stored by the capacitor.

From Kirchhoff's voltage law, we can derive that ($V = V_1 + V_2 + V_3$) and from Eq. 12-1, ($Q = CV$) we obtain:

$$\frac{Q_T}{C_T} = \frac{Q_1}{C_1} + \frac{Q_2}{C_2} + \frac{Q_3}{C_3} \qquad (12\text{-}13)$$

Since Q is constant, it follows that the total capacitance for capacitances in series is given by the equation:

$$\frac{1}{C_T} = \frac{1}{C_1} + \frac{1}{C_2} + \frac{1}{C_3} \qquad (12\text{-}14)$$

This equation is similar to the equation used for the total resistance of a purely resistive parallel circuit. It should be obvious from Eq. 12-14 that when capacitances are connected in series the total capacitance will decrease. This effect from combining capacitances in series is similar to that of increasing the thickness of the dielectric of a capicitor.

The total capacitive reactance for capacitances connected in series is readily obtained from Eq. 12-14.

$$\frac{1}{2\pi f C_T} = \frac{1}{2\pi f C_1} + \frac{1}{2\pi f C_2} + \frac{1}{2\pi f C_3} \quad (12\text{-}15)$$

$$X_{C_T} = X_{C_1} + X_{C_2} + X_{C_3} \quad (12\text{-}16)$$

Note that Eq. 12-16 is similar to the equation used for combining resistances in series. When reactances are connected in series, the total reactance increases. This is logical, because when capacitors are connected in series, the total capacitance decreases and, hence, from Eq. 12-11, the total reactance will increase.

Example 12-5: Assume in Fig. 12-12, $C_1 = 20\,\mu F$, $C_2 = 30\,\mu F$, $C_3 = 40\,\mu F$, and $V = 120$ V, 60 Hz. Determine (a) the reactance of each capacitor, (b) the total capacitance, (c) the total reactance, and (d) the circuit current.

(a) $X_{C_1} = \dfrac{1}{2\pi f C_1}$

$\quad = \dfrac{10^6}{2 \times 3.14 \times 60 \times 20}$

$\quad = 132.70\ \Omega$

$\quad X_{C_2} = \dfrac{1}{2\pi f C_2}$

$\quad = \dfrac{10^6}{2 \times 3.14 \times 60 \times 30}$

$\quad = 88.46\ \Omega$

$\quad X_{C_3} = \dfrac{1}{2\pi f C_3}$

$\quad = \dfrac{10^6}{2 \times 3.14 \times 60 \times 40}$

$\quad = 66.35\ \Omega$

(b) $\dfrac{1}{C_T} = \dfrac{1}{C_1} + \dfrac{1}{C_2} + \dfrac{1}{C_3}$

$\quad = \dfrac{1}{20} + \dfrac{1}{30} + \dfrac{1}{40}$

$\quad = \dfrac{13}{120}$

$\quad C_T = 9.23\ \mu F$

(c) $X_{C_T} = X_{C_1} + X_{C_2} + X_{C_3}$

$\quad = 132.70 + 88.46 + 66.35$

$\quad = 287.51\ \Omega$

or

$\quad X_{C_T} = \dfrac{1}{2\pi f C_T}$

$\quad = \dfrac{10^6}{2 \times 3.14 \times 60 \times 9.23}$

$\quad = 287.51\ \Omega$

(d) $I = \dfrac{V}{X_{C_T}}$

$\quad = \dfrac{120}{287.51}$

$\quad = 0.42$ A

CAPACITANCES IN PARALLEL

When capacitors are connected in parallel, as in Fig. 12-13, the charge which moves to each capacitance is independent of each other because each capacitor forms a separate loop across the supply voltage. Therefore, the total charge stored by capacitors in parallel is given by the equation:

$$Q_T = Q_1 + Q_2 + Q_3 \quad (12\text{-}17)$$

From Kirchhoff's voltage law

$$(V = V_1 = V_2 = V_3),$$

and from Eq. 12-1 ($Q = CV$), we obtain:

$$C_T V = C_1 V + C_2 V + C_3 V \quad (12\text{-}18)$$

Since V is constant, it follows from Eq. 12-18 that the total capacitance for capacitors in parallel is given by the equation:

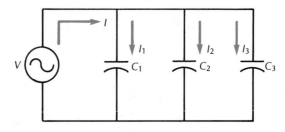

Figure 12-13 Capacitances in parallel.

$$C_T = C_1 + C_2 + C_3 \qquad (12\text{-}19)$$

This equation is similar to the equation used for finding the total resistance of a purely resistive series circuit. It should also be obvious from Eq. 12-19 that when capacitors are connected in parallel, the total circuit capacitance will increase. Therefore, the effect of connecting capacitors in parallel is similar to that of increasing the plate area of a capacitor.

The total capacitive reactance for capacitances connected in parallel is readily obtained from Eq. 12-11 and Eq. 12-14:

$$2\pi f C_T = 2\pi f C_1 + 2\pi f C_2 + 2\pi f C_3 \qquad (12\text{-}20)$$

Therefore

$$\frac{1}{X_{C_T}} = \frac{1}{X_{C_1}} + \frac{1}{X_{C_2}} + \frac{1}{X_{C_3}} \qquad (12\text{-}21)$$

Note that Eq. 12.21 is similar to the equation used for combining resistances in parallel. When reactances are connected in parallel, the total reactance decreases. This is logical, because when capacitors are connected in parallel, the total capacitance increases and, hence, from Eq. 12-12, the total reactance will decrease.

Example 12-6: Assume in Fig. 12-13, $C_1 = 20\ \mu F$, $C_2 = 30\ \mu F$, $C_3 = 40\ \mu F$, and $V = 120$ V, 60 Hz. Determine (a) the total circuit capacitance, (b) the circuit capacitive reactance, (c) the current in each branch, and (d) the total circuit current.

(a) $C_T = C_1 + C_2 + C_3$
$= 20\ \mu F + 30\ \mu F + 40\ \mu F$
$= 90\ \mu F$

(b) From Ex. 12-5, $X_{C_1} = 132.70\ \Omega$, $X_{C_2} = 88.46\ \Omega$, and $X_{C_3} = 66.35\ \Omega$

$$\frac{1}{X_{C_T}} = \frac{1}{X_{C_1}} + \frac{1}{X_{C_2}} + \frac{1}{X_{C_3}}$$
$$= \frac{1}{132.70} + \frac{1}{88.46} + \frac{1}{66.35}$$
$$= 7.53 \times 10^{-3} + 11.30 \times 10^{-3}$$
$$+ 15.07 \times 10^{-3}$$
$$= 33.90 \times 10^{-3}$$

$$= \frac{10^3}{33.90}$$
$$= 29.50\ \Omega$$

or

$$X_{C_T} = \frac{1}{2\pi f C_T}$$
$$= \frac{10^6}{2 \times 3.14 \times 60 \times 90}$$
$$= 29.50\ \Omega$$

(c) $I_1 = \dfrac{V}{X_{C_1}}$
$$= \frac{120\text{ V}}{132.70\ \Omega}$$
$$= 0.90\text{ A}$$

$I_2 = \dfrac{V}{X_{C_2}}$
$$= \frac{120\text{ V}}{88.46\ \Omega}$$
$$= 1.36\text{ A}$$

$I_3 = \dfrac{V}{X_{C_3}}$
$$= \frac{120\text{ V}}{66.35\ \Omega}$$
$$= 1.81\text{ A}$$

(d) $I = I_1 + I_2 + I_3$
$$= 0.90 + 1.36 + 1.81$$
$$= 4.07\text{ A}$$

or

$$I = \frac{V}{X_{C_T}}$$
$$= \frac{120}{29.50}$$
$$= 4.07\text{ A}$$

Example 12-7: If in the circuit of Fig. 12-14, $C_1 = 20\ \mu F$, $C_2 = 30\ \mu F$, $C_3 = 40\ \mu F$, and $V = 120$ V, 60 Hz, determine (a) the total circuit capacitance, (b) the total circuit reactance, (c) the currents I, I_2, and I_3, and (d) the voltages V_1, V_2, and V_3.

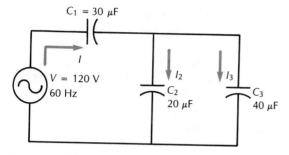

Figure 12-14 Capacitances in series-parallel.

(a) Since C_2 and C_3 are in parallel, the total capacitance (C_A) for these two capacitances is:

$$C_A = C_2 + C_3$$
$$= 30 \,\mu F + 40 \,\mu F$$
$$= 70 \,\mu F$$

C_1 is connected in series with C_A. Therefore, the total circuit capacitance (C_T) is:

$$\frac{1}{C_T} = \frac{1}{C_1} + \frac{1}{C_A}$$
$$= \frac{C_1 \times C_A}{C_1 + C_A}$$
$$= \frac{20 \times 70}{90}$$
$$= 15.56 \,\mu F$$

(b) $X_{C_T} = \dfrac{1}{2\pi f C_T}$

$$= \frac{10^6}{2 \times 3.14 \times 60 \times 15.56}$$
$$= 170.56 \,\Omega$$

or

$$X_{C_T} = X_{C_1} + \left(\frac{X_{C_2} \times X_{C_3}}{X_{C_2} + X_{C_3}}\right)$$

(Recall the values for X_{C_1}, X_{C_2}, and X_{C_3} from Ex. 12-5)

$$X_{C_T} = 132.70 + \frac{88.46 \times 66.35}{88.46 + 66.35}$$
$$= 170.6 \,\Omega$$

(c) $I = \dfrac{V}{X_{C_T}}$

$$= \frac{120}{170.56}$$
$$= 0.70 \text{ A}$$

(d) $V_1 = I X_{C_1}$

$$= 0.7 \times 132.7$$
$$= 92.89 \text{ V}$$

$V_2 = V_3$
$$= V - V_1$$
$$= 120 - 92.89$$
$$= 27.11 \text{ V}$$

$I_2 = \dfrac{V_2}{X_{C_2}}$

$$= \frac{27.11}{88.46}$$
$$= 0.30 \text{ A}$$

$I_3 = \dfrac{V_3}{X_{C_3}}$

$$= \frac{27.11}{66.35}$$
$$= 0.40 \text{ A}$$

PHASE RELATIONSHIP IN A PURELY CAPACITIVE CIRCUIT

It was mentioned in Chapters Ten and Eleven that the current is in phase with the voltage in a purely resistive circuit and that the current lags the applied voltage by 90° in a purely inductive circuit. It will now be shown that the current leads the voltage by 90° in a purely capacitive circuit.

It was shown that when a capacitor is connected across a dc supply, there is a momentary flow of charge to the capacitor. This movement of charge to the capacitor *decreases* exponentially, and ceases when the capacitor's voltage becomes equal to the applied voltage. The current is at a maximum the instant the switch is closed, and it decreases exponentially to zero as the capacitor builds up its charge. If the charged capacitor is connected across a resistance, the capacitor will now discharge exponentially from its maximum charge to zero. During this period, a reverse current will flow through the circuit.

When a capacitor is connected across an ac source, it will charge and discharge during each half cycle. It will charge when the applied voltage is increasing and then discharge when the applied voltage is decreasing. The charge stored ($Q = CV$) at any instant will be proportional to the applied voltage at that instant. Therefore, the waveforms for the applied voltage and the capacitor's charge will be in phase as shown in Fig. 12-15.

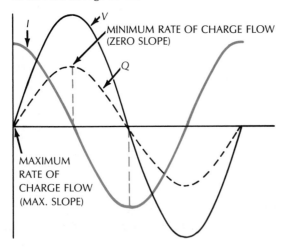

Figure 12-15 Voltage and current phase relationship in a purely capacitive circuit.

The current flow in the circuit at any instant will be equal to the rate of flow of charge to the capacitor at that instant. In Fig. 12-15, the rate of charge flow is maximum at 0°, 180°, and 360°. Therefore, at these points the current flow will be maximum. At 90° and 270°, the rate of charge flow will be instantaneuously zero (capacitor is at maximum charge); therefore, at these instances the current flow will be zero.

From 0° to 90°, the capacitor is charging and the current will decrease from a maximum at 0° to zero at 90°. At 90°, the capacitor is at full charge and the current will be zero at this instant. From 90° to 180°, the capacitor is discharging, and therefore the current will now be opposite (negative) to the charging current. From 180° to 270°, the capacitor is again charging, but in the opposite direction. Therefore, the current will remain negative during this period, but it will also decrease from a maximum at 180° to zero at 270°. From 270° to 360°, the capacitor is discharging. Therefore, the current will be positive (opposite to the charging current between 180° and 270°) during this period. The current flow for a complete cycle of applied ac voltage is shown in Fig. 12-15.

From this discussion, it is obvious that when an ac voltage is applied across a capacitor, it will cause an ac current flow through the circuit, and this current will lead the applied voltage by 90°. The waveforms for the applied voltage, capacitor charge, and the resulting circuit current are shown in Fig. 12-15. The phasor representation for the voltage and current in a purely capacitive circuit is shown in Fig. 12-16.

Figure 12-16 Current leads voltage by 90°.

POWER IN A CAPACITOR (REACTIVE POWER)

In the previous section, the current in a purely capacitive circuit was shown to lead the applied voltage by 90°. The power in watts taken by this circuit at any instant is equal to the product of the voltage and current at that same instant. The power waveform obtained for a complete cycle of applied ac voltage is shown in Fig. 12-17.

At positions, 0°, 90°, 180°, 270°, and 360°, either the voltage or the current is instantane-

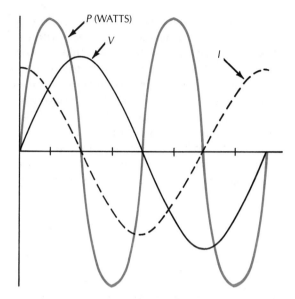

Figure 12-17 The power taken by a pure capacitance.

ously zero. Therefore, at these instances the power in the pure capacitance will be zero.

From 0° to 90°, both the current and the voltage are positive and, hence, the power pulse during this period will be positive, as shown. Since the capacitor is being charged during this period, the positive power pulse will represent energy taken from the source and stored by the capacitor. This energy is stored in an electrostatic field.

From 90° to 180°, the current is increasing negatively while the voltage is positive and decreasing from a maximum to zero. Therefore, the power pulse will be negative, as shown. Since the capacitor is being discharged during this period, the negative power pulse will represent energy returned from the capacitor to the circuit.

From 180° to 270°, both the current and the voltage are negative and, hence, the power pulse will again be positive (the product of two negatives). Note that during this period, the capacitor is again charging but in the opposite direction. Therefore, the capacitor will be taking energy from the source.

From 270° to 360°, the current is increasing

positively while the voltage is negative and decreasing to zero. Therefore, the power pulse during this period will be negative. During this period, the capacitor is being discharged and the stored energy is being returned to the source.

From this discussion and the waveforms shown in Fig. 12-17, it is seen that one cycle of ac voltage will produce two positive power pulses and two equal negative power pulses. The positive power represents energy taken from the source, and the negative power represents energy returned to the source. Therefore, the average watts taken by a pure capacitance is equal to zero per ac cycle of applied voltage.

In a purely resistive circuit, the current and voltage are in phase, and the true power (watts) is equal to the product of the current and the voltage. In a purely capacitive circuit, the current leads the voltage by 90°, and the watts consumed is zero. However, the product of the voltage and current for this capacitive circuit will not be mathematically zero. This is similar to a purely inductive circuit in which the current lags the voltage by 90°. Therefore, the product of the voltage and current for a pure capacitance will give a kind of reactive power similar to that for a pure inductance. This reactive power for a capacitance is also measured in VARs.

For a pure inductance, the current lags the voltage by 90° and the reactive power in VARs is considered to be positive. However, for a pure capacitance, the current leads the voltage by 90°. Therefore, the reactive power in VARs for a pure capacitance will be opposite to that for a pure inductance. Hence, the reactive power for a pure capacitance is considered to be negative. An inductance is often said to consume VARs, while a capacitance is said to produce VARs. In the next chapter it will be shown how this property of a capacitor may be used to cancel the inductive effect of an ac circuit (power-factor correction).

When a pure resistance is connected across

an ac supply, the current is in-phase with the voltage and the true power in watts may be calculated by any of the equations:

$$W = VI, \; W = I^2R, \; W = \frac{V^2}{R} \qquad (12\text{-}22)$$

When a pure inductance is connected across an ac supply, the current lags the voltage by 90°. The reactive power consumed may be calculated by any of the equations:

$$\text{VARs} = VI = I^2X_L = \frac{V^2}{X_L} \qquad (12\text{-}23)$$

When a pure capacitance is connected across an ac supply, the current leads the voltage by 90°. The reactive power produced may be calculated by any of the equations:

$$\text{VARs} = -VI = -I^2X_C = -\frac{V^2}{X_C} \qquad (12\text{-}24)$$

R-C SERIES CIRCUIT

In an R-C series circuit as shown in Fig. 12-18, the total opposition to current flow in the circuit is due both to the circuit's resistance, R, and its capacitive reactance, X_C. This total opposition to current flow in an ac circuit is called *impedance*. This is the same for a circuit containing both resistance and inductive reactance. The symbol for impedence is Z, and the unit of measurement is the ohm (Ω). From Ohm's law, the impedance of a circuit will be equal to the source voltage divided by the circuit's current.

$$Z = \frac{V}{I} \qquad (12\text{-}25)$$

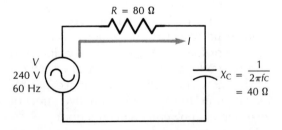

Figure 12-18 An ac circuit with R and X_C in series.

It was shown that the current flowing through a pure resistance is in phase with the voltage across the resistance and that the current through a capacitor leads the voltage across the capacitor by 90°. Therefore, in a series circuit, the current will be in phase with the resistance voltage drop, V_R, and it will lead the capacitance voltage drop, V_C, by 90°. This phase relationship is illustrated by the phasor diagram of Fig. 12-19. From Ohm's law, these two voltage drops may be calculated from the equations:

$$V_R = IR \qquad (12\text{-}26)$$

$$V_C = IX_C \qquad (12\text{-}27)$$

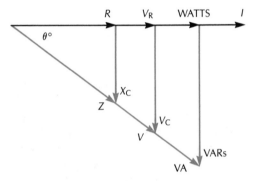

Figure 12-19 Power, voltage, and impedance phasor triangles for the circuit shown in Figure 12-18.

It is obvious from Fig. 12-19 that the voltage V_R leads the voltage, V_C, by 90°. Therefore, from Kirchhoff's voltage law, the source voltage in Fig. 12-18 will be equal to the phasor sum of the voltage drops, V_R and V_C. Since V_R and V_C are 90° out-of-phase, they will form a right triangle with the voltage, V, as shown in Fig. 12-19. Therefore, the phasor sum of V_R and V_C simplifies to:

$$V = \sqrt{(V_R)^2 + (V_C)^2} \qquad (12\text{-}28)$$

This equation may be expanded to the form:

$$IZ = \sqrt{(IR)^2 + (IX_C)^2} \qquad (12\text{-}29)$$

and cancelling I from each term gives the impedance equation:

$$Z = \sqrt{(R)^2 + (X_C)^2} \qquad (12\text{-}30)$$

The impedance phasor triangle will be similar to the voltage phasor triangle. These two triangles are shown in Fig. 12-19. From the impedance equation (Eq. 12-30), it is obvious that the impedance, Z, of a series R-C circuit is equal to the phasor sum of the circuit's resistance, R, and its reactance, X_C. These two quantities will always be 90° out-of-phase and, hence, their phasor addition is sometimes expressed as:

$$Z = R - jX_C \qquad (12\text{-}31)$$

This expression for impedance is similar to that shown for a R-L circuit (Eq. 11-20) in Chapter Eleven. Note, however, that the minus "j" operator is used in expressing a capacitive reactance. The "(−j)" operator implies that X_C lags R by 90°, and the "(+ j)" operator implies that X_L leads R by 90°. More details on the "j" operator is given in the Appendix at the back of the text.

Example 12-8: If in the series R-C circuit of Fig. 12-18, $R = 80\ \Omega$, $X_C = 40\ \Omega$, and $V = 240$ V, find the following: (a) Z, (b) I, (c) V_R, (d) V_C, and (e) draw the phasor diagram.

(a) $Z = R - jX_C$
$$= \sqrt{(R)^2 + (X_C)^2}$$
$$= \sqrt{(80)^2 + (40)^2}$$
$$= 89.44\ \Omega$$

(b) $I = \dfrac{V}{Z}$
$$= \dfrac{240\ V}{89.44\ Z}$$
$$= 2.68\ A$$

(c) $V_R = I \times R$
$$= 2.68\ A \times 80\ \Omega$$
$$= 214.40\ V$$

(d) $V_C = I \times X_C$
$$= 2.68\ A \times 40\ \Omega$$
$$= 107.20\ V$$

check: $V = V_R + V_C$ (phasor addition, V_R and V_C at right angle)
$$= \sqrt{(V_R)^2 + (V_C)^2}$$
$$= \sqrt{(214.40)^2 + (107.20)^2}$$
$$= 240\ V \text{ (agrees with the given value)}$$

(e)   The phasor diagram is given in Fig. 12-19.

APPARENT POWER (VOLT-AMPERE OR VA)

The total power for an R-C circuit is similar to that for a R-L circuit. The R-C circuit total power will contain both a true power (watts) component, and a reactive power (VARs) component. This power combination in a circuit is called *apparent power*, and it is measured in *volt-amperes*, (VA). The apparent power for a circuit may be determined from the equations:

$$\text{apparent power (VA)} = VI = I^2 Z \qquad (12\text{-}32)$$

The phasor triangle formed by the power phasors will be similar to the voltage and impedance triangles for the circuit as shown in Fig. 12-19. Since the true power and reactive power components are 90° out-of-phase, the circuit's apparent power can also be obtained from the phasor addition of these two power components.

$$\text{apparent power (VA)}$$
$$= \sqrt{(\text{watts})^2 + (\text{VARs})^2} \qquad (12\text{-}33)$$

Example 12-9: Determine (a) the true power, (b) the capacitive power, and (c) the apparent power for the circuit shown in Fig. 12-18.

(a)   From Ex. 12-8, the current through this circuit is $I = 2.68$ A.

true power $= I^2 R$
$$= (2.68)^2 \times 80$$
$$= 574.59\ W$$

(b) capacitive reactive power $= I^2 X_C$
$$= (2.68)^2 \times 40$$
$$= 287.30\ VARs$$

(c) apparent power (VA) = $V \times I$
$$= 240 \times 2.68$$
$$= 643.20 \text{ VA}$$
or apparent power (VA)

$$= \sqrt{(\text{watts})^2 + (\text{VARs})^2}$$
$$= \sqrt{(574.50)^2 + (287.30)^2}$$
$$= 643.0 \text{ VA}$$

POWER FACTOR AND PHASE ANGLE

It was seen that the current is in phase with the voltage in a purely resistive circuit, and that the current leads the voltage by 90° in a purely capacitive circuit. An R-C circuit, however, as in Fig. 12-18, is neither purely resistive nor purely capacitive because it contains both a resistive and a capacitive component. Therefore, the current in this circuit will lead the source voltage by an angle, $\theta°$, which lies somewhere between 0° (purely resistive) and 90° (purely capacitive) as shown in Fig. 12-19. This leading angle, $\theta°$, is commonly called the *phase angle* or *power factor angle* of the circuit, and the cosine function of this phase angle is termed the *power factor* of the circuit. Since the current is leading the voltage, the power factor is more accurately described as a *lead power factor*.

$$\text{power factor} = \cos \theta° \qquad (12\text{-}34)$$

From Fig. 12-19, it is obvious that a series R-C circuit power factor may be found from any of the relationships:

$$\text{P.F.} = \cos \theta°$$
$$= \frac{R}{Z}$$

$$\text{or} = \frac{\text{true power (watts)}}{\text{apparent power (VA)}}$$

$$\text{or} = \frac{V_R}{V} \qquad (12\text{-}35)$$

Of course, the phase angle, $\theta°$, can be found first by using any of the other trigonometric functions, and then the power factor can be

determined by taking the cosine of this phase angle. For example:

$$\theta° = \tan^{-1} \frac{X_C}{R} \qquad (12\text{-}36)$$

Because of the phase relationship between the current and voltage in an R-C circuit, many of the equations used in the analysis of this type of circuit will be related to the phasor addition and trigonometric functions of a right triangle. Consequently, the practice of first sketching the phasor diagram of a circuit will invariably help in establishing many of the equations required for the solution of the question. For example, in the phasor diagram shown in Fig. 12-19, if the voltage (V), the current (I), and the circuit power factor are known, we can write the equations:

$$\text{true power} = VI \cos \theta° \qquad (12\text{-}37)$$
$$= \text{apparent power}$$
$$\times \text{power factor}$$

$$\text{reactive power} = VI \sin \theta° \qquad (12\text{-}38)$$

and others.

Example 12-10: A capacitive circuit draws 1.4 A when it is operated on a 120-V ac supply. If the circuit's power factor is 0.5 lead, determine: (a) the true power; (b) the reactive power; (c) circuit impedance; (d) circuit resistance; (e) circuit capacitive reactance.

Since the circuit has a lead power factor of 0.5, it will contain both a resistive and a capacitive component. That is, the circuit is capacitive with a lead phase angle:

$$\theta° = \cos^{-1} 0.5 = 60°$$

The phasor triangle for this circuit can be easily sketched, as shown in Fig. 12-19.

(a) true power = $VI \cos \theta°$
$$= 120 \times 1.4 \times 0.5$$
$$= 84 \text{ W}$$

(b) reactive power = $VI \sin \theta°$
$$= 120 \times 1.4 \times 0.866$$
$$= 145.49 \text{ VARs}$$

(c) circuit impedance $(Z) = \dfrac{V}{I}$

$$= \frac{120 \text{ V}}{1.4 \text{ A}}$$

$$= 85.71 \ \Omega$$

(d) circuit resistance $(R) = Z \cos \theta°$

$$= 85.71 \times .5$$

$$= 42.86 \ \Omega$$

(e) capacitive reactance $(X_C) = \sqrt{Z^2 - R^2}$

$$= Z \sin \theta°$$

$$= 85.71 \times 10.866$$

$$= 74.22 \ \Omega$$

You may have noticed that the equations used in this chapter are relatively similar to the equations on inductance developed in the previous chapter (Chapter Eleven). One major difference, however, must be emphasized. Whereas in a purely inductive circuit the current lags the source voltage by 90°, in a purely capacitive circuit, the current leads the source voltage by 90°. This means although the phase shift caused by either a pure inductance or a pure capacitance is 90°, the phase difference between capacitive and inductive quantities will be 180°. For this reason, for example, inductive VARs will be opposite to capacitive VARs. Similarly, X_L and V_L will be opposite to X_C and V_C, respectively. Therefore, capacitive effects will cancel inductive effects in a circuit. This opposing property will be important in the next chapter where circuits containing both L and C are examined.

SUMMARY OF IMPORTANT POINTS

The Capacitor:

1. —is formed by two metal plates separated by a dielectric, an insulating material.
2. —is basically an ac device. Special dc capacitors are called electrolytics.
3. —will store a charge in an electrostatic field. This property is called capaci-

tance and it is measured in micro farads (μF).

4. —the charge stored Q is given by: $Q = CV$.
5. —capacitance may be increased by (a) increasing the plate area, (b) decreasing thickness of dielectric, and (c) use a dielectric with a higher dielectric constant.
6. —the dielectric constant measures the material's effectiveness as a dielectric (insulator). Air is the standard with a given dielectric constant of one.
7. —Dielectric strength describes the voltage required to break down a piece of material which is 1 cm thick.

The Capacitive Circuits

8. —will block dc with the exception of the momentary charging current.
9. —charges exponentially $V_C = V(1 - e^{-t})$.
10. —discharges exponentially $V_C = VC^{-t}$
11. —time constant, $t = R \times C$. This is the time required for capacitor to charge to 63.2 percent of the source voltage.
    —for R-L circuits, the time constant is $t = L/R$.

The ac Capacitive Circuits

12. The current limiting property of the capacitor voltage is called capacitance:

$$X_C = \frac{1}{2\pi f c}$$

13. When connected in series:

$$\frac{1}{C_T} = \frac{1}{C_1} + \frac{1}{C_2} + \cdots\cdots$$

$$X_{C_T} = X_{C_1} + X_{C_2} + \cdots\cdots$$

14. When connected in parallel:

$$C_T = C_1 + C_2 + \cdots\cdots$$

$$\frac{1}{X_{C_T}} = \frac{1}{X_{C_1}} + \frac{1}{X_{C_2}} + \cdots\cdots$$

15. Capacitive power is reactive power (VARs).
16. Current leads V_C by 90°.

Series R-C Circuit Equations

17. $Z = \dfrac{V}{I} = \sqrt{R^2 + X_C^2}$

true power $= I^2 R$
$$= VI \cos \underline{/\theta}$$

$$\text{P.F.} = \cos \underline{/\theta} = \frac{R}{Z} = \frac{\text{watts}}{\text{VA}}$$

apparent power $= V \times I$
$$= I^2 Z$$
$$= \sqrt{(\text{watts})^2 + (\text{VARs})^2}$$

$V = I \times Z = \sqrt{V_R^2 + V_C^2}$
reactive power $= T^2 X_C$
$$= VI \sin \underline{/\theta}$$

ESSAY AND PROBLEM QUESTIONS

1. Describe the composition of a capacitor.
2. List four applications for capacitors.
3. Describe a method used for increasing the capacitance of a capacitor while keeping its physical size relatively small.
4. (a) List three properties of a capacitor which determine its capacitance. (b) State how each property will affect capacitance.
5. Briefly describe the construction and operation of a variable capacitor.
6. Briefly describe the construction and circuit connection of an electrolytic capacitor.
7. (a) The charge that a capacitor will store depends on two circuit properties. Name these two properties. (b) Why is it important to discharge a capacitor before handling it?
8. An R-C series circuit contains a resistance, $R = 20\ \Omega$, a capacitance, $C = 40\ \mu F$, and a supply voltage, $V = 120$ V dc. Determine the charge stored by the capacitor when it is fully charged.
9. (a) Draw the curves to illustrate the charging current and the capacitor voltage for the circuit in Q. 8. (b) What type of curves are these? (c) Write the general equation for each of these two curves.
10. Define (a) capacitance and (b) capacitive reactance. State the unit of measure for each quantity.
11. (a) What is the difference between the

charging and discharging current of a R-C series circuit. (b) Draw the graphs to illustrate these two currents.
12. Define (a) dielectric constant and (b) dielectric strength.
13. Name four types of dielectrics used in the construction of capacitors.
14. Describe the effect of a capacitance on the circuit current for, (a) a dc series R-C circuit and (b) an ac series R-C circuit.
15. What effect does increasing the resistance, R, and increasing the capacitance, C, have on the rate of charging or discharging of a capacitor in a R-C series circuit?
16. A series dc circuit has, $R = 600\ \Omega$, $C = 20\ \mu F$, and $V = 300$ V. Determine (a) the circuit's time constant, (b) the voltage V_R at the moment the circuit is energized, (c) the voltage V_R at the end of one time constant, and (d) the voltage V_R at the end of two time constants.
17. For the circuit values given in Q. 16, what is the circuit current (a) the moment the circuit is energized and (b) one time constant after the circuit is energized?
18. A series R-C circuit has a resistance, $R = 400\ \Omega$, a capacitance, $C = 60\ \mu F$, and a source voltage, $V = 200$ V. Determine (a) the circuit's time constant, (b) the capacitor's voltage, V_C, the moment the circuit is energized, and (c) the voltage, V_C, one time constant after the circuit is energized.
19. What will be the approximate value for (a) the circuit's current, I_C, (b) the voltage, V_C, and (c) the voltage, V_R, in Q. 18, after five time constants?
20. A discharging R-C circuit contains a resistance, $R = 500\ \Omega$ and a capacitance, $C = 40\ \mu F$. The capacitor is charged to 400 V. Determine (a) the circuit's time constant, (b) the voltage, V_R, at the moment the circuit is closed, and (c) the current, I, at this same instant.
21. For the circuit given in Q. 20, determine the circuit current (a) one time constant,

and (b) two time constants after the circuit is closed.

22. Define the time constant for (a) an R-C series circuit and (b) an R-L series circuit.

23. A series R-C circuit has a resistance, $R = 80\ \Omega$, a capacitance, $C = 50\ \mu F$, and a source voltage, $V = 240$ V. Determine (a) the circuit time constant and (b) the capacitor voltage, V_C, at a time 12 seconds after the circuit is energized.

24. For the circuit given in Q. 23, determine the current at the times (a) 8 s, and (b) 12 s after the circuit is energized.

25. A series R-L circuit has a resistance, $R = 40\ \Omega$, an inductance, $L = 60$ H, and a source voltage, $V = 240$ V. Determine (a) the circuit time constant, (b) the circuit current after one time constant, and (c) after 2.7 s.

26. Draw a circuit to show three capacitors connected in series and across an ac voltage, $V = 200$ V, 60 Hz. If $C_1 = 30\ \mu F$, $C_2 = 60\ \mu F$, and $C_3 = 90\ \mu F$, determine: (a) the total capacitance, (b) the reactance of each capacitance, and (c) the total capacitive reactance.

27. For the circuit given in Q. 26, determine: (a) the circuit current and (b) the voltage drop across each capacitance.

28. Draw the waveforms and phasor diagram to show the circuit current and source voltage for Q. 27.

29. Draw a circuit to show three capacitors connected in parallel and across an ac voltage, $V = 200$ V, 60 Hz. If $C_1 = 30\ \mu F$, $C_2 = 60\ \mu F$, and $C_3 = 90\ \mu F$, determine (a) the total capacitance, (b) the reactance of each capacitor, and (c) the total capacitive reactance.

30. For the circuit given in Q. 29, determine: (a) the current through each capacitance, and (b) the total circuit current.

31. Draw the waveforms and phasor diagram to show the circuit current and source voltage for Q. 30.

32. If in Fig. 12-14, $C_1 = 30\ \mu F$, $C_2 = 15\ \mu F$, $C_3 = 25\ \mu F$ and $V = 240$ V, 60 Hz, determine (a) the total capacitance, (b) the reactance of each capacitor, and (c) the total capacitive reactance.

33. For the circuit given in Q. 32, determine (a) the currents I, I_2 and I_3, and (b) the voltages V_1, V_2 and V_3.

34. Draw the waveforms to illustrate (a) the voltage (b) the current and (c) the charge for a purely capacitive circuit. Indicate where the rate of charge flow is at a maximum.

35. Draw the waveform to illustrate the instantaneous power for the current and voltage in Q. 34. What is the average value for this power waveform?

36. An R-C series ac circuit has a resistance, $R = 50\ \Omega$, a capacitive reactance, $X_C = 30\ \Omega$, and source voltage, $V = 240$ V, 60 Hz. Determine: (a) the circuit impedance, Z, (b) the circuit current, (c) the voltage drop V_R across the resistance, (d) the voltage drop V_C across the capacitor, (e) the circuit true power, (f) the circuit reactive power, and (g) the circuit apparent power.

37. Determine: (a) the power factor, (b) the phase angle, and (c) draw the phasor diagram to show the impedance, voltage, and power phasors for Q. 36.

38. A capacitive circuit has a power factor of 0.8 lead, and operates on a supply voltage of 240 V, 60 Hz. The circuit draws a current of 2.5 A. (a) Sketch a phasor diagram for the circuit to show the impedance, voltage, and power phasors. (b) Determine: (i) the circuit apparent power (ii) the true power, (iii) the reactive power, (iv) the impedance, (v) the circuit resistance, and (vi) the circuit capacitive reactance.

39. A circuit has a lead power factor of 0.6, a resistance, $R = 750\ \Omega$, and uses 6400 watts of power. (a) Sketch a phasor diagram for the circuit to show the impedance, voltage, and power phasors. (b) Determine: (i) the circuit current, (ii) the source voltage, (iii) the impedance, (iv) the capacitive reactance (v) the apparent power, and (vi) the reactive power.

SINGLE-PHASE *R-L-C* CIRCUITS

In this chapter, the student will study the basic principles of series *R-L-C* circuits, series circuit resonance, parallel *R-L-C* circuits, parallel circuit resonance, power factor correction, and series-parallels *R-L-C* circuits.

SERIES *R-L-C* CIRCUITS

In the previous two chapters, the properties of the simple *R-L* and *R-C* circuits were discussed. It was shown that the circuit current is in phase with the voltage drop across a pure resistance, that the current lags the voltage drop across a pure inductance by 90°, and that the current leads the voltage drop across a pure capacitance by 90°. In a series *R-L-C*

circuit, as shown in Fig. 13-1, these relationships remain true, but they are now all combined into a single circuit.

Figure 13-2 shows the component phasor diagram for the *R-L-C* series circuit of Fig. 13-1.

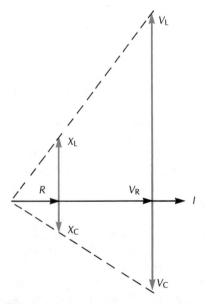

Figure 13-2 Phasor diagram showing the current, voltage, drops, and impedance phasor for Figure 13-1.

Figure 13-1 Series *R-L-C* circuit.

The circuit current is in phase with the voltage drop, V_R, it lags the voltage, V_L, by 90°, and it leads the voltage V_C, by 90°. The resistance and reactance phasors which cause the voltage drops, V_R, V_L, and V_C, are also shown. These impedance phasors will be in phase with their respective voltage drop.

THE NET CIRCUIT REACTIVE VOLTAGE AND REACTANCE

It should be obvious from Fig. 13-2, that the inductive voltage V_L, and the capacitive voltage, V_C, are 180° out-of-phase. Therefore, these two reactive voltage drops will oppose each other, and the net reactive voltage, V_X, in the circuit may be obtained from the equation:

$$V_X = V_L - V_C \tag{13-1}$$

The relationship between the reactance phasors is very similar and, therefore, the net circuit reactance, X_T, may be obtained by:

$$X_T = X_L - X_C \tag{13-2}$$

The resultant reactive voltage, phasor, V_X, and reactance phasor, X_T are shown in Fig. 13-3. For this circuit, the net reactive effect is inductive because X_L is greater than X_C (or $V_L > V_C$). When X_L is less than X_C, the net circuit effect will be capacitive, and when X_L is equal to X_C, the circuit will act as if it is purely resistive. These properties will be illustrated in later examples in the chapter.

CIRCUIT IMPEDANCE, CURRENT, AND VOLTAGES

The total impedance, Z, of Fig. 13-1 is equal to the phasor sum of the circuit's resistance and reactances. From the circuit phasor diagram shown in Fig. 13-3, it should be evident that this phasor addition simplifies to:

$$Z = \sqrt{(R)^2 + (X_L - X_C)^2} \tag{13-3}$$
$$= \sqrt{(R)^2 + (X_T)^2}$$

From Kirchhoff's voltage law, the source voltage, V, in Fig. 13-1 is equal to the phasor

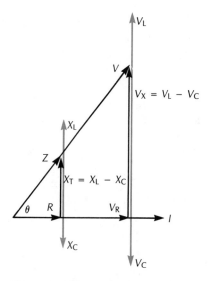

Figure 13-3 Phasor diagram showing the resultant voltage and impedance phasors for Figure 13-1.

sum of the voltage drops, V_R, V_L, and V_C. Also, as shown in Fig. 13-3, the voltage phasor diagram is similar to the impedance phasor diagram. Thus, the source voltage, V, may be expressed by the equation:

$$V = \sqrt{(V_R)^2 + (V_L - V_C)^2} \tag{13-4}$$
$$= \sqrt{(V_R)^2 + (V_X)^2}$$

The current flow, I, and the voltage drops in the series circuit may be obtained by using Ohm's law:

$$I = \frac{V}{Z} \tag{13-5}$$

$$V_R = I \times R \tag{13-6}$$

$$V_L = I \times X_L \tag{13-7}$$

$$V_C = I \times X_C \tag{13-8}$$

Example 13-1: Determine the impedance, current, and voltage drops in the circuit of Fig. 13-1.

(a) $Z = \sqrt{(R)^2 + (X_L - X_C)^2}$
$\qquad = \sqrt{(30)^2 + (60-20)^2}$
$\qquad = \sqrt{(30)^2 + (40)^2}$
$\qquad = 50 \ \Omega$

(b) $I = \dfrac{V}{Z}$

$\qquad = \dfrac{240}{50}$

$\qquad = 4.8$ A

(c) $V_R = I \times R$

$\qquad = 4.8 \times 30$

$\qquad = 144$ V

(d) $V_L = I \times X_L$

$\qquad = 4.8 \times 60$

$\qquad = 288$ V

(e) $V_C = I \times X_C$

$\qquad = 4.8 \times 20$

$\qquad = 96$ V

(f) $V = \sqrt{(V_R)^2 + (V_L - V_C)^2}$

$\qquad = \sqrt{(144)^2 + (288-96)^2}$

$\qquad = \sqrt{(144)^2 + (192)^2}$

$\qquad = 240$ V

PHASE ANGLE AND POWER FACTOR

The circuit's phase angle, $\theta°$, is always the angle that separates the circuit's current and the source voltage, and the circuit's power factor is given by the equation:

$$\text{power factor} = \cos \theta° \qquad (13\text{-}9)$$

Figure 13-3 shows that the impedance and voltage phasors for a circuit will form similar right triangles. Hence, the power factor of a series circuit may also be obtained from trigonometric relationships such as:

$$\text{power factor} = \cos \theta°$$

$$\text{or P.F.} = \dfrac{R}{Z} \qquad (13\text{-}10)$$

$$\text{or P.F.} = \dfrac{V_R}{V}$$

When the net circuit effect is inductive ($X_L > X_C$, or $V_L > V_C$), the circuit current will lag the source voltage, resulting in a lagging power

factor. Figure 13-3 shows a phasor diagram of a net inductive circuit. When the net circuit effect is capacitive ($X_L < X_C$, or $V_L < V_C$), the circuit current will lead the source voltage, resulting in a leading power factor. It may be recalled that a circuit's power factor can lie anywhere between 0 and 1, inclusive.

CIRCUIT POWER

TRUE POWER COMPONENT

In a series R-L-C circuit, there will be a true power component and a reactive power component. It may be recalled that the true power (watts) component will be caused by the circuit resistance and it will be in-phase with the circuit current. The circuit true power may be obtained by:

$$\text{true power} = I^2 \times R \qquad (13\text{-}11)$$

$$\text{or} \qquad V \times I \cos \theta° \qquad (13\text{-}12)$$

$$\text{or} \qquad \dfrac{V_R^2}{R} \qquad (13\text{-}13)$$

REACTIVE POWER COMPONENT

The reactive power in the circuit will be caused by the circuit inductive and capacitive reactances. In the previous two chapters, an inductance was considered to consume reactive VARs and a capacitance was considered to produce reactive VARs.

As shown in Fig 13-4, the inductive and capacitive reactance phasors are 180° out-of-phase; therefore their respective reactive power will also be 180° out-of-phase. It is customary to consider inductive VARs to be positive and capacitive VARs to be negative. It should be obvious now that the net circuit reactive power will be given by:

net reactive power (P_x)
= inductive VARs (P_L)
− capacitive VARs (P_C)

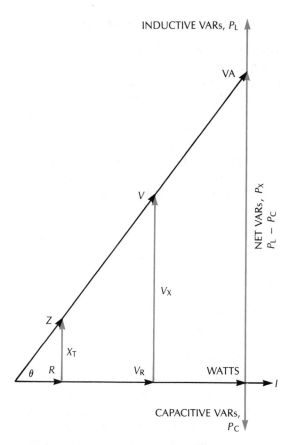

Figure 13-4 Phasor diagram showing the power, voltage, and impedance phasors for a *R-L-C* series circuit.

$$I^2 X_T = I^2 X_L - I^2 X_C \qquad (13\text{-}14)$$

In Fig. 13-4, the net circuit effect is inductive, hence, the net reactive power will be inductive or positive. When the net circuit effect is capacitive, the net reactive power will be capacitive or negative.

APPARENT POWER

The total circuit power is called the *apparent power* and it is equal to the phasor sum of the circuit's true power and reactive power components. From Fig. 13-4, this phasor addition simplifies to:

apparent power (VA)

$$= \sqrt{\text{(true power)}^2 + \text{(net reactive power)}^2}$$
$$\text{(VA)} = \sqrt{\text{(watts)}^2 + \text{(VARs)}^2} \qquad (13\text{-}15)$$

Using Ohm's law, the circuit apparent power may also be expressed by the equation:

$$\text{apparent power (VA)} = V \times I \qquad (13\text{-}16)$$

From the phasor diagram of Fig. 13-4 and the use of trigonometric relationships, other useful circuit equations should easily be recognized. Two of these equations are:

$$\begin{aligned}
\text{true power} &= \text{apparent power} \times \cos \underline{/\theta} \\
&= V \times I \times \cos \underline{/\theta} \\
&= V \times I \times \text{P.F.} \qquad (13\text{-}17)
\end{aligned}$$

$$\begin{aligned}
\text{net reactive power} &\\
&= \text{apparent power} \times \sin \underline{/\theta} \\
&= V \times I \times \sin \underline{/\theta} \qquad (13\text{-}18)
\end{aligned}$$

Example 13-2: Determine the following for the circuit of Fig. 13-1: (a) power factor, (b) true power, (c) net reactive power, and (d) the apparent power.

From Ex. 13-1, $Z = 50\ \Omega$, $I = 4.8\ \text{A}$, $V_R = 144\ \text{V}$, $V_L = 288\ \text{V}$, and $V_C = 96\ \text{V}$.

(a) $\text{P.F.} = \cos \underline{/\theta}$

$$= \frac{R}{Z}$$

$$= \frac{30}{50}$$

$$= 0.60\ \text{lag}$$

Note P.F. is lagging because net circuit effect is inductive.

or $\text{P.F.} = \dfrac{V_R}{V} = \dfrac{144}{240}$

$$= 0.60\ \text{lag}$$

circuit phase angle $\underline{/\theta} = \cos^{-1}(0.60)$
$$= 53.13°\ \text{lag}$$

(b) true power $= I^2 \times R$
$$= (4.8)^2 \times 30$$
$$= 691.2\ \text{W}$$

or true power $= V \times I \times \cos \underline{/\theta}$
$$= 240 \times 4.8 \times 0.6$$
$$= 691.2\ \text{W}$$

(c) inductive reactive power $= I^2 \times X_L$
$= (4.8)^2 \times 60$
$= 1382.4$ VARs

capacitive reactive power $= -I^2 \times X_C$
$= -(4.8)^2 \times 20$
$= -460.8$ VARs

Note: capacitive VARs are designated negative. Therefore,

net reactive power $= 1382.4 - 460.8$
$= 921.6$ VARs

Note also that when the net circuit effect is inductive, the net reactive power is also inductive.

or net reactive power $= V \times I \times \sin\underline{/\theta}$
$= 240 \times 4.8 \times 0.8$
$= 921.6$ VARs

(d) apparent power
$= \sqrt{(\text{watts})^2 + (\text{net VARs})^2}$
$= \sqrt{(691.2)^2 + (921.6)^2}$
$= \sqrt{1\ 327\ 104}$
$= 1152.00$ VA

or

apparent power $= V \times I$
$= 240 \times 4.8$
$= 1152$ VA

The student should be aware that the sketching of a phasor diagram of a circuit, whether it is asked for or not in the question, will be of invaluable assistance. Many of the equations required in the solution will become more apparent because they will, in general, be an extension of Ohm's and Kirchhoff's laws, using trigonometric relationships and phasor additions.

Note that for a series circuit, the current is constant and, hence, the current phasor is used as the reference phasor.

SERIES CIRCUIT RESONANCE

In an inductive circuit, the current was shown to lag the source voltage, whereas in a capacitive circuit the current was shown to lead the source voltage. When the current in a series R-L-C circuit is in phase with the source voltage, a condition known as *series resonance* exists. At resonance, the series curcuit will act *as if* it is purely resistive, because under this condition the inductive and capacitive effect completely cancel each other. Therefore, at resonance of a series R-L-C circuit:

(a) $X_L = X_C$

(b) $Z = \sqrt{(R)^2 + (X_L - X_C)^2} = R$

(c) $V_L = V_C$

(d) $V = V_R$

(e) $P_L = P_C$, and

(f) power factor $= 1$

Since the net reactance is zero at resonance, the circuit impedance will be at a minimum and it will be equal to the circuit resistance. Hence, the circuit current will be at a maximum. The concept of series resonance is very important in communications applications.

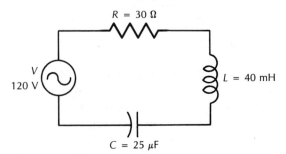

Figure 13-5 Resonant R-L-C series circuit.

The resonant condition of a circuit can be achieved by varying either the inductance, the capacitance, or the source frequency in the circuit. The frequency at which resonance occurs is called the *resonant frequency*, f_r, of the circuit. At resonance:

$$X_L = X_C$$

$$2\pi fL = \frac{1}{2\pi fC}$$

(13-19)

$$f_r = \frac{1}{2\pi\sqrt{L\,C}}$$

(f)   the net reactive power
(g)   the true power
(h)   circuit power factor, and
(i)   sketch the phasor diagram

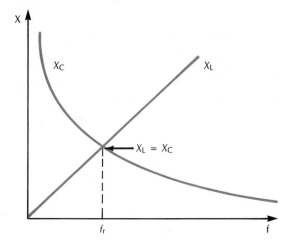

Figure 13-6(a) X_L and X_C versus frequency.

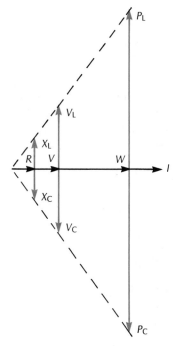

Figure 13-7 Phasor diagram for Figure 13-5.

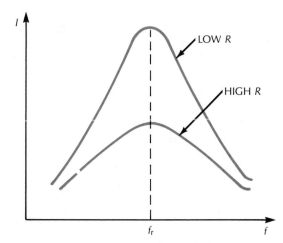

Figure 13-6(b) I_T versus frequency.

Example 13-3: Determine the following for the circuit shown in Fig 13-5, when at resonance:
(a)   resonant frequency
(b)   the circuit impedance
(c)   the circuit current
(d)   the reactive voltage, V_L and V_C
(e)   the resistive voltage V_R

(a) $f_r = \dfrac{1}{2\pi\sqrt{LC}}$

$$= \frac{1}{2\pi\sqrt{(40 \times 10^{-3}) \times (25 \times 10^{-6})}}$$

$$= \frac{1}{2\pi\sqrt{10^{-6}}}$$

$$= 159.15 \text{ Hz}$$

(b)   At resonance, $Z = R = 30\ \Omega$

(c) $I = \dfrac{V}{Z} = \dfrac{120}{30} = 4$ A

(d) $V_L = I \times X_L$
      $= I \times 2\pi f_r L$
      $= 4 \times 2 \times 3.14 \times 159.15 \times 40 \times 10^{-3}$
      $= 160$ V
   $V_C = V_L = 160$ V

or $V_C = I \times X_C$

$= \dfrac{I}{2\pi f_r C}$

$= \dfrac{4}{2 \times 3.14 \times 159.15 \times 25 \times 10^{-6}}$

$= 160 \text{ V}$

(e) $V_R = I \times R$

$= 4 \times 30$

$= 120 \text{ V}$

Note: V_L and V_C are greater than the source voltage in magnitude. This is often called the resonant voltage rise. In fact, in some circuits these voltages can become dangerously high.

(f)   net reactive power $P_X = P_L - P_C$

$= I^2 X_L - I^2 X_C$

$= 0$

(g)   true power $= I^2 \times R$

$= (4)^2 \times 30$

$= 480 \text{ W}$

(h)   power factor $= \cos \underline{/\theta}$

$= \dfrac{R}{Z}$

$= \dfrac{30}{30} = 1$

PARALLEL *R-L-C* CIRCUITS

Parallel circuits have many more applications in alternating current work than do series circuits. Parallel circuits are used in the majority of residential and industrial installations because source voltage must be maintained across the various loads. An important point to remember when analyzing parallel circuits is the fact that the branch currents may not necessarily be in phase with each other. Therefore, phasor addition must be used whenever these branch currents are being added. Of course, phasor addition can be done either by the trigonometric component method or by the graphical method. Both of these methods were discussed in Chapter Ten.

Another important point to note is in the construction of the phasor diagram. In a series circuit, the current is constant and, hence, the current phasor is used as the reference phasor. However, in a parallel circuit, the voltage across the parallel branches is constant. Therefore, when constructing the phasor diagram of a parallel circuit, the voltage phasor is used as the reference phasor.

All other principles and laws of parallel circuits and on inductance and capacitance remain the same. The following examples should present little or no difficulty.

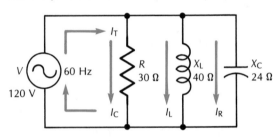

Figure 13-8 A parallel *R-L-C* circuit.

Example 13-4: Determine the following for the circuit of Fig. 13-8:

(a)   the branch currents and their phase angles
(b)   sketch the branch and circuit current phasors
(c)   the total circuit current
(d)   the circuit power factor
(e)   the circuit impedance
(f)   the circuit true and net reactive power
(g)   the circuit apparent power

(a) $I_R = \dfrac{V}{R}$ (I_R in phase with V)

$= \dfrac{120}{30}$

$= 4 \text{ A } \underline{/0°}$

$I_L = \dfrac{V}{X_L}$ (I_L lags V by 90°)

$= \dfrac{120}{40}$ (note negative sign)

$= 3 \text{ A } \underline{/-90°}$

$$I_C = \frac{V}{X_C} \qquad (I_C \text{ leads } V \text{ by } 90°)$$
$$= \frac{120}{24}$$
$$= 5 \text{ A } \underline{/90°}$$

(b)

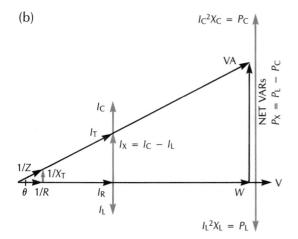

Figure 13-9 Phasor diagram for the parallel *R-L-C* circuit of Figure 13-10.

(c) The total circuit current can be obtained by using Kirchhoff's current law. From the phasor diagram of Fig. 13-9, it is obvious that I_T is equal to the phasor sum of I_R, I_L, and I_C

$$I_T = \sqrt{I_R^2 + (I_C - I_L)^2}$$
$$= \sqrt{I_R^2 + (I_X)^2}$$
$$= \sqrt{(4)^2 + (5-3)^2}$$
$$= \sqrt{16 + 4}$$
$$= 4.47 \text{ A}$$

(d) Since I_C is greater than I_L, the net reactive current in the circuit will be capacitive. Hence, the circuit will have a net capacitive effect and leading power factor. The circuit phase angle is shown in the phasor diagram.

$$\text{power factor } = \cos \underline{/\theta}$$
$$= \frac{I_R}{I_T}$$

$$= \frac{4}{4.47}$$
$$= 0.894 \text{ lead}$$

$$\text{phase angle } = \cos^{-1}(0.894)$$
$$= 26.57° \text{ lead}$$

(e) The circuit impedance, Z, can be found using either Ohm's law or the reciprocal method for a parallel circuit. Although the reciprocal method can be used effectively, it is generally more complex for a parallel *R-L-C* circuit. This is because it will involve the phasor addition of the reciprocals of out-of-phase quantities. The phasor diagram shows these reciprocal quantities. This method is illustrated in the appendix at the back of the text. In this chapter we will limit ourselves to questions for which the impedance may be readily found using Ohm's law.

$$Z = \frac{V}{I_T}$$
$$= \frac{120}{4.47}$$
$$= 26.84 \ \Omega$$

(f) The true power in any circuit is consumed only by the circuit's effective resistance. In this circuit, resistance appears only in branch one.

$$\text{true power } = I_R^2 \times R$$
$$= (4)^2 \times 30$$
$$= 480 \text{ W}$$

$$\text{or true power } = \frac{V^2}{R}$$
$$= \frac{(120)^2}{30}$$
$$= 480 \text{ W}$$

$$\text{or true power } = V \times I_R$$
$$= V \times I_T \cos \theta°$$
$$= 120 \times 4.47 \times 0.894$$
$$= 480 \text{ W}$$

The circuit also contains both an inductive and a capacitive power component.

These two reactive power components are 180° out-of-phase, as shown in the phasor diagram. Therefore, the net reactive power is given by:

net reactive power
$= I^2 X_L - I^2 X_C$
$= [(3)^2 \times 40]$
$\quad - [(5)^2 \times 24]$
$= -240$ VARs

When the net reactive power is negative, it indicates that the circuit has a net capacitive effect and is supplying magnetizing VARs to the source. This net capacitive effect of the circuit was also indicated by the circuit currents.

or net reactive power
$= - V \times I_T \sin \underline{/\theta}$
$= - 120 \times 4.47 \times 0.447$
$= - 240$ VARs

(g) apparent power
$= \sqrt{(\text{watts})^2 + (\text{net VARs})^2}$
$= \sqrt{(480)^2 + (240)^2}$
$= 536.6$ VA

or circuit apparent power $= V \times I_T$
$\qquad = 120 \times 4.47$
$\qquad = 536.6$ VA

PARALLEL CIRCUIT RESONANCE

In Ex. 13-4, the circuit analyzed had a net capacitive effect. In this circuit, I_C was greater than I_L and the total line current, I_T led the source voltage. For an inductive circuit, I_L will be greater than I_C, and the total line current, I_T will lag the supply voltage.

What effect will $I_L = I_C$ have on a parallel R-L-C circuit? When I_L equals I_C, in a parallel circuit, the net reactive component in the circuit disappears and the circuit acts as if it is purely resistive. Under this condition, the circuit is said to be in parallel resonance.

Under parallel resonance, therefore, the line current will be in phase with the source

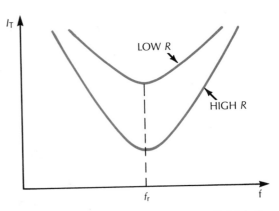

Figure 13-10 I_T versus frequency for a parallel R-L-C circuit.

voltage and the circuit will have a power factor of one. Since only an in-phase component of current is now taken from the source, the line current and power under parallel resonance will be at a minimum, and the circuit impedance will be at a maximum. This is somewhat opposite to that for series circuit resonance. Recall that with series circuit resonance, the line current was at maximum and the impedance was at minimum.

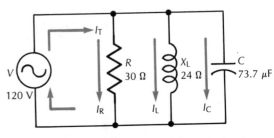

Figure 13-11(a) A resonant parallel R-L-C circuit.

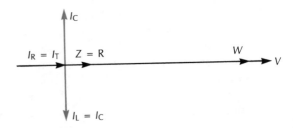

Figure 13-11(b) Phasor diagram for resonant parallel circuit.

The following properties may be summarized for a parallel resonant circuit:

(i) $I_L = I_C$ (ii) P.F. = 1
(iii) $X_L = X_C$ (iv) $I_T = I_R$
(v) $P_L = P_C$ (vi) phase angle = 0°
(vii) $Z = R$

Example 13-5: The parallel circuit shown in Fig. 13-11 is at resonance. Determine the following:
(a) the resonant frequency, f_r
(b) each of the branch currents, I_R, I_L, and I_C
(c) the line current, I_T
(d) the circuit impedance, Z
(e) the net reactive power, P_X
(f) the circuit power factor
(g) the true power
(h) the circuit apparent power
(i) sketch the circuit phasor diagram

(a) At parallel resonance, $X_L = X_C$,
     therefore: $X_C = 24$

$$= \frac{1}{2\pi f_r C}$$

therefore $f_r = \frac{10^6}{2 \times 3.14 \times 24 \times 73.7}$
            $= 90$ Hz

(b) $I_R = \frac{V}{R}$

     $= \frac{120}{30} = 4$ A

     $I_L = \frac{V}{X_L}$

     $= \frac{120}{24} = 5$ A

     $I_C = \frac{V}{X_C}$

     $= \frac{120}{24} = 5$ A

     (Note, $I_L = I_C$ at resonance)

(c) $I_T = \sqrt{(I_R)^2 + (I_L - I_C)^2}$
     $= \sqrt{(4)^2 + (5 - 5)^2}$
     $= 4$ A
     (Note, $I_T = I_R$ at resonance)

(d) $Z = \frac{V}{I_T}$

     $= \frac{120}{4} = 30\ \Omega$

     (Note, $Z = R$ at resonance)

(e) net reactive power
     $= I_L^2 X_L - I_C^2 X_C$
     $= [(5)_2 \times 24] - [(5)_2 \times 24]$
     $= 600 - 600$
     $= 0$ VARS
     (Note $P_L = P_C$ at resonance)

(f) power factor = $\cos{\underline{/\theta}}$

     $= \frac{I_R}{I_T}$

     $= \frac{4}{4} = 1$

     phase angle, $\underline{/\theta} = \cos^{-1}(1)$
     $= 0°$

(g) true power $= V \times I \cos{\underline{/\theta}}$
     $= 120 \times 4 \times 1$
     $= 480$ W

(h) apparent power $= V \times I$
     $= 120 \times 4$
     $= 480$ VA
     (Note, true power = apparent power at resonance)

(i) The phasor diagram for a resonant circuit is shown in Fig. 13-11 (b).

POWER FACTOR CORRECTION

Practically all commercial and residential power installations are in parallel. Many of these parallel circuits must supply power to motors, transformers, and other loads which operate with a lagging power factor. When a load is reactive, it will take more current to develop the required power as compared to a load with zero net reactance. As a result, when larger currents flow through a line, the power

250   ELECTRICAL SYSTEMS TECHNOLOGY

losses and line voltage drop increase, and also heavier wires must be installed to carry the current. Therefore, the overall cost of the installation increases. For this reason, it is often more economical to modify a system so that it will operate under a zero net reactive condition.

In the previous section on parallel resonance, it was shown that when a parallel circuit was resonant, its net reactive component was zero, its line current was in phase with the supply voltage, and its power factor was unity. When a circuit is modified to the point that the power factor approaches unity, the modification is called *power factor correction*. Since most loads are inductive, power factor correction usually involves the use of capacitances to nullify the inductive component in the circuit (Fig. 13-12).

Figure 13-12 KVAR power factor correction capacitor.   *Courtesy: Steelman Electric Mfg. Co. Ltd.*

Example 13-6: If the wattmeter to the inductive circuit in Fig. 13-13 reads 600 W and the ammeter reads 15 A, determine the following:

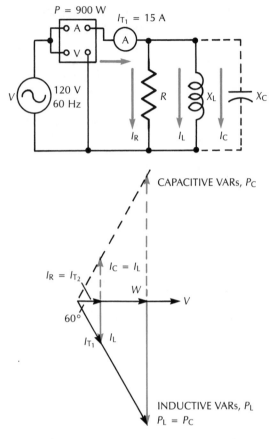

Figure 13-13 An inductive circuit with power factor correction.

(a) the circuit power factor
(b) the inductive power in the circuit
(c) the inductive current, I_L
(d) the capacitance necessary for power factor correction
(e) the new line current with the capacitor connected into the circuit

Solution:
(a)   power factor $= \cos\theta°$

$$= \frac{\text{true power}}{\text{apparent power}}$$

$$= \frac{900 \text{ W}}{V \times I_{T_1}}$$

$$= \frac{900}{120 \times 15}$$

$$= 0.5 \text{ lag}$$

(b)   inductive circuit power

$= V \times I_T \sin \theta°$

$= 120 \times 15 \times \sin 60°$

$= 1558.85$ VARs

(c)   from (b) inductive power is also equal to $V \times I_L$ or $I_L^2 X_L$

therefore, $I_L = \dfrac{\text{inductive power}}{V}$

$= \dfrac{1558.85}{120} = 12.99$ A

or from the phasor diagram of Fig. 13-12,

$I_L = I_{T_1} \sin \theta°$

$= 15 \times 0.866$

$= 12.99$ A

(d)   For power factor correction, $I_L = I_C = 12.99$ A, and $P_L = P_C = 1558.85$ VARs.

since $P_C = I_C^2 X_C$

$X_C = \dfrac{P_C}{I_C^2}$

$= \dfrac{1558.85}{(12.99)^2} = 9.24\ \Omega$

or

$X_C = \dfrac{V}{I_c}$

$= \dfrac{120}{12.99} = 9.24\ \Omega$

$C = \dfrac{1}{2\pi f X_C}$

$= \dfrac{1}{2 \times 3.14 \times 60 \times 9.24}$

$= 287.2\ \mu F$

(e)   The new line current, I_{T_2}, for the power factor corrected circuit, will be the in-phase current, I_R, because I_C will cancel out I_L. This is illustrated in the phasor diagram.

$I_{T_2} = I_R$

$= I_{T_1} \cos \underline{/\theta}$

$= 15 \times 0.5 = 7.5$ A

Note that the line current for the circuit,

when its power factor has been corrected to unity, is reduced from 15 A to 7.5 A.

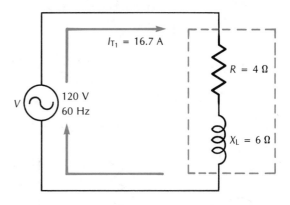

Figure 13-14(a) Circuit with an *R-L* series load.

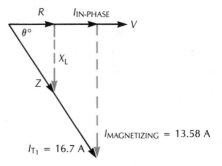

Figure 13-14(b) Phasor diagram for Figure 13-14(a).

Example 13-7: Figure 13-14 shows a circuit with an *R-L* series inductive load. Determine (a) the line current, (b) the power factor, and (c) the true power.

$Z = \sqrt{(R)^2 + (X_L)^2}$ (*R* in series with X_L)

$= \sqrt{(4)^2 + (6)^2}$

$= 7.2\ \Omega$

(a) $I_{T_1} = \dfrac{V}{Z}$

$= \dfrac{120}{7.2}$

$= 16.67$ A

(b)   power factor $= \cos \underline{/\theta}$

$= \dfrac{R}{Z}$

$$= \frac{4}{7.2}$$
$$= 0.556 \text{ lag}$$
$$\underline{/\theta} = \cos^{-1} 0.556$$
$$= 56.2° \text{ lag}$$

(c) true power = $V \times I_{T_1} \times \cos \underline{/\theta}$ or $(I_{T_1})^2 \times R$
$$= 120 \times 16.7 \times 0.556$$
$$= 1114.2 \text{ W}$$

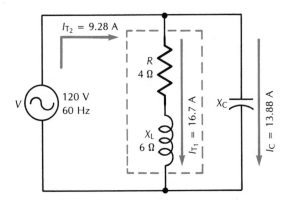

Figure 13-15(a) Power factor correction circuit.

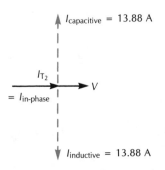

Figure 13-15(b) Phasor diagram for power factor corrected circuit of Figure 13-15(a).

To correct the inductive effect of the load in Fig. 13-14, a capacitor is connected in parallel across the load as shown in Fig. 13-15. The capacitance used must be just large enough to cancel the magnetizing current component of the inductive load. Under this condition, the circuit will have the required unit power factor.

Determine the following:

(d) the inductive magnetizing current component which the capacitance must oppose
(e) the size of capacitance required
(f) the new line current
(g) the net reactive power
(h) the true power

(d) From the phasor diagram of Fig. 13-15(b):

$$I_{magnetizing} = - I_{T_1} \sin \underline{/\theta}$$
$$= - 16.7 \times \sin 56.2°$$
$$= - 16.7 \times 0.831$$
$$= - 13.88 \text{ A}$$
(inductive current component)

(e) The capacitive current will oppose the inductive current component as shown in the phasor diagram of Fig. 13-15(b). For unity power factor correction, the net reactive current must be reduced to zero. Therefore, the capacitive current must be equal in magnitude to the inductive current component.

$$I_C = 13.88 \text{ A}$$
$$X_C = \frac{V}{I_C}$$
$$= \frac{120}{13.88}$$
$$= 8.66 \text{ }\Omega$$

$$C = \frac{1}{2\pi f X_C}$$
$$= \frac{1}{2 \times 3.14 \times 60 \times 8.66}$$
$$= 306.5 \text{ }\mu\text{F}$$

Therefore, the parallel capacitance required to correct the power factor of the series R-L inductive load is 306.8 μF.

(f) The new line current, I_{T_2}, is equal to the phasor sum of I_{T_1} and I_C. Since I_C cancels the inductive component of I_{T_1}, this phasor addition will simply be equal to the in-phase component of I_{T_1}.

$$I_{T_2} = I_{in\text{-}phase}$$

$$= I_{T_1} \cos \underline{/\theta}$$
$$= 16.7 \times 0.556$$
$$= 9.28 \text{ A}$$

Note that the current through the inductive load remains at $I_{T_1} = 16.7$ A $\underline{/- 56.2°}$. However, the net line current, I_{T_2}, has been greatly reduced. Under the corrected power factor condition, the new line current, which is the phasor sum of I_{T_1} and I_C, is now 9.28 A. This new line current is in phase with the supply voltage. In fact, it is equal to only the in-phase component of the load current.

(g) The net reactive power in the modified circuit should be zero. (Circuit with unity power factor.)

$$P_X = P_L - P_C$$
$$= (I_{T_1})^2 X_L - I_C^2 X_C$$
$$= [(16.67)^2 \times 6] - [(13.88)^2 \times 8.65]$$
$$= 1666 - 1666$$
$$= 0 \text{ VARs}$$

(h) The true power taken by the load should remain the same as for the unmodified circuit.

$$\text{true power} = V \times I_{T_2} \times \cos \theta \text{ or } (I_{T_1})^2 \times R$$
$$= 120 \times 9.28 \times 1$$
$$= 1114 \text{ W}$$

Note that the modified circuit takes only in-phase or true power from the supply. In the unmodified inductive circuit, the power taken included both the required true power and also an inductive power component. In the modified circuit, the inductive power requirement is now supplied by the capacitance.

It may have been noticed that for both the parallel R-L load and series R-L load as described in examples 13-6 and 13-7, the power factor correcting capacitor was connected in parallel across the load. Connecting the capacitor in series with the load may also be used to reduce the circuit power factor to unity. This, however, will produce a series resonant circuit, and under series resonance, the line current is at a maximum. Since the intent

of correcting a circuit power factor is to reduce the line current, the series arrangement of the capacitor is obviously not suitable.

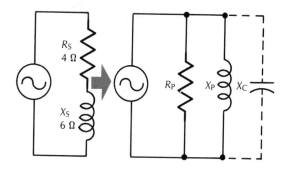

Figure 13-16 Series circuit to parallel circuit conversion.

Another method of determining how to correct the power factor of a series R-L load is by first converting the series circuit into its parallel equivalent. The equations for converting a series circuit (Fig. 13-16) into its parallel equivalent circuit are given below. The remaining solution for finding the required capacitance will then be the same as described in Ex. 13-6.

$$R_P = \frac{R_S^2 + X_S^2}{R_S} = \frac{Z_S^2}{R_S} \qquad (13\text{-}20)$$

$$X_P = \frac{R_S^2 + X_S^2}{X_S} = \frac{Z_S^2}{X_S} \qquad (13\text{-}21)$$

SERIES-PARALLEL CIRCUITS

Several methods may be used in the solutions of series-parallel ac circuits. However, the examples discussed in this chapter will be limited to circuits requiring only a knowledge of basic mathematics. It should also be noted that the fundamental rules discussed in previous examples in this chapter and on dc circuits in Chapter Three remain basically the same. The first example given below analyzes a series-parallel circuit in terms of its impedances and currents, and the second example uses the terms conductance (G), susceptance (B), and admittance (Y). These terms are defined in the example.

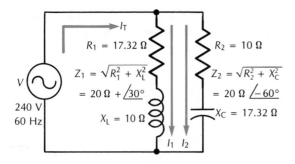

Figure 13-17 Series-parallel impedances.

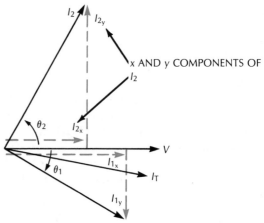

Figure 13-18 Phasor diagram for Figure 13-17.

Example 13-8: Figure 13-17 shows a circuit with two impedances. The value of each impedance and their respective phase angles can easily be determined. These values are indicated on the circuit. Determine the following:

(a) the current, I_1, I_2, and I_T
(b) the circuit impedance
(c) total true power
(d) circuit apparent power
(e) circuit power factor

(a) $I_1 = \dfrac{V}{Z_1}$

$= \dfrac{240}{20}$

$= 12 \text{ A}\underline{/-30°}$ (I_1 lags V by 30°)

$I_2 = \dfrac{V}{Z_2}$

$= \dfrac{240}{20}$

$= 12 \text{ A}\underline{/+60°}$ (I_2 leads V by 60°)

From the phasor diagram of Fig. 13-18, it is obvious that I_T will be equal to the phasor sum of I_1 and I_2. Using the phasor component method for phasor addition:

$I_{T_x} = I_{1_x} + I_{2_x}$ (this is the net in-phase current)
     $= I_1 \cos\underline{/\theta_1} + I_2 \cos\underline{/\theta_2}$
     $= (12 \times 0.5) + (12 \times 0.866)$
     $= 16.39 \text{ A}$

$I_{T_y} = I_{1_y} + I_{2_y}$
     $= I_1 \sin\underline{/\theta_1} + I_2 \sin\underline{/\theta_2}$
     $= (-412 \times 0.866) + (12 \times 0.5)$
     $= -4.39 \text{ A}$(net reactive current is inductive)

$I_T = \sqrt{(I_{T_x})^2 + (I_{T_y})^2}$
     $= \sqrt{(16.39)^2 + (4.39)^2}$
     $= 16.97 \text{ A lag}$

(b) $Z = \dfrac{Y}{I_T}$

     $= \dfrac{240}{16.97}$

     $= 14.14 \text{ }\Omega$

(c)   total true power
     $= (I_1^2 \times R_1) + (I_2^2 \times R_2)$
     $= [(12)^2 \times 17.32] + [(12)^2 \times 10]$
     $= 2492.08 + 1440$
     $= 3934.08 \text{ W}$

(d)   circuit apparent power $= V \times I$
                      $= 240 \times 16.97$
                      $= 4072.8 \text{ VA}$

(e)   circuit power factor $= \dfrac{\text{watts}}{\text{VA}}$

                           $= \dfrac{2934.08}{4072.8}$

                           $= 0.966 \text{ lag}$

or power factor $= \dfrac{I_{T_x}}{I_T}$

                      $= \dfrac{16.39}{16.97}$

                      $= 0.966 \text{ lag}$

The next example uses the terms conductance, susceptance, and admittance.

Conductance is the ease with which electrons flow through a resistance. In a purely resistive circuit, it is equal to the reciprocal of the resistance, but in a reactive circuit it is equal to the resistance divided by the squared value of the impedance. Conductance is measured in siemens (S).

$$\text{conductance, } G = \frac{R}{Z^2} \qquad (13\text{-}22)$$

Susceptance is the ease with which electrons flow through a reactance. It is also measured in siemens (S) and is equal to the reactance divided by the square of the impedance.

$$\text{susceptance, } B = \frac{X}{Z^2} \qquad (13\text{-}23)$$

Admittance is the ease with which electrons flow through an impedance. It is equal to the reciprocal of the impedance, and like conductance and susceptance, it too is measured in siemens (S).

$$\text{admittance, } Y = \frac{1}{Z} \qquad (13\text{-}24)$$

$$\text{since } Z = \frac{V}{I}$$

$$\text{therefore } Y = \frac{I}{V} \qquad (13\text{-}25)$$

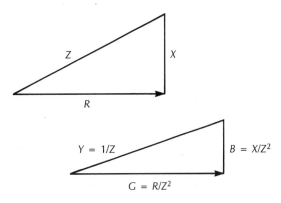

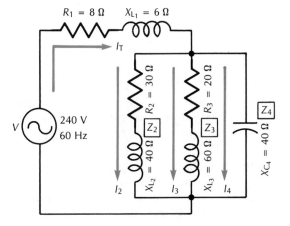

Figure 13-19 Series-parallel circuit.

Figure 13-20 Impedance and admittance phasor triangles.

The relationship between the impedance and admittance phasor triangles is the same as shown in Fig. 13-20. Therefore, the admittance is also equal to:

$$Y = \sqrt{(G)^2 + (B)^2} \qquad (13\text{-}26)$$

Example 13-9: For the circuit shown in Fig. 13-19 determine the following:

(a) the total impedance
(b) the circuit current
(c) the power factor for the circuit
(d) the total circuit true power

This problem will be solved in sections. First the impedance, conductance, and susceptance for each parallel branch section are found as follows:

(a) $Z_2 = \sqrt{R_2^2 + X_{L_2}^2}$
$= \sqrt{(30)^2 + (40)^2}$
$= 50 \ \Omega$

$G_2 = \frac{R_2}{Z_2^2}$
$= \frac{30}{(50)^2}$
$= 0.012 \text{ S}$

$B_2 = \frac{X_2}{Z_2^2}$
$= \frac{40}{(50)^2}$
$= 0.016 \text{ S}$

(b) $Z_3 = \sqrt{R_3^2 + X_{L_3}^2}$
$= \sqrt{(20)^2 + (60)^2}$
$= 63.25 \ \Omega$

$G_3 = \dfrac{R_3}{Z_3^2}$
$= \dfrac{20}{(63.25)^2}$
$= 0.005 \ S$

$B_3 = \dfrac{X_3}{Z_3^2}$
$= \dfrac{60}{(63.25)^2}$
$= 0.015 \ S$

(c) $Z_4 = \sqrt{R_4^2 + X_C^2}$
$= \sqrt{0 + 40^2}$
$= 40 \ \Omega$

$G_4 = \dfrac{R_4}{Z_4^2}$
$= \dfrac{0}{(40)^2}$
$= 0 \ S$

$B_4 = -\dfrac{X_4}{X_4^2}$
$= \dfrac{-40}{(40)^2}$
$= -0.025 \ S$

Note that capacitive susceptance opposes inductive susceptance in the same way as capacitive reactance is opposite to inductive reactance. Therefore, the capacitive susceptance, B_4, will be negative and the susceptances B_2 and B_3 will be positive.

Next, the net conductance and the net susceptance for the parallel portion of the circuit is obtained as follows:

$G_P = G_2 + G_3 + G_4$ (13-27)
$= 0.012 + 0.005 + 0.0$
$= 0.017 \ S$

$B_P = B_2 + B_3 + B_4$ (13-28)
$= 0.016 + 0.015 - 0.025$
$= 0.006 \ S$

This net conductance and susceptance can now be used to determine the net admittance, Y_P, and net impedance, Z_P, for the parallel portion of the circuit. From equations 13-26 and 13-24:

$Y_P = \sqrt{(G_p)^2 + (B_p)^2}$
$= \sqrt{(0.017)^2 + (0.006)^2}$
$= 0.018 \ S$

$Z_P = \dfrac{1}{Y_P}$
$= \dfrac{1}{0.018}$
$= 55.47 \ \Omega$

To sum up, the parallel portion of the circuit has been reduced to a single equivalent impedance, Z_P. A look at the circuit will indicate that this equivalent impedance, Z_P, is in series with the remaining part of the circuit. From equations 13-22 and 13-23 this equivalent impedance can be expressed in terms of an equivalent resistance, R_P, and a reactance, X_P.

$G_P = \dfrac{R_P}{Z_P^2}$
$R_P = G_P \ Z_P^2$
$= 0.017 \ (55.47)^2$
$= 52.31 \ \Omega$

$B_P = \dfrac{X_P}{Z_P^2}$
$X_P = B_P \ Z_P^2$
$= 0.006 \ (55.47)^2$
$= 18.46 \ \Omega$

The total circuit resistance, R_T, reactance, X_T and impedance, Z_T, can now be easily obtained. The remaining solution should be obvious from the equivalent circuit of Fig. 13-21, and it is left as an exercise for the reader.

The conductance- susceptance- admittance type of solution is sometimes abbreviated as the G-B-Y method. Although somewhat long, the G-B-Y method requires only basic mathe-

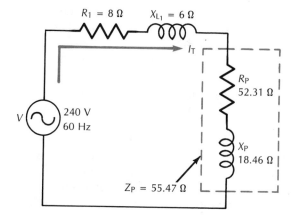

Figure 13-21 Equivalent circuit for Figure 13-19.

matics and can be effectively used to analyze many types of ac parallel and series-parallel circuits.

SUMMARY OF IMPORTANT POINTS

In a series R-L-C circuit

1. Current (I), resistance (R), true power (W), and voltage drop (V_R) are in phase.
2. I lags V_L by 90° and leads V_C by 90°.
3. Impedances R, X_L and X_C are in phase with their respective voltage drop V_R, V_L, and V_C.
4. Net reactive voltage $V_x = V_L - V_C$.
5. Net reactance $X_T = X_L - X_C$.
6. Net reactive power $P_x = P_L - P_C = I^2 X_T = VI \sin \underline{/\theta}$.
7. True power $= I^2 R = VI \cos \underline{/\theta}$.
8. Apparent power $= I^2 Z = VI$
   $= \sqrt{(\text{watt})^2 + (\text{VARs})^2}$.
9. Power factor $= \cos \theta = \dfrac{R}{Z} = \dfrac{V_R}{V} = \dfrac{\text{watts}}{\text{VA}}$.
10. Other useful equations:
    $V_L = I \times X_L$ $V = I \times Z = \sqrt{V_R^2 + (V_x)^2}$
    $V_C = I \times X_C$
    $V_R = I \times R$ $Z = \sqrt{R^2 + X_T^2}$

11. At series resonance:
    $X_L = X_C$, $V_L = V_C$, $R = Z$, P.F. $= 1$,
    $f_r = \dfrac{1}{2\pi\sqrt{LC}}$, and I is at a maximum and in phase with V.
12. The current (I) is used as the reference phasor.
13. In a net inductive circuit: $V_L > V_C$, $X_L > X_C$, $P_L > P_C$ and I lags V by the angle $\underline{/\theta}$.
14. Inductive VARs is considered positive and capacitive VARs negative.

In a parallel R-L-C circuit

1. $V = V_R = V_L = V_C$
2. $I_R = \dfrac{V}{R}$ $I_L = \dfrac{V}{X_L}$ $I_C = \dfrac{V}{X_C}$
3. $I = \dfrac{V}{Z} = I_R + I_L + I_C$ (phasor addition)
4. For a parallel branch:
   conductance $G = \dfrac{R}{Z^2}$; susceptance $B = \dfrac{X}{Z^2}$
5. Total conductance of all branches:
   $G_T = G_1 + G_2 + G_3 +--$.
6. Total susceptance of all branches:
   $B_T = B_1 + B_2 + B_3 +--$.
7. Circuit admittance $Y_T = \dfrac{1}{Z_T} = \sqrt{G_T^2 + B_T^2}$
8. Power factor $= \cos \underline{/\theta} = \dfrac{I_R}{I_T} = \dfrac{\text{watts}}{\text{VA}}$
9. True power $= VI \cos \underline{/\theta} = I_1^2 R_1 + I_2^2 R_2 +---$.
10. Net reactive power $= VI \sin \underline{/\theta} = I_1^2 X_1 + I_2^2 X_2 +---$.
11. The voltage phasor is used as the reference phasor.
12. At parallel resonance $I = I_R$, $I_L = I_C$, $Z = R_T$, $P_L = P_C$, P.F. $= 1$, watts $= V \times I$, impedance is at a maximum, current is at a minimum and in phase with V.

Series to parallel circuit conversion
The subscripts "P" denotes parallel equivalent and "S" series equivalent.

$$R_P = \frac{R_S^2 + X_S^2}{R_S} \qquad X_P = \frac{R_S^2 + X_S^2}{X_S}$$

ESSAY AND PROBLEM QUESTIONS

Where not stated otherwise, assume the supply voltage to be 60 Hz.

1. A series R-L-C circuit was determined to have a lag power factor. Comment on the properties of: (a) the circuit current, (b) the circuit reactance, (c) the circuit reactive voltage drops, and (d) the circuit reactive power.

2. In a series R-L-C circuit, $R = 20\ \Omega$, $X_L = 60\ \Omega$, $X_C = 40\ \Omega$, and $V = 240\ V$. (a) Sketch the phasor diagram for the circuit and determine the following:
    (b) the circuit impedance
    (c) the circuit phase angle and power factor
    (d) the circuit current
    (e) the voltage drops, V_R, V_L, and V_C
    (f) the true and net reactive power components
    (g) the circuit apparent power

3. In a series R-L-C circuit, $R = 30\ \Omega$, $X_L = 20\ \Omega$, $X_C = 40\ \Omega$, and $V = 120\ V$.
    (a) Sketch the circuit phasor diagram and determine the following:
    (b) the circuit impedance
    (c) the circuit current
    (d) the circuit power factor
    (e) the resistance voltage drop and net reactive voltage drop
    (f) the true and net reactive power components
    (g) the circuit apparent power

4. The true power to a series R-L-C circuit is 950 W, and $R = 38\ \Omega$, $X_L = 50\ \Omega$, and $X_C = 30\ \Omega$.
    (a) Sketch the phasor diagram for the circuit and determine the following:
        (i)   the circuit current
        (ii) the voltage drops V_R, V_L, and V_C
        (iii) the source voltage
        (iv) the circuit impedance
        (v)   the circuit power factor

5. A R-L-C series circuit has a source voltage of 220 V, a current of 8 A, and a lag power factor of 0.6:
    (a) Sketch the circuit phasor diagram and determine the following:
        (i)   the circuit impedance
        (ii) the circuit resistance and net reactance
        (iii) the circuit true and net reactive power components
        (iv) the circuit apparent power

6. Describe the properties of a series R-L-C resonant circuit. Comment on the following:
    (a) circuit current
    (b) circuit impedance and reactance
    (c) circuit power factor
    (d) net reactive circuit power
    (e) true power transferred to the load resistance

7. A series R-L-C circuit has $R = 40\ \Omega$, $L = 20\ mH$, and $C = 32\ \mu F$. Determine the frequency at which this circuit will be resonant.

8. (a) A series R-L circuit has $R = 25\ \Omega$, $X_L = 60\ \Omega$, and $V = 220\ V$. Determine the following:
        (i)   circuit impedance
        (ii) the circuit current, and
        (iii) the true power taken by the circuit.
    (b) This circuit is to be made series resonant. Determine the following:
        (i)   the series capacitance required
        (ii) the new circuit impedance
        (iii) the new circuit current, and
        (iv) the true power taken by the circuit

9. A parallel R-L-C circuit was determined to have a lagging power factor.
    (a) Sketch the circuit phasor diagram and comment on the following:
        (i)   the net reactive circuit current
        (ii) the net reactive circuit power

10. In a parallel R-L-C circuit as shown in Fig.

13-8, $R = 40\ \Omega$, $X_L = 30\ \Omega$, $X_C = 24\ \Omega$, and $V = 120$ V.

(a) Sketch the phasor diagram for this circuit and determine the following:
   (i) the branch currents, I_R, I_L, and I_C
   (ii) the line current
   (iii) the circuit power factor and phase angle
   (iv) the circuit true and net reactive power
   (v) the circuit apparent power
   (vi) the circuit impedance

11. If in the parallel circuit of Fig. 13-8, $I_T = 12$ A, $V = 360$ V, and the circuit phase angle is 25° lag, determine:
   (a) the circuit true power
   (b) the circuit in-phase current, I_R
   (c) the circuit net reactive current, I_X
   (e) sketch the circuit phasor diagram

12. If in the parallel circuit of Fig. 13-8, $I_R = 8$ A, $R = 30\ \Omega$, $X_L = 60\ \Omega$, and $X_C = 80\ \Omega$, determine the following:
   (a) the supply voltage, V
   (b) the branch currents I_L and I_C
   (c) the total circuit current
   (d) the circuit impedance
   (e) the circuit power factor

13. Assume the parallel *R-L-C* circuit of Fig. 13-11(a) is in parallel resonance, and comment on the following circuit properties:
   (a) the line current I_T and in-phase current I_R
   (b) the reactive currents I_L and I_C
   (c) the circuit reactances X_L and X_C
   (d) the inductive and reactive power components
   (e) the circuit power factor

14. (a) What change(s) must be made to the capacitive branch in the circuit of Q. 12, so that the circuit will become resonant?
   (b) What will be the line current for this circuit when it is made resonant?

15. Draw the graphs to illustrate the current versus frequency for:
   (a) a series *R-L-C* circuit
   (b) a parallel *R-L-C* circuit
   What will happen to the current in the series circuit at resonance if the resistance is made very small?

16. An industrial load contains a resistance of 80 Ω in parallel with an inductance of 60 Ω, and is connected across 480 V. Determine:
   (a) the capacitance required to correct the circuit power factor
   (b) the circuit current before it is modified, and
   (c) the circuit current after it is modified

17. (a) An industrial load contains a resistance of 40 Ω in series with an inductance of 60 Ω and connected across a 300-V supply. Determine:
   (i) the circuit impedance
   (ii) the line current
   (iii) the circuit power factor
   (iv) the circuit reactive power
   (b) The power factor of this load is to be corrected by connecting a capacitance in parallel across the load. Determine:
   (i) the reactive power which this capacitance must produce
   (ii) the reactance of this capacitor
   (iii) the current through this capacitor

18. Convert the *R-L* series circuit load in Q. 17
   (a) into its parallel equivalent circuit, and
   (b) then determine the capacitive reactance required to correct the power factor of this equivalent circuit. Compare the result with that obtained in Q. 17 (b).

19. An industrial load is operated on a voltage of 600 V and draws a line current of 15 A with a lag angle of 40°. Determine the following:
   (a) the in-phase and reactive current components for this circuit

(b) the true and reactive power components for the circuit
(c) the capacitance required to correct the power factor of this circuit

20. An industrial motor operates on a voltage of 400 V, and draws a true power of 2000 W at 12 A. Determine the following:
(a) the circuit power factor
(b) the circuit reactive power
(c) the capacitance required to correct the circuit power factor
(d) the new line current to the modified circuit

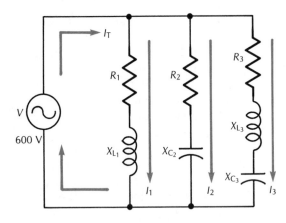

Figure 13-22 A series-parallel circuit.

21. In the series-parallel circuit shown in Fig. 13-22, $R_1 = 40\ \Omega$, $X_{L_1} = 30\ \Omega$, $R_2 = 60\ \Omega$, $X_{L_2} = 80\ \Omega$, $R_3 = 20\ \Omega$, $X_{L_3} = 70\ \Omega$, and $X_{C_3} = 60\ \Omega$,
(a) determine the impedance for each of the three branches
(b) determine the current and phase angle for each of the branches
(c) sketch the phasor diagram for these three branch currents
(d) determine the total line current and its phase angle
(e) determine the circuit impedance
(f) determine the circuit true, reactive, and apparent power

22.
(a) Define the following terms for a series R-L-C branch of a circuit:

(i)   conductance
(ii)   susceptance
(iii) admittance
(iv) impedance
(b) How will you find the total susceptance in a parallel circuit containing three branches?

23. For the circuit values given in Q. 21, determine the following:
(a) the impedance of each branch
(b) the conductance of each branch
(c) the net susceptance of each branch
(d) the admittance for the complete circuit
(e) the total circuit impedance
(f) the net equivalent resistance in the circuit
(g) the net equivalent reactance in the circuit

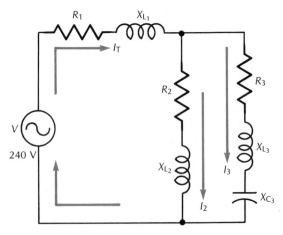

Figure 13-23 A series-parallel circuit.

24. In the series-parallel circuit shown in Fig. 13-23, $R_1 = 25\ \Omega$, $X_{L_1} = 40\ \Omega$, $R_2 = 60\ \Omega$, $X_{L_2} = 80\ \Omega$, $R_3 = 50\ \Omega$, $X_{L_3} = 30\ \Omega$ and $X_{C_3} = 70\ \Omega$.
Using the G-B-Y method, determine:
(a) the net circuit resistance
(b) the net circuit reactance
(c) the circuit impedance
(d) the circuit current
(e) the circuit power factor

CHAPTER
FOURTEEN

TRANSFORMERS

A transformer is an electrical device to which electrical energy is supplied at a definite voltage and from which electrical energy is obtained at a different voltage. A transformer is thus a voltage-changing device and is *not* a device in which electrical energy is converted into another form of energy, such as occurs in generators and motors.

A transformer is an efficient, relatively simple and quiet device that has no moving parts. Transformers can be used with very high voltages because there are no rotating parts. Efficiencies obtained in transformers range as high as 99 percent.

Large transformers are mainly used to step-up the voltage output of generators, which is generally 13.8 kV to 500 kV for transmission across long distances. When electrical energy is transmitted at relatively low voltages, there is considerable power loss as energy is converted to heat energy as the electrons overcome the resistance of the circuit conductors. This power loss in watts is given as:

$$P = I^2R$$

where P is the power in watts

    I is the current value in amperes

    R is the resistance of the line conductors in ohms

It is not practical to reduce line power losses by decreasing the resistance of the conductors. Since a conductor's resistance is inversely proportional to its cross-sectional area, the size of the line conductors would have to be increased substantially to accomplish a major decrease of line power losses. Economically and practically this is not feasible where long distances are involved.

The I^2R losses are proportional to the square of the current. When the current is reduced by one-half, the line power losses are decreased by a factor of four. When the current is decreased to one-third, the line power losses are decreased by a factor of nine.

The total power transmitted is not basically affected by a lowered current if the voltage is increased proportionally. The transmission, then, of a constant amount of power with minimum line power losses is accomplished by increasing the voltage V. Since $P = VI$, any decrease in I is only possible with an increase in V. If the current is to be decreased by a factor of four, the voltage must be increased by a factor of four. The above is the *main* reason for the use of transformers.

The voltage of 13.8 kV generated by an ac generator is increased to 500 kV (or to another value) by a *step-up* transformer. The voltage transmission lines may be many hundreds of kilometres in length before reaching the area where the electrical energy is to be used. It is obvious that the various energy consumers cannot make use of this energy at the voltage at which it is being transmitted across the long distances. The high voltage is, therefore, lowered by *step-down* transformers to a value which is suitable for the consumers. To accommodate the needs of consumers, transformers are built in a great variety of voltage values and capacities.

TRANSFORMER CONSTRUCTION

Transformers are classified into three basic types. These are: the core-type, the shell-type, and the H-type, sometimes called the distributed-core type.

Figure 14-1 shows a single-phase core built up of thin laminations. This core forms a closed magnetic circuit through which the main magnetic flux circulates. The core is built up with thin laminations of silicon steel insulated from one another. The core not only serves as a main magnetic circuit, but its side legs also mechanically support the primary and secondary coils. These coils are supported in bobbins or coil forms and are impregnated with insulation. The leads from the coils are carefully supported and are

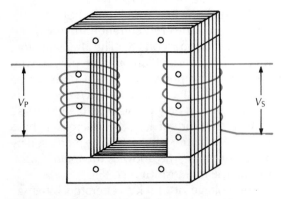

Figure 14-1 A laminated transformer core with primary and secondary windings.

brought out to a terminal board or through insulating bushings to specially designed terminals. This whole core assembly with primary and secondary coils is supported and clamped very carefully in an upright position.

The clamping is necessary to prevent the laminations from moving under the magnetic forces developed. The whole assembly is placed in a steel tank, except for the very smallest size transformers.

Larger transformers are filled with an insulating transformer oil that serves also as a coolant to transmit the heat produced in the coils to the outside surface.

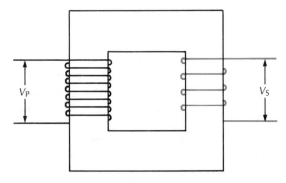

Figure 14-2(a) Primary and secondary coil mounted on a laminated core-type transformer.

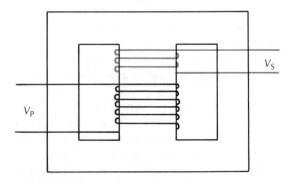

Fig. 14-2(b) Primary and secondary coils mounted on a laminated shell-type transformer coil.

The core-type transformer (Fig. 14-2(a)) has a longer length of core and therefore a longer magnetic circuit than the shell-type transformer. The core-type transformer also has a lighter core of smaller cross-sectional area. An advantage of the core-type transformer is that its relatively large winding space is better suited for high voltages which require many coil turns and heavier insulation.

In the *shell-type transformer core*, the legs are placed horizontally. In this type of core, the leakage flux is minimal because the coil windings surround the centre core and the outer core surrounds the coils. This core is also constructed of thin silicon steel laminations which are insulated from one another.

The primary and secondary voltage coils for the shell-type transformer core are intermixed in order to reduce magnetic leakage. These intermixed primary and secondary coils are both wound on the centre leg of the shell-type core. (See Fig. 14-2(b)) The shell-type core is used when better provision is required for mechanically supporting and bracing the primary and secondary coils.

A third type of transformer core used is the *H-type or distributed-core type*. This type of core is a modification of the two types of cores discussed above.

The transformer core consists of four smaller cores whose magnetic paths are in parallel. The core is cruciform, as shown in Fig. 14-3. The primary and secondary coil windings are wound on the centre leg of the core, and the outside legs of the core structure surround the coils. As a result, the leakage flux in this core is very small.

This type of transformer has two high-voltage coils and two low-voltage coils. Figure 14-3 indicates how the two high-voltage coils are wound between the two low-voltage coils. This type of arrangement ensures that the insulation requirements are kept to a minimum. Considerable space is left between the primary and secondary coils. This allows for the insulating and cooling oil to circulate from the bottom to the top of the transformer tank. The heat produced in the coils and core is thus conducted away from the coils to the tank surface.

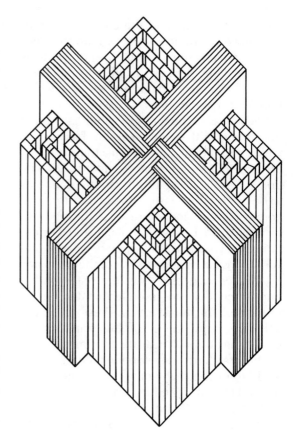

Figure 14-3 The H-type transformer core with double coils.

The H-type transformer is sometimes called a wound-core transformer and is not as often used today as the core-type and shell-type transformers. Several modifications of the above types of transformer cores are available and the use of a specific type will depend on its particular application. The core-type transformer is generally used for high voltages with relatively small output, and the shell-type transformer is often used for lower voltages but with higher output.

The construction of the primary and secondary coil windings is very complex and many factors must be considered by the engineers designing the coil structures. A few of these factors are: the type of insulation required for specific voltages, the spacing available on the core for the location of the coils, whether to use single- or double-layer cylindrical windings, and many others. The types of windings used are dependent on the differing mechanical and electrical characteristics of specific transformer applications.

Regardless of the types of windings used in transformers, there are certain basic rules which must be followed when a transformer is assembled. The Canadian Standards Association (CSA) has established various standards to be used in the construction of transformers. Some of these standards relate to the methods of cooling to be used, voltage and current ratings of coils and leads, power ratings, and various insulation levels. One of the standards established concerns how the leads of the transformer coils are to be terminated on the outside of the transformer. These leads are generally brought out of the transformer's steel casing through insulating bushings.

The standards developed state that the leads from the high-voltage coils are to be marked H_1 and H_2 and the leads from the low-voltage coils are to be marked X_1 and X_2. The actual polarity markings of these leads depend on several factors. The polarity of a transformer refers to the voltage phasor relations of the transformer leads as brought outside the transformer tank. There are two methods of terminating the high- and low-voltage leads on transformers. Figure 14-4 indicates the two types of arrangement possible.

TRANSFORMER POLARITY

In general, most larger transformers have their leads terminated so that they have what is called *subtractive polarity* (Fig. 14-4(b)). When standing behind the low-voltage side of a transformer, the left-side low-voltage lead will be X_1 and the high-voltage lead across from it will be H_1. Transformers having a power rating of 200 kVA or less and a voltage rating of 7.5 kV or less have what is called *additive polarity* (see Fig. 14-4(a)). Transformer polarity is important

when they are to be connected in parallel or for connections discussed in Chapter Fifteen.

TESTING FOR TRANSFORMER POLARITY

By convention, the top-left terminal when the transformer is looked at from its low voltage side is always labelled H_1. To determine the X polarity markings, first measure the output voltage for a given input voltage. Use a low voltage for safety reasons. Secondly, connect one of the input and one output lead, as shown in Fig. 14-4. Now measure the voltage across the other two terminals. The results for

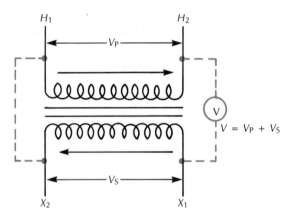

(a) ADDITIVE POLARITY

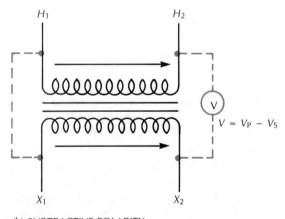

(b) SUBTRACTIVE POLARITY

Figure 14-4 Transformer polarity markings.

additive and subtractive polarities are shown in Fig. 14-4.

PRINCIPLES OF TRANSFORMER OPERATION

The coil of a transformer to which an ac voltage is applied is called the *primary* coil. This coil may be the high- or low-voltage coil. The *secondary* coil delivers a voltage and current to a load. Any transformer may be used as a step-up or a step-down transformer, depending on how the coil connections are made.

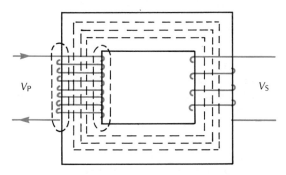

Figure 14-5 The magnetic flux in a core-type transformer.

When a 60-Hz ac voltage is connected to a transformer's primary coil, an ac current begins to flow through the primary coil whose resistance is low. During the first alternation of the ac sine wave voltage, the amplitude of the voltage and the current increases to its peak during the first 1/240 s. As the current increases in amplitude, its magnetic field also expands perpendicular to the turns of the coil. This expanding flux cuts, first of all, through the turns of the current-carrying primary coil. This action produces a voltage across the primary coil counter to the applied voltage. This voltage is called the voltage of *self-induction* or counter voltage or back voltage.

The voltage of self-induction is not quite equal in value to the applied voltage. As a result, assuming that no load is connected to the transformer, a small current will flow

through the primary coil from the voltage source. This current is called the magnetizing current or the no-load current.

During the same 1/240 s, the expanding magnetic flux flows through the iron core and also cuts through the turns of the secondary coil. Therefore, a voltage will also be induced in this coil. This voltage induced in the secondary coil is called the voltage of *mutual induction* because the magnetic flux set up by current flowing in one coil induces a voltage in another coil which is not linked electrically to the first coil.

When the applied voltage has reached its peak value after 1/240 s, the induced voltages are also at their maximum value. During the next 1/240 s, the applied voltage begins to decrease to zero. The current from the source will also begin to decrease to zero during this time. As a result, the flux which had expanded maximally, now begins to contract or to collapse.

The collapsing magnetic field also cuts through the turns of the primary coil in a direction opposite to when the flux was expanding. The previously induced voltage of self-induction begins to decrease in value until at the end of the 1/240 s, the voltage of self-induction is zero just as the applied voltage is at zero. The applied voltage now changes its polarity, and the total induction process is repeated during the next alternation of the cycle. Of course, since the current from the source will now flow in opposite direction, the polarities of the voltage of self-induction and of mutual induction will also be reversed.

In Fig. 14-6, the phasors indicate the relationships in the primary coil with no load applied. V (induced) does not quite equal V (applied) but is 180° out of phase with it. I (source) will lag the voltage by 90° due to the inductive reactance of the coil (resistance is considered negligible). The I (induced) does not completely cancel I (source) and the resultant current is the magnetizing or no-load current in the primary coil.

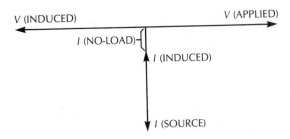

Figure 14-6 Primary coil voltage and current phasors (no-load).

The above discussion has assumed an ideal transformer. However, certain factors prevent it from being ideal, such as the resistance of both coils, and the leakage flux which does occur in all transformers. These factors will result in certain losses which will be discussed later. For most calculations, however, these losses are neglected because they are relatively very small. Hence, the primary self-induced voltage is assumed to equal the applied voltage, and the primary input energy is assumed to equal the secondary output energy.

The total induced voltage of the secondary coil is equal to the volts induced per turn of the primary coil multiplied by the number of turns of the secondary coil. The ratio of the induced voltages of the primary and secondary coils is directly proportional to the ratio of the number of turns in the primary and secondary coils. Therefore

$$\frac{V_P}{V_S} = \frac{N_P}{N_S} \qquad (14\text{-}1)$$

where V_P and V_S are respectively the induced voltages of the primary and secondary coils and N_P and N_S are respectively the number of turns in the primary and secondary coils.

Example 14-1: A transformer has a primary coil of 2500 turns and a secondary coil of 250 turns. 2400 V ac is applied to the primary coil. The voltage induced in the primary coil will almost be 2400 V and the volts induced per turn of the primary coil is $\frac{2400}{2500} = 0.96$ V. The voltage induced in the secondary coil is:

$$V_S = V_P \times \frac{N_S}{N_P} \qquad (14\text{-}2)$$

$$V_S = 2400 \times \frac{250}{2500}$$

$$V_S = 240 \text{ V}$$

The volts induced *per turn* are the same for the secondary coil and for the primary coil and, therefore, the voltage induced in the secondary coil is equal to the volts induced per turn in the primary coil times the number of turns in the secondary coil, or 0.96 x 250 = 240 V.

The turns ratio of a transformer's primary and secondary coils is extremely important as it will determine the factor by which a voltage is stepped-up or is stepped-down. This turns ratio is sometimes called the *transformation ratio* of a transformer.

THE TRANSFORMER UNDER LOAD

When the secondary coil circuit of a transformer is connected to a load through switches or circuit breakers, or both, the transformer is said to be operating under load. The operating characteristics of the transformer change considerably from when there is a no-load condition. The induced secondary coil voltage causes a load current to flow through the load and through the secondary coil. This current is, of course, an ac current and its amplitude and polarity vary at the same rate as the secondary coil voltage varies.

The ac load current flowing through the secondary coil sets up a magnetic flux in the iron core that opposes the magnetic flux produced by the magnetizing current in the primary coil. As a result, the voltage induced in both secondary and primary coils would be lowered. As the induced voltage in the primary coil begins to decrease, the applied voltage sends more current through the primary coil because there is now less voltage of self-induction.

The amount of additional current flowing in the primary coil when a load current flows in the secondary coil is the amount required to strengthen the primary coil's magnetic flux and overcome the effects of the opposing secondary coil's magnetic flux due to the load current. The resultant magnetic flux in the iron core is sufficient to maintain the level of induced voltages in the primary and secondary coils. Any change in load current will result in a corresponding change in primary coil current and the required induced voltage levels will be maintained. This may also be stated as follows: the current in the primary coil varies exactly as the current in the secondary coil.

Neglecting the small magnetizing current in the primary coil, the magnetizing force of the primary coil which is due to the load current in the secondary coil is equal to the magnetizing force of this load current. The magnetizing force is normally expressed in ampere-turns. The ampere-turns, $I_P N_P$, of the primary coil are equal to the ampere-turns, $I_S N_S$, of the secondary coil. This may be stated as follows:

$$I_P N_P = I_S N_S \qquad (14\text{-}3)$$

or

$$\frac{I_P}{I_S} = \frac{N_S}{N_P} \qquad (14\text{-}4)$$

From Eq. 14-2, 14-3, and 14-4, it is evident that the ratio of the number of turns in the primary coil to the number of turns in the secondary coil is equal to the ratio of the induced voltage in the primary coil to the induced voltage in the secondary coil. However, the ratio of the current in the primary coil to the current in the secondary coil is equal to the ratio of the number of turns in the *secondary* coil to the number of turns in the primary coil. Note that this ratio is the reverse of the numbers ratio in Eq. 14-1.

The ratio $\dfrac{V_S}{V_P}$ can be substituted for the ratio $\dfrac{N_S}{N_P}$ in Eq. 14-4 and the ratio of Eq. 14-4 has then been changed to:

$$\frac{I_P}{I_S} = \frac{V_S}{V_P} \qquad (14\text{-}5)$$

This equation, when cross-multiplied, produces the following equation:

$$V_P I_P = V_S I_S \qquad (14\text{-}6)$$

Since power is expressed as $P = VI$, Eq. 14-6 indicates that the output power of a transformer equals the input power to the transformer. This would be true in an ideal transformer and this transformer would have an efficiency of 100 percent. However, due to factors mentioned previously, there are certain losses in a transformer and they will cause the efficiency to be from about 85 percent to 99 percent.

Example 14-2: A transformer rated at 4 kVA has 1800 turns in its primary coil and 200 turns in its secondary coil. An ac voltage of 2400 V, 60 Hz is applied to the primary coil. A load current of 12 A flows in the secondary coil.

Determine:
(a) the voltage induced in the secondary coil
(b) the current flowing in the primary coil
(c) the power supplied by the primary coil (input)
(d) the power supplied by the secondary coil (output)

Solutions:

(a) From Eq. 14-2, $V_S = V_P \times \dfrac{N_S}{N_P}$

$$V_S = 2400 \times \frac{200}{1800}$$
$$V_S = 2400 \times 0.11$$
$$V_S = 264 \text{ V}$$

(b) From Eq. 14-4, $I_P = I_S \times \dfrac{N_S}{N_P}$

$$I_P = 12 \times \frac{200}{1800}$$

$$I_P = 12 \times 0.11$$

$$I_P = 1.32 \text{ A}$$

(c) $V_P \times I_P = 2400 \times 1.32$ (assume P.F. = 1)

$$= 3168 \text{ W}$$

(d) $V_S \times I_S = 264 \times 12$

$$= 3168 \text{ W}$$

Since efficiency $= \dfrac{\text{output watts}}{\text{input watts}} \times 100,$

the above example indicates that the transformer is 100 percent efficient. In practice, this never occurs due to losses resulting from the following:

1. There are some magnetic leakage losses in all transformers.
2. There are eddy-current losses.
3. The alternating current causes the magnetic circuit to be magnetized first in one direction and then in the other direction. Energy is required to compensate for the hysteresis effects resulting from this.
4. Both windings have some resistance and as a result I^2R losses occur.

As mentioned above, transformer efficiency varies from about 85 percent to 99 percent, and the larger a transformer is, the higher its efficiency is generally.

PHASOR DIAGRAMS OF A TRANSFORMER

When the efficiency of a transformer is close to 100 percent, one may consider such a transformer as an ideal transformer and to understand better the voltage, current, and flux relationships in this transformer, phasor diagrams are used. Figure 14-7 is a phasor diagram of an ideal transformer with no load applied.

The no-load or magnetizing current in the primary coil, (I_M), is the phasor sum of the induced current (I_n) and the in-phase source current (I_R). This no-load current produces the

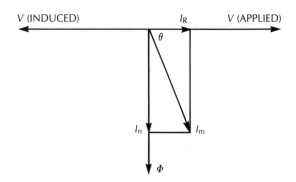

Figure 14-7 No-load transformer phasor diagram.

flux (ϕ) which lags the applied voltage by 90°. The no-load current lags the applied voltage by $\underline{/\theta}$. The cosine of this angle would give the no-load power factor for the transformer.

It may happen occasionally that a transformer is used to supply a purely resistive load. Figure 14-8 shows the phasor diagram for the phasor relationships that exist under this condition.

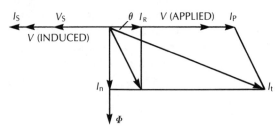

Figure 14-8 Phasor diagram for transformer with resistive load.

Note that since there is a resistive load, load current I_S is in phase with the secondary coil voltage V_S, as well as with the voltage of self-induction in the primary coil (V_P). Due to the load current, there is an increased flow of current in the primary coil and this total primary coil current I_T is equal to the phasor sum of the in-phase current I_P from the voltage source and the magnetizing current I_n. Note that the resistive load current I_S causes the angle of lag between the total primary current and the applied voltage to become much smaller. The power factor of the transformer has thus improved.

Most loads connected to transformers have resistive and reactive components, and the transformer coils also have resistance and reactance components. Figure 14-9 is a phasor diagram for such a condition. For the sake of simplicity, it has been assumed that there is a 1:1 ratio of voltage transformation.

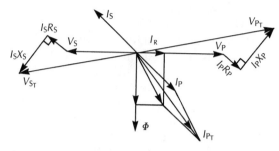

Figure 14-9 Phasor diagram of a transformer connected to a resistive and reactive load (inductive).

To make the diagram more legible, only a few of the normally required construction lines have been drawn. When current I_P flows in the primary coil, it causes voltage drop $I_P R_P$ across its resistance and voltage drop $I_P X_P$ across its inductive reactance. These voltage drops are 90° out of phase with each other. It is obvious that the coil's inductive reactance is much greater than its resistance.

When current I_S flows in the secondary coil, it causes a voltage drop $I_S R_S$ across its resistance and a voltage drop $I_S X_S$ across its inductive reactance. Note that these voltage drops are also 90° out-of-phase with each other.

Voltage phasor V_{P_T} represents the total applied voltage and is the phasor sum of phasors $I_P R_P$, $I_P X_P$, and V_P. The voltage drop $I_P R_P$ is in phase with the total current in the primary coil, I_{P_T}. The voltage drop $I_P X_P$ is 90° out of phase with current I_{P_T}.

When load current I_S flows in the secondary coil, the phasor V_{S_T} represents the total induced voltage of the secondary coil and V_S represent the secondary terminal voltage under the inductive load.

The phasor diagrams for transformers vary tremendously because the impedance of the load will determine the amount of secondary coil current and, as a result, the amount of primary coil current. These phasor diagrams tend to be very detailed and complex.

TRANSFORMER LOSSES AND EFFICIENCY

Since a transformer has no rotating parts, it has no mechanical losses. The power losses in a transformer consist of copper losses and iron losses. Both losses are heat losses. The copper losses consist of I^2R losses in both primary and secondary coils. Several methods may be used to determine the amount of I^2R losses in a transformer. One of these methods is called the *Short-Circuit Test* (Fig. 14-10).

In this test, the *secondary coil* is short-circuited and a wattmeter, ammeter, and voltmeter are connected as shown. A rheostat is con-

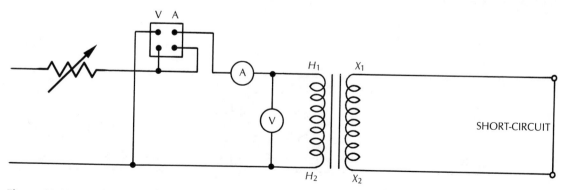

Figure 14-10 Transformer short-circuit test.

nected in series with the primary coil so that the input current may be controlled. Prior to applying the ac voltage, the rheostat is adjusted so that maximum resistance is inserted in the circuit. After the ac voltage is applied, the rheostat is slowly adjusted until the ammeter reading is equal to the *rated current value of the primary coil*. At this time, the ammeter reading, voltmeter reading, and wattmeter reading are recorded, and the rheostat is turned so that its maximum resistance is again in series with the primary coil. The voltage supply is then disconnected from the transformer.

The impedance Z is then determined from $Z = \frac{V}{I}$. It will be found that Z is very low. The voltage recorded on the voltmeter is only about 3 to 5 percent of the rated voltage for the primary coil. This voltage, which causes the rated coil current flow is sometimes called the *impedance voltage*.

The wattmeter in Fig. 14-10 will indicate the total copper losses of the transformer. The wattmeter reading indicates the input power, during the short circuit test, that is used to overcome the total copper losses.

The iron losses in a transformer may be determined through an *Open-Circuit Test* (Fig. 14-11).

During the open-circuit test there are almost no copper losses in the secondary coil and none in the open primary coil. The wattmeter reading accounts for the iron losses in the magnetic circuit. These losses include eddy-current losses and hysteresis losses discussed earlier.

The open-circuit test must be done very carefully. Note that the ac voltage will be supplied to the *secondary coil* and that the voltage induced in the open-circuit primary coil may be very high, depending on the turns ratio of the coils.

In addition to the meters shown in Fig. 14-11, a rheostat is connected in series with the secondary coil. Also, the open high-voltage terminals should be insulated. Prior to applying the ac voltage, the rheostat is adjusted so that maximum resistance is inserted in the circuit. After the ac voltage is applied, the rheostat is adjusted slowly until the voltmeter indicates the *rated voltage of the secondary* coil. At this voltage level, the ammeter reading, the voltmeter reading, and the wattmeter reading are recorded. The rheostat is then adjusted until its maximum resistance is again in series with the secondary coil and then the applied ac voltage is disconnected from the circuit.

The power reading on the wattmeter represents the power necessary to force the required flux to overcome the reluctance of the magnetic circuit and the power necessary to supply the eddy-current and hysteresis losses.

The maximum current flowing during the open-circuit test is only about 1 to 7 percent of the normal full-load current in the secondary coil. During an open-circuit test on a trans-

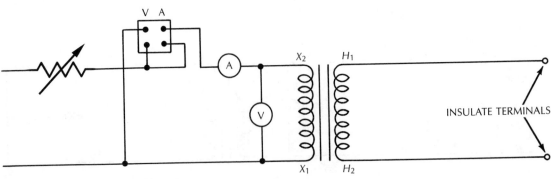

Figure 14-11 Transformer open-circuit test.

former, about 0.25 of the power is used compared to that used during the short-circuit test.

The efficiency of a transformer is very high and may be found as follows:

$$\text{efficiency } (\eta) = \frac{\text{output watts}}{\text{output watts + total losses}}$$

The total losses consist of the measured I^2R losses and the iron losses.

Example 14-3: The total losses in a transformer were found to be 46 W. The output power was 860 W. Determine the transformer efficiency.

$$\eta = \frac{860}{860 + 46}$$
$$\eta = 94.9\ \%$$

The discussion of transformer operation may have left the impression that the transformer's only purpose is to transform a given voltage value to a higher or lower value. Although this is basically the purpose of all transformers, except the isolation transformer, transformers are generally classified in terms of the type of specific service they provide, other than voltage transformation. Some of the transformer classifications are:

– high-voltage power transformers
– distribution transformers
– low-power transformers
– instrument transformers consisting of current
  transformers and potential transformers
– street-lighting transformers
– autotransformers
– regulating transformers
– grounding transformers
– display sign transformers
– controls and signals transformers
– chimes and bells transformers

INSTRUMENT TRANSFORMERS

High voltages and high currents cannot be recorded directly through the use of voltmeters and ammeters. Instrument transformers step-down the voltage and current of a circuit to a value that can be safely used for the efficient operation of voltmeters, ammeters, wattmeters, varmeters, telemetering equipment, and relay equipment. There are basically two types of instrument transformers. *The current transformer* is used to reduce high currents, and *the potential or voltage transformer* is used to reduce high voltages to voltages suitable for metering equipment.

Instrument transformers have a small VA capacity. These transformers are manufactured very carefully so that they are highly accurate in voltage and current transformation. The construction of instrument transformers is quite different from other types of transformers and they are usually quite small in size.

CURRENT TRANSFORMERS

Current transformers have a primary coil of sufficient capacity to carry the line current, which may be very high. In very large current applications, the transformer may not have a primary coil, but the main line conductor is made to go through an iron core on which the secondary coil is wound. The secondary coil is almost always designed for 5 A maximum. This transformer is designed with various standardized current ratios. It is sometimes called a series transformer because its primary coil must be connected in series with the line conductor.

The current ratings of the primary coil of a current transformer depend on the value of the current to be carried. If the current rating of the primary coil is 300 A and of the secondary coil is 5 A, there will be a ratio of turns of 300:5. This means that the secondary coil will have 60 times as many turns as the primary coil. The ratio of primary coil current to secondary coil current is inversely proportional to the ratio of the primary coil turns to the secondary coil turns, that is

$$\frac{I_P}{I_S} = \frac{N_S}{N_P}$$

For example, in the current transformer above, the primary coil would carry a current of 300 A and the secondary coil would carry 5 A. If the primary coil current was decreased to 250 A, the secondary coil current would be 4.17 A. The ratio of 60:1 would apply to any value of load current less than 300 A.

The fact that current transformers consist of turns of insulated wire introduces the factors of resistance and inductive reactance. If nothing were done to compensate for these factors, there would be considerable power losses and the readings obtained would not be very accurate. These factors are mainly compensated for by:

(a) using a steel for the magnetic circuit that has a very high permeability
(b) purposely keeping the flux density low so that the negative effects of the magnetizing currents are kept low
(c) decreasing the number of turns of the secondary coil slightly so that more current will flow in this coil than there usually is

When connecting a current transformer in a circuit, great care must be taken. If no external load is connected immediately but the transformer is to be tested first, the secondary coil *must* be short-circuited. When a large line current flows through the primary coil, its strong magnetic flux will induce a very high voltage in the secondary coil which will endanger persons working with the transformer. Also, the insulation of the turns of the secondary coil will not be rated for such a high voltage and rupturing of the insulation may occur.

All current transformers are wound to have subtractive polarity, and the secondary coil circuit is always grounded for reasons of safety. There are three types of current transformers, the wound type, the window type, and the bar type.

The wound-type current transformer, as shown in Fig. 14-12, has separate primary and secondary coils mounted on a common laminated iron core. The primary coil consists of

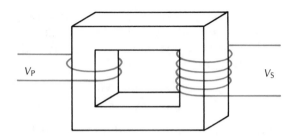

Figure 14-12 A wound-type current transformer.

one or more turns of heavy wire conductor which is connected in series in the circuit whose current is to be measured.

A window-type current transformer (Fig. 14-13) is sometimes called a through-type transformer. This transformer has a cylindrical iron core, which is doughnut-shaped, built of thin laminations as a magnetic circuit. There is no primary coil and the core fits around a high-current conductor. These transformers are often used around the bushings of large circuit breakers. The current flowing in the conductor which leaves a circuit breaker through insulating bushings induces a current in the secondary coil. This secondary coil consists of insulated copper turns of conductor which are wound around the iron core. The total core and coils are often enclosed in an insulated mold. These transformers are also often used to obtain low currents for test equipment.

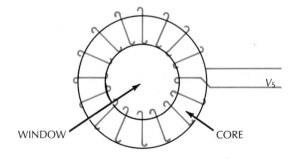

Figure 14-13 A window-type current transformer.

The third type of current transformer is the bar-type. *A bar-type current transformer* is basically a window-type transformer with a solid copper bar placed permanently through the hole. These transformers are completely enclosed except for the ends of the copper bar and the terminals of the secondary coil. The primary connections are made from the bar to the high-current circuit in which the transformer is to be connected. These transformers are designed to have a large primary current rating.

The type of current transformer to be used will be determined by the specific application. Current transformers are classified as dry, oil-filled, or compound-filled for indoor or outdoor use.

POTENTIAL OR VOLTAGE TRANSFORMERS

A potential transformer is used for the measurement of high voltages of circuits indirectly. Voltmeters cannot measure these voltages directly. This transformer operates on the same principles as the transformers discussed previously. The power ratings of these transformers vary from 100–500 VA. The secondary coil is generally wound to supply 120 V to instruments. The primary coil may be wound for connection to very high-voltage circuits (Fig. 14-14).

Just as with the current transformer, one side of the secondary coil is grounded to ensure safety for persons working on it. These

transformers are also wound for subtractive polarity. There are various types of potential transformers. For circuit voltages greater than 25 kV, the transformers are designed for outdoor use. The high-voltage transformer will have one insulating bushing (Fig. 14-15(a)) for the lead to the primary coil if the voltage to be measured is between line and ground. A double-bushing transformer is used when the leads from the primary coil are to be connected to two lines for the measurement of the voltage between lines (Fig. 14-15(b)).

Potential transformers are designated by

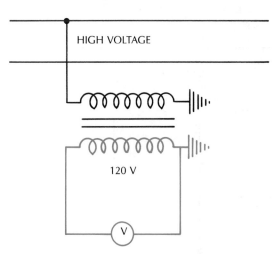

Figure 14-15(a) A one-bushing potential transformer.

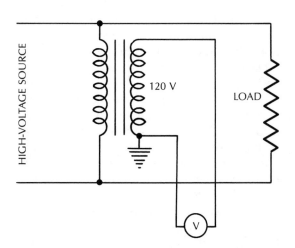

Figure 14-14 Potential transformer connections.

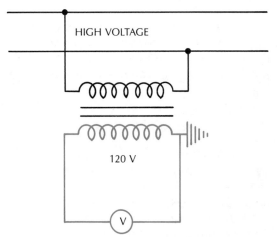

Figure 14-15(b) A two-bushing potential transformer.

voltage ratio. For instance, a 13.8-kV line would require a 13 800:120 = 115 or a 115:1 voltage ratio potential transformer. These transformers are manufactured in standard voltage ratios for all possible voltages available.

Potential transformers for voltages above 100 kV are now generally made to incorporate both capacitors and primary and secondary coils. *The capacitor potential transformer* consists of a capacitive voltage divider and a voltage transformer. Several high-voltage capacitors are connected in series with the primary coil, as shown in Fig. 14-16.

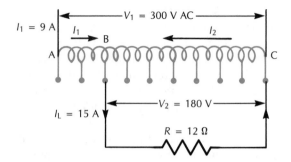

Figure 14-17 A step-down autotransformer (5:3 ratio).

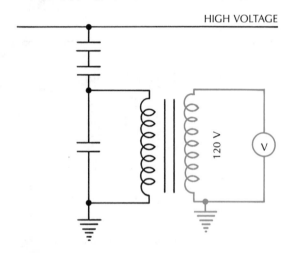

Figure 14-16 A capacitor potential transformer

This type of transformer is used to decrease the voltage connected to the primary coil. This is more economical as the primary coil does not have to be insulated for the very high voltages.

AUTOTRANSFORMERS

An autotransformer is a special transformer in which the primary and secondary circuits share a common winding. Autotransformers are used in the starting of rotating machinery such as synchronous and induction motors. The synchronous motor is started at about

one-third of its normal operating voltage.

Figure 14-17 shows how the primary and secondary coils are connected in series. If a voltage is applied across the coils connected in series, a voltage and current may be supplied from either end of the series-connected coils and a tap which may be brought out anywhere along the coil. Autotransformers often have quite a series of percentage taps available so that different voltages may be obtained from the transformer.

The ratio of voltage transformation, as in a two-coil transformer, is equal to the ratio of primary coil turns to secondary coil turns.

The single coil with various percentage taps is wound on a laminated silicon steel core and the magnetic circuit is common to both the primary and secondary turns of the coil. When an ac voltage is applied to the coil, an ac current will flow through the coil and its magnetic flux will include a voltage of self-induction or counter voltage equal to the applied voltage. When a load is connected across B and C, as in Fig. 14-17, the counter voltage induced in this section of the coil will send a current through the load and through section B to C of the coil. This section of the coil can be considered to be the secondary coil of the transformer. The currents in the two sections of the whole coil are the same as those in a two-coil transformer.

Autotransformers may be used as step-up or step-down transformers. The ratio between the sections of the coil is seldom greater than

5:1. The reason for this is that a higher ratio would result in having to insulate the total coil for the highest voltage applied to the coil. This extra cost for an autotransformer would result in it being less economical than a two-coil transformer.

An input voltage of 300 V is applied to the autotransformer in Fig. 14-17. Since the resistive load of 12 Ω is connected between the bottom side of the coil and a tap at 60 percent of the total turns, the voltage induced in the secondary coil part of the autotransformer is 300 × 0.60 = 180 V. This induced voltage across the load opposes the applied voltage and causes a current of 15 A to flow through the load.

$$I_L = \frac{180}{12}$$

$$I_L = 15 \text{ A}$$

Recall that $P = V \times I \times \cos \underline{/\theta}$.

In this exercise, therefore, the power output would be 180 × 15 × 1.0 = 2700 W. Assuming that there are no power losses and that therefore the power input equals the power output, the power input would also be 2700 W. The current flowing from the voltage source would be:

$$I_1 = \frac{P \text{ (input)}}{V \text{ (input)}}$$

$$I_1 = \frac{2700}{300}$$

$$I_1 = 9 \text{ A}$$

The source current of 9 A flows through winding A - B. The load requires 15 A, and thus a current of 6 A will flow through coil C - B. The 9 A and the 6 A combine to flow through the load. The secondary load current is then $I_1 + I_2$. The ratio of voltage transformation in the autotransformer is

$$\frac{V_1}{V_2} \text{ or } \frac{300}{180} = 1.67{:}1 \text{ (5:3)}$$

In Fig. 14-17, the 300 V applied causes a current of 9 A to flow from A to B. At point B,

the voltage level is 180 V. There is then a voltage drop of 120 V between A and B and a power value of 120 × 9 = 1080 W. The voltage and current in this section of the coil are responsible for the flux that induces the 180 V from C to B. The power in the section C - B is 1080 W. The load has a power requirement of 180 × 15 = 2700 W.

The 9 A from point A of the coil is forced through the load at 180 V. The power for this is 180 × 9 = 1620 W. Thus, the load equals the power from section C - B of the coil (180 × 6) and the power supplied by the current in the primary coil (180 × 9).

$$2700 \text{ W} = 1080 \text{ W} + 1620 \text{ W}$$

The total load power is equal to the product of the voltage across the secondary coil and the sum of the coil currents flowing through the load.

The autotransformer in Fig. 14-17 is a step-down transformer. Autotransformers may also be used as step-up transformers.

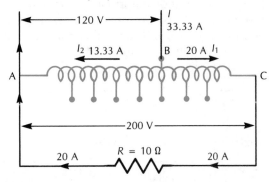

Figure 14-18 A step-up autotransformer (3:5 ratio).

The resistive load of 10 Ω will have a current of 20 A flow through it as a result of a voltage of 200 V between A and C. The power output to the load is $P = V \times I \times \cos \underline{/\theta}$.

$$P = 200 \times 20 \times 1.0$$
$$P = 4000 \text{ W}$$

Assuming no power losses, the input power will also be 4000 W. The current from the voltage source will then be

$$I = \frac{P \text{ (input)}}{V \text{ (input)}}$$

$$I = \frac{4000}{120}$$

$$I = 33.33 \text{ A}$$

This means that 20 A flows through section B - C of the coil and 13.33 A flows from B to A of the coil. The 13.33 A flows at a voltage of 120 V through coil A - B. The power of this section of the coil is 120 × 13.33 = 1600 W. The 20 A flowing from B to C is flowing at a voltage of 120 V. The power associated with this current is 120 × 20 = 2400 W. The total power of the load consists of this 2400 W and the 1600 W of the primary coil. Thus the total auto-transformer power output is 4000 W.

An autotransformer has two very important limitations. The first one concerns the effects of the transformer's impedance. The impedance of an autotransformer is much less than that of the two-coil transformer. This low impedance would result in a very high current flow in the event of a short-circuit of the transformer. These transformers cannot be built economically to withstand this short-circuit current. Sometimes external impedances are provided for autotransformers.

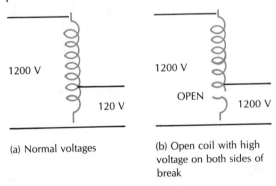

(a) Normal voltages

(b) Open coil with high voltage on both sides of break

Figure 14-19 An open coil in an autotransformer.

Another limitation of an autotransformer concerns a voltage hazard resulting from an open coil. It is possible that the section of the total coil which is common to both secondary and primary may become open. When this happens, the total source voltage will exist across the secondary and the load connected to it. The danger from electrocution would be very real because either side of the break in the coil would be at a very high potential energy level. Figure 14-19 shows the diagrams for this dangerous situation.

TAP-CHANGING TRANSFORMERS

When voltages are transmitted over long distances, there will be a voltage drop along the length of the conductors. The voltage applied to a transformer will then be somewhat lower than the rated primary coil voltage. The voltage obtained from the secondary side of the transformer will then also be lower than its rated voltage value. Some transformers have an internal arrangement whereby the voltage ratio between primary and secondary coils may be changed to ensure that the output voltage is of the correct rated value. This change in voltage ratio is made by changing the connections of taps, usually from the primary coil, brought out to a *tap changer* within the transformer.

Most transformers in use today have the primary coil in two parts with taps taken from the primary coils. The simplest means of changing the voltage ratio of primary to secondary coils is by changing tap connections on a terminal board mounted inside the transformer. To do this, the transformer must be isolated from circuits and the transformer oil must be lowered. A manhole cover is then removed from the top or the side of the transformer tank. Taps are changed by changing the position of one or more shorting bars which can be bolted down on the bolts or studs on the terminal board. The taps from the primary coil are connected to these bolts or studs. The instructions for changing the taps are usually found on the transformer's nameplate diagram.

This method of changing taps is usually used

on distribution and small power transformers. Figure 14-20 indicates how this manual method of tap changing will change the turns ratio of the coils by a precise percentage.

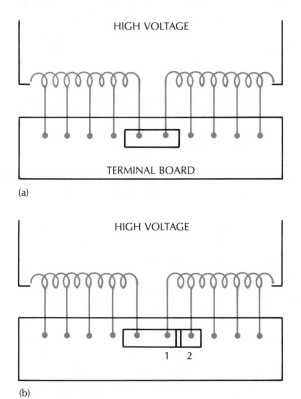

Figure 14-20 Manual tap changing.

Figure 14-20 (a) shows the primary coil taps connected to bolts or studs on a terminal board. With the connections as shown, the full primary coil is being used. In Fig. 14-20 (b), a short-circuit bar has been bolted on the studs 1 and 2 thereby shorting out a known percentage of the primary coil. The new turns ratio will ensure that the voltage induced in the secondary coil will be of the rated value.

This manual method of changing taps has certain disadvantages. It is possible to install the short-circuit bar across the incorrect taps or neglect to tighten down the bar so that too much resistance is introduced. It is also possi-

ble to drop tools or small parts into the transformer tank.

Another method to change the taps involves a manually operated tap-changing switch. This method is only used when the load is disconnected from the transformer. In this method, the leads from the taps of the primary coil are brought out to heavy contacts of a switch. This switch is generally operated with a handle mounted on the side of the transformer which is connected to the switch mechanism through a shaft. This method of changing taps is shown in Fig. 14-21.

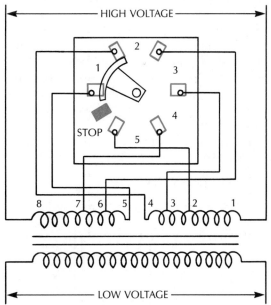

Figure 14-21 Manually operated tap-changing switch.

Note that with the handle as shown in position 1, all turns of the primary coil are used. By turning the shorting bar to position 2, the section 5–6 is removed form the primary coil. Other changes are made by changing the handle setting in a similar way. The position to be selected depends on by how much the secondary coil voltage needs to be increased. This type of tap-changing switch is generally used with distribution transformers and has the following advantages over the previous manual tap-changing method:

(a) no incorrect connections can be made
(b) no loose connections, having high resistance, can be made
(c) no parts or tools can be dropped inside the transformer tank
(d) the oil does not have to be lowered

In the operation of manually operated tap-changing equipment, the transformer has to be disconnected from its load. In large utility systems it is not always possible to remove transformers from service in order to change taps. Equipment has been developed whereby the taps of a transformer's primary coil may be changed while the transformer is not disconnected from service but continues to supply a load. Several methods have been developed for changing taps under load, but one method, called the *single-winding method*, is used most often today and is discussed below.

In the single-winding method, an auto-transformer is used to "bridge" taps on the

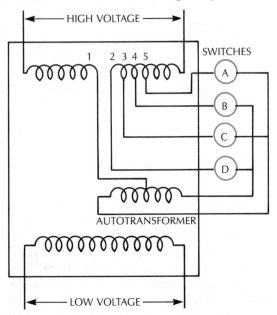

Figure 14-22 The single-winding method of changing taps under load.

primary coil of the transformer. The auto-transformer used gives also an additional voltage step between taps. Externally operated switches are connected, as shown in Fig. 14-22. There is a definite sequence of operating positions for the switch mechanism, and for each position certain switch contacts are open while others are closed. For each of these positions, certain percentage of the primary coil's turns are taken out of service, thereby increasing the secondary coil voltage.

The following discussion is not intended to give all details about the changing voltage ratios for each position but to show how sections of the primary coil are taken out of service while the load is being maintained. Note carefully what occurs at each position.

Position 1: Main contact A is closed while contacts B, C, and D are open. In this position, the full primary coil is connected across the supply voltage and all taps of the coil are inoperative and also half of the auto-transformer is in the circuit.

Position 2: Main contacts A and B are closed and contacts C and D are open. In this position, the autotransformer is connected across taps 4 and 5 and half of the load current flows through each half of the autotransformer coil.

Position 3: Main contact B is closed while contacts A, C, and D are open. Half of the autotransformer coil is in series with the primary coil.

Position 4: Main contacts B and C are closed and contacts A and D are open. In this case, one-half of the autotransformer is connected to tap 3 while the other half is connected to tap 4.

Position 5: Main contact C is closed while contacts A, B, and D are open. In this case half of the autotransformer is connected in series with the primary coil and the section 2–3 is not in the circuit.

Position 6: Main contacts C and D are closed and contacts A and B are open. The auto-transformer is connected across taps 2 and 3 of the primary coil.

Position 7: Main contact D is closed and contacts A, B, and C are open. Half of the autotransformer is again connected in series with the primary coil.

Normally, the tap-changing mechanism is connected to taps in the primary coil. However, if the voltage level is too high to economically insulate all tap-changer parts, the mechanism may be connected to taps from the secondary coil of the transformer.

VOLTAGE REGULATING TRANSFORMERS (REGULATORS)

When it is desirable to maintain the voltage from the secondary coil of a transformer to a load at a constant value, a voltage regulator may be used. A voltage regulator is often used when it is necessary that a load current be maintained at an exact level. Of course, any change in secondary coil voltage will result in a change in load current.

An induction voltage regulator automatically increases or decreases a secondary coil's voltage to a load, thereby maintaining a constant current value. An example of this is the constant current required for a street-lighting circuit. Several types of voltage regulators are available. One type is the induction voltage regulator.

Induction voltage regulators are made for single-phase and three-phase applications. The regulator discussed below is for single-phase voltage operation. The voltage induction regulator consists of a stator or stationary part and a rotor or moving part similar to those of an induction motor, as discussed in Chapter Seventeen. The construction of this type of voltage regulator is shown in Fig. 14-23.

The induction voltage regulator is basically a transformer whose voltage ratio may be adjusted automatically. The secondary turns of this transformer are wound on the nonrotating stator. The turns of the primary coil of this transformer are wound on the rotor which

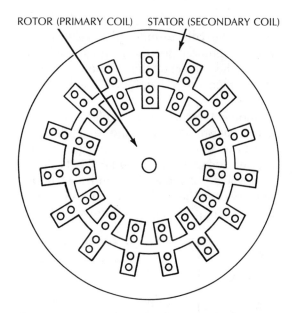

ROTOR (PRIMARY COIL)   STATOR (SECONDARY COIL)

Figure 14-23 An induction voltage regulator.

may be rotated with respect to the stator.

When the primary coil is fully aligned with the secondary coil, as shown, maximum voltage is induced in the secondary coil. By rotating the rotor, the number of lines of magnetic force cutting the secondary coil on the stator is reduced and the induced voltage is decreased.

The rotor is turned by a motor to which current is supplied through a closed relay contact. The relay has two sets of contacts. When the secondary coil voltage is too low, a contact-making voltmeter causes one set of contacts to close on the relay. The motor will now turn the rotor in one direction for a certain number of degrees so that the number of lines of force from the primary coil is such that it induces the exact voltage required for the load.

When the secondary coil voltage is too high, the contact-making voltmeter causes a different set of contacts to close on the relay. The motor will now turn in the opposite direction and cause the rotor to also turn a certain number of degrees in opposite direction.

Fewer lines of force will cut the secondary coil so that the voltage induced is again exactly what is required for the load. In each case above, the contact on the voltmeter is opened again when the voltage level has been corrected.

Another type of voltage regulator consists of a mechanism in which the secondary coil moves with respect to the primary coil. The moving-coil regulator, shown in Fig. 14-24, has a movable secondary coil and a stationary primary coil. The secondary coil is suspended from a set of pulleys and counterweights.

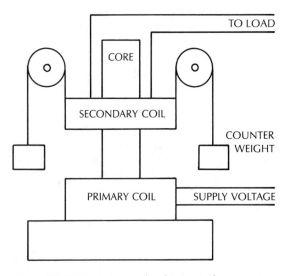

Figure 14-24 A moving-coil voltage regulator.

This type of regulator is also used on circuits where a constant value current flow is required. This regulator is often mounted on poles for use with street-lighting circuits. The primary coil of this regulator is supplied with the voltage from the secondary coil of a transformer.

The insulated primary and secondary coils are mounted on the same vertical magnetic core. Although not shown in Fig. 14-24, the magnetic core also surrounds both coils and the secondary coil is suspended within the iron core frame and may be raised or lowered.

When the primary coil is energized, its mag-netic field cuts across the turns of the secondary coil, inducing a certain level of voltage which is connected to a set of street lights. The load current flowing through the secondary coil is opposite in direction to the current in the primary coil. Therefore, the magnetic fields set up around the coils will oppose each other.

The counterweights which keep the secondary coil suspended is adjusted so that it *almost* balances the mass of the secondary coil. With a given amount of load current, the repelling magnetic field is sufficient to push the secondary coil upward so that it is now balanced perfectly. In this position, the magnetic fields are such that the voltage induced in the secondary coil is exactly the value which is required for the load current.

Assume that the load current has increased. This will result in a larger current in the secondary coil and a corresponding increase in current in the primary coil. This change in current causes an increase in the repelling magnetic forces between the coils and the secondary coil is pushed upward. The upward movement of the secondary coil continues until the voltage induced is exactly what is required for the load current. The secondary coil is then perfectly balanced in that position.

When the secondary coil current is decreased, the magnetic fields in both coils are weakened and the repelling forces are also weaker. As a result, the secondary coil begins to move downward. This continues until the voltage induced in the secondary coil is exactly what is required for the load current. This moving action between the coils will be such as to maintain the current in the load circuit.

THREE-PHASE TRANSFORMERS

Throughout this chapter, the transformer action between a single primary coil and a single secondary coil has been stressed. However, almost all voltages produced by generators today are produced in three-phase generators. These generators have three sets of coils (see also Chapter Sixteen) wound on their stator. The voltages induced in these coils are 120 electrical degrees out of phase with each other. For practical purposes, the three coils in an ac generator are connected in one of two methods, namely the *delta* connection or the *star* or *Wye* connection. With both of these connections, only three leads will terminate from the generator, although a fourth lead may be terminated from the Wye connection.

The generated voltage is usually 13.8 kV, and this needs to be stepped-up for transmission purposes to much higher values, as discussed earlier in this chapter. For this purpose, either three single-phase transformers may be connected in parallel or the three leads from the generator may be connected to a three-phase transformer.

The actual transformer action in a three-phase transformer is similar to what occurs in three single-phase transformers and as has been discussed in this chapter. In Chapter Fifteen, much attention will be paid to the actual connections required for single-phase and three-phase transformers. Figure 14-25 shows how the primary and secondary coils are wound in a three-phase transformer.

Note that the magnetic circuit of a three-phase transformer consists of insulated laminations formed as shown above. The magnetic circuit consists of three legs held together by yokes at the top and bottom. Each leg supports one primary coil and one secondary coil. As mentioned earlier, the primary coils may be connected in star, or Wye, or in delta. The secondary coils may also be connected in this manner. For all practical purposes, each of the three legs of the magnetic circuit and its associated primary and secondary coils may be considered as a single-phase transformer.

SUMMARY OF IMPORTANT POINTS

1. A transformer is used to step-up or step-down an ac voltage. There is no change in the frequency.
2. A transformer has a very high efficiency: 85 to 99 percent.
3. Ac voltages are stepped up for transmission in order to minimize I^2R line losses. The line current is reduced by the same factor as the voltage is increased.
4. Three types of transformer cores are:
   (a) core-type used mainly for high voltages with relatively small output
   (b) the shell-type used for lower voltages with high output
   (c) the H-type with very low flux leakage
5. There are CSA rules for transformer current, voltage, cooling and insulating ratings, and terminal polarities.
6. When observed from the low-voltage side, the left high-voltage terminal is always labelled as H_1. If the low-voltage terminal directly below is X_1, the transformer is subtractive. If this terminal is X_2, then the transformer is additive.
7. For a rating greater than 200 kVA and a voltage more than 7.5 kV, transformers are constructed with subtractive polarity.
8. The input side is called the primary and the output side the secondary.
9. The self-induced primary voltage is almost equal to the applied voltage.
10. Under no-load, the small resulting pri-

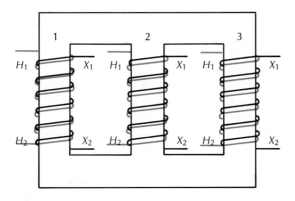

Figure 14-25 Primary and secondary coils of a three-phase transformer.

mary current is called the magnetizing current. This current produces the magnetizing flux which induces a voltage in the secondary.

11. When a load is applied, the secondary flux opposes the magnetizing flux which causes the primary current to increase proportionally.

12. Useful transformer equations assuming no losses:

(a) $V_P I_P = V_S I_S$

(b) $N_P I_P = N_S I_S$

(c) $\dfrac{V_P}{V_S} = \dfrac{N_P}{N_S}$

13. Transformer losses include: magnetic leakage, eddy currents, hysteresis, and I^2R losses.

14. Transformer phasor diagrams depend on winding impedance and load power factor.

15. For the rated I in the short-circuit test, V is about 3 to 5 percent of the rated value.

16. The open-circuit test is used to determine the iron losses. At rated voltage, the current is between 1 and 7 percent of rated value.

17. A current transformer is always subtractive. It usually has a maximum secondary rating of 5 A. It should be always short-circuited unless a load is connected to it when in a circuit.

18. Three types of current transformers are the wound, window, and bar type. The secondary is always grounded for safety reasons.

19. Potential transformers are used for measuring very high voltages. For above 100 kV, this transformer is incorporated with a capacitor divider circuit.

20. In an autotransformer, the primary and secondary share a single winding. There may be many taps for step-up or step-down operation.

21. In a step-down autotransformer $I_L = I_P + I_S$. In a step-up autotransformer $I_L = I_P - I_S$.

22. In an autotransformer (a) a short circuit will cause a dangerously high current and (b) an open coil may cause full voltage across the secondary.

23. Tap changers are used in transformers to compensate for line voltage losses. To change taps under load, an autotransformer and a number of switches are incorporated with the tap circuit. The taps are used to vary the primary winding in small precise steps in order to correct the secondary voltage.

24. A voltage regulator is used to maintain a constant output voltage under a varying load condition. Two types of regulators are (a) the induction and (b) the moving coil.

25. In the induction regulator, a motor turns the primary coil to change its position with respect to the secondary in order to maintain a constant output voltage.

26. In the moving-coil regulator, the secondary coil is suspended with counterweights. Variations in the magnetic force due to change in load will change the setting of the secondary coil so that a constant voltage is maintained.

27. To transform a three-phase voltage, either three single-phase transformers or a single three-phase transformer may be used. A three-phase transformer is basically three single transformers combined into a single housing.

REVIEW QUESTIONS

Essay and problem questions

1. Explain why there are lower power losses in transmission lines whose voltage is of very high value.

2. State precisely what the function is of a step-up transformer and of a step-down transformer.

3. Discuss the differences in three types of cores used in transformers. Indicate when each type of core is used and name some

of its advantages and disadvantages.

4. State at least four factors regarding the construction of transformers that have been standardized by the Canadian Standards Association.

5. Explain the difference between subtractive and additive transformer polarity. State when each of the polarities is used in transformer winding.

6. Assume that no load is connected to a transformer. A voltage is applied to the primary coil. Explain:
   (a) what is the voltage of self-induction?
   (b) what is the voltage of mutual induction?
   (c) why is the secondary coil voltage an ac voltage?
   (d) what is no-load current?

7. Which coil of the transformer is called: (a) primary coil, (b) secondary coil, (c) high-voltage coil, (d) low-voltage coil?

8. Explain why the induced secondary coil voltage is equal to the number of turns of this coil times the voltage induced per turn in the primary coil.

9. A transformer, with no load connected, has a primary coil of 3000 turns and a secondary coil of 265 turns. An ac voltage of 2400 V is applied to the primary coil. Determine:
   (a) the voltage of self-induction per turn of the primary coil
   (b) the voltage of mutual induction in the secondary coil
   (c) the voltage of mutual induction per turn in the secondary coil

10. A transformer with no load connected has a turns ratio of 20:1 between its coils. Its secondary coil has 200 turns and the voltage of mutual induction is 6 V per coil turn. Determine:
   (a) the total voltage of mutual induction
   (b) the number of turns of the primary coil
   (c) the total voltage of self-induction
   (d) the voltage of self-induction per turn

in the primary coil

11. What is the transformation ratio of a transformer and why is it so important?

12. Explain in some detail why and how an increase in load current causes a proportional increase in primary coil magnetizing current.

13. A transformer has a load current of 20 A and its primary coil has 4000 turns, while its secondary coil has 100 turns. How much magnetizing current will flow in the primary coil?

14. The current flowing in a primary coil is 2 A while the load current is 20 A. The applied voltage is 2000 V. Determine the voltage of mutual induction.

15. State the factors responsible for having a transformer obtain an efficiency of less than 100 percent. Explain why each of these factors is responsible for a percentage of the total power losses.

16. Assume that in Q. 14 the transformer losses amount to 10 percent of the transformer input power. Determine the efficiency of this transformer.

17. The phasor diagram of Fig. 14-8 shows the phasor relationships that exist when a resistive load is connected to the transformer. Discuss the relationships between the phasors shown in this diagram.

18. Why are phasor diagrams for loaded transformers generally very extensive and involve diagrams?

19. State carefully the procedure to be followed to perform a short-circuit test on a transformer. What precautions should be taken? Explain the values of the meter readings obtained during this test.

20. The iron losses of a transformer may be determined by doing an open-circuit test. Explain:
   (a) the procedures to be followed for doing this test
   (b) the precautions to be taken during this test
   (c) the values of the meter readings

obtained

21. State the standard secondary coil ratings for current transformers and potential transformers. Why is the secondary coil always grounded on these transformers ?

22. A current transformer has a rated current ratio of 40:1. A load current of 3.6 A is indicated on an ammeter. Determine the primary coil current.

23. Explain carefully why a current transformer must never be energized unless a short-circuit is connected across its secondary coil or a load is connected to it.

24. State the means used to compensate for current transformer meter errors due to inductive reactance in the coils.

25. Discuss three types of current transformers and indicate how they differ and why they differ and when a specific type is used.

26. Draw a diagram of a high-voltage circuit and make the correct connections for a current transformer and a potential transformer.

27. State and explain the reasons why potential transformers of over 100 kV have capacitors connected in series with the primary coil.

28. Explain why the sections of the coil of an autotransformer seldom have a higher ratio that 5:1.

29. In a step-down autotransformer, the applied voltage is 400 V and a load of 10 Ω is connected across 60 percent of the coil. Determine:
    (a) the secondary coil voltage
    (b) the load current
    (c) the output power and input power
    (d) the input current from the source
    (e) the current supplied to the load by the secondary coil
    (f) the voltage transformation ratio

30. In a step-up autotransformer, 120-Vac is connected across 66.7 percent of the coil. A 12-Ω resistor is connected across the secondary coil. Determine:
    (a) the secondary coil voltage
    (b) the load current
    (c) the output power and input power
    (d) the input line current
    (e) the secondary coil current
    (f) the primary coil current
    (g) the voltage transformation ratio

31. Explain why a break in the section of the coil common to both primary and secondary in an autotransformer is very hazardous and damaging.

32. State details of how a voltage ratio change may be accomplished by the use of a tap changer and why the secondary coil voltage increases.

33. Explain how a manual tap-changing switch operates to change the voltage ratio on a transformer disconnected from its load.

34. Describe in some detail how voltage ratio changes are made on a transformer under load.

35. Explain why voltage regulators are used on certain circuits.

36. Describe the construction and operation of an induction voltage regulator. Which main factor accomplishes a change in amount of mutual induction in this regulator ?

37. Describe the construction of a moving-coil voltage regulator.

38. Explain carefully when and how the secondary coil voltage of a moving-coil voltage regulator is increased or decreased.

39. Discuss the construction of a three-phase transformer and explain why this transformer may be treated exactly as three single-phase transformers.

CHAPTER
FIFTEEN

THREE-PHASE (POLYPHASE) CIRCUITS

INTRODUCTION TO THREE-PHASE CIRCUITS

Almost all of the discussions of ac circuits presented thus far in the text were on single-phase systems. A single-phase voltage is represented by a single voltage waveform, as shown in Fig. 15-1(a). This is the kind of voltage that is supplied to all residential homes and many commercial institutions.

The service supplied to a home consists of a single-phase voltage supply with a centre tap, as shown in Fig. 15-1(b). The centre tap allows for two different voltages (120/240 V) from the

single-phase supply. This type of voltage supply is adequate for most lighting and power appliances in the home.

Most industrial and heavy electrical equipment, on the other hand, requires three-phase voltage. A three-phase voltage system, as will be shown, is actually a combination of three, single-phase voltages. In fact, the single-phase voltage supplied to residential homes is simply one of the phases taken from a three-phase distribution system. In this chapter, the properties of three-phase (poly-phase) voltage connections and distribution will be discussed.

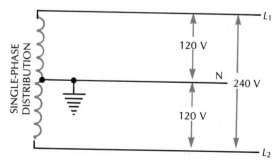

Figure 15-1(a) Single-phase ac.

Figure 15-1(b) Single-phase voltage supply with centre tap.

ADVANTAGES

Practically all voltage generation and distribution by power utilities is done with a three-phase network system. There are several reasons why three-phase is preferred to single-phase service for so many applications.

1. A balanced three-phase distribution system requires only three or four wires, whereas three single-phases for the same kVA capacity will require six wires. A three-phase balanced system with equal voltage between the lines will also require only about 75 percent of the copper used in one single-phase system for the same kVA and voltage ratings.
2. The power delivered by a single-phase source is pulsating. This property was illustrated in Chapter Ten. The power delivered by a three-phase system, however, is relatively constant at all times. This means that even though the power in each phase is pulsating, the total power at any instant will be relatively constant. Therefore, the operating characteristics of three-phase machines will be superior to single-phase devices with similar ratings.
3. For the same physical size, three-phase generators and motors generally have larger kVA and (kW) ratings than single-phase machines.
4. A three-phase distribution system can be used to supply both three-phase and single-phase service.

PHASE ROTATION

A three-phase alternator contains three sets of coils positioned 120° apart and its output is a three-phase voltage, as shown in Fig. 15-2. A three-phase voltage consists of three voltage waves 120 electrical degrees apart, and the order in which these voltages succeed one another is called the *phase rotation or phase sequence*. In Fig. 15-2, phase A starts to rise in a positive direction at zero electrical degrees,

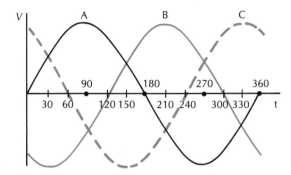

Figure 15-2 A three-phase voltage waveform.

120 electrical degrees later phase B starts to increase in a positive direction, and then phase C follows at another interval of 120°. Therefore the phase sequence for this system is...A,B,C,A,B,C....The phase sequence of a system may be reversed by either reversing the direction of rotation of the alternator, or by interchanging the connections of any two of the three wires used for the transmission of the three-phase voltage.

Correct phase sequence is important in the construction and connection of various three-phase systems. For example, correct phase rotation is important when the outputs of three-phase alternators must be paralleled into common voltage system, or for that matter, when the three coils of an alternator must be connected to form a three-phase system. The phase sequence of a three-phase system may be determined from the use of an oscilloscope or from the lamps of a phase sequence indicator.

THE WYE OR STAR CONNECTION

The construction of the alternator will be discussed in more detail in Chapter Sixteen. However, simply, the three-phase alternator contains three sets of coils positioned 120° apart. The output from these coils is three, separate voltages with the same frequency and magnitude, but 120 electrical degrees apart. The three windings of the alternator are connected so that only three or four wires, instead of six wires, are required for the transmission of the three-phase voltage. There are two methods, (a) Wye or star and (b) delta, for connecting the three coil windings in the alternator to form a three-phase system.

Most alternators have their three single-phase coils interconnected in the Wye or star connection. The Wye connection is so named because the three coils form the shape of the letter "Y", as shown in Fig. 10-3. In the Wye-connection, the end of each of the three coils is connected together and the beginning of each coil is brought out to the three line wires. The connection may be reversed by connecting the beginning of each of the three coils together and then bringing out the end of each coil to the three line wires. In either case, however, the sequence of the phases must be maintained as indicated. It is important that the turns of the three coils be wound in the same direction.

A fourth wire is often used in the Wye connection. This wire is called the neutral, and is connected to ground and to, the junction of the three coils, as shown in Fig. 15-3. The neutral wire provides protection against lightning and, also, it permits two voltage values, a line voltage and a phase voltage, to be obtained from the system. *The line voltage* is the voltage measured across any two of the three lines, and it is sometimes called the line-

to-line voltage. *The phase voltage*, also called the coil voltage, is measured from any of the three lines to the neutral.

The three-phase voltages in a three-phase system may be represented by three voltage phasors displaced 120°, as shown in Fig. 15-4(a). It should be noted from Figs. 15-4(a) and 15-4(b) that the line voltage in a Wye connection is equal to the phasor sum of the two-phase voltages between the two lines. For simplicity only, the phase voltages, V_{AO} and

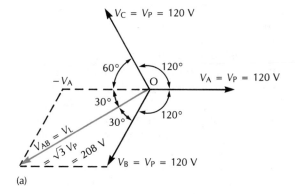

(a)

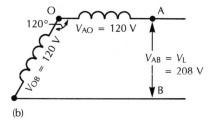

(b)

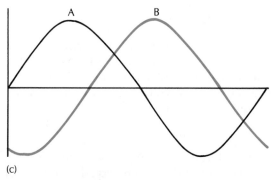

(c)

Figure 15-4 Determination of the line and phase voltage in a Wye system.

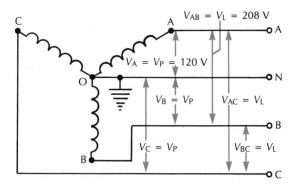

Figure 15-3 The phase and line voltages for a Wye or Star connection.

V_{OB}, between the lines A and B are shown in Fig. 15-4(b).

The two-phase voltages across any two lines in a Wye-connected system are not only 120° out of phase, but also when one phase voltage is positive the other voltage is negative. This property can be illustrated using either Fig. 15-4(b) or Fig. 15-4(c). In Fig. 15-4(b), if the subscripts "AO" in V_{AO} represent a foward positive voltage, then the reverse subscripts "OB" in V_{OB} will indicate a reverse negative voltage. Similarly, in the wave diagram of Fig. 15-4(c), observe that when the wave V_{AO} is positive, the second wave is negative for most of the same interval, and vice versa. Therefore, it can be concluded that the line voltage in a Wye-connected system will be equal to the phasor sum of the positive value of one phase voltage and the negative value of the other phase voltage that exist across the two lines. This will be the same as connecting the two phasors at 60° instead of at 120°.

The phasor addition for determining the line voltage across the lines A and B is illustrated in Fig. 15-4(a). The phasor parallelogram in the diagram shows a positive phase voltage, V_B, combined with the negative phase voltage, $(-V_A)$. From simple geometry or from phasor addition by the component method, it can be shown that the line voltage, V_L, in a Wye connection is equal to $\sqrt{3}$ times the phase voltage, V_P.

$$V_L = \sqrt{3}\, V_P \qquad (15\text{-}1)$$
$$= 1.73\, V_P$$

It should be apparent from Fig. 15-3 that in the Wye connection the phase voltage is the same as the coil voltage. Figure 15-5 shows the phase and line voltage phasors for a Wye-connected system.

$$V_P = V \text{ coil} \qquad (15\text{-}2)$$

In a three-phase Wye-connected system, the line current, I_L, will be equal to the coil or phase current, I_P, because each phase winding is connected directly in series with one of the three line wires.

$$I_L = I_P \qquad (15\text{-}3)$$

Also, it will be recalled from Kirchhoff's current law that the sum of the currents at a junction in a circuit is equal to zero. Therefore, the phasor sum of the three coil currents in a Wye system will be zero.

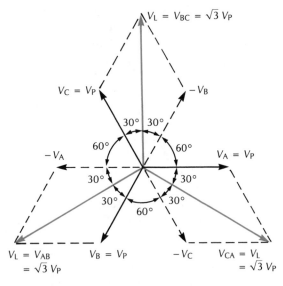

Figure 15-5 The phase and line voltage phasors for a three-phase Wye-connected voltage system.

Example 15-1: A balanced three-phase Wye-connected system has a phase voltage, V_P, of 120 V and phase or coil current of 10 A. Determine the line voltage and line current for the system.

$$V_L = \sqrt{3} \times V_P$$
$$= 1.73 \times 120 = 208 \text{ V}$$
$$I_L = I_P = 10 \text{ A}$$

THE DELTA CONNECTION

The second standard connection for the three, single-phase windings of an alternator is called

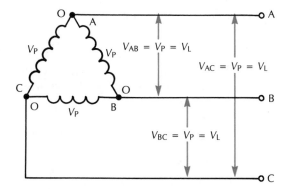

Figure 15-6 The delta connection.

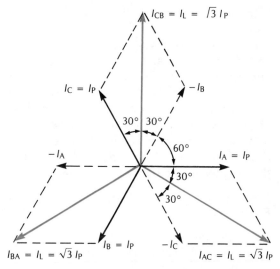

Figure 15-7 The coil and line current phasors for a delta-connected system.

the delta connection, so named because when the three coils are interconnected they form the shape of the Greek letter "delta (Δ)." In this delta connection, the end of one coil is connected to the beginning of another coil, and a line wire is connected to the junction of these two coil ends. This method of connection is continued for the remaining coil ends. The complete connection is shown in Fig. 15-6. Note that the delta connection requires only three line wires.

It should be evident from Fig. 15-6 that the coil or phase voltage, V_P, in a delta connection is equal to the line voltage, V_L, because the line wires are connected directly across the phase winding.

$$V_P = V_L \qquad (15\text{-}4)$$

The line current in a delta connection, however, will be supplied from two phase windings 120° apart. Therefore, the line current in a delta connection will be equal to the phasor addition of two phase currents 120 electrical degrees apart. This phasor addition is very similar to the phasor addition for determining the line voltage in a Wye-connected system.

Using the same reasoning, the line current, I_L, in a delta-connected system will be found to equal $\sqrt{3}$ times the phase or coil current, I_P. Figure 15-7 shows the line and coil current phasors for a delta-connected system.

$$I_L = \sqrt{3}\, I_P \qquad (15\text{-}5)$$

Example 15-2: A delta-connected system has a phase voltage of 120 V, and a phase current of 15 A. Determine the line voltage and line current for this three-phase system.

For the delta connection:

$$V_L = V_P = 120 \text{ V}$$

$$I_L = \sqrt{3}\, I_P$$
$$= 1.73 \times 15 = 25.95 \text{ A}$$

POWER FACTOR OF A THREE-PHASE SYSTEM

The power factor angle in either a Wye or a delta-connected system is always the angle between the *coil voltage* and *coil current*. Therefore, in a balanced three-phase system, the power factor will be equal to the cosine of this angle. Since in a three-phase system the power factor is defined in terms of a single coil voltage and coil current in the system, it follows then that any of the appropriate power factor equations developed in Chapter Thirteen for single-phase circuits may also be used here.

If a three-phase system is slightly unbalanced, that is, there are small differences either between the line voltages or between the line currents, then the power factor of the system may be determined from average line or phase values. For a badly unbalanced system, however, power factor will have practically no meaning. All the circuits described in this chapter will be balanced systems.

Example 15-3: For the Wye system shown in Fig. 15-8(a):

(a) Determine the following:
   (i) the alternator phase current
   (ii) the line current
   (iii) the line voltage, and
   (iv) the circuit power factor
(b) Draw the circuit's voltage and current phasors.

Solutions:

(a)   (i) Phase current, $I_P = I_A$

$$= \frac{V_A}{R_A}$$

$$= \frac{120}{6} = 20 \text{ A}$$

Note that in the circuit of Fig. 15-8(a), the phase current in the three-phase Wye-

connected load will be the same as the phase current in the Wye-connected alternator.

(ii) The line current, I_L is equal to the phase current, I_P, for a Wye connection. Therefore,

$I_L = I_P$
   $= 20 \text{ A}$

(iii) For a Wye connection,

$V_L = \sqrt{3} \, V_P$
   $= 1.73 \times 120$
   $= 208 \text{ V}$

Note that in this circuit the line voltage for the Wye-connected alternator will be the same as the line voltage for the Wye-connected load.

(iv) Since the three-phase load is purely resistive, it should be obvious that the circuit power factor will be unity.

(b) The current and voltage phasors for this circuit are shown in Fig. 15-8(b). The phase voltage and phase current for each of the three coils in the system are in phase because the circuit power factor is unity. Also, since the load is connected in Wye, the line current, I_L, leads the line voltage, V_L, by 30°, as shown in the phasor diagram of Fig. 15-8(b).

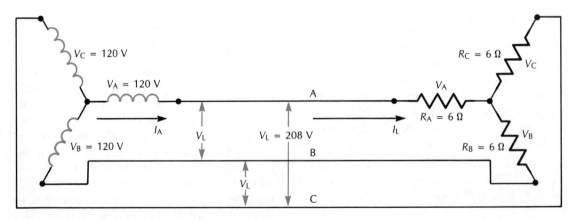

Figure 15-8(a) A Wye system supplying a purely resistive load.

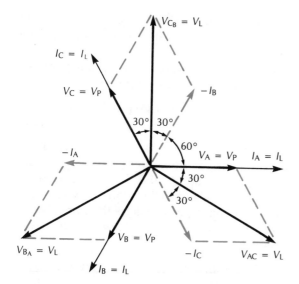

Figure 15-8(b) The current and voltage phasors for a Wye system with unity power factor.

Example 15-4: A line voltage of 208 V is connected to a balanced three-phase, delta-connected inductive load as shown in Fig. 15-9(a).

(a) Determine:
  (i) the phase impedance
  (ii) the circuit power factor
  (iii) the load phase voltage
  (iv) the phase current
  (v) the line current, and
(b) Draw the circuit voltage and current phasors.

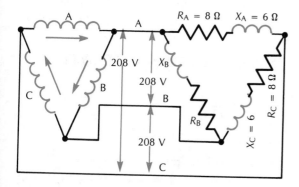

Figure 15-9(a) A balanced three-phase delta-connected inductive circuit.

Solutions:

(a) (i) The phase impedance is the same for each phase in a balanced load.

$$Z_P = Z_A = \sqrt{(R_A)^2 + (X_A)^2}$$
$$= \sqrt{(8)^2 + (6)^2}$$
$$= 10 \ \Omega$$

(ii) In a balanced load, the phase power factor is the same as the three-phase circuit power factor.

$$P.F. = \cos \underline{/\theta} = \frac{R_A}{Z_A}$$
$$= \frac{8}{10}$$
$$= 0.8 \ \text{lag}$$
$$\underline{/\theta} = \cos^{-1} 0.8$$
$$= 36.87° \ \text{lag}$$
(phase current lags phase voltage)

(iii) For a delta connection, the line voltage is equal to the phase voltage.

$$V_P = V_L = 208 \ \text{V}$$

(iv) The phase current, $I_P = \dfrac{V_P}{Z_P}$

$$= \frac{208}{10}$$
$$= 20.8 \ \text{A}$$

(v) The line current for a delta connection is:

$$I_L = \sqrt{3} \ I_P$$
$$= 1.73 \times 20.8$$
$$= 35.98 \ \text{A}$$

(b) The phasor diagram for the three-phase, delta-connected inductive load is shown in Fig. 15-8(b).

Since the current and voltage phasors for one phase will be similar to those of the other two phases in a balanced three-phase system, for simplicity, the phasor diagram for only one phase need be drawn. Also, when constructing the phasor diagram for either a Wye-connected or delta-connected system, the student will find it easier first, to construct the three coil voltage phasors and, then, to insert the remaining phasors for the system. Recall that the three, phase voltages for either a Wye or delta connection will always be 120° apart. However, in the case of the Wye connection the line voltage is equal to the phasor sum of two of the phase voltages, whereas in the delta system it is the line current that is equal to the phasor sum of two phase currents. Hence, for the Wye connection, the phasor parallelogram will use two coil voltage phasors, but for the delta connection, the phasor parallelogram will use two coil current phasors.

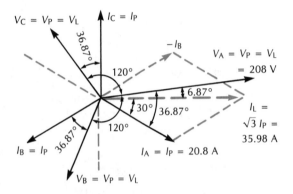

Figure 15-9(b) Phasor diagram for an inductive three-phase, delta-connected circuit.

In the phasor diagram of Fig. 15-9(b), the phasor parallelogram for only one phase is completed. In Ex. 15-4, the phase current for the circuit was found to lag the phase voltage by 36.87° and, hence, the coil current phasors were drawn as shown. The line current for this circuit is illustrated in the phasor parallelogram for the coil currents I_{AO}, and $(-I_{OB})$. From this diagram, it can be seen that when the coil current, I_A, lags the coil voltage, V_A, by 36.87°, then the line current, I_{AB}, will lag the line voltage, V_{AB}, by 6.87°.

POWER IN A THREE-PHASE SYSTEM

In Chapter Thirteen, it was shown that the apparent power in a single-phase circuit is equal to the product of the single-phase circuit voltage and the circuit current. Therefore, the volt-amperes in each phase of a three-phase system will be equal to the product of the phase voltage and the phase current.

$$\text{apparent power (VA) (single-phase)} = V_P \times I_P \qquad (15\text{-}6)$$

In a three-phase system, the total circuit volt-amperes is equal to the sum of the volt-amperes in each of the three phases. However, in a balanced three-phase system, the power in each phase will be the same, therefore, the total three-phase apparent power will be equal to:

$$\text{total apparent power (VA) (three-phase)} = 3 \times V_P \times I_P \qquad (15\text{-}7)$$

Note that Eq. 15-7 applies for any balanced three-phase system, that is, for both Wye and delta connections. Also, it is common practice to determine power in terms of the circuit's line voltage and line current because it is generally easier to measure the line voltage and line current in a three-phase system.

IN A WYE-CONNECTED SYSTEM

$$V_L = \sqrt{3}\, V_P \text{ and}$$
$$I_L = I_P$$

therefore Eq. 15-7 becomes:

$$\text{total volt-amperes} = 3 \times V_P \times I_L$$
$$= \sqrt{3}\, V_L \times I_L \qquad (15\text{-}8)$$

IN A DELTA-CONNECTED SYSTEM

$$V_L = V_P \text{ and}$$
$$I_L = \sqrt{3}\, I_P$$

therefore Eq. 15-7 becomes:

$$\text{total volt-amperes} = 3 \times V_P \times I_P$$
$$= 3 \times V_L \times \sqrt{3}\, I_L$$
$$= \sqrt{3}\, V_L \times I_L \qquad (15\text{-}9)$$

It should be evident from equations 15-7, 15-8, and 15-9 that the equation for determining the total volt-amperes in either a balanced three phase Wye-connected or delta-connected system is the same. If the circuit power factor angle is known, then the total *true power* in a balanced three-phase system is given by:

total true power (watts)

$$= 3 \times V_P \times I_P \times \cos \underline{/\theta} \qquad (15\text{-}10)$$
$$= \sqrt{3} \times V_L \times I_L \times \cos \underline{/\theta} \qquad (15\text{-}11)$$

Similarly, the net three-phase *reactive power* will be equal to:

net reactive power (VARs)

$$= 3 \times V_P \times I_P \times \sin \underline{/\theta} \qquad (15\text{-}12)$$
$$= \sqrt{3} \ V_L \times I_L \times \sin \underline{/\theta} \qquad (15\text{-}13)$$

Example 15-5: A three-phase Wye-connected generator is connected to an industrial motor load. The generator has an output coil voltage of 120 V, a coil current of 15 A, and a lag power factor angle of 40°. Determine the circuit's:

(a) line voltage, V_L
(b) line current, I_L
(c) total apparent power
(d) total true power
(e) total reactive power

Solutions:

(a) for a Wye connection,

$$V_L = \sqrt{3} \ V_P$$
$$= 1.73 \times 120$$
$$= 208 \text{ V}$$

(b) for a Wye connection,

$$I_L = I_P$$
$$= 15 \text{ A}$$

(c) total apparent power $= 3 \times V_P \times I_P$
$$= 3 \times 120 \times 15$$
$$= 5400 \text{ VA}$$
$$= 5.4 \text{ kVA}$$

or   apparent power $= \sqrt{3} \ V_L \times I_L$
$$= 1.73 \times 208 \times 15$$
$$= 5400 \text{ VA}$$
$$= 5.4 \text{ kVA}$$

(d) total true power $= 3 \times V_P \times I_P \cos \underline{/\theta}$
or
$$= \sqrt{3} \ V_L \times I_L \cos \underline{/\theta}$$
or
$$= \text{apparent power}$$
$$\times \cos \underline{/\theta}$$
$$= 5400 \times 0.766$$
$$= 4136.4 \text{ W}$$

(e) total reactive power $= 3 \times V_P \times I_P \sin \underline{/\theta}$
or
$$= \sqrt{3} \ V_L \times I_L \sin \underline{/\theta}$$
or
$$= \text{apparent power}$$
$$\times \sin \underline{/\theta}$$
$$= 5400 \times 0.643$$
$$= 3472.2 \text{ VARs}$$

Example 15-6: A delta-connected three-phase generator is connected to an industrial network. The generator has a rated phase voltage of 8000 V and rated current of 425 A, and at full load its power factor is 80 percent lag.

Determine:

(a) the line voltage
(b) the line current
(c) the generator kVA rating
(d) the full-load power in kilowatts

Solutions:

(a) for a delta connection,

$$V_L = V_P$$
$$= 8000 \text{ V}$$

(b) for a delta connection,

$$I_L = \sqrt{3} \ V_P$$
$$= 1.73 \times 425$$
$$= 735.25 \text{ A}$$

(c) kVA rating $= \dfrac{3 \times V_P \times I_P}{1000}$

$$= \frac{3 \times 8000 \times 425}{1000}$$
$$= 10 \ 200 \text{ kVA}$$

or   kVA rating $= \dfrac{\sqrt{3} \ V_L \times I_L}{1000}$

$$= \frac{1.73 \times 2000 \times 735.25}{1000}$$
$$= 10 \ 200 \text{ kVA}$$

(d) full-load kilowatts $= \dfrac{\sqrt{3}\ V_L \times I_L \times \cos\ \underline{/\theta}}{1000}$

$= 10\ 200\ \text{kVA} \times 0.8$

$= 8160\ \text{kW}$

THREE-PHASE POWER MEASUREMENT

Two wattmeters can be used to measure the three-phase true power in any three-phase system which contains only three line wires. The so-called "two-wattmeter method" therefore can be used to measure the power in either a three-wire delta system or Wye system. Three wattmeters can also be used to measure three-phase true power, but this method is limited for use only in four-wire, Wye-connected systems. These two power measurement methods are discussed below.

THE TWO-WATTMETER METHOD

Figure 15-10 shows the standard connections for the two-wattmeter method used in a three-wire, three-phase system. The load in Fig. 15-10, may also be delta-connected. It is important to observe the polarity marks (±) on the voltage and current coils of the wattmeters and make the connections exactly as shown. Observe that the current coils of the two

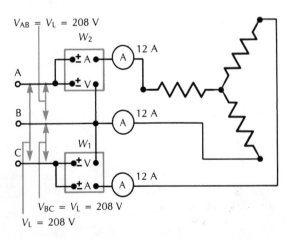

$V_{AB} = V_L = 208\ \text{V}$

$V_{BC} = V_L = 208\ \text{V}$

$V_L = 208\ \text{V}$

Figure 15-10 The two-wattmeter method.

wattmeters are connected in series with two of the three line wires. The voltage coil of each wattmeter is connected between the line wire to which its current coil is connected and the third line wire.

The readings of the two wattmeters may not necessarily be the same for different loads even though the loads may be balanced (equal line voltages and the same current in each line). Any difference in readings between the two wattmeters is directly related to the power factor of the balanced three-phase load. The following is a summary of the wattmeter readings for various power factor conditions:

(a) When the three-phase system is balanced, and the circuit power factor is unity, then the two wattmeters will have the same readings. The total circuit power will be equal to the sum of the two wattmeter readings, W_1 and W_2.

(b) If the circuit is balanced and has a power factor of less than one but greater than 0.5 lag, and the same phase sequence A-B-C-A-, then the wattmeter, W_2, will read less than W_1. The total three-phase circuit power is still, however, the sum of W_1 and W_2.

(c) If the circuit is balanced and has a power factor of exactly 0.5 lag, and the same phase sequence A-B-C-A-, then W_2 will read zero. The total circuit power will now be equal to that of W_1.

(d) If the circuit is balanced and has a power factor of less than 0.5 lag, and the same phase sequence A-B-C-A-, then wattmeter, W_2, will read backwards. Therefore, it will be necessary to reverse either the current coil connections or the voltage coil connections to this wattmeter in order to obtain its negative reading. The total power under this power factor condition will be equal to the reading of W_2 subtracted from the reading of W_1.

The question which must now be answered is: "Why does a change in the power factor of a three-phase circuit affect the wattmeter readings even though the circuit is kept balanced?" The answer to this question lies in the method of connection of the two wattmeters. Note that the voltage to the potential coil of each wattmeter is equal to the line voltage and the current to each current coil is equal to the line current. Even though the magnitude of the voltage and current to both wattmeters is the same, the phase angle between the line current and line voltage to one wattmeter can be different from the phase angle between the line current and line voltage to the other wattmeter. Since each wattmeter will use the phase angle that it senses and not the circuit power factor angle in computing its reading, it follows that their readings can be different. The phase angle between the line current and line voltage to a wattmeter is dependent on the circuit power factor angle and the circuit phase rotation. The following examples should help to further explain the reasons for any difference between the two wattmeter readings.

Example 15-7: Assume the three-phase power factor to the circuit shown in Fig. 15-10 is unity. Determine the wattmeter reading of W_1 and W_2.

Figure 15-10 shows that the line voltage and line current to wattmeter, W_1, are V_{CB} and I_C, respectively, and the voltage and current to wattmeter, W_2, are, respectively, V_{AB} and I_A.

These two line voltages and currents are illustrated in the phasor diagram of Fig. 15-11. This is the phasor diagram for a balanced Wye-connected unity power factor circuit. From the phasor diagram, it is obvious that the phase angle between the line voltage and line current is the same to both wattmeters and is equal to 30°. Therefore, both wattmeters will have the same reading when the circuit power factor is unity.

$$W_1 = V_{CB} \times I_C \times \cos 30°$$
$$= 208 \times 12 \times 0.866$$
$$= 2161.5 \text{ W}$$

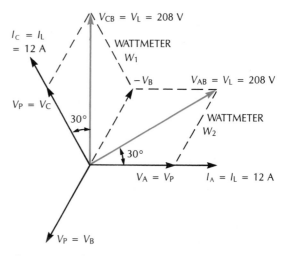

Figure 15-11 The voltage, current, and phase angle to each wattmeter for a unity power factor Wye-connected circuit.

$$W_2 = V_{AB} \times I_A \times \cos 30°$$
$$= 208 \times 12 \times 0.866$$
$$= 2161.5 \text{ W}$$

Example 15-8: Assume that the power factor for the circuit shown in Fig. 15-10 is now 70° lag with the line voltages and currents remaining unchanged. Determine the new wattmeter readings for W_1 and W_2.

The circuit phasor diagram has been redrawn in Fig. 15-12 (see page 296) to show the circuit's new power factor angle. In this diagram, the phase current now lags the phase voltage by 70°. However, notice that the phase angle between the line current and line voltage to wattmeter, W_2, is $(70 + 30) = 100°$, whereas the phase angle between the line current and line voltage to wattmeter, W_2, is $(70 - 30) = 40°$. Therefore, the respective readings for wattmeter W_1 and W_2 will be:

$$W_1 = V_{CB} \times I_C \times \cos 40°$$
$$= 208 \times 12 \times 0.766$$
$$= 1911.94 \text{ W}$$

$$W_2 = V_{AB} \times I_A \times \cos 100°$$
$$= 208 \times 12 \times (- 0.174)$$
$$= - 434.3 \text{ W}$$

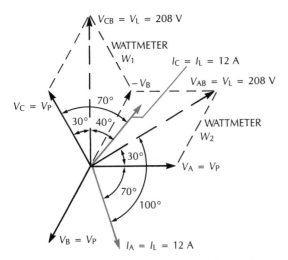

Figure 15-12 The voltage and current phasors to the two wattmeters in a 70° lag power factor Wye-connected circuit.

In this example, the circuit power factor was less than 0.5 lag (P.F. = cos 70° = 0.342 lag) and, hence, as indicated in the calculation, the wattmeter, W_2, will read negatively. This agrees with the statement made previously about the wattmeter readings when the circuit power factor is less than 0.5 lag. Example calculations to determine the two wattmeter readings for a balanced three-phase circuit for

other lag power factors are left as an exercise to the student.

If the circuit power factor is made leading instead of lagging, then the conditions at the two wattmeters will be interchanged. That is, when the power factor is less than one and more than 0.5 lead, then W_1 will read less than W_2. At exactly 0.5 lead power factor, W_1 will read zero, and at below 0.5 lead power factor, W_1, will read negatively.

THE THREE-WATTMETER METHOD

Most Wye-connected three-phase systems have a neutral wire as well as the three line wires. This neutral wire is connected to the junction of the Wye system, as shown in Fig. 15-13, and the system is described as four-wire, Wye-connected, three-phase system. The neutral wire makes available the two different voltages, the phase voltage and the line voltage. A second important property of the neutral wire is that it helps to maintain relatively constant voltages across the three line wires when the line currents are unbalanced.

The three-wattmeter method for power measurement can be used only on a four-wire system and the connections are made, as shown in Fig. 15-13. Each wattmeter is con-

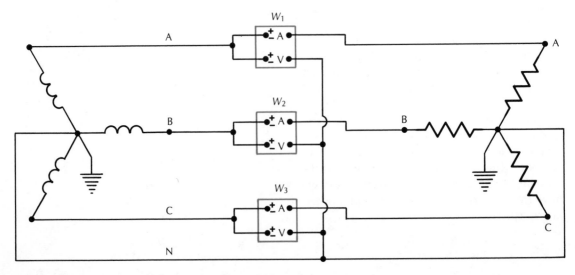

Figure 15-13 Power measurement in a three-phase, four-wire system.

nected across a phase voltage and in series with a phase current. Hence, each wattmeter will measure the power of a particular phase in the system. Note, too, that each wattmeter will sense the power factor angle $\underline{/\theta}$ for the phase to which it is connected. In a balanced system, the phase angle for each phase will be identical and, hence, the three wattmeters will all indicate similar readings. The reading of each wattmeter will be equal to:

true power per phase
$$= V_P \times I_P \times \cos \underline{/\theta} \qquad (15\text{-}14)$$

Obviously, if the system is unbalanced the readings of the wattmeters will be different. In any case, the wattmeter readings will always be positive regardless of the circuit power factor and load conditions. At all times, the total system power will be equal to the sum of the power in each of the three phases.

$$W_T = W_1 + W_2 + W_3 \qquad (15\text{-}15)$$

THREE-PHASE TRANSFORMER CONNECTIONS

Since most electrical energy is generated and transmitted over a three-phase network, it is important to be able to raise or lower a three-phase voltage. For example, the three-phase voltage output from a generating station is usually stepped up to a very high value prior to transmission. This increase in line voltage causes a decrease in current for the same kVA capacity, and this smaller current results in lower line voltage and power losses during transmission. At the other end, the three-phase must then be stepped down to a much lower value for connection to the various consumer systems.

A three-phase voltage may be raised or lowered with the use of single-phase transformers connected to form a three-phase transformer bank. A three-phase transformer bank may be formed by any one of five standard connections. They are: the delta-delta, the Wye-Wye, the delta-Wye, the Wye-delta, and the open-

delta (or V) connections. Each of these connections, except the open-delta method, use three single-phase transformers in its bank. The open-delta (or V) connection uses only two single-phase transformers.

When connecting transformers to form a three-phase bank, close attention must be given to the transformer polarity. The polarity of a transformer was discussed in Chapter Fourteen. It was explained that the high voltage polarity markings, H_1 and H_2, were fixed on any transformer, with the H_1 lead always positioned on the left when the transformer is looked at from the low voltage side. The polarity markings, X_1 and X_2, on the low voltage side, however, are dependent on whether the transformer is additive or subtractive. If the transformer polarity markings are known, then the formation of any three-phase bank is a simple matter of following the proper connection procedure for that bank. If, on the other hand, the transformer polarity is unknown, additional testing will be necessary during the connection of the transformers. The various transformer connections are described below.

THE DELTA-DELTA CONNECTION

The delta-delta (Δ-Δ) connection gets its name from the way in which the transformer coils are connected. The first delta symbol tells how the primary windings are connected and the second tells how the secondary windings are connected. The delta-delta connection of three single-phase transformers is shown in Fig. 15-14(a), and a simplified drawing is shown in Fig. 15-14(b). (See page 298.)

Observe in Fig. 15-14, that for the delta-delta transformer connection, each primary coil end, H_2, is connected to the beginning, H_1, of another primary coil, and that to each of these junctions an input line wire is also connected. Since in this connection each primary coil is connected directly across the line voltage, it follows that the transformer primary windings must have the same voltage rating as the line voltage.

Recall that the high voltage polarity mark-

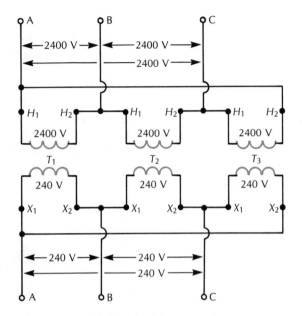

Figure 15-14(a) The delta-delta connection.

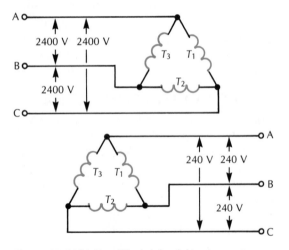

Figure 15-14(b) Simplified delta-delta connection.

ings are fixed on any transformer; therefore, the primary delta connection is straightforward. The delta connection of the output low voltage coils is very similar to the high voltage coils, but to avoid the possibility of a dangerous error, especially since the transformers may be of different polarity or their polarity may be unknown, the following procedure for

connecting the low voltage coils is strongly recommended:

1. Check to see that the output voltage for each of the transformers is the same.
2. Connect the end (X_2) of one secondary coil to the beginning (X_1) of another secondary coil, as shown in Fig. 15-15. If the connection is correct, the voltage across the open ends of the two windings will be equal to the output voltage of each secondary winding. The phasor addition for two delta-connected coil voltages is shown in Fig. 15-16(b). In this example if the coil voltage, $V_1 = V_2 = 240$ V, then the phasor addition of V_1 and V_2 will also equal 240 V.

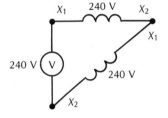

Figure 15-15(a) Correct delta-connection. Voltage across open ends is equal to coil voltage.

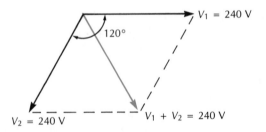

Figure 15-15(b) Phasor addition for two delta-connected coil voltages.

If the connections of one of the two secondary coils is reversed, as shown in Fig. 15-16, the incorrect delta connection will result. The voltmeter will now read $\sqrt{3}$ times the secondary coil voltage. The phasor addition for the incorrect delta connection of two coil voltages is shown in Fig. 15-16(b). In this example, if $V_1 = V_2 = 240$ V, then the phasor addition of V_1 and V_2 will yield $\sqrt{3}$ times 240 V, or 416 V.

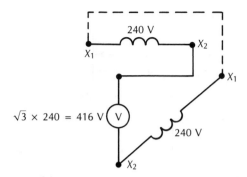

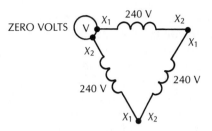

Figure 15-17(a) Correct delta-connection. Voltage across last pair of open ends equals zero.

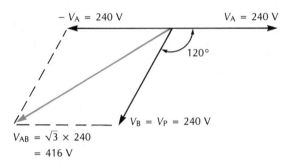

Figure 15-16(a) Incorrect delta-connection for two secondary coils–voltmeter reads $\sqrt{3}$ times coil voltage.

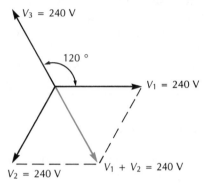

Figure 15-17(b) Phasor sum of V_1 and V_2 is equal but directly opposite to V_3. Phasor sum of V_1, V_2 and V_3 equals zero.

Figure 15-16(b) Phasor addition for the incorrect delta-connection of two secondary coils.

This faulty delta connection may be corrected by simply reversing the connections of any one of the two secondary coils. In the next section, it will be seen that this incorrect delta connection is actually the proper Wye connection for two secondary coils.

One end of the third transformer's secondary coil is now connected to the correct arrangement of the two previous transformer secondary coils, as shown in Fig. 15-17. If the voltmeter across the last pair of open ends reads zero volts, the connection of the third coil is correct, and these last two open coil ends can now be connected together to complete the delta connection of the three secondary coils. The phasor addition for the three delta-connected coil voltages is shown in Fig. 15-17(b). In this delta connection, the phasor

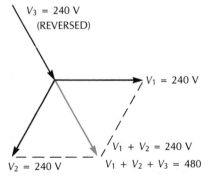

Figure 15-17(c) Coil voltage, V_3, connected in reverse in a delta-connection. The phasor sum of V_1, V_2 and V_3 equals two times the coil's voltage.

addition of V_1 and V_2 gives a phasor voltage of equal magnitude, but directly opposite to the phase voltage, V_3. Hence, the phasor sum of V_1, V_2 and V_3 will be equal to zero.

If the third transformer's coil was connected in reverse, the voltmeter across the remaining two open ends will read two times the coil's voltage. **Caution**: Under this incorrect connection, never attempt to connect together these two open coils ends because this will constitute a dead short across a voltage. Connection of these two leads can be made only when the voltage between them is zero. It is obvious that the connection of the third transformer's coil will have to be reversed to correct the circuit. The phasor diagram shown in Fig. 15-17(c) explains why when the third secondary coil is connected in reverse, the voltmeter reads two times the coil's voltage. In this connection, the phase voltage, V_3, will be equal to and also in the same direction with the phasor sum of V_1 and V_2.

The voltage, current, and kVA capacity in a three-phase delta-connected transformer bank should be obvious from previous discussions in this chapter.

(a)   In the delta-connected primary:

$$V_L = V_P \quad \text{and} \quad I_L = \sqrt{3}\, I_P \qquad (15\text{-}16)$$

where V_L and I_L = respective primary line voltage and line current

V_P and I_P = respective primary phase or coil voltage and current

(b)   Similarly, in the delta-connected secondary:

$$V_L = V_P \quad \text{and} \quad I_L = \sqrt{3}\, I_P \qquad (15\text{-}17)$$

(c)   The total kVA capacity of a delta-connected, three-phase transformer bank is equal to the sum of the three transformers' kVA capacity.

three-phase kVA
= kVA$_1$ + kVA$_2$ + kVA$_3$

$$(15\text{-}18)$$

Example 15-9: Three single-phase transformers are connected in a delta-delta connection, as shown in Fig. 15-18. Each transformer is rated at 75 kVA, 2400/240 V.

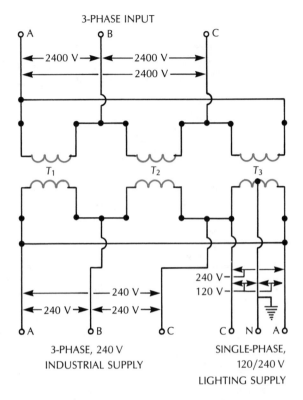

Figure 15-18 A delta-delta transformer bank supplying voltage to a single-phase three-wire load and to a three-phase three-wire load.

The power factor of the three-phase balanced load connected to the transformer bank is 0.8 lag, and the secondary line voltage and line current are, respectively, 240 V and 300 A. Assume negligible transformer losses. Determine:

(a)   the rated transformer bank kVA capacity
(b)   the load kVA on the transformer bank
(c)   the kW load on the transformer bank
(d)   the secondary coil voltage and current
(e)   the primary line voltage and line current
(f)   the primary coil voltage and current
(g)   the input kVA to the transformer bank

Solutions:

(a)   rated bank kVA capacity
= kVA$_1$ + kVA$_2$ + kVA$_3$

$$= 3 \times 75 \text{ kVA}$$
$$= 225 \text{ kVA}$$

(b) load kVA on bank $= \dfrac{\sqrt{3} \times V_L \times I_L}{1000}$

(secondary)
$$= \dfrac{1.73 \times 240 \times 300}{1000}$$
$$= 124.56 \text{ kVA}$$

Note that the bank is operating at only $\dfrac{124.56}{225} \times 100 = 55.36$ percent of its rated capacity.

(c) kW load on bank $= \dfrac{\sqrt{3} \times V_L \times I_L}{1000} \cos \underline{/\theta}$
$$= \text{load kVA} \times \cos \underline{/\theta}$$
$$= 124.56 \times 0.8$$
$$= 99.65 \text{ kW}$$

(d) In a delta-connected secondary:
$$V_L = V_P = 240 \text{ V}$$
$$I_P = \dfrac{I_L}{\sqrt{3}} = \dfrac{300}{1.73}$$
$$= 173.41 \text{ A}$$

(e) Assuming no transformer losses, load kVA on primary will equal load kVA on secondary. Also in a delta-connected primary, $V_L = V_P = 2400$ V.

primary load kVA $= \dfrac{\sqrt{3} \times V_L \times I_L}{1000}$
$$= 124.56 \text{ kVA}$$
$$I_L \text{ (primary)} = \dfrac{124.56 \times 1000}{1.73 \times 2400}$$
$$= 30 \text{ A}$$

Since the transformer turns ratio is 10:1, note that in a delta-delta transformer connection, the primary line current may also be found from:

$I_L \text{ (primary)} = \dfrac{I_L}{10} \text{ (secondary)}$
$$= \dfrac{300}{10}$$
$$= 30 \text{ A}$$

This result agrees with the value determined above.

(f) $I_P \text{ (primary)} = \dfrac{I_L}{\sqrt{3}} \text{ (primary)}$
$$= \dfrac{30}{1.73}$$
$$= 17.34 \text{ A}$$

or using the turns ratio method,

$I_P \text{ (primary)} = \dfrac{I_P}{10} \text{ (secondary)}$
$$= \dfrac{173.41}{10}$$
$$= 17.34 \text{ A}$$

(g) If no losses are assumed, then the input kVA must equal the load kVA. This load kVA was determined to be 124.56 kVA. This value may also be determined using input conditions.

input kVA $= \dfrac{\sqrt{3} \times V_L \times I_L}{1000} \text{ (primary)}$
$$= \dfrac{1.73 \times 2400 \times 30}{1000}$$
$$= 124.56 \text{ kVA}$$

The delta-delta transformer connection is used by some public utilities for moderate voltages. One of the main advantages of this connection is that if one transformer fails, then the other two can be operated as an open-delta connection. The open-delta will continue to supply three-phase voltage with the same phase sequence except that its kVA capacity will be only 57 percent of the original kVA. The open-delta connection will be described in a later section.

Whenever the closed-delta bank is used, it is often modified so that it can supply both three-phase, three-wire loads, and single phase, three-wire lighting loads. The only addition required to the basic delta connection of the transformers is a grounded centre tap or neutral wire connected to the middle point of one of the transformers' secondary coil. The transformer with the centre tap is usually of a higher kVA rating because of the extra load demand placed on it. In Fig. 15-18, the grounded neutral wire is shown attached to the secondary coil of T_3. Therefore, the lines B-N-C will supply single-phase 120/240 V for lighting loads, and the line wires A-B-C will supply three-phase 240 V for industrial loads. There is, however, one possible safety hazard with this system; whereas the voltage from either B to ground or C to ground is 120 V, the voltage from A to ground is 208 V.

THE WYE-WYE CONNECTION

A second standard connection for three-phase voltage transformation is the Wye-Wye connection, sometimes called the star connection. In this three-phase bank, the three primary transformer coils and the three secondary transformer coils are connected in Wye, as shown in Fig. 15-19.

In the Wye connection of the high voltage transformer coils, either the coil ends with

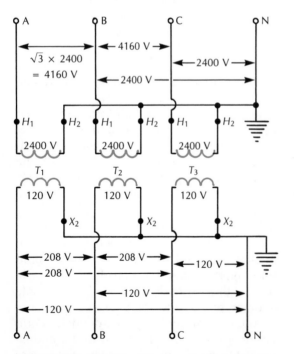

Figure 15-19(a) The Wye-Wye transformer bank with neutral.

polarity, H_2, are connected together as shown in Fig. 15-19, or the coil ends with polarity, H_1, may be commoned together. A line wire is then connected to each of the three remaining open coil ends. This Wye connection of the primary coils should be relatively simple because, as was mentioned before, the polarity of the high voltage coil of any transformer is fixed.

The Wye connection of the low voltage secondary coils is similar to the primary coil connection. However, the following procedure is recommended in order to avoid the possibility of error during connection.

1. Check to see that the output secondary voltage of each transformer is the same.
2. Connect one end of one secondary coil to one end of a second coil. If the connection is correct, a voltmeter across the two open ends should read $\sqrt{3}$ times the coil's voltage. Recall that this Wye connection of the two coils is actually the incorrect delta connection shown in Figs. 15-16(a) and (b). It should be obvious that if the Wye connection is incorrect, the voltmeter will read the same as the coil's voltage, and that to correct this would involve simply reversing the connection of one of the two coils. When the correct end of the third secondary coil is attached to the junction of the first two coils to complete the Wye connection, the voltmeter across any two of the three open ends should read $\sqrt{3}$ times the coil voltage.

The voltage, current, and kVA capacity in a three-phase Wye-connected transformer bank should be obvious from previous discussion in this chapter.

(a) In the Wye-connected primary:

$$V_L = \sqrt{3}\, V_P \qquad \text{and } I_L = I_F \qquad (15\text{-}19)$$

where V_L and V_P = respective primary line voltage and phase voltage

current I_L and I_P = respective primary line current and phase or coil current

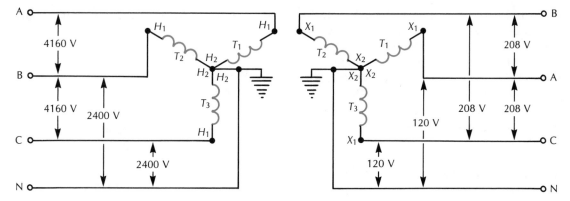

Figure 15-19(b) Simplified Wye-transformer bank with neutral.

(b) Similarly, in the Wye-connected secondary:

$$V_L = \sqrt{3}\, V_P \quad \text{and} \quad I_L = I_P \qquad (15\text{-}20)$$

(c) The total kVA capacity of a Wye-connected transformer bank is found the same way as for a delta-connected bank.

three-phase kVA
$$= kVA_1 + kVA_2 + kVA_3 \qquad (15\text{-}21)$$

The Wye-Wye connected transformer bank suffers from many disadvantages. For example, when the Wye transformer bank is supplying a load, it may cause severe interference in nearby communication circuits. Also, if the load is unbalanced, the three-wire Wye transformer bank will not operate satisfactorily. Therefore, for these and other reasons, the three-wire, Wye-Wye connected transformer bank is not as commonly used as the other transformer connections.

The Wye-connected system, however, could be improved to remedy any voltage imbalance by use of a fourth wire called the *neutral* wire. This neutral wire is grounded and connects to the junction point of the three coils in the Wye connection, as shown in Fig. 15-19. Therefore, the system may now be described as a three-phase, four-wire Wye-Wye connection.

The neutral wire will maintain a relatively constant coil voltage by conducting any imbalance current in the system when the load is unbalanced. Since the neutral wire is also grounded, it will provide some measure of safety from lightning surges. Notice, too, that the neutral wire is connected to the common point in the Wye connection and as a result, it will make available the two voltages, the line, and the phase voltage to the system. The coil or phase voltage will be the voltage from any one of the three lines to neutral wire, and the line voltage will be the voltage across any two lines.

It was already shown that the line voltage in a Wye connection will be equal to $\sqrt{3}$ times the coil voltage. Therefore, in the Wye connection shown in Fig. 15-19, the primary line and phase voltage will be respectively 4160/2400 V, and in the secondary Wye connection, the secondary line and phase voltage will be, respectively, 208/120 V. It should also be obvious from the diagram that the four-wire Wye-connected secondary can supply both three-phase, 208-V voltage for industrial loads, and single-phase, 120-V voltage for lighting loads.

If one transformer in a Wye-connected transformer bank should fail, it must be replaced before the system is re-energized. The Wye-connected bank cannot be used to supply three-phase voltage with only two transformers. This is possible, only with the delta-connected transformer bank.

Example 15-10: Three single-phase transformers, each rated at 2400/120 V, 25 kVA, are connected to form a three-phase four-wire, Wye-Wye transformer bank. The bank supplies a balanced load having a power factor of 0.7 lag with a line current of 200 A. Determine:

(a) the secondary line and phase voltage
(b) the three-phase bank kVA capacity
(c) the load kVA
(d) the load kW
(e) the input kVA (assume negligible losses)
(f) the primary phase and line voltages
(g) the primary phase and line currents

Solutions:

(a) In the four-wire, Wye-connected secondary:

$$V_P = V_{coil}$$
$$= 120 \text{ V}$$
$$V_L = \sqrt{3} \times V_P$$
$$= 1.73 \times 120$$
$$= 208 \text{ V}$$

(b) total kVA capacity $= kVA_1 + kVA_2 + kVA_3$
$$= 3 \times 25 \text{ kVA}$$
$$= 75 \text{ kVA}$$

(c) load kVA $= \dfrac{\sqrt{3} \, V_L \times I_L}{1000}$
$$= \dfrac{1.73 \times 208 \times 200}{1000}$$
$$= 72.0 \text{ kVA}$$

or load kVA $= \dfrac{3 \times V_P \times I_P}{1000}$
$$= \dfrac{3 \times 120 \times 200}{1000}$$
$$= 72.0 \text{ kVA}$$

(d) load kW $= \sqrt{3} \times V_L \times I_L \times \cos \underline{/\theta}$
$$= 72 \text{ kVA} \times 0.7$$
$$= 50.4 \text{ kW}$$

(e) if the losses are negligible, then

input kVA = load kVA = 72.0 kVA

(f) in the four-wire, Wye-connected primary:

$$V_P = V_{coil}$$
$$= 2400 \text{ V}$$
$$V_L = \sqrt{3} \times V_P$$
$$= 1.73 \times 2400$$
$$= 4160 \text{ V}$$

(g) primary input kVA = 72 kVA
$$= \dfrac{3 \, V_P \times I_P}{1000}$$

$$I_P = \dfrac{72 \times 1000}{3 \times 2400}$$
$$= 10 \text{ A}$$

or from transformer turns ratio of 20 : 1,

$$I_P \text{ (primary)} = \dfrac{I_P}{20} \text{ (secondary)}$$
$$= \dfrac{200}{20} = 10 \text{ A}$$

In a balanced four-wire, Wye-connected primary:

$$I_L = I_P = 10 \text{ A}$$

THE DELTA-WYE TRANSFORMER CONNECTION

The delta-Wye connection of three single-phase transformers is shown in Fig. 15-20. In the delta-Wye connection, the transformer primary coils are connected in delta, and the transformer secondary coils are connected in Wye.

The procedure for making the primary delta connection and the secondary Wye connection for the delta-Wye transformer bank follows the same rules previously described for the delta and Wye connections. Note, however, that the transformer bank shown in Fig. 15-20 is for a step-up voltage transformation. Hence, the primary coils will be the low voltage coils and their polarity will be $X_1, X_2 \ldots$. Similarly, the secondary coils will be the high voltage coils and the polarity will be $H_1, H_2 \ldots$. Recall, too, that the high voltage polarity of

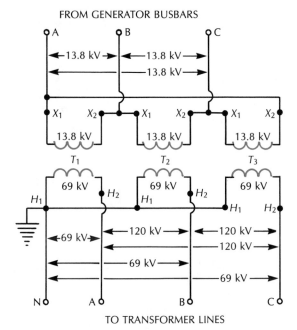

FROM GENERATOR BUSBARS

TO TRANSFORMER LINES

Figure 15-20 The delta-Wye transformer bank.

any single-phase transformer is fixed.

It should also be evident from previous discussions that in this delta-Wye bank,

(a) for the delta-connected primary:

$$V_L = V_P \quad \text{and} \quad I_L = \sqrt{3}\, I_P \qquad (15\text{-}22)$$

(b) for the Wye-connected secondary:

$$V_L = \sqrt{3}\, V_P \quad \text{and} \quad I_L = I_P \qquad (15\text{-}23)$$

and

(c) three-phase kVA capacity
$$= kVA_1 + kVA_2 + kVA_3 \qquad (15\text{-}24)$$

The delta-Wye transformer connection may be used for both step-up and step-down voltage transformation. However, this connection is especially adapted for step-up voltage transformation because the voltage is stepped up by the transformer ratio and is further increased by the factor of $\sqrt{3}$. Since the secondary coil voltage is only $\dfrac{1}{\sqrt{3}}$ or 57.7 percent of

the secondary output line voltage, this also means that the insulation requirements of the secondary windings are reduced. This is particularly useful when the output voltage is very high. This type of transformer bank is commonly used at generating stations where, for example, a three-phase generator output of 13 800 V is stepped-up to 120 000 V prior to transmission.

The use of the neutral wire on the high voltage secondary output also allows this system to have balanced voltages even when the load current is unbalanced. Note, too, that the neutral wire is grounded, and hence the high voltage secondary windings will be protected from lightning surges.

THE WYE-DELTA TRANSFORMER CONNECTION

The Wye-delta transformer connection is the opposite of the delta-Wye transformer connection. In this three-phase bank, as shown in Fig. 15-21, the transformer primary coils are connected in Wye and the secondary coils are connected in delta. The current, voltage, and kVA capacity relationships are the same as previously described for Wye and delta connections.

The Wye-delta transformer bank may be used for both step-up and step-down transformation, but it is better adapted for stepping down relatively high transmission line voltages at the load centre. There are two important reasons for using this bank for step-down transformation. First, the line voltage is reduced by the step-down turns ratio multiplied by the factor $\sqrt{3}$. Secondly, the insulation requirements for the high voltage windings are reduced because the primary coil voltage is only 57.7 percent of the primary line voltage.

The use of the neutral wire in the high voltage Wye-connected primary will also maintain balanced three-phase voltages even though the load currents may be unbalanced. Since the neutral wire is also grounded, it will pro-

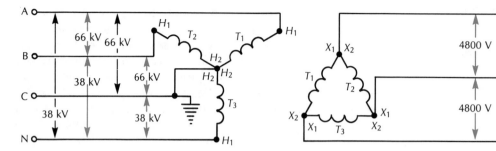

Figure 15-21 The Wye-delta transformer bank.

vide a measure of protection from lightning surges.

Example 15-11: A Wye-delta transformer bank as shown in Fig. 15-23 is connected to a 66-kV input three-phase transmission line. Each transformer is rated at 1000 kVA and has a step-down ratio of 8:1. The bank supplies a balanced load having a power factor of 0.8 lag with a load line current of 70 A. Assume negligible transformer losses. Determine:

(a) the primary line and phase voltages
(b) the secondary line and phase voltages
(c) the secondary line and phase currents
(d) the load kVA and kW
(e) the input kW
(f) the total kVA capacity
(g) the primary line and phase currents

Solutions:

(a) primary line voltage, $V_L = 66$ kV for a Wye-connected primary:

$$V_P = \frac{V_L}{\sqrt{3}}$$

$$= \frac{66\ 000}{1.73}$$

$$= 38.15\ \text{kV}$$

(b) Step-down ratio is 8 : 1, therefore the secondary phase voltage is:

$$V_P\ (\text{secondary}) = \frac{V_P}{8}\ (\text{primary})$$

$$= \frac{38\ 150}{8}$$

$$= 4.8\ \text{kV}$$

for a delta-connected secondary:

$$V_L = V_P = 4.8\ \text{kV}$$

(c) Secondary load line current, $I_L = 70$ A for a delta-connected secondary:

$$I_P = \frac{I_L}{\sqrt{3}}$$

$$= \frac{70}{1.73}$$

$$= 40.5\ \text{A}$$

(d) load kVA $= \dfrac{\sqrt{3} \times V_L \times I_L}{1000}$

$$= \frac{1.73 \times 4800 \times 70}{1000}$$

$$= 581.3\ \text{kVA}$$

load kW $= \dfrac{\sqrt{3} \times V_L \times I_L \times \cos\ \underline{/\theta}}{1000}$

$$= \text{load kVA} \times \cos\ \underline{/\theta}$$

$$= 581.3\ \text{kVA} \times 0.8 = 465\ \text{kW}$$

(e) Assuming no losses, then

input kW = output kW = 465 kW

(f) total kVA capacity $= kVA_1 + kVA_2 + kVA_3$

$$= 3 \times 1000\ \text{kVA}$$

$$= 3000\ \text{kVA}$$

(g) input load kVA $= \sqrt{3} \times V_L \times I_L$

$$= 581.3\ \text{kVA}$$

$$I_L \text{ (primary)} = \frac{581\ 300}{1.73 \times 6600}$$
$$= 5.1 \text{ A}$$

For a Wye-connected primary:

$$I_P = I_L = 5.1 \text{ A}$$

or from step-down ratio 8 : 1,

$$I_P \text{ (primary)} = \frac{I_P}{8} \text{ (secondary)}$$
$$= \frac{40.5}{8}$$
$$= 5.1 \text{ A}$$

THE OPEN-DELTA TRANSFORMER CONNECTION

It was mentioned in an earlier section that if one of the transformers in a delta-delta connected three-phase bank burns out, then the remaining two transformers may be used to continue the three-phase supply. This two-transformer connection is called the open-delta or V-connection and may be used to replace only a delta-delta connected system. This connection could be very useful in the case of an emergency where one of the transformers in the delta-delta bank becomes defective. However, note, as will be shown below, that the capacity of this open-delta bank will be only 57.7 percent of the capacity for the closed-delta connection.

For economical reasons, the open-delta system is also used sometimes in communities where the present demand for power is low, but where there is a strong possibility of future growth. When the community grows and the power demand increases, the system capacity can be easily increased by adding a third transformer to convert the open-delta bank to a closed-delta bank.

Figure 15-22 shows three single-phase transformers. By disconnecting all the leads to the defective transformer as shown, the closed-delta transformer connection becomes an open-delta connection. The question

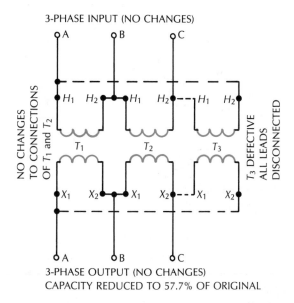

3-PHASE INPUT (NO CHANGES)

3-PHASE OUTPUT (NO CHANGES)
CAPACITY REDUCED TO 57.7% OF ORIGINAL

Figure 15-22 The open-delta connection.

which must now be answered is, why does the capacity for the two connected transformers reduce to only 57.7 percent instead of 66.7 percent (2/3) of the original capacity for the three transformers? Fig. 15-23 will be used to answer this question.

For simplicity, only the secondary coil connections of the transformers are shown in Figs. 15-23(a) and 15-23(b) for the respective closed-delta and open-delta transformer banks. The transformers are all assumed to be operating at their rated coil voltage of 120 V, and coil current of 15 A.

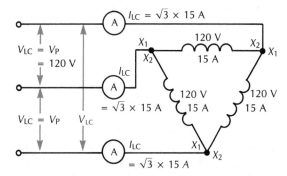

Figure 15-23(a) Secondary voltages and currents in a closed-delta transformer bank.

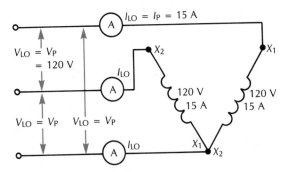

Figure 15-23(b) Secondary voltages and currents in an open-delta bank.

For the closed-delta connection of Fig. 15-23(a), the line current $I_{LC} = \sqrt{3} \times 15 = 25.95$ A, and the line voltage, $V_{LC} = V_P = 120$ V. Therefore,

$$\text{closed-delta VA capacity} = \sqrt{3} \times V_{LC} \times I_{LC}$$
$$= 1.73 \times 120$$
$$\times 25.95$$
$$= 5387.2 \text{ VA}$$

For the open-delta connection of Fig. 15-23(b), the line current, for a balanced three-phase load, is equal to the coil current, $I_{LO} = I_P = 15$ A, and the line voltage, $V_{LO} = V_P = 120$ V.
Therefore, open-delta VA capacity

$$= \sqrt{3} \times V_{LO} \times I_{LO}$$
$$= 1.73 \times 120 \times 15$$
$$= 3114.0 \text{ VA}$$

The ratio of the open-delta capacity to the closed-delta capacity is:

$$\frac{\sqrt{3} \times V_{LO} \times I_{LO}}{\sqrt{3} \times V_{LC} \times I_{LC}}$$
$$= \frac{3114}{5387.2} = 0.557 \text{ or } 57.7 \text{ percent}$$

Therefore, when a closed-delta bank is changed to operate as an open-delta bank, the capacity of the open-delta bank will be only 57.7 percent of the capacity for the closed-delta bank. That is, each transformer in the open-delta bank can now only be loaded to 86.6 percent of its actual rating.

Example 15-12: A defective transformer is removed from a delta-delta transformer bank and the new open-delta bank continues to supply voltage to the system. The rated line voltage and line current for the closed-delta system were 240 V, and 17.3 A, respectively, and the VA capacity of each transformer is 20 VA.
Determine:

(a) the total KVA capacity for the closed-delta bank

(b) the KVA capacity for the open-delta bank, and

(c) the rated line current for the open-delta bank

Solutions:

(a) Closed-delta bank capacity $= 3 \times \text{kVA}_1$
$$= 3 \times 20$$
$$= 60 \text{ kVA}$$

(b) Open-delta bank capacity $= \dfrac{57.7}{100} \times 60$
$$= 34.68 \text{ kVA}$$

(c) Rated line current for open-delta bank is equal to the rated transformer coil current I_{LO} (open-delta) $= I_P$

$$= \frac{I_{LC}}{\sqrt{3}} \text{ (closed-delta)}$$
$$= \frac{17.3}{1.73}$$
$$= 10 \text{ A}$$

THREE-PHASE TRANSFORMERS

A three-phase voltage can also be transformed with the use of a three-phase transformer instead of a bank of three single-phase transformers. The advantages of this system are that the unit is more compact and occupies less space. The initial installation cost is also less than for three single-phase transformers, and the wiring connections are usually simpler. The main disadvantage is that if a coil burns out, then the complete unit must be replaced.

Figure 15-24 A three-phase power transformer.
Courtesy: Reliance Electric Limited.

Basically, a three-phase transformer contains a core consisting of three legs with the primary and secondary coils for one phase wound on each of the three legs. The assembly is housed in a single tank filled with an insulating transformer oil. A three-phase transformer is shown in Fig. 15-24. The fluxes in each of the three coil legs are 120° out-of-phase with each other and will therefore reach their maximum values at different instants. At any instant at least one of the legs will act as the return path for the fluxes of the other phases.

The connections between the coils are made inside the bank and brought to the outside through insulated bushings. When the coils are connected in Wye, there will be four external terminals, and when the coils are connected in delta, there will be only three external terminals on the bank.

SUMMARY OF IMPORTANT POINTS

Three–phase voltage
1. Consists of three voltage phases 120 electrical degrees apart.
2. Requires three or four wires for transmission instead of six wires.
3. Its power is relatively constant compared to the pulsating power for a single-phase voltage.
4. Can supply both three-phase and single-phase loads.
5. Its phase sequence may be reversed by either reversing the alternator's rotation or by interchanging the connections to any two of the three line wires.

The Wye connection
6. Either the end or the beginning of each of the three coils is connected together. The three line wires are connected to other coil ends. A fourth wire, the grounded neutral, is connected to the junction of the Wye-connection.
7. $V_P = V_C$ = voltage from line to neutral.
8. $V_L = \sqrt{3}\, V_P$ = line-to-line voltage.
9. $I_L = I_P$ = phase and line current.

The delta connection
10. The end of one coil is connected to the beginning of the other. A line wire is connected to each of three junctions.
11. $V_L = V_P$ and $I_L = \sqrt{3}\, I_P$.

Three-phase power
12. The angle between a phase voltage and its phase current is the three-phase power factor angle.
13. Three-phase VA $= 3\, V_P \times I_P = \sqrt{3}\, V_L I_L$

Three-phase watts = $3 V_P \times I_P \cos \underline{/\theta} = \sqrt{3} V_L \times I_L \cos \underline{/\theta}$.

Three-phase VARs = $3 V_P \times I_P \sin \underline{/\theta} = \sqrt{3} V_L \times I_L \sin \underline{/\theta}$.

14. Three-phase power measurement by two wattmeters: meter reading is dependent on the voltage phase rotation and the load power factor $W_T = W_1 + W_2$.

    One meter will read negatively when P.F. < 0.5.

    Meter reading = $V_L I_L \cos \underline{/\theta_L}$ where $\underline{/\theta_L}$ is the phase angle between the line voltage and line current for that particular meter.

15. Three-phase power measurement by three wattmeters: can only be used on four-wire systems. Each meter measures the power of a particular phase $W_T = W_1 + W_2 + W_3$.

Three-phase transformer connections

16. Types of connections are: (a) delta-delta, (b)Wye-Wye, (c) delta-Wye, (d) Wye-delta, and (e) open delta.

17. In delta-delta, Wye-Wye, delta-Wye, etc, the first term refers to the primary coil connections and the second term to the secondary connection.

18. Particular attention must be paid to the transformer polarity when making any transformer connections.

19. The H_1, H_2 markings are for the high-voltage connections and the X_1, X_2 markings are for the low-voltage connections.

20. For the delta-connection, H_1 of one coil is connected to H_2 of another coil, etc. Similar connections are used for X_1 and X_2.

21. For the Wye connection, either all three H_1 or the three H_2 terminals are connected together. The Xs are connected in the same manner.

22. The open-delta connection can be used to replace only the delta-delta connection. Its capacity will be only 57.7 percent of the delta-delta bank capacity.

23. Three-phase bank kVA = $kVA_1 + kVA_2 + kVA_3$.

24. Bank equations are the same as for the delta and Wye connections given previously.

25. The three-phase transformer is slightly more compact and efficient than a three-phase transformer bank.

ESSAY AND PROBLEM QUESTIONS

1. Draw the waveform for a three-phase voltage and indicate the phase sequence of the voltages.

2. Draw a diagram to illustrate how a 240-V single-phase supply can be made into a three-wire 120/240-V supply.

3. Describe the advantages of a three-phase generation and distribution system over a single-phase system.

4. (a) Describe what is meant by "phase sequence of a polyphase system."
   (b) Describe two methods whereby the phase sequence of a three-phase system may be reversed.

5. (a) Give two examples where proper phase sequence connections are important.
   (b) How can the phase sequence of a system be determined?

6. Explain how the three coils of a three-phase alternator may be connected in (a) Wye and (b) delta. Which one of the connections generally uses four wires?

7. In a four-wire Wye connection, what is the voltage from (a) line to neutral and (b) line to line called? What is the relationship between these two voltages?

8. What are the advantages of using a grounded neutral wire in a Wye connection?

9. Explain why, when three coils are connected in Wye, the voltage across any two open ends is equal to $\sqrt{3}$ times the coil voltage.

10. A Wye-connected three-phase circuit has a power factor angle of 60° lag. Draw the

phasor diagram to show (a) the phase voltages, (b) the line voltages, (c) the phase currents, and (d) the line currents.

11. Explain why, when three coils are connected in delta, the line current is equal to $\sqrt{3}$ times the coil current.

12. A delta-connected three-phase circuit has a power factor angle of 30° lag. Draw the phasor diagram to show (a) the phase voltages, (b) the line voltages, (c) the phase currents, and (d) the line currents.

13. If a balanced three-phase Wye-connected load, as shown in Fig. 15-8, has a phase impedance, $Z_P = 20\ \Omega$, and a phase voltage, $V_P = 240$ V, determine:
(a) the phase current, (b) the line current, (c) the line voltage, and (d) the load apparent power.

14. A balance delta-connected three-phase load, as shown in Fig. 15-9(a), has a phase resistance, $R_P = 12\ \Omega$, in series with an inductive reactance, $X_P = 8\ \Omega$. If the phase voltage to the load is equal to 360 V, determine: (a) the phase impedance, (b) the load power factor (c) the phase current, (d) the line current, and (e) the load true power.

15. A three-phase delta-connected industrial motor load is supplied with a three-phase line voltage of 480 V, and draws a line current of 20 A. If the load power factor is 0.7 lag, determine: (a) the circuit phase voltage, (b) the phase current, (c) the load apparent power, and (d) the load true power.

16. (a) Draw a diagram showing how two wattmeters must be connected in a three-wire, three-phase circuit to measure the circuit power.
(b) Compare and comment on the reading from the two wattmeters for each of the following circuit power factor conditions, and indicate how the total circuit power will be determined: (a) P.F. = 1, (b) P.F. = 0.8, (c) P.F. = 0.5, and (d) P.F. = 0.3.

17. Describe the voltage, current, and phase angle for each wattmeter in the two-wattmeter connection for determining its reading.

18. Two wattmeters are used to measure the power in a balanced three-phase delta-connected circuit with a power factor angle of 45° lag. Assume the circuit phase sequence is...A.B.C.A..., with wattmeter, W_1, connected across the line voltage $V_{AB} = 240$ V, and wattmeter, W_2, connected across the line voltage, $V_{CB} = 240$ V. Wattmeter, W_1, is in series with line current, $I_A = 8$A, and W_2 is in series with line current, $I_C = 8$ A.
(a) Draw a phasor diagram to show the voltage, current, and phase angle that each wattmeter senses.
(b) Determine the reading of each wattmeter.

19. Assume the same circuit connections and values as given in Q. 18, except that the circuit power factor angle is now 60° lead.
(a) Draw the phasor diagram to show the voltage, current, and phase angle that each wattmeter senses.
(b) Determine the reading of each wattmeter.

20. Draw a diagram to show the connections for the three-wattmeter method for power measurement in a three-phase circuit.

21. (a) Describe the voltage, current, and phase angle for each wattmeter in the three-wattmeter power measurement method.
(b) Compare and comment on the wattmeter readings for three-phase loads that are balanced but have a different power factor.

22. A three-phase voltage may be transformed using a bank of single-phase transformers. Name five connection methods that may be used to construct a three-phase transformer bank. Which of

these banks uses only two single-phase transformers?

23. Name two important transformer properties which must be taken into account when connecting them into a three-phase bank.

24. Describe the method of connecting three single-phase transformers into a delta-delta bank. Assume that the polarity of each transformer is unknown.

25. Draw a diagram to show how an delta-delta three-phase transformer bank may be modified to supply both three-wire, three-phase, and three-wire, single-phase loads.

26. The secondary coils of three transformers are to be connected in delta. What important precaution should be taken before attempting to close the final two open ends of the delta formation?

27. Three transformers are connected into a delta-delta bank. Each transformer is rated at 50 kVA, 4800/240 V. The bank is connected to a balanced load which has a power factor of 0.6 lag, and draws a line current of 100 A at 240 V. Assume negligible transformer losses. Determine, (a) the rated bank kVA capacity (b) the load kVA on bank, (c) the kW load, (d) the secondary phase current, (e) the primary phase and line current, and (f) the input kVA.

28. Describe the method for connecting three single-phase transformers into a Wye-Wye bank. Assume that the polarity of each transformer is unknown.

29. Describe three advantages for using a grounded fourth wire (grounded neutral) in a Wye-connected system.

30. Three single-phase transformers are connected into a four-wire Wye-Wye three-phase bank. Each transformer is rated at 4800/240 V, 100 VA, and the bank supplies a balanced load having a power factor of 0.8 lag with a line current of 300 A. Assume negligible transformer losses. Determine: (a) the load VA and kV, (b) the

input VA, (c) the rated VA, (d) the primary phase and line currents, and (e) the primary phase and line voltages.

31. Give two reasons why the delta-Wye transformer bank is best adapted for step-up voltage transformation.

32. A delta-Wye transformer bank is used at a generating plant to step up a generator output voltage of 5 kV to a transmission line voltage of 69 kV. If each transformer is rated at 200 VA, determine: (a) the rated primary phase and line current, (b) the rated secondary phase and line current, (c) the rated primary phase and line voltage, and (d) the three-phase VA capacity of the bank.

33. (a) Draw a diagram to show the connections for three single-phase transformers connected into a Wye-delta bank. (b) Give two reasons why the Wye-delta transformer bank is best adapted for three-phase step-down voltage transformation.

34. A delta-Wye transformer bank is used to step-down a three-phase three-wire primary voltage of 4800 V to a three-phase, four-wire secondary voltage of 120/208 V. Each transformer is rated at 25 VA, 4800/120 V. If the bank supplies a balance three-phase load in which the impedance of each phase is 6 Ω, and the power factor is 0.8 lag., determine: (a) the secondary coil and line current, (b) the load VA and kV, (c) the three-phase VA capacity of the bank, and (d) the primary coil and line load currents.

35. An Wye-delta transformer bank is used to step-down a three-phase, four-wire primary voltage of 4160/2400 V to a three-phase, three-wire voltage of 240 V. Each transformer in the bank is rated at 15 VA, 2400/240 V. If the bank supplies a balanced load with a line current of 30 A, determine:
    (a) the secondary coil current
    (b) the load VA on the bank

(c) the three-phase VA capacity of the bank, and

(d) the primary coil and line current

36. (a) Draw a diagram to show two single-phase transformers connected into an open-delta transformer bank. (b) Which of the other transformer bank connections could this open-delta bank be used to replace in an emergency? (c) What will be the system's new VA capacity if such a substitution is made?

37. Two single-phase transformers, each rated at 30 VA, 2400/120 V, are connected to form an open-delta, three-phase bank. Determine: (a) the VA capacity for this bank, and (b) the rated secondary line current.

38. Give three advantages and one disadvantage of the three-phase transformer as compared with a three-phase bank consisting of three single-phase transformers.

39. A three-phase transformer has its primary low voltage coils connected in delta, and its secondary high voltage coils connected in Wye. The transformer has a total VA capacity of 5000 VA and each set of phase windings has a voltage ratio of 1 to 8. If the primary input line to the transformer is 2400 V, determine: (a) the secondary phase and line voltage, (b) the rated primary coil and line currents, and (c) the rated secondary line current.

CHAPTER
SIXTEEN

AC GENERATORS (ALTERNATORS)

It was noted in Chapter Fourteen that the *major* function of transformers is their ability to step-up a generated voltage to a very high voltage, for instance 500 kV, for the purpose of long-distance voltage transmission. At this very high voltage, the line current is very low and the line power losses can be kept to an acceptable level.

In dc generators, the voltage generated can be only of a few thousand volts due to the limitations of the generator's commutator and brushes. The generated dc voltage cannot be increased in value conveniently, and it is for these reasons that during the second half of the twentieth century ac voltage generation and transmission have become nearly universal. Charles Steinmetz was the scientist who originally was deeply involved in the development of ac generators because he understood the necessity of producing a voltage which could be transformed to higher values for transmission purposes.

CONSTRUCTION OF AC GENERATORS

The voltage induced in most ac generators is 13.8 kV. This voltage is induced in coils wound on a stationary *stator*. Leads from the stator coils are connected to terminals located on the bottom or side of the stator. There are no rotating parts involved with the stator and, therefore, there will be no arcing at brushes as no slip rings are required.

The armature of an ac generator is fixed and the windings in which the ac voltage is induced are located in stator slots (Fig. 16-1) located on its inner surface. The coils to be used are formed-wound and are of large cross-sectional area and are insulated prior to being placed in the parallel-side slots of the stator. The iron of the stator is built up of thin insulated laminations to minimize eddy-current losses due to the varying magnetic flux. These laminations of silicon steel alloy are usually 0.3- or 0.4-mm thick in 60-Hz generators.

Note that the sides of the slots have been machined parallel with each other so as to be able to accommodate the large area turns of the coils. The teeth on the stator, adjacent to the parallel-sided slots, are wide and strong near the bottom of the slots. The centrifugal forces developed in a generator are very large, and unless the teeth are made wide and strong, they may possibly break off due to the mechanical stresses placed on them. Not

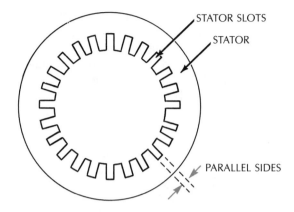

Figure 16-1 The stator slot structure of an ac generator.

shown in Fig. 16-1 is the provision made in the sides of the slots for a covering wedge to keep the coils tightly in the slots.

The stator coil windings installed in a large three-phase generator may be either lap or wave windings. The lap winding is the most often used type of stator winding because of its shorter coil connections. Since single-phase windings are seldom used in ac generators, the three-phase generator requires three identical groups of windings spaced 120 electrical degrees apart. Electrical degrees refers to the degrees of a complete sine wave where each cycle consists of 360 electrical degrees.

It is very important that the induced voltage in ac generators be of sinusoidal wave form. To obtain this form of induced voltage, the location and connection of the three separate coils in the stator are critical.

The three single-phase windings may be connected in either delta or Wye (star). The Wye connection is used most often because the terminal voltage will be 1.732 times the induced coil voltage. This factor reduces the coil's current and hence less insulation is required by the coils. Since the three groups of windings must be 120 electrical degrees apart, and since the magnetic poles to be developed must exist in pairs, there are certain rules that must be observed regarding the coil winding and the spacing of the slots in the stator.

An ac generator is designed and built to produce a specific voltage at a specific frequency. This frequency of the induced voltage in North America is 60 Hz. The actual winding of the stator coils is dependent on the number of magnetic poles of the rotor, to be discussed later, and the speed of rotation of these magnetic poles. For each of the three sets of coils wound on the stator, the frequency of the induced voltage must be 60 Hz regardless of whether the coils are connected in delta or in Wye. The frequency of the induced voltage is expressed in the equation

$$f = \frac{P\omega}{4\pi} \qquad (16\text{-}1)$$

where f = frequency in Hz
P = number of poles
ω = radians per second

In terms of revolutions per minute (S), this equation may also be expressed as $f = \frac{PS}{120}$.

Since one mechanical rotation equals 2π radians and one mechanical rotation induces one voltage cycle for every two poles, it follows that the frequency (f) of the induced voltage for a generator with (P) poles and rotating at (ω) radians per second is given by:

$$f = \frac{\omega}{2\pi} \times \frac{P}{2}$$
$$= \frac{P\omega}{4\pi} \text{ Hz}$$

Example 16-1: Determine the frequency of an induced generator voltage whose rotor has six poles and turns at 125.664 rad/s.

$$\text{where } f = \frac{P\omega}{4\pi}$$
$$= \frac{6 \times 125.664}{12.5664}$$
$$= 60 \text{ Hz}$$

From the above it can be seen that the rotative speed and the number of magnetic poles will determine the frequency of the induced voltage in *each* of the three coils located on the stator. Students are also reminded that in the metric system the speed of rotation is not expressed in revolutions per minute (r/min) but in radians per second (1 r/min = 2π rad/s).

The maximum voltage induced along a sine wave occurs every 180°. This voltage is induced at each successive pole. Therefore, the number of electrical degrees in one complete mechanical revolution equals 180 times the number of poles.

electrical degrees in one revolution
= 180 P   (16-2)

Example 16-2: Assume that there are 48 slots in the stator of a three-phase, four-pole ac generator. This means that there are $\frac{48}{4}$ = 12 slots per pole position and $\frac{12}{3}$ = 4 slots per phase per pole. Since the total number of electrical degrees for one revolution is 180 × 4 = 720, and since the frequency of the voltage is to be 60 Hz, the number of slots per pole position must be equal to $\frac{720}{60}$ = 12.

In the above example, each coil with the correct number of turns must be wound on 12 slots so that the voltage may be of a certain value and have a frequency of 60 Hz. In a 48-slot, four-pole stator, each coil will have spanned 90° of the stator. This stator is said to be wound with *full-pitch* coils (Fig. 16-2) because each side of the coil will be under one of the rotor poles. However, very often the stator coils are wound to cover fewer slots. The stator is then said to be wound *fractional pitch*. For instance, if the coil enters slot 1 and returns from slot 9 of 12 slots, the coil covers 9 – 1 = 8 slots. The pitch is then $\frac{8}{12}$ = 66.7 percent.

The fractional-pitch wound stators have the following advantages over full-pitch wound stators:

1. The copper used in the coils is less and therefore the I^2R loss is less.

2. It is easier for the assemblers to form the coils in a more common form.
3. There will be cancellation of higher harmonics produced in the generated voltages and this will result in a more pure sinusoidal wave form of voltage.

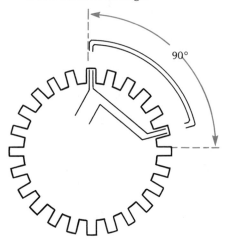

Figure 16-2 A four-pole generator with full pitch coils.

There are several methods which can be used to wind the three-phase coils on a stator, but no attempt is made here to discuss details of coil winding. Students who are interested in the details of coil winding are advised to consult generator winding manuals.

As mentioned earlier, the three coils of a stator may be connected in delta or in Wye (star), as shown in the diagrams below.

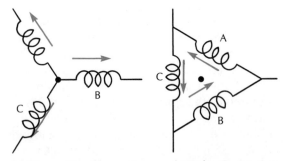

(a) Wye or Star Connection   (b) Delta Connection

Figure 16-3 Three-phase ac generator connection.

For full-pitch windings in each of the three-phases, per pole groups, the rms voltage induced pole coil may be determined from the following equation

$$V_{rms} = 4.4428 \; \phi Nnf \qquad (16-3)$$

where ϕ = number of Wb
  N = number of turns per coil
  n = number of coils per phase per pole
  f = frequency

Equation 16-3 is a fundamental generator equation and needs to be remembered.

Please note that Eq. 16-3 refers to the voltage induced per pole per phase. This voltage may then be multiplied by the number of pole-phase groups that are connected in series to obtain the voltage induced per phase. Keep in mind that, for instance, a four-pole generator would have all four-pole coils connected in series to obtain the desired voltage level. One other point to remember is that the terminal voltage is equal to the coil voltage if the coils are connected in delta, but that the terminal voltage equals 1.732 times the coil voltage if the coils are connected in Wye (star).

Also note that Eq. 16-3 is only used to determine the rms voltage for coil groups with full-pitch windings. However, in a fractional pitch-wound coil, the voltage induced is not the same value at each side of the coil at exactly the same time. Therefore, Eq. 16-3 must be multiplied by a constant, k_P, which takes account of the pitch of the coils. This constant, factor k_P, is equal to the sine of half the coil span in electrical degrees (full pitch = 180°) or

$$k_P = \sin\left(\frac{S}{2}\right) \qquad (16-4)$$

where S = coil span in electrical degrees

Equation 16-3 is then changed to read

$$V_{rms} = 4.4428 \; \phi Nnf k_P \qquad (16-5)$$

There is, however, one other factor to be considered. When coils are wound in the slots of the stator, the voltages induced in turns cannot be added arithmetically to obtain the coil voltage. There will be a difference of phase angle between the voltages in the various turns. A constant, called distribution factor k_d, is added to the above equation. This distribution factor is determined as follows:

$$k_d = \frac{\text{phasor sum of coil voltages per phase}}{\text{arithmetic sum of coil voltages per phase}}$$

The final equation used to determine the rms voltage induced in a fractional pitch coil is then

$$V_{rms} = 4.442\,8 \; \phi Nnf k_{k_P} k_d \qquad (16-6)$$

Example 16-3: The three-phase coils of an ac generator are connected in Wye (star). The stator is wound as a four-pole stator and consists of 48 slots. Each of the coils consists of 20 turns and spans 9 slots out of 12 possible slots, that is, slots 1 to 10. The field coils will produce 0.008 Wb per pole. The rotor's flux turns at 188.496 rad/s. Determine the generator's terminal voltage.

Solution: The number of slots per pole is $\frac{48}{4}$ = 12. The number of slots per pole per phase (n) is $\frac{12}{3}$ = 4. There are $\frac{180}{12}$ = 15 electrical degrees per slot, and with a span of 9 slots, there are 9 × 15 = 135 electrical degrees per coil. From a table in a generator winding manual, it may be determined that the pitch factor k_P is 0.923 88 for a $\frac{9}{12}$ pitch and that the distribution factor k_d is 0.957 66. Using Eq. 16-6, the voltage induced per pole coil group will be

$$V_{rms} = 4.4428 \times 0.008 \times 20 \times 4 \times 60 \times$$
$$0.9238\,8 \times 0.957\,66$$
$$\left(\text{Note that } f = 60 \text{ Hz is found using} \right.$$
$$\left. f = \frac{P \times \omega}{4\pi} \right)$$
$$V_{rms} = 150.943 \text{ V per pole group}$$
$$\text{or pole pair}$$

Since there are two pole groups (four poles), the total coil voltage is 150.9436 × 2 = 301.8872 V. Since these coils are connected in Wye (star), the terminal voltage is 301.8872 × 1.732 = 522.87 V.

THE ROTOR

Throughout the above discussion the assumption has been made that a rotating magnetic field will be cutting through the turns of the coils wound on the stator. This rotating magnetic field is produced by a rotor which is located inside the stator and can rotate freely inside the stator. The rotor structure is built also of thin insulated laminates of silicon steel alloy to minimize eddy-current losses.

There are two types of rotors, the *salient-pole* type and the *cylindrical* type. The salient-pole rotor is generally used for low-speed generators. Figure 16-4 shows a simplified diagram of a salient-pole rotor consisting of four electromagnetic poles.

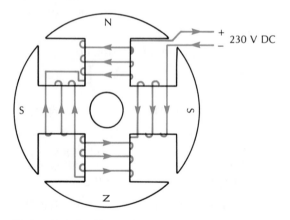

Figure 16-4 A salient-pole ac generator rotor.

The circuit for the field coils wound on the pole structures of this rotor is connected to two slip rings. These slip rings are copper alloy rings that are mounted on the rotor shaft but are insulated from it. The excitation current necessary to provide the magnetic flux is generally provided from an up to 500-V dc source. This current from the dc source flows through

carbon brushes, supported by brush rigging, to the slip rings and the coils of the rotor poles. The field coils are connected in series and are wound so that alternate magnetic poles are produced when current flows through the coils.

The whole pole structure is mounted on and keyed to a rotor shaft. The prime mover for slow-speed rotors provides the mechanical energy to produce the rotor torque so that its magnetic flux rotates across the stator coils. The salient-pole rotors are used on ac generators whose rotative speed may not exceed 190 rad/s.

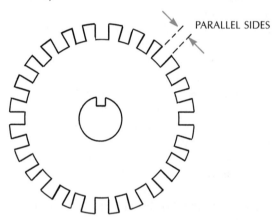

Figure 16-5 End view of a cylindrical rotor.

Cylindrical-type rotors (Fig. 16-5) are used on most steam turbine-driven generators which develop higher speeds. On cylindrical rotors, the coils required to produce the magnetic poles are located in slots of the laminated rotor core.

The coils are wound in the rotor slots in such a manner that the desired number of magnetic poles will be developed when the dc current flows through these coils. The coils wound in the slots are held in place with wedges. Also in this type of rotor, the field pole circuit is connected to two slip rings. The excitation voltage of up to 500 V dc is often supplied by a small motor-generator or *exciter*, mounted on the same generator shaft as the rotor.

ROTOR FIELD DISCHARGE CIRCUIT

When the dc voltage supply from the exciter to the rotor is to be disconnected, a special field discharge circuit is required. If the dc voltage were disconnected from the exciter by the opening of a two-pole switch, the collapsing magnetic field would cut through the rotor coils and induce a large enough voltage that could damage the insulation and possibly the equipment. To prevent this from happening, a special *field discharge switch* is normally used.

Figure 16-6 shows the connections of this switch between the dc exciter and the rotor field coils when the switch is closed and current from the exciter flows through a control rheostat and the rotor field coils.

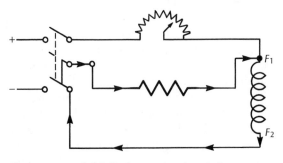

Figure 16-6 A field discharge circuit switch.

When the field discharge switch is closed, an auxiliary switch blade is in the open position.

When the dc voltage supply is to be disconnected, the field discharge switch is opened, thereby closing the auxiliary blade. This blade actually closes before the main switch blades are fully open.

When the main switch blades are in the open position (Fig. 16-7), a circuit still exists through the closed auxiliary switch blade. As the magnetic field collapses in the rotor coils, a voltage will be induced which decreases very quickly as the current flows through the field discharge resistor and is dissipated as heat. This arrangement ensures that no one is endangered when the switch is opened, and the coils' insulation is protected from being damaged by a high voltage.

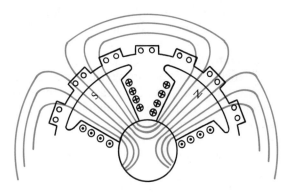

Figure 16-7 Magnetic flux of one pair of salient rotor poles cutting through turns of stator coils.

VENTILATION AND COOLING

When the salient-field pole rotor rotates, the poles act as fans to circulate the air throughout the generator, and no problems are encountered with the heat present from all coils. However, cylindrical rotors are normally used in large generators. These rotors are quite smooth and cannot be used as fans to circulate the collected heat away from the stator coils.

Elaborate systems of cooling ducts are provided in the larger ac generators and the whole generator is sealed in a gas-tight casing. Hydrogen is used in the larger generators as a cooling medium. The hydrogen is maintained in the generator at a slight pressure above atmospheric pressure and any leakage will then be in an outward direction. This is done so that an explosive hydrogen-air mixture cannot develop inside the generator. The hydrogen pressure may be varied to accomplish changes in the power rating of the generators. At higher pressure, the cooling effect is increased and the power rating of a generator increases.

Hydrogen has some major advantages over air as a cooling medium in generators. The density of hydrogen is about 0.07, that of air at the same temperature and pressure. As a result, the windage losses of the generator are much less. Hydrogen also serves as a much better transfer agent for the heat developed in the generator than air. Its heat conductivity is about 7.5 times the heat conductivity of air.

When hydrogen is used as a cooling agent, the danger of fires occurring is eliminated. As long as the hydrogen content is more than 70 percent, a hydrogen-air mixture will not explode. Hydrogen is circulated throughout the generator by a blower system, and the heated hydrogen is blown through a water-cooled heat exchanger and is then recirculated through the generator.

The use of hydrogen as a cooling agent in a generator ensures that the insulation of all coils does not deteriorate. Maintenance costs also decrease because there will be no dirt, moisture, and oxygen in the generator. Generators that are cooled with hydrogen kept at a certain pressure can be built of smaller size, since the hydrogen acts as a good medium for the transfer of heat from all coils and, therefore, the power rating may be increased for a specific size generator.

Improvements in the cooling of generators have been made over the years. In some generators, cooling ducts are installed in both stator and rotor and the heat produced is forced through these ducts by the hydrogen. In other cases, metal ducts are installed in the stator and liquids are then circulated through them causing transfer of heat directly from the coils of the stator.

Some generators are made in which stator cooling is accomplished through the forced flow of water through metallic tubes in the stator while the rotor is cooled by hydrogen at a definite pressure.

INDUCED VOLTAGES IN AN AC GENERATOR

When the prime mover, such as steam or water, causes rotation of the generator shaft and the rotor, the dc excitation current is applied to the rotor coils. The current flowing through these coils produces the number of pairs of magnetic poles for which the rotor has been wound. The voltage to be induced in the three stator coils is in practice controlled by varying the amount of dc voltage applied to the rotor windings. As a result, the dc current flow is controlled, as well as its magnetic flux.

Since the rotor turns at a controlled number of radians per second, the magnetic flux of its poles is also rotating at the same number of radians per second. Note well that the flux density of each pair of poles is dependent on the amount of dc current flowing. Since the induced ac voltage must be of a predetermined value, the dc current flow is regulated extremely carefully; this regulation process will be discussed later.

It was seen earlier that the fixed armature or stator is wound for three complete phases. The coils for these phases are located 120° apart. As a result of the rotating flux, voltages are induced in these coils in such a manner that they are 120° apart. Without entering into all possible details for the induced voltage per coil, it needs to be stressed that the number of turns of the coils, the pitch used, and the number of poles of the rotor will determine the actual amplitude of the induced voltage at any instant, as well as its polarity. It was stressed earlier that the winding of these coils is very complex since it is very necessary that the coil formation be such that the induced voltage be of sinusoidal wave form.

Assuming that the stator coils have been designed and wound correctly, and assuming that the density of the rotor's magnetic flux is as desired, the amplitude of the voltage induced during each of the 360 electrical degrees for each phase will be of the value for which the generator was designed.

The sequencing of the induced voltages in the three phases is critically important and this will be discussed further when the paralleling of generators is discussed. Figure 16-8 illustrates the sequencing of the three phase vol-

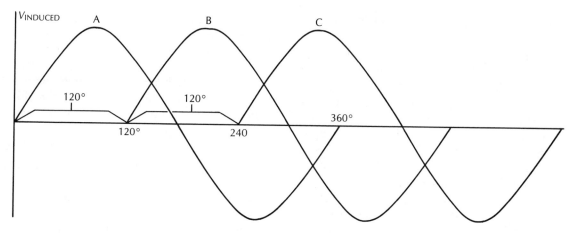

Figure 16-8 Three-phase voltage induced in an ac generator with phase sequence A.B.C...

tages of sinusoidal wave form and the fact that they are 120° apart with respect to one another.

The terminal voltage of the generator will be of the same value as the induced voltage in each main coil if the three coils have been connected in delta. The terminal voltage will be 1.732 times the induced voltage per coil if the three coils have been connected in Wye (star). Regardless of the value of the terminal voltage, it must be remembered that the three voltages are 120° out of phase with one another and that this fact is very important when connecting three phase loads to the voltage supply.

The frequency of the induced voltage in each of the three phases is directly related to the number of field poles of the rotor and stator and to the rotative movement of the rotor. These factors are responsible for this type of generator often being called a *synchronous* generator. The most common frequency of generated voltages in North America is 60 Hz, although certain other frequencies are also standard frequencies. These frequencies are 25 Hz, 50 Hz, and 400-Hz. Aircraft ac power supplies are almost always 400-Hz supplies. Generators of voltages at this frequency are high-speed units which do not require much space and are therefore very suitable for aircraft.

When no electrical load is connected to the generator, the terminal voltage will remain of the value which the generator has been designed to produce, taking into account that the stator phase coils have been connected in delta or Wye (star). The application of a load to a generator will have considerable influence on the value of the terminal voltage of the generator.

The load current flowing through the stator phase coils will cause changes in the terminal voltage due to the following factors:

(a) stator coil *resistance*
(b) stator coil *reactance*
(c) generator *armature reaction*
(d) load *power factor*

Each of the stator phase coils has a certain amount of resistance. The coils have been wound of sufficiently large cross-sectional area so that the resistance of the coils is relatively small. The load current flowing through the stator coils causes a voltage or *IR* drop, thus decreasing the voltage at the generator terminals.

The rotating magnetic flux of the rotor causes a continuously varying current flow in the stator phase coils. This varying induced current and the load current result in considerable inductive reactance (X_L), whose value is generally considerably larger than the value of the resistance. As a result, there will be a higher voltage drop across X_L and the terminal voltage will be decreased.

A third factor affecting the terminal voltage of an ac generator is the so-called armature reaction. When no load is connected to the generator, the magnetic flux of the rotor is distributed very evenly across the gap between rotor and stator and across the stator coils. However, when a load current flows in the stator phase coils, and the load is inductive, the magnetic flux produced by the current flow is such that it opposes the rotor's flux. As a result, the voltage induced in the stator coils is decreased. On the other hand, when the load has a leading power factor, its current will produce a flux which aids the rotor's flux and results in an increased terminal voltage.

The ac generator may have a considerable voltage drop at its terminals or a considerable voltage increase at its terminals, depending on the power factor of the load connected to the generator. Of course, such variation in ac terminal voltage is highly undesirable, and it is necessary to control these voltage variations by external means, which will be discussed later.

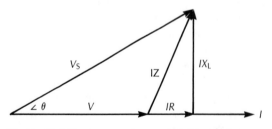

Figure 16-9 Generator voltage relations at unity power factor.

Figure 16-9 indicates the voltage relations when a resistive load is connected to a generator and a unity power factor exists.

Due to the unity power factor, the current phasor I is shown as being in phase with terminal voltage phasor V and resistance voltage drop phasor IR. Current phasor I lags the phasor for the voltage drop across the coil's inductive reactance (IX_L) by 90°. Phasor IZ indicates

the phasor sum of IR and IX_L. Phasor V_S indicates the value of the stator coil voltage and the angle between phasor I and phasor V_S indicates the angle by which the current lags the stator coil voltage when a unity power factor load is connected to the generator. The relationship between the stator phase voltage (V_C) and the terminal voltage V at *unity power factor* may be expressed as follows:

$$V_S = \sqrt{(V + IR)^2 + (IX_L)^2} \qquad (16\text{-}7)$$

In most cases, the power factor of the load connected to a generator is lagging. Figure 16-10 indicates the voltage relations for this condition.

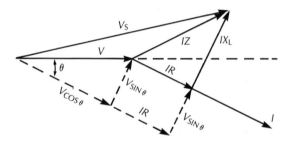

Figure 16-10 Generator voltage relations when a load with lagging power factor is applied.

When the load current lags the voltage, phasor I and phasor IR are in phase, but both lag terminal phasor voltage (V) by a certain angle. Phasor I lags phasor IX_L by 90°, and phasor IZ represents the phasor sum of the stator coil's voltage drops. The value of the voltage V_s induced in the stator coils may be determined from the following equation:

$$V_S = \sqrt{(V_{\cos\theta} + IR)^2 + (V_{\sin\theta} + IX_L)^2} \qquad (16\text{-}8)$$

When the load current leads the voltage and there is a leading power factor, the voltage relations may be determined from a phasor diagram, as shown in Fig. 16-11.

The current in a load with leading power factor causes phasor IR to lead terminal voltage V by an angle θ. The phasor diagram indicates that the relationship between the stator phase coil voltage and the terminal voltage may be stated in equation:

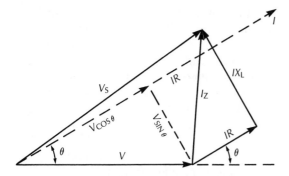

Figure 16-11 Generator voltage relations when a load with a leading power factor is applied.

$$V_S = \sqrt{(V_{\cos \theta} + IR)^2 + (V_{\sin \theta} + IX_L)^2} \quad (16\text{-}9)$$

The above phasor diagrams illustrate the effects of lagging and leading load currents on the value of the generator terminal voltage and the induced stator coil voltage in an ac generator.

VOLTAGE REGULATION

The process used in an ac generator to keep the terminal voltage from varying due to the factors discussed above, is called *voltage regulation*. A generator itself does not have the ability to regulate its own terminal voltage. In most cases, the generator's terminal voltage is controlled by automatically varying the rotor current so that the voltage at the stator terminals may remain constant, regardless of the power factor of the load. This automatic variation of the amount of rotor current is accomplished by the use of very sensitive control and relay equipment.

Regulation of a generator is generally expressed as a percentage. Regulation is often defined as the generator terminal voltage at *no-load* less the terminal voltage at *rated-load* V_F divided by the terminal voltage at *rated-load*, when the rotor speed and the excitation current are held constant.

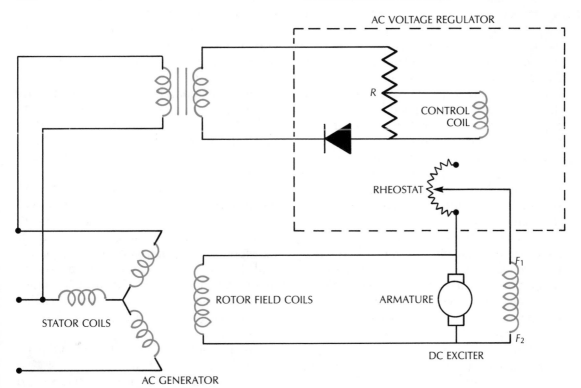

Figure 16-12 An ac generator voltage regulation circuit.

percent voltage regulation

$$= \frac{V_{NL} - V_{FL}}{V_{FL}} \times 100 \qquad (16\text{-}10)$$

where V_{NL} = terminal voltage at no-load
V_L = terminal voltage at rated-load

Example 16-4: The terminal voltage of a generator at no-load is 550V, while its terminal voltage with an inductive load applied is 523.6 V.

percent voltage regulation

$$= \frac{550 - 523.6}{523.6} \times 100$$

$$= 5.04\%$$

The regulation of the generator's terminal voltage is accomplished with a device called an ac *generator voltage regulator*. The simplified regulator circuit diagram illustrates one method used to accomplish generator voltage regulation.

The voltage from the ac generator terminals is connected to a transformer whose voltage output is proportional to the generator terminal voltage. The transformer output is connected to the voltage regulator's main resistance. The dc exciter's voltage is not only connected to the ac generator's rotor coils, but this dc voltage is also supplied to the exciter's own field coils. Note that the exciter's field coils are connected across the armature, and that the amount of current flowing through the shunt field coils is dependent on the setting of its control rheostat. The control rheostat is physically connected to the core of the regulator's control coil.

The voltage from the transformer is connected across regulating resistance *R*. This rectified voltage (a bridge rectifier is often used) is proportional to the stator terminal's voltage. The strength of the control coil's magnetic flux is dependent on the voltage across regulating resistance *R*.

When the generator's terminal voltage begins to decrease, the control coil's magnetic field weakens and the wiper arm of the rheostat moves downward. This means that less

resistance is then connected in series with the dc shunt field coils. As a result, more dc current flows through the shunt field coils. In turn, the output voltage of the dc generator (exciter) increases. Since this voltage is applied to the ac generator's rotor coils, its magnetic field is strengthened and the stator's induced voltage is increased. This causes the terminal voltage to remain near or at its rated voltage value. This process of voltage regulation requires very voltage-sensitive equipment.

PARALLEL OPERATION OF AC GENERATORS

The tremendous use of electrical energy in the world today has made it necessary that ac generators be connected in parallel to produce the voltages and energy required. The voltage output of many generators in large geographical areas are connected together in parallel so that the customers may be assured of reliable service and the provision of uniform voltages. For instance, the very large power grid in Ontario is connected to power grids in Michigan and New York, as well as to other provinces' grids. This common power grid is supplied by hundreds of ac generators which are all connected in parallel. A very extensive control system is necessary to ensure that the paralleling of generators is done according to the requirements for the paralleling of ac generators.

The voltages of the ac generators to be paralleled need not be identical. These voltages can be made identical by having the generator output voltage connected to transformers. The output of the transformers must be identical. There are several requirements that have to be met before ac generator output voltages may be operated in parallel.

1. The voltages to be paralleled must be equal in value.
2. The frequency of the voltages to be paralleled must be identical.
3. As discussed earlier, the phase sequence of the three-phase generator voltage must be

the same as the phase sequence of the voltage to which a generator is to be paralleled. At any point in time, the voltage of phase A must parallel the voltage of phase A of the grid voltage. The same must be true of B phase and C phase.

4. The generator voltage to be paralleled with the grid voltage must be in phase with the grid voltage. This means that for each phase, zero voltage and maximum voltage for every alternation must occur at the same time for the generator voltage and for the grid voltage to which it is to be paralleled. A closer look will now be taken at each of the above very important requirements for the parallel operation of ac generators.

When ac generators are operating in parallel, they are said to be operating *in synchronism*. This synchronous operating condition can only occur when the above requirements are met. As mentioned above, the voltages to be connected in parallel must be identical. If this were not the case, currents would circulate between generators. These circulating currents would produce voltage drops in the various transformer coils, as well as I^2R power losses. Such losses would produce more heat in the generators.

Since the voltage grids in North America extend over large areas, the voltage transmission is also over long distances. As discussed in Chapter Fourteen, long-distance voltage transmission will result in smaller line power losses if the voltage is high. Since the output of ac generators is normally 13.8 kV, transformers are used to increase the voltages of the generators to higher values so that they may be paralleled with the voltage grid. These grid voltages are not identical throughout North America. There are 115-kV grids, 230-kV grids, and more recently 500-kV grids.

The frequency of the voltages to be paralleled must be the same also. The frequency of a voltage depends on the rotative speed of the rotor and this can be changed by changing the speed produced by the prime mover. A

change of frequency occurs very seldom since most ac generators are manufactured to produce a three-phase voltage at a frequency of 60 Hz. Frequency synchronism between a generator's voltage and a grid voltage with which it is to be paralleled is critical.

A critically important requirement for operating ac generators in parallel with each other and a large voltage grid is the phase sequence of the three-phase voltage. Since the three-phase coils of the stator are located 120° apart, the three-phase voltages are 120 electrical degrees apart, as shown in Fig. 16-8.

When a generator's voltage is to be connected in parallel and its phase sequence is A-B-C, as shown in Fig. 16-8, it is critical that the grid voltage be exactly in the same phase sequence at exactly the same time. A tremendous short-circuit situation would develop if the phase sequence of the two voltages were not the same. For example, at 270° in Fig. 16-8, the voltage induced in phase A is at its maximum value. At this exact point in time the voltages of phases B and C are opposite in polarity to the phase A voltage. If phase A were at this instant paralleled with another voltage whose polarity was similar to phases B and C above, there would be tremendous flows of circulating currents whose values would be those of short-circuit currents.

Tremendous damage could be done to generators if no protection existed to prevent this. However, elaborate control and relay systems exist whereby equipment will be automatically isolated if an out-of-sequence situation occurs between voltage systems. These elaborate control and relay systems prevent the development of very large short-circuit currents, which will cause rotor speeds that will destroy generators.

As mentioned earlier, not only must the voltages to be paralleled have a matching phase sequence, the voltages must also be *in phase*. For the voltages to be paralleled and be in phase, their instantaneous voltages at any point in time must be the same. Since these voltages are of sinusoidal-wave pattern, their zero values and maximum values for each alternation must occur at the same time. This also holds true for the values between zero value and maximum value. Figure 16-13 shows one voltage which lags the base voltage by 45 electrical degrees.

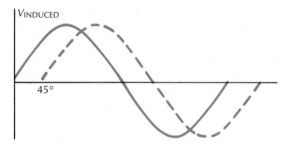

Figure 16-13 A lagging voltage condition.

If the above voltages were being paralleled, there would be considerable circulating current between generators, which in turn would result in increased torque and speed as well as heating up of the generators.

The process of paralleling an ac generator voltage with a voltage of a voltage grid is called *synchronization*. The purpose of this process is to make certain that the voltages to be paralleled are in correct phase sequence (e.g., A-B-C) and that the two voltages are in phase at every point in time for each complete cycle.

To accomplish the process of synchronizing, a synchronizing device is used. When large generators are to be paralleled, a *synchroscope* is used, in addition to a dark lamp circuit or a bright lamp circuit. For the paralleling of smaller generators, the dark lamp and bright lamp circuits are generally used.

Although only the paralleling switch is shown in Fig. 16-14 connecting the generator voltage to the grid voltage, the normal procedure is to use a potential transformer for each side so that two voltages are obtained across which a lamp may be safely connected.

With the paralleling switch open and the ac generator not yet turning, the lamps will glow very dimly and steadily. When the generator is

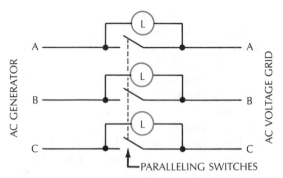

Figure 16-14 Dark lamp synchronizing circuit.

started and reaches its rated speed and induces the rated voltage, the three lamps become bright together and then dim together. This indicates that the phase sequence is correct. If, however, the lamps become bright and then dim, the one after the other but not together, the phase sequence is incorrect. Any pair of terminal leads from the generator may then be reversed to obtain the correct phase sequencing.

In addition to determining whether the phase sequence is correct, the dark lamp synchronizing circuit may also be used to determine the in-phase condition. When the generator to be connected in parallel is not turning yet, the lamps will all glow very dimly and steadily. As the generator is started, its voltage and frequency are not of rated value and the three lamps glow brightly and steadily. As the generator voltage and frequency approach rated value, the lamps become bright together, and then dark together. At first this change from brightness to darkness occurs quite quickly. The closer the voltage and frequency are to rated value, the longer the periods of brightness and darkness become. Finally, the lamps stay dark for quite a few seconds. The middle of this dark period indicates the point when the two voltages are perfectly in phase.

Having determined earlier that the phase sequence is correct and having also determined when the voltages are in phase, the paralleling switch between the two voltages may be closed. When this is done, the lamps remain dark, since the energy level of each voltage will be identical and no current will flow through the lamps.

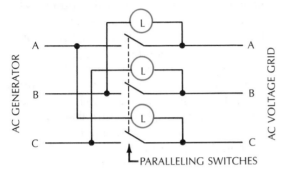

Figure 16-15 Bright lamp synchronizing circuit.

Figure 16-15 shows the circuit used for bright lamp synchronizing. It should be noted here that the person who is doing the paralleling must know whether dark lamp or bright lamp synchronizing is being used. If he or she does not know this, serious conditions will result when the paralleling switch is closed.

When the generator has not been started, the three lamps will glow very dimly and steadily. When the generator is started and induces its rated terminal voltage, all three lamps become bright together and then dim together, and this continues. This pattern indicates that the phase sequence is correct. If the lamps become bright and then dim, the one after the other but not together, the phase sequence is incorrect and may be corrected by reversing any pair of generator terminal leads.

The bright lamp synchronizing circuit has certain disadvantages in determining whether the voltages are in phase. The phases A, B, and C should be 120° out-of-phase when the voltages are being paralleled, but the lamps are brightest at maximum voltage values at 180° positions, thereby introducing an error of 60°. Bright lamp synchronizing circuits may be used effectively for the paralleling of single-phase voltages, but are seldom used with three-phase systems.

For both of the above synchronizing methods, the lamps are dark when the paralleling switch is closed. However, even with the dark lamps, there may be considerable voltage between the two sides of the switch, and this can create considerable disturbance in high-speed generators. A circuit has been designed which will partially overcome this difficulty and this circuit is shown in Fig. 16-16.

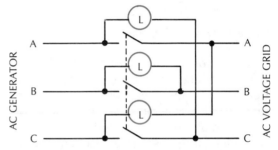

Figure 16-16 Two bright lamps, one dark lamp synchronizing circuit.

Note that the lamps from A and C phases have been cross-connected. When this circuit is used, one lamp will be dark while the other two lamps will be bright when the phase sequence is correct. The lamps actually twinkle when the phase sequence is correct. When the phase sequence is incorrect, the lamps glow and become dark together. When the voltages and frequencies of the two voltage systems are being close to being matched, the time involved in the changing from bright to dark becomes longer.

The correct synchronizing point occurs when one lamp is dark and the other two lamps are equally bright. At this point, the paralleling switch is closed.

As noted earlier, large generators are generally paralleled by using one of the above lamp systems but in addition, by using a *synchroscope*. A synchroscope shows the relative phase angle and frequency difference between the voltages that are to be paralleled. The synchroscope is used only after the correct phase sequence has been established, using one of the above lamp synchronizing circuits.

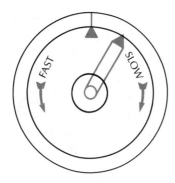

Figure 16-17 A synchroscope.

A synchroscope has a rotating hand whose speed indicates the difference between the frequency of the grid voltage and the frequency of the generator's voltage. The slower the speed of the rotating hand, the closer the two voltage frequencies are. In addition, when the hand turns clockwise, it indicates that the generator's frequency is higher than that of the grid voltage frequency. When the hand turns counter-clockwise, the hand indicates that the frequency of the generator voltage is lower than the frequency of the grid voltage.

When the frequencies of the two voltage systems are almost identical, the hand will be at or within a few degrees of an index mark located at the top of the synchroscope. When the hand is stopped at this mark or is almost stopped within a few degrees of the mark, the paralleling switch is closed.

HUNTING OF ALTERNATING CURRENT GENERATORS

It should not be assumed that because a generator was synchronized and paralleled with a voltage grid, no further difficulties may occur. The prime mover connected to the shaft of the generator rotor normally supplies a very uniform torque. The controls used ensure that with large generators, the torque is constant. When the rotors of various generators connected in parallel are turned by various types of prime movers, the torque developed in the

generators may vary slightly over time. As a result, the rotative speed of the rotor's magnetic field will vary somewhat. The voltage induced in the stator's coils will then also change. The rotative speed of the rotor may be ahead or behind its normal running position. This moving faster or slower of the rotor is pulsating and is called *hunting*.

The effect of hunting is that the induced voltage in the three stator coils will vary from the grid voltage with which it is paralleled. When this happens, large currents may circulate between the generator and the grid.

When the hunting is quite severe, the generator may actually be pulled out of synchronism with the grid. Generators are protected by overload devices and when the hunting condition causes large circulating currents between the generator and the grid, the overload devices will function, thereby disconnecting the generator from being in parallel with the grid. This action prevents serious damage being done to the generator.

The hunting in ac generators may be reduced by the use of several methods, (a) by using a heavy flywheel on the shaft (the heavier flywheel helps to dampen the oscillations caused by changes in prime mover speed), and (b) a special winding, called an amortisseur winding, is often wound in slots on the pole faces of the rotor poles. This amortisseur winding consists of heavy conductors which are welded to end rings on the poles.

When hunting develops, the normal path of

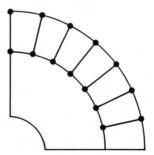

Figure 16-18 Amortisseur's windings in a rotor's pole face.

the flux changes and it now cuts through the short-circuited amortisseur windings. A voltage is then induced in the amortisseur winding, and this causes current to flow in the winding. The magnetic flux of this current aids the main flux and the effects of hunting are cancelled. As a result, the voltage induced in the stator coils remains constant, and the development of circulating currents is prevented. The generator then remains connected in parallel with the voltage grid.

AC GENERATOR LOSSES AND EFFICIENCY

The power losses in ac generators are very similar to the power losses in dc generators. A consideration of these losses is important for several reasons:

1. Losses determine the efficiency of a generator and may influence its operating costs considerably.
2. Losses determine the amount of heating of a generator and thus the power output that may be obtained without overheating of the coils' insulation.
3. Power losses (W) involve both voltage drops (V) and current flow (I) and need to be considered when determining a generator's efficiency.

There are basically four types of power losses in ac generators.

COPPER LOSSES OF I^2R LOSSES

Copper losses occur in all windings of a generator. When electric currents flow through the resistances of the various coils, a certain amount of electrical energy is converted to heat energy. This loss to the system is expressed in watts (W) and equals I^2 times the resistance of the various coils.

MECHANICAL LOSSES

These losses consist of bearing friction loss, brush friction loss at the slip rings, windage loss, and the loss of power due to the total efforts required to ventilate and cool a generator.

CORE LOSS

This power loss consists of the eddy-current and hysteresis losses due to the changing flux in the iron of the generator. These losses occur mainly in the iron of the stator. In a loaded generator, the stator core losses increase considerably. These losses will also vary with rotative speed of the generator.

STRAY LOAD LOSS

This is a loss due to the nonuniform distribution of current in the copper coils, and also when a load current tends to distort the stator's magnetic flux and causes more core losses.

GENERATOR EFFICIENCY

The efficiency (η) of a generator may be determined when it is operating at rated capacity. The input and output power can be measured and the efficiency is expressed as follows:

$$\text{percent efficiency } (\eta)$$
$$= \frac{\text{power output (W)}}{\text{power input (W)}} \times 100 \qquad (16\text{-}11)$$

However, a generator is rated in kVA and the efficiency is normally expressed as follows:

$$\eta = \frac{\text{output kVA} \times \cos\underline{/\theta}}{\text{output } kVA \times \cos\underline{/\theta} + \text{total losses}}$$
$$\times 100 \qquad (16\text{-}12)$$

Example 16-15: The output of an ac generator is 200 kVA at 80 percent power factor. The total losses of the generator are 2.5 kW. Determine the generator's efficiency.

$$\eta = \frac{200\,000 \times 0.80}{200\,000 \times 0.80 + 2\,500} \times 100$$
$$= \frac{160\,000}{162\,500} \times 100$$
$$= 98.5\,\%$$

The maximum efficiency of a generator is obtained when it operates at or near its rated full-load capacity.

SUMMARY OF IMPORTANT POINTS

1. In ac generators, the voltage is induced in coils wound on a stationary stator.
2. Stator coils are either lap or wave windings.
3. Ac generators are built to produce a specific voltage at a specific frequency.
4. The frequency (f) of the induced voltage for a generator with (P) poles and rotating at (ω) radians per second is:

$$f = \frac{p\omega}{4\pi}$$

5. Ac generators basically have one of two types of rotors: (a) salient-pole, used for low speeds, (b) cylindrical, used for high speeds.
6. A special rotor field discharge circuit is required to disconnect the dc voltage supply of the exciter from the rotor.
7. The coils in a three-phase generator are located 120° apart; thus, the three-phase voltages of sinusoidal wave form are 120° apart with respect to one another.
8. The application of a load to a generator will cause changes in the terminal voltage due to the following factors; (a) stator coil resistance, (b) stator coil reactance, (c) generator armature reaction, and (d) load power factor.
9. Voltage regulation is the process used in an ac generator to keep the terminal voltage from varying as a varying load is applied.
10. Regulation percent voltage regulation =

$$\frac{V_{NL} - V_{FL}}{V_{FL}} \times 100$$

11. Generators may be connected in parallel to supply large loads. To operate generators in parallel the following requirements must be met:

(a) voltages to be paralleled must be equal in value
(b) the frequency of the voltages to be paralleled must be identical
(c) the voltages paralleled must have the same phase as well as the same phase sequence

12. When ac generators are operating in parallel, they are said to be operating in synchronism.
13. When large generators are to be paralleled, a synchroscope is used in addition to a dark lamp or bright lamp circuit.
14. When the rotors of generators connected in parallel vary in speed, the voltage induced in the stator's coils will also change. The rotative speed of the rotor may be ahead or behind its normal running position. This moving faster or slower of the rotor is pulsating and is called *hunting*.
15. Hunting may be reduced by: (a) using a heavy flywheel on the shaft, (b) using a special winding called an amortisseur winding.
16. Ac generator losses: (a) copper (I^2R) losses, (b) mechanical losses, (c) core losses, (d) stray load losses.
17. Percent efficiency = $\dfrac{\text{power output}}{\text{power input}} \times 100$.

REVIEW QUESTIONS

1. Discuss why ac voltage generation is so widespread when compared with dc voltage generation.
2. Explain why the iron of the stator and rotor of ac generators consists of thin insulated iron laminations.
3. Explain why the three-phase coils of ac generators are usually connected in Wye or star.
4. Determine the frequency of the voltage induced in an ac generator whose rotor has four poles and whose rotative speed is 251.328 radians per second.

5. The frequency of a voltage to be induced in a three-phase ac generator is 60 Hz. The generator rotor has six poles, 48 slots, and the generator is wound full pitch. Determine:
   (a) the number of electrical degrees per revolution
   (b) the number of stator slots per pole
   (c) the number of stator slots per phase per pole
   (d) the total number of stator slots
6. Compare full-pitch winding of stators with fractional-pitch winding. Name three advantages of fractional-pitch winding.
7. Explain why the basic voltage formula (Eq. 16-3) must be multiplied by a pitch constant and a distribution constant.
8. A three-phase ac generator has a stator which is wound as a four-pole machine and consists of 60 stator slots. Each coil has 20 turns and spans 11 slots (1 to 12). The field coils will produce 0.01 Wb per pole. The pitch constant is 0.913 55 and the distribution constant is 0.956 68. The rotative speed is 188.496 radians per second. Determine:
   (a) the number of slots per pole
   (b) the number of slots per pole per phase
   (c) the total number of electrical degrees per rotation
   (d) the number of electrical degrees per slot
   (e) the number of electrical degrees per coil
   (f) the voltage induced in one coil group
   (g) the total induced coil voltage
        Assume that the coils are connected in Wye. Determine the terminal voltage of this generator.
9. Discuss the construction of a salient-pole rotor, its pole coils, its poles, the means whereby it receives its current, and its current source.
10. The rotative speed of a rotor is 335.104 radians per second. Determine the number of revolutions this rotor makes (a) per

second, (b) per minute. Would this be a salient-pole rotor or a cylindrical rotor?
11. Draw a diagram of the rotor field discharge circuit. Explain the purpose of this circuit for
    (a) closed condition
    (b) open condition
12. Discuss the use of hydrogen as a cooling means for large generators with reference to:
    (a) pressure of the hydrogen
    (b) power rating of the generator
    (c) its advantages over air as a cooling means
    (d) its heat conductivity
    (e) maintenance costs
    (f) explosive air-hydrogen mixtures
13. Name three factors which determine the amplitude of the induced voltage in a stator's coils.
14. Discuss the correct sequence of the three-phase voltage induced in a stator and explain why the correct sequence is so important.
15. The application of a load to an ac generator will have a major effect on its terminal voltage. Name the factors responsible for causing changes in the terminal voltage and explain how each of these factors contributes to the changes in the terminal voltage.
16. Draw a voltage phasor diagram for an ac generator when a unity power factor load is connected to the generator. Explain all aspects of every phasor of this diagram.
17. Draw a voltage phasor diagram for an ac generator when a lagging power factor load is connected to the generator. Explain all aspects of every phasor of this diagram.
18. Draw a voltage phasor diagram for an ac generator when a leading power factor load is connected to the generator. Explain all aspects of every phasor of this diagram.
19. Explain what voltage regulation is and why it is so important.

20. The terminal voltage of an ac generator is 13.8 kV, and without voltage regulation this voltage would drop 8 percent when operating with rated load. Determine the percent voltage regulation of this generator.

21. Give a detailed explanation of the construction and operation of an ac generator voltage regulator. The voltage regulation process is sometimes called a "cyclic process." Discuss this.

22. Explain the importance of transformers for the paralleling of ac generator voltages to high-voltage grids.

23. Voltage systems to be paralleled must not only have a correct phase sequence but the voltages must also be in phase. Explain why an in-phase condition is imperative before the voltages are paralleled. Discuss the consequences of paralleling voltage systems with their voltages out of phase.

24. Discuss the details of the synchronization process used to parallel two voltage systems using a dark lamp synchronization circuit. The correct phase sequence is to be determined.

25. The in-phase condition of two voltage systems to be paralleled is to be determined. Discuss the process for this when the dark lamp synchronization circuit is to be used.

26. Discuss the bright lamp synchronizing circuit as it is used to parallel different voltage systems. Explain why this method is not used very often with three-phase voltage systems.

27. Why must the person doing the paralleling of two ac voltage systems know whether dark lamp or bright lamp synchronizing is to be used?

28. Explain how the two bright lamp, one dark lamp synchronizing circuit functions during the paralleling operation of two voltage systems. Why is this circuit used instead of the dark lamp or the bright lamp circuits?

29. Discuss the functioning of a synchroscope during the paralleling of two voltage systems with regard to:
    (a) speed of the hand
    (b) direction of the hand
    (c) closing of the paralleling switch

30. Explain what hunting of generators is and how it is caused.

31. Explain why generators may be destroyed by severe hunting and how generators are protected from this.

32. Discuss the construction and operation of the rotor pole amortisseur winding. Discuss how this winding is used to overcome the effects of hunting.

33. State why a consideration of the power losses in an ac generator is important when discussing a generator's efficiency.

34. State the various losses of an ac generator and explain how each is caused.

35. The rated output of an ac generator is 180 kVA at 75 percent power factor. The losses of the generator amount to 3 kW. Determine the generator efficiency.

CHAPTER
SEVENTEEN

THREE-PHASE INDUCTION MOTORS

Three-phase induction motors may be classified into two groups, the *squirrel-cage motor* and the *wound-rotor motor*. Both types of induction motors, as will be shown, operate on very much the same basic principle. They both have similar stators which produce a rotating magnetic field but they differ in the constructions of their rotors. These two types of induction motors are described in this chapter. Another group of induction motors which are designed to operate on single-phase voltage will be discussed in Chapter Nineteen.

SQUIRREL-CAGE INDUCTION MOTOR

The three-phase squirrel-cage induction motor is a simple and very rugged ac machine. It has good speed regulation under varying load conditions and it requires little maintenance. The physical size for this type of motor for a given power rating is also relatively small when compared with other types of motors. Given these advantages, the squirrel-cage induction motor is commonly used for many industrial applications.

Figure 17-1 The parts of a three-phase squirrel-cage induction motor.

Courtesy: Century Electric, Inc.

CONSTRUCTION

The squirrel-cage induction motor, as shown in Fig. 17-1 contains three basic parts: the stator, the rotor, and a pair of end shields. Note the absence of any brushes in this motor, which is one of the reasons for its ruggedness.

A typical stator of an induction motor resembles the stator in a revolving field three-phase alternator. The stator or stationary part consists of three phase windings held in place in the slots of a laminated steel core with the core supported by a cast-iron or steel frame. The phase windings are placed 120 electrical degrees apart and connected in either Wye or delta, and the three leads are brought out to a terminal box mounted on the frame of the motor. When the stator is energized from a three-phase voltage, it will produce a rotating magnetic field on the inside surface of the stator core.

The rotor for the squirrel-cage motor contains no windings. Instead, it is a cylindrical core, constructed of steel laminations or punchings with conductor bars mounted parallel or approximately parallel to the shaft, and embedded near the surface of the rotor core. These conductor bars are short-circuited by an end ring at either end of the rotor core. On large machines, these conductor bars and two end rings are made of copper, with the bars brazed or welded to the end rings. On small machines, the conductor bars and end rings are sometimes made of aluminum, with the bars and rings cast in one piece on the rotor core.

The rotor or rotating part is not connected electrically to the power supply, but has a voltage induced in it by transformer action from the stator. For this reason, the stator is sometimes called the primary and the rotor is referred to as the secondary of the motor. Since this motor operates on the principles of induction and the construction of the rotor with the bars and end rings resembles a squirrel cage, the name squirrel-cage induction motor is used.

The rotor bars are not insulated from the rotor core but because they have less resistance than the core, the induced current will flow mainly in them. Also, the bars are usually not quite parallel to the rotor shaft, but are mounted in a slightly skewed direction. This design feature tends to produce a more uniform rotor field and torque and, also, it helps to reduce some of the internal magnetic noise when the motor is running.

The function of the two end shields is to support the rotor shaft. They are fitted with bearings and attached to the stator frame.

THE ROTATING FIELD FROM A THREE-PHASE STATOR

The operation of the induction motor is dependent on the presence of a rotating stator field. It may be recalled that the stator contains three phase windings, placed 120 electrical degrees apart with the windings wound on the stator core to form nonsalient-stator field poles. When the stator is energized from a three-phase voltage supply, each phase winding will set up a pulsating field. However, by virtue of the spacing between the windings, and the phase difference between the phase currents, the flux from the three phase windings will combine to produce a magnetic field which rotates at a constant speed around the inside surface of the stator core. This resultant flux is called the *rotating magnetic field* and its speed is called the *synchronous speed*.

The manner in which the rotating field is set up may be described by considering the direction of the phase currents at successive instants during a cycle. Figure 17-2 (a) shows a simplified Wye-connected three-phase stator winding. The winding shown has two nonsalient field poles per phase, and hence the motor will be described as a two-pole induction motor. Figure 17-2 (b) shows the phase currents for the three phase windings. The phase currents will be 120 electrical degrees apart, as shown. Figure 17-2 (c) illustrates the resultant magnetic field at increments of 60° for one cycle of applied voltage.

It is assumed that when the phase current is in a positive direction, the direction of the line current to the coil is as shown in Fig. 17-2(a), and when the phase current is negative, the

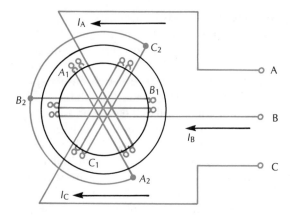

Figure 17-2(a) Simplified three-phase stator winding with two poles per phase.

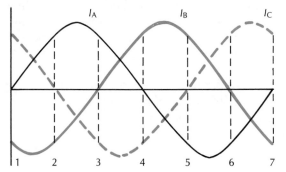

Figure 17-2(b) The three-phase current in the winding.

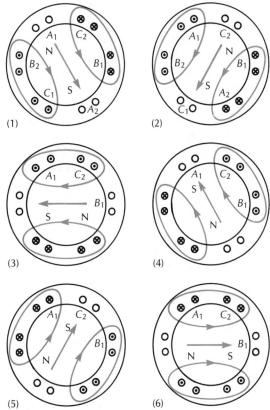

Figure 17-2(c) Rotating magnetic field of a three-phase induction motor.

direction of the current is opposite to the direction shown in Fig. 17-2(a).

At position (1) in Fig. 17-2(b), the phase current, I_A is zero and, hence, coil A will be producing zero flux. However, the phase current I_C is positive and I_B is negative. Therefore, from these current directions in Fig. 17-2(a) and as shown in Fig. 17-2(c)(1), the current will be flowing into the conductors at coil ends C_2 and B_1 and out of the conductors at coil ends C_1 and B_2. Using the left-hand rule, it should be evident that the flux produced by the conductors at C_2, B_1 and at C_1, B_2 will be as illustrated in Fig. 17-2(c)(1). The resultant of these two phase fluxes will therefore form magnetic poles in the stator core, as shown in the same diagram.

At position (2), 60 electrical degrees later, phase current, I_C, is now zero, the current I_A is the positive and the current I_B is negative. From Fig. 17-2 (a), the current is now observed to be flowing into the conductors at coil ends A_2 and B_1, and out of the conductors at coil ends A_1 and B_2. Therefore, as shown in Fig. 17-2(c)(2), the resultant magnetic poles are now at a new position in the stator core. In fact, the poles have also rotated 60° from position (1).

Using the same reasoning as above for the current wave positions (3), (4), (5), (6), and (7), it will be seen that for each successive increment of 60 electrical degrees, the resultant stator field will rotate a further 60°, as shown in Fig. 17-2(c). Note that the resultant flux for position (7) is not shown because it will be the same as for position (1). It should now be obvious that for each cycle of applied voltage, the field from the two-pole stator will also rotate one revolution around its core.

This rotating magnetic field from a three-phase stator can also be demonstrated with the use of a compass. If the rotor and end shields are removed from an induction motor and a compass held over the open stator, the compass needle will be observed to rotate when the stator is energized from a three-phase supply. The rotating compass needle indicates the presence of a revolving stator field. **Caution**: Do not energize a stator winding in this manner for too long a period because it can be damaged due to the excessive heat produced in the windings.

The description for the rotating stator field indicated that when the phase sequence of the applied voltage is A.B.C, as shown in Fig. 17-2(a) and (b), the stator field will rotate in a clockwise direction. *To reverse the direction of rotation* of this field, and consequently reverse the direction of rotation of the motor, simply requires the interchange of any two of the three line wire connections. The reversed rotating field may be shown using the same reasoning previously given for the phase sequence A.B.C, and is left as an exercise for the student.

SYNCHRONOUS SPEED

It was mentioned that the speed of the rotating stator field is called the synchronous speed. This synchronous speed is dependent on both the number of stator poles and on the frequency of the supply. The relationship is the same as that between the speed and number of poles and frequency for a rotating field alternator.

In the two-pole stator of Fig. 17-2, it was shown that the field makes one revolution for one cycle of applied voltage. If the stator is constructed to have four poles by connecting

two groups of coils in series in each phase, it may be shown that the stator field will make only one-half revolution for one cycle of applied voltage. That is, doubling the stator poles would reduce the synchronous speed by one-half. Similarly, if the stator has six poles, the rotating field will make only one-third revolution for one cycle of applied voltage.

It may be easily reasoned that the synchronous speed is inversely proportional to the stator pole pairs and directly proportional to the supply frequency. This relationship was previously stated in Chapter Sixteen. It may be expressed in terms of either revolutions per minute (rpm) or radians per second (rad/s). Both forms are given below. The student should note the relative simple relationship between revolutions and radians, namely, 1 revolution equals 2π radians.

$$\text{synchronous speed in r/min} \quad (17\text{-}1)$$
$$= \frac{60 \times \text{supply frequency}}{\text{pole pairs}}$$
$$S = \frac{60 \times f}{P}$$

$$\text{or synchronous speed in rad/s} \quad (17\text{-}2)$$
$$= \frac{2\pi \times \text{supply frequency}}{\text{pole pairs}}$$
$$\eta = \frac{2\pi f}{P}$$

Example 17-1: If a six-pole three-phase induction motor is operated on a 60-Hz supply, determine the synchronous speed of the rotating stator field.

$$S = \frac{60 \times f}{P} \qquad \eta = \frac{2\pi f}{P}$$
$$= \frac{60 \times 60}{3} \qquad = \frac{2 \times 3.14 \times 60}{3}$$
$$= 1200 \text{ r/min} \qquad = 125.6 \text{ rad/s}$$

where $f = 60$ Hz
$P = 3$ pole pairs

INDUCTION MOTOR ACTION PRINCIPLE

In Chapter Eight, it was shown that armature motion in a dc motor is produced from the interaction between two magnetic fields. It was also shown that the windings for these two fields are electrically connected to an external supply.

In an induction motor, there are also two magnetic fields interacting to produce rotor motion. These are the stator and rotor fields. The stator, it may be recalled, produces a rotating field when it is connected to a three-phase supply. The rotor, however, is not connected electrically to any external supply. Instead, the rotor field is produced by currents induced in the rotor bars by transformer action as the stator field sweeps across the rotor.

The induced rotor current will produce a magnetic field having the opposite polarity of the field that produces the induced current (Lenz's law). For example, at the instant the field polarity is north, the induced rotor polarity will be south. When the applied field voltage reverses, the field polarity will be south and the rotor field polarity will now be north. Therefore, the stator field and rotor field will always have opposite polarities and so they will be attracted to each other. To put this another way, the induced rotor currents tend to oppose any movement between the stator field and the rotor conductors. Since the stator field is rotating, it should now be obvious that the rotor would tend to rotate along or follow this rotating stator field.

Figure 17-3 illustrates another way of showing the force responsible for rotor motion. As is well known, a current-carrying conductor in a magnetic field will have a force exerted on it, tending to move it at right angle to the magnetic field. This tangential force is exerted around the circumference of the rotor and causes the rotor to turn in the same direction as the rotating field.

In Fig. 17-3, only one rotor conductor is

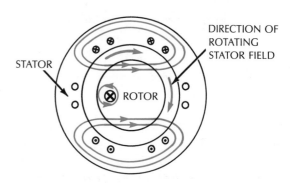

Figure 17-3 Induction motor action.

shown and the stator field is assumed to be rotating in a clockwise direction. An application of the left-hand generator rule gives the direction of the induced current in the rotor conductor as moving away from the reader for the flux direction and flux motion shown. The right-hand motor rule shows that the force on this current-carrying conductor is upward. This force will tend to turn the rotor in the same clockwise direction as the stator field. The other rotor bars under other stator poles may be shown likewise tending to produce similar clockwise rotation.

ROTOR SLIP

It was shown that the torque developed by the rotor is dependent on the current induced in the rotor conductors. If the rotor were to turn at the same speed (synchronous speed) as the rotating stator field, then the relative motion between the rotor conductors and stator field would be zero. Under this condition, no stator flux can cut the rotor conductors, there will be no induced rotor currents, and the rotor will not develop any torque. Therefore, the rotor of an induction motor will always turn at a speed that is less than synchronous speed. In this way, the stator flux can now cut the rotor conductors and induce the necessary rotor currents.

The difference between a motor's synchronous speed and its rotor speed equals the number of rpm the rotor slips behind the rotating stator field, and is called the *rotor slip*.

$$\text{rotor slip = synchronous speed} \quad (17\text{-}3)$$
$$\text{– rotor speed}$$

$$\text{slip} = S_s - S_R$$

Rotor slip is more commonly expressed as a percentage of the synchronous speed. This value, called *percent slip*, is usually used to describe the speed performance of a squirrel-cage motor. The smaller the percent slip, the better the speed regulation of the motor. Most squirrel-cage induction motors at rated load have a percent slip of between 2 and 5 percent.

$$\text{percent slip} \quad (17\text{-}4)$$
$$= \frac{\text{synchronous speed – rotor speed}}{\text{synchronous speed}} \times 100$$

$$= \frac{\text{rotor slip} \times 100}{\text{synchronous speed}}$$

Example 17-2: If a three-phase, four-pole, 60-Hz induction motor has a full-load speed of 1760 r/min (184.2 rad/s), what is its: (a) synchronous speed, (b) rotor slip, and (c) percent slip?

(a) $S_s = \dfrac{60 \times f}{P}$ $\eta = \dfrac{2\pi f}{P}$

$\quad = \dfrac{60 \times 60}{2}$ $= \dfrac{6.28 \times 60}{2}$

$\quad = 1800 \text{ r/min}$ $= 188.4 \text{ rad/s}$

where S_s = synchronous speed
$\qquad P$ = 2 pole pairs
$\qquad f$ = 60 Hz

(b) $\text{slips} = S_s - S_R$

$\quad = 1800 - 1760$ | or $= 188.4 - 184.2$
$\quad = 40 \text{ r/min}$ | $= 4.2 \text{ rad/s}$
where S_R = rotor speed

(c) $\text{percent slip} = \dfrac{\text{rotor slip} \times 100}{\text{synchronous speed}}$

$\qquad\qquad = \dfrac{40 \times 100}{1800}$

$\qquad\qquad = 2.2 \%$

ROTOR FREQUENCY

It was stated previously that the rotor slip is equal to the revolutions per minute that the rotor slips back of the stator field. That is, the rotor slip is equal to the number of times the flux of a pair of stator poles passes over a given rotor conductor in one minute. In Chapter Seven, it was shown that one cycle of voltage is induced each time a conductor passes over a pair of unlike field poles. Therefore, the frequency of the current induced in the rotor conductors, called the *rotor frequency*, will be proportional to the rotor slip. This relationship is similar to that between the frequency, r/min (rad/s), and pole pairs of an alternator and may be expressed as

$$f_r = \frac{PS}{60} \quad (17\text{-}5)$$

where f_r = rotor frequency
$\qquad P$ = stator pole pairs
$\qquad S$ = rotor slip

This equation may also be written as

$$f_r = f_s \times \frac{(\text{percent slip})}{100} \quad (17\text{-}6)$$

where f_r = rotor frequency
$\qquad f_s$ = stator frequency

The rotor frequency is a very important property, because when the rotor frequency changes, the rotor inductive reactance ($X_L = 2\pi f_r L$) will also change. The rotor reactance, as will be shown, affects the starting and running characteristics of the motor.

Example 17-3: A 220-V, three-phase, eight-pole, 60-Hz squirrel-cage motor has a rated full-load speed of 860 r/min (90 rad/s).

Determine:
(a) the synchronous speed
(b) the rotor slip at rated load
(c) the percent slip at rated load
(d) the percent slip at the instant of startup

(e)   the rotor frequency at rated load
(f)   the rotor frequency at the instant of startup

Solutions:

(a)   synchronous speed $= \dfrac{60 \times f}{\text{pole pairs}}$

$\qquad = \dfrac{60 \times 60}{4}$

$\qquad = 900 \text{ rpm}$

or

synchronous speed
$= \dfrac{2\pi f}{P} = \dfrac{6.28 \times 60}{4}$
$= 94.2 \text{ rad/s}$

(b)   slip at rated load $= 900 - 860 = 40 \text{ r/min}$

or

$\qquad\qquad = 94.2 - 90 = 4.2 \text{ rad/s}$

(c)   percent slip at rated load

$= \dfrac{\text{slip} \times 100}{\text{synchronous speed}}$

$= \dfrac{40 \times 100}{900}$

$= 4.4\%$

(d)   At the instant of startup, the rotor speed is zero, and hence the percent slip will be 100 percent.

(e)   rotor frequency at rated load

$f_r = \dfrac{\text{pole pairs} \times \text{slip}}{60}$

$\quad = \dfrac{4 \times 40}{60}$

$\quad = 2.66 \text{ Hz}$

or f_r (at rated load) $= \dfrac{f_s \times (\text{percent slip})}{100}$

$\qquad\qquad = \dfrac{60 \times 4.44}{100}$

$\qquad\qquad = 2.66 \text{ Hz}$

(f)   At the instant of startup, the slip is 100 percent. Therefore, at this instant, the rotor frequency will be equal to the stator frequency.

$f_r \text{ (at start-up)} = f_s$
$\qquad\qquad\quad = 60 \text{ Hz}$

SPEED AND TORQUE CHARACTERISTICS

It was shown that the rotor speed of a squirrel-cage motor will always lag behind the synchronous speed of the stator field. The rotor slip is necessary in order to induce the rotor currents required for motor torque. At no-load, only a small torque is required to overcome the motor's mechanical losses, and the rotor slip will be very small, about 2 percent. As the mechanical load is increased, however, the rotor speed will decrease and, hence, the slip will increase. This increase in slip will cause an increase in the induced rotor currents, and the increased rotor current, in turn, will produce the higher torque required by the increase in load.

Since the squirrel-cage rotor is constructed basically of heavy copper bars shorted by two end rings, the rotor impedance will be relatively low and, hence, a small increase in rotor-induced voltage will produce a relatively large increase in rotor current. Therefore, as the squirrel-cage motor is loaded from no-load to full-load, only a small decrease in speed is required to cause a relatively large increase in rotor current. For this reason, the speed regulation of a squirrel-cage motor is very good and the motor is often classified as a constant speed device.

It was stated earlier that torque is produced in an induction motor by the interaction of the stator and rotor fluxes. The amount of torque produced is dependent on the strength of these two fields and the phase relations between them. This may be expressed mathematically as:

$$T = K\phi_s I_R \cos \angle \theta_R \qquad (17\text{-}7)$$

where T = torque

$\quad K$ = a constant

$\quad \phi_S$ = stator flux

$\quad I_R$ = rotor current

$\cos\angle\theta_R$ = rotor power factor

From no-load to full-load, the torque constant (K), the stator flux (ϕ_S) and the rotor power factor ($\cos\angle\theta_R$) for a squirrel-cage motor will be practically constant. Hence, the motor's torque will vary almost directly with the induced rotor current (I_R). Since the rotor current, in turn, will vary almost directly with its slip, the torque variation of a squirrel-cage motor is often plotted against its rotor slip, as shown in Fig. 17-4.

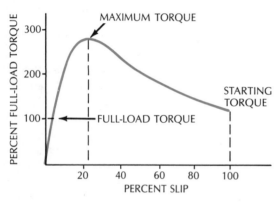

Figure 17-4 Torque variation versus slip for a standard squirrel-cage motor.

The increase in rotor current, and hence the increase in rotor torque for a given increase in rotor slip, is dependent on the rotor impedance (resistance and reactance) and the rotor power factor. The rotor resistance for a squirrel-cage motor will be constant. However, an increase in slip will increase the rotor frequency and the resulting inductive reactance of the rotor. From no-load to full-load and even up to about 125 percent of rated load, the amount of rotor slip for a standard squirrel-cage motor is relatively small and the rotor frequency will seldom exceed 2 to 5 Hz. Therefore, for the above range of load, the effect of frequency change on impedance will

be negligible, and as shown in Fig. 17-4, the rotor torque will increase in almost a straight-line relationship with slip.

At between 10 to 25 percent slip, the squirrel-cage motor will attain its maximum possible torque. This torque is referred to as the motor's *breakdown torque* and it may reach between 200 and 300 percent of rated torque, as shown in Fig. 17-4. At this maximum torque, the rotor's resistance will be equal to its reactance.

However, when the load and the resulting slip are increased much beyond the rated full-load values, the increase in rotor frequency, and hence the increase in rotor reactance and impedance, becomes appreciable. This increase in rotor inductive reactance and the resulting decrease in rotor power factor will have two effects: first, the increase in impedance will cause a decrease in the rate at which rotor current increases with an increase in slip. Second, the lagging rotor power factor means that the rotor flux will reach its maximum at some time after the stator peak flux has swept by it. The out-of-phase relationship between these two fields will reduce their interaction and their resulting torque. Hence, if the motor load is increased so that the slip exceeds that at the breakdown point, the loss in torque due to the above two effects will exceed any gain due to the increase in slip and the motor will stall.

STARTING CHARACTERISTICS

The torque, input current, and rotor power factor at the instant of starting a squirrel-cage motor can be considerably different from those values when, for example, the motor is operating under rated load conditions.

The *startup torque* for a squirrel-cage motor, also called the locked rotor torque, can be considerably less than the motor's rated full-load torque. At the instant of startup, the rotor's slip will be 100 percent and its frequency will equal the supply frequency. Therefore, at startup, the rotor's reactance will be large compared with its resistance and the rotor will have a low lagging power factor. As described previously, a lagging rotor flux will produce poor interaction with the stator flux and, hence, the starting torque of the squirrel-cage motor will be low. In a later section it will be shown how the startup torque of an induction motor may be improved by increasing its rotor resistance.

As the motor's speed increases from startup towards its rated speed, the rotor frequency and its reactance will decrease. This decrease in rotor impedance and rotor power factor angle will, in turn, cause an increase in torque, as shown in Fig. 17-4. At about 20 percent slip, the motor will have attained its maximum torque. The motor will continue to accelerate, but beyond this maximum torque point, a decrease in slip will now produce a net decrease in torque. The final speed of the motor will be where the motor barely produces the required torque. The rotor slip at this final speed will be about 2 to 10 percent.

The *startup line current* to an induction motor may also be considerably greater than its rated current. During the brief instant of startup, the induction motor resembles a static transformer with the stator winding as the primary and the rotor as the secondary. At this instant, the stator flux will cut the rotor bars at a faster rate than when the rotor is turning and, hence, the rotor bars will carry a relatively high induced current. Consequently, the stator current will also be high at this instant. In fact, when the motor is started with full-line rated voltage, the starting surge of current may momentarily be as high as three to five times the rated full-load current.

Generally most squirrel-cage motors are started with full-rated voltage, which means that the starting protective devices used in their installation circuits must be rated to carry this high-starting surge current. Sometimes some large induction motors are started with starters which will apply a reduced voltage at the instant of startup. Under this condition, however, the startup torque will be considerably reduced.

For a given slip, the torque produced by an induction motor may be shown to vary as the square of the applied voltage. For example, if the stator voltage is reduced by one-half, then the stator flux and the induced rotor current will also be reduced by one-half. From the torque equation, Eq. 17-7, it is obvious that when both the stator flux and rotor current are reduced by one-half, the motor torque will be only one-quarter its original value.

Some of the devices and circuits used for the startup and control of three-phase motors will be described in Chapter Twenty-One.

MOTOR TORQUE CALCULATION

Since the stator flux and induced rotor current for an induction motor are not easily measured, the torque equation given in Eq. 17-7 is not the most practical equation to use for determining a motor's torque. Instead, the prony brake torque equation described in Chapter Eight, may be used, providing the motor's output power and r/min are known.

$$\text{output power (watts)} = \frac{2\pi \times \text{torque} \times \text{r/min}}{60}$$
$$(17\text{-}8)$$

$$\text{torque (newton metres)} = \frac{60 \times \text{output watts}}{2\pi \times \text{r/min}}$$
$$(17\text{-}9)$$

$$= \frac{9.55 \times \text{output watts}}{\text{r/min}}$$

or

$$\text{output power (watts)} = \text{torque} \times \text{rad/s}$$

$$\text{torque (newton-metres)} = \frac{\text{output watts}}{\text{rad/s}}$$

A motor's power may also be stated in horsepower (hp). In this case, the output power in watts will be equal to the output horsepower multiplied by 746 (1 hp = 746 W).

Example 17-4: Determine the torque in newton-metres produced by a 5-hp squirrel-cage motor rotating at 1740 r/min (182.12 rad/s).

$$\text{output power (watts)} = \text{hp} \times 746$$
$$= 5 \times 746$$
$$= 3730 \text{ W}$$

$$\text{torque (newton-metres)} = \frac{60 \times \text{output watts}}{2\pi \times \text{r/min}}$$
$$= \frac{60 \times 3730}{2 \times 3.14 \times 1740}$$
$$= 20.48 \text{ N.m}$$

or

$$\text{torque (newton-metres)} = \frac{3730 \text{ W}}{182.12 \text{ rad/s}}$$
$$= 20.48 \text{ N.m}$$

THE SPEED, TORQUE, AND SLIP CHARACTERISTIC CURVES

Figure 17-5 shows the typical speed, torque, and slip characteristic curves for a standard squirrel-cage motor. The speed curve shows that a standard squirrel-cage motor will operate at a relatively constant speed from no-load to full-load. This is due to the extremely small rotor impedance, which will require only a small decrease in speed to cause a large

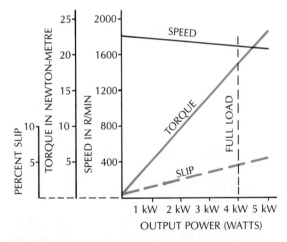

Figure 17-5 Speed, slip and torque curves for a standard squirrel-cage motor.

increase in the induced rotor current. The small drop in speed is also evident from the low slip from no-load to full-load. Since the torque will increase in almost direct proportion to the rotor slip, the torque graph, like the slip graph, will also have a straight line characteristic, as shown.

MOTOR POWER FACTOR

The magnetic circuit for the stator field of an induction motor consists of the stator core, the rotor core, and the air gap separating the stator from the rotor. Due to this air gap, the reluctance of the stator magnetic circuit will be high, and as a result, the input current will always contain a large magnetizing component. Therefore, the induction motor will have a lagging power factor both at startup and during operation.

At the instant of startup, the in-phase current component is negligible, and the high starting surge current is practically all magnetizing current. Hence, the stator input phase current will lag the applied phase voltage by a large angle, and the squirrel-cage motor will have a poor power factor at the instant it is started.

When the motor is running at no-load, the in-phase stator current will be due mainly to the motor losses, and this current will be still relatively small compared with the stator magnetizing current. Therefore, at no-load, the squirrel-cage motor will also have a low lagging power factor.

As the load on the motor is increased, the in-phase current to the motor increases, but the magnetizing component remains about the same. Consequently, the power factor of a squirrel-cage motor will improve as its load is increased. At the instant of startup and at no-load operation, the power factor of a squirrel-cage motor may be as low as 5 to 10 percent lagging, and at rated load it may be as high as 90 percent lagging. The following example illustrates one method for determining the power factor of an induction motor circuit under a given load.

Example 17-5: The following readings were taken from a three-phase induction motor circuit while it was operating under a given load: $V_L = 208$ V, $I_L = 20$ A, wattmeter (W_1) = 3150 W, and wattmeter (W_2) = 1230 W. Determine the circuit power factor.

$$\text{power factor} = \frac{\text{circuit true power (W)}}{\text{circuit apparent power (VA)}}$$

$$= \frac{W_1 + W_2}{\sqrt{3}\, V_L \times I_L}$$

$$= \frac{3450 + 1230}{1.73 \times 208 \times 20}$$

$$= 0.65 \text{ or } 65\% \text{ lag}$$

MOTOR LOSSES AND EFFICIENCY

The power losses of a squirrel-cage motor consist of the motor's copper losses and fixed losses. The *copper losses* are due to the resistance in the stator and rotor windings. This loss, also called the I^2R loss, will vary with the motor's load and load power factor. The *fixed losses* are made up of the motor's core, friction, and windage losses. The fixed losses will vary with speed, however, because the squirrel-cage motor is a relatively constant speed device. This loss is usualy considered to be constant. The core loss in induction motors consists of the losses due to eddy currents and hysteresis. These two core effects were discussed previously in the chapters on dc generators and dc motors. The example below describes a method for determining the copper losses, fixed losses, and efficiency of an induction motor.

Example 17-6: A Wye-connected, three-phase, four-pole, 60-Hz squirrel-cage motor is rated at 4 kW, 220 V, 14.5 A, and 1740 r/min. Determine the motor's copper losses, fixed losses, and efficiency at rated full-load.

In order to determine the copper losses at any load, it will be necessary to find the "effective resistance" of the motor windings. For this test, sometimes called *the blocked rotor test*, the rotor is secured stationary and the applied ac voltage is slowly increased until rated current flows to the motor. The rotor may be secured with a prony brake.

The following readings were taken from the blocked rotor test of the above motor: $V_L = 50$ V, $I_L = 14.5$ A, $W_1 = 660$ W, and $W_2 = -10$ W.

Observe that only 50 V is required to cause full-load stator current under the blocked rotor condition. The core loss will be negligible at this low voltage. Also, there will be no friction and windage losses. Therefore, the watts taken by the motor under the blocked rotor test can be assumed to be basically copper losses. The motor's effective resistance (R_e) per phase is now determined as follows:

$$\text{total watts (blocked rotor)} = 3(I^2 R_e) \qquad (17\text{-}10)$$
$$660 - 10 = 3(14.5)^2 \times R_e$$

effective phase resistance (R_e)

$$= \frac{650}{3(14.5)^2}$$
$$= 1.03 \ \Omega$$

The motor's copper losses at any load can now be easily determined. The copper losses for all three phases at rated full load will be:

$$\text{full-load copper losses} = 3\, I_L^2 R_e$$
$$= 3\,(14.5)^2 \times 1.03$$
$$= 650 \text{ W}$$

To determine the motor's fixed losses, the motor must be tested under no-load and rated voltage conditions. The following readings were taken from the no-load test of this motor: $V_L = 220$ V, $I_L = 4.8$ A, $W_1 = 620$ W, and $W_2 = -330$ W.

The total watts taken by the motor under this no-load test will contain the motor's fixed losses and also a small cooper loss component.

fixed losses = total watts at N.L.   (17-11)
$$- \text{copper loss at N.L.}$$
$$= (620 - 330) - (3\ I_L^2 R_e)$$
$$= 290 - [3(4.8)^2 \times 1.03]$$
$$= 290 - 71.2$$
$$= 218.8\ \text{W}$$

To determine the motor's efficiency at rated full-load, it will be necessary to know the motor's input power and total losses at rated-load. The motor's losses at full-load have been determined above, and the following readings were taken from a full-load test of the motor: $V_L = 220$ V, $I_L = 14.5$ A, $W_1 = 3076.2$ W, and $W_2 = 1792.06$ W.

efficiency at F.L.   (17-12)
$$= \frac{\text{input power at F.L.} - \text{total losses at F.L.}}{\text{input power at F.L.}}$$
$$= \frac{(3076.2 + 1792.6) - (650 + 218.8)}{3076.2 + 1792.6} \times 100$$
$$= \frac{4000 \times 100}{4868.8}$$
$$= 82.2\ \%$$

POWER FACTOR AND EFFICIENCY CHARACTERISTIC CURVES

Figure 17-6 shows the typical power factor and efficiency characteristic curves for a standard squirrel-cage motor. Due to the air gap in the stator magnetic circuit, the stator magnetizing current is very much greater than the stator in-phase current at no-load. Therefore, the squir-rel-cage motor will have an extremely low lagging power factor at no-load. The in-phase current, however, will increase as the load on the motor is increased. Therefore, the motor's power factor will improve and it is usually about 0.9 lag at rated-load.

The efficiency of a squirrel-cage motor is dependent on its power losses. At light loads, the fixed losses form a large part of the motor's input power and, as a result, this motor will have a low efficiency when operating under light loads. However, as the load is increased, the useful output becomes much greater than

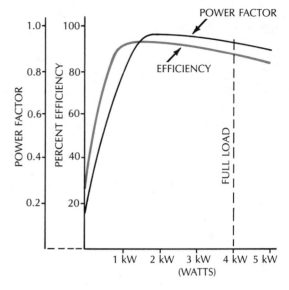

Figure 17-6 Power factor and efficiency curves for a standard squirrel-cage motor.

the motor losses, and the motor's efficiency will increase, as shown in Fig. 17-6. When the load exceeds rated load, the copper losses will have increased sufficiently to cause the efficiency to decrease.

STATOR VOLTAGE RATING AND CONNECTION

Many three-phase squirrel-cage motors are designed to operate at two different voltage ratings. For example, the name plate of a motor may be stamped for operation at 220/440 V and 14.5/7.3 A. This means that when the motor is operated at 220 V, the full-load current will be 14.5 A, and when the motor is operated at 440 V, the full-load cur-rent will be 7.5 A.

This double voltage rating of a three-phase, squirrel-cage motor is made possible by designing each stator phase winding with two equal sections. When the two sections in each phase are connected in series, the voltage rating will be double the voltage rating of the two sections in each phase connected in par-allel. Figures 17-7(a) and (b) show the Wye

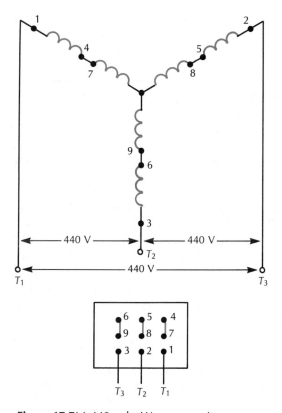

Figure 17-7(a) 440 volts Wye-connection.

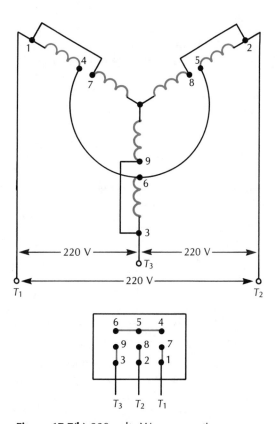

Figure 17-7(b) 220 volts Wye-connection.

connections of the phase sections for 440- and 220-V operation, respectively. When the stator is connected in Wye, there will be nine leads brought out from the stator to the terminal box on the motor. These terminal are numbered and they permit easy connection for either 440-V or 220-V operation, as shown in Fig. 17-7. The delta-connections for dual voltage operation are also used, and the connections follow a similar series and parallel arrangement of the stator phase sections.

SINGLE-PHASE OPERATION

The three-phase induction motor may not start if one of its three line wires is disconnected. Although the stator and induced rotor poles set up by a single-phase supply are of opposite polarity, they will be in complete

alignment and thus no torque can be produced. However, if the motor is running when one of its leads is disconnected, then the motor will continue to operate, but at a considerably reduced load capacity. The induced current from the transformer action will produce rotor poles that are displaced from the stator poles, and thus some rotor torque is maintained.

TEMPERATURE RATING AND MOTOR ENCLOSURES

A squirrel-cage motor is designed to have a specific temperature rating and for use in a particular type of environment. The temperature rating of a motor is usually given on its name plate. For example, a squirrel-cage motor may be designed to have a temperature rating of 50°C temperature rise continuous duty. This means that the motor can be operated continuously and safely at a temperature of up to 50°C above the ambient temperature of 40°C. The temperature rating of a squirrel-cage motor is dependent on the kind of insulation used to separate and insulate the stator windings.

Protection of the windings from the environment depends on the type of housing which encloses the motor. The various enclosures in use may be classified into (a) the open type and (b) the closed type.

The *open type of motor enclosure* is one in which air is drawn in from the surrounding atmosphere and circulated through the interior of the motor. The air circulation is achieved by fans or fins mounted on the rotor assembly. Open motors are further divided into (a) drip-proof, (b) splash-proof, and (c) flood-proof motors.

The *totally closed motor* is one in which there is no exchange of air between the interior of the motor and the surrounding atmosphere. Some types of totally enclosed motors are (a) totally enclosed fan-cooled (b) explosion proof, (c) pipe ventilated.

TYPES OF INDUCTION MOTOR — CODE IDENTIFICATION

In the section above, it was shown that induction motors may be classified according to the design of their enclosures. A motor's housing will dictate the type of environment for which the motor is best suited. Another important method used for classifying squirrel-cage motors is based on the motor's startup torque, and on the ratio between the startup and full-load currents. This property is identified by a code letter and it is usually stamped on the motor's name plate. A motor's code, for example, may be used to quickly determine the fuse and breaker ratings for the motor's startup circuit.

In a previous section, it was shown that because of its low rotor resistance, the standard squirrel-cage motor has a poor startup torque and a high starting surge current. This motor, however, has good speed regulation and very small rotor copper losses.

If the rotor resistance is increased by constructing the rotor cage with relatively small conductors of an alloy material, the starting current will decrease. The in-phase current component, however, will increase and, consequently, the starting power factor will improve, and the startup torque will increase. This motor, however, will have a higher slip, and the increased rotor resistance losses will cause a decrease in efficiency. Therefore, squirrel-cage motors with high rotor resistance are suitable only for start-stop operations. The high rotor loss may cause the rotor to overheat if this motor is operated continuously.

It should be apparent now that the actual design of a squirrel-cage motor will be a compromise of these two extremes just described. The rotor design selected will depend on the application for which the motor is to be used. The three most commonly used squirrel-cage motors are those coded CEMA B, CEMA C, CEMA D. CEMA is the abbreviation for Canadian Electrical Manufacturers' Association.

THE CEMA B DESIGN

The rotor stamping and the normal starting torque curve for a CEMA B motor are shown in Fig. 17-8. This is the most widely used design. Although these motors have relatively low rotor resistance, their starting torques are suitable for a variety of applications and their high starting current is usually at an acceptable value. The locked rotor torque for this design can vary between 50 and 120 percent of full-load torque and the full-load slip is usually between 1 and 3 percent.

CEMA B squirrel-cage motors are used in fans, blowers, machine tools, crushers,

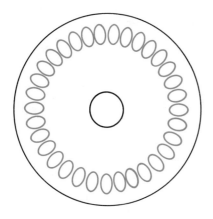

Figure 17-8(a) The rotor cage of a CEMA B motor.

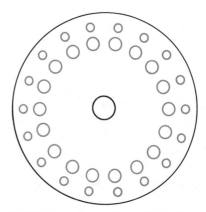

Figure 17-9(a) The rotor cage for a CEMA C motor (the double-squirrel-cage motor).

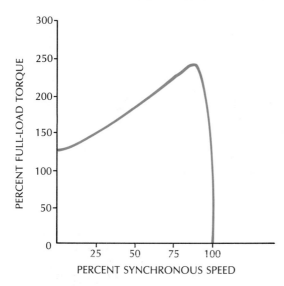

Figure 17-8(b) The normal starting torque for a CEMA B motor.

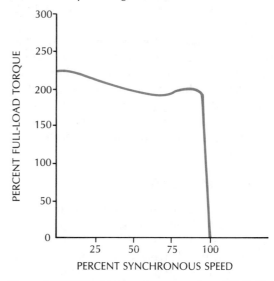

Figure 17-9(b) The high starting torque for a CEMA C motor.

chippers, centrifugal pumps, and various types of mills.

THE CEMA C DESIGN (DOUBLE SQUIRREL-CAGE)

The rotor cage and the starting torque curve for a CEMA C motor are shown in Fig. 17-9. This design is an attempt to combine the good properties of a low-resistance rotor and a high-resistance rotor. These motors have a *double*

squirrel-cage rotor, as shown in Fig. 17-9(a). The inner cage is made of large, low-resistance bars and is surrounded by a large amount of iron. Therefore, the inner cage will have a high inductive reactance at startup. Because of this high inductive reactance, there will be very little current in these inner bars at startup, and the inner cage will have negligible effect when this motor is being started.

The outer cage, however, is constructed of high-resistance bars, and they have less iron surrounding them. Therefore, the outer cage will have a low reactance and high resistance at startup. This results in a high startup torque with a low starting current.

When these motors are at operating speeds, the rotor slip frequency will be low and, as a result, the low resistance inner cage will now have a low reactance. Therefore, at operating speed, most of the induced rotor current will now flow through the low resistance, low reactance inner cage, and these motors will now operate with relatively low slip and high efficiency. The slip at full-load is normally between 1.5 and 5 percent and the startup torque may be over 200 percent of full-load torque.

The high startup torque, low starting current, relatively low operating slip, and good efficiency make these motors suitable for applications which must sometimes be started under load. They are used for operating such systems as: conveyors, elevators, crushers, compressors, and positive displacement pumps.

THE CEMA D DESIGN

The rotor cage and the starting torque curve for a CEMA D motor are shown in Fig. 17-10. Since the rotor for the CEMA D motor is constructed with relatively small conductor bars, the rotor will have a high resistance. As a result, these motors will have a high blocked rotor (startup) torque and a low starting current. However, they will also have a high percent slip and high resistance power loss. Because of these properties, the CEMA D motor is best suited for operations which must be frequently stopped and started under load, and for loads having high intermittent peaks.

When used on loads with intermittent peaks, the motor is usually equipped with a flywheel. At light loads, the motor's slip will be small, but under peaks loads, the slip will be high. The flywheel helps to carry the load through its peaks, and this results in a more constant power demand from the electrical

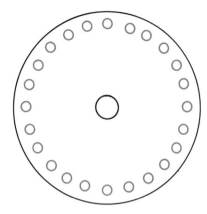

Figure 17-10(a) The rotor cage for a CEMA D motor.

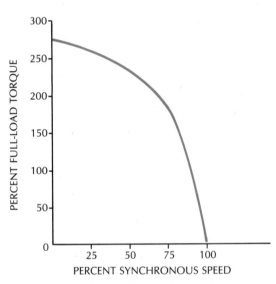

Figure 17-10(b) The high starting torque, high slip for a CEMA D motor.

system. CEMA D motors are used for operating chippers, crushers, shears, punch presses, and reciprocating pumps.

SPEED-CONTROL OF SQUIRREL-CAGE MOTORS

The speed of a squirrel-cage motor was shown to depend on the supply frequency, the number of stator poles, and the rotor resistance. The motor's speed is proportional to the supply frequency, but this frequency is usually limited to the only available line frequency of 60 Hz. Therefore, supply frequency variation for speed control is usually not practical. The motor's speed can also be decreased by increasing the number of stator poles or by increasing the rotor resistance, but these two factors are constant by design for a given squirrel-cage motor. Therefore, the squirrel-cage motor is usually not suitable for applications which require variable speed operation.

The advent of solid state technology, however, has somewhat improved the versatility of the squirrel-cage motor. Semi-conductor variable-frequency supplies are now available. They can be used to vary the supply frequency to a motor and thus vary the motor's speed. These devices, however, can be complex and expensive. Variable speed operation of a squirrel-cage motor is also possible if the motor is powered by an independent generator. The output frequency of the generator can be easily altered by changing the speed of the generator's prime mover. The wound-rotor induction motor described in the next section was developed as a means of overcoming this disadvantage of squirrel-cage motor.

THE WOUND-ROTOR INDUCTION MOTOR

It is usually not practical to vary the speed of a squirrel-cage motor. In addition, some of the motor's operating efficiency and low slip must be traded off in order to improve the motor's startup torque. The wound-rotor induction motor was developed as a means of achieving variable speed, high starting torque, and high operating efficiency characteristics.

The wound-rotor induction motor is shown in Fig. 17-11. The stator for this motor is very much the same as for the squirrel-cage motor. However, the construction of its rotor is very different. The wound rotor contains (instead

Figure 17-11 Rotor and stator of wound-rotor induction motor.

Courtesy: John Dubiel

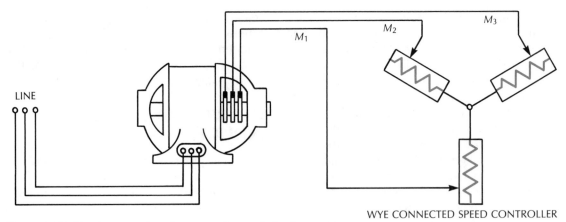

LINE

M_1 M_2 M_3

WYE CONNECTED SPEED CONTROLLER

Figure 17-12 Circuit connections for a wound-rotor induction motor.

of rotor bars), three phase windings similar to the stator; hence, the name wound-rotor. The rotor is wound to have the same number of poles as the stator and the phase windings are connected in Wye (or delta.)

The open ends of these windings are connected to three slip rings mounted on the rotor shaft. This rotor circuit is, in turn, connected by means of three brushes to an external Wye-connected variable resistance, as shown in Fig. 17-12. The rotor terminals are marked M_1, M_2, and M_3, with the M_3 terminal being the one closest to the rotor winding.

When the stator of this motor is energized, it produces a rotating magnetic field in the same way as in the squirrel-cage motor. The rotating field will induce voltages in the rotor windings, and the resulting induced currents will flow through the closed rotor circuit formed by the rotor windings, the slip rings, the brushes, and the external speed controller.

When this motor is started, the speed controller is set to its maximum resistance. Therefore, the rotor resistance will be high. The high rotor resistance will limit the startup current to a low value and at the same time it will cause the motor to have an excellent starting torque. Note that even though the starting current is reduced, the startup torque is stronger because the high resistance will cause the rotor currents to be practically in-phase with

the stator flux. If the motor is started with a reduced resistance, the starting current will increase and the startup torque will decrease.

As the motor accelerates, the resistance of the Wye-connected speed controller is gradually cut out in steps. When all the controller resistance is cut out, the three slip rings will in effect now be short-circuited, and the motor will operate identically to a squirrel-cage motor with full speed and low slip. The low rotor resistance at this controller setting also means a high operating efficiency. The torque-speed curves and current-speed curves for three resistance settings of the speed controller are shown in Figs. 17-13 (a) and (b). Examples of resistance controllers will be described in Chapter Twenty-One.

The wound-rotor motor can be operated at relatively heavy loads with controller resistance cut in the rotor circuit to obtain the desired torque. However, the motor's slip will now be greater and it will have a poorer speed regulation and greater I^2R losses.

The external resistance controller may also be used to vary the speed of the wound-rotor motor when it is operating under a given load. As shown by the speed curves of Figs. 17-13 (a) and (b), when the controller resistance is increased, the motor's speed will decrease to a new level at which the torque produced is again equal to the load. If the load for this new

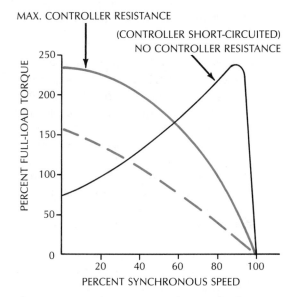

Figure 17-13(a) The torque-speed curves for the settings of the speed controller.

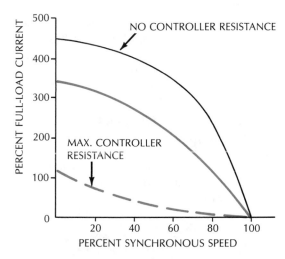

Figure 17-13(b) The current-speed curves for three settings of the speed controller.

controller setting was now decreased, then the motor's speed will increase to another speed-torque level. From this it can be seen that an increase in rotor resistance will decrease the motor's speed but, also, this increase in resistance will cause wider changes in speed for variations in the load. When used for speed control, the wound-rotor motor is generally operated between 50 and 100 percent of rated speed. The motor is considered unstable below 50 percent of rated speed.

It should be obvious that the efficiency of the wound-rotor motor depends on the I^2R losses in the rotor circuit. This power loss consists of heat energy which is dissipated in the controller. Therefore, a high-resistance controller setting will also cause a poor operating efficiency.

The properties of the wound rotor motor make it suitable for applications which require high starting torque and/or variable speed control. Some of these applications are: compressors, pumps, conveyors, cranes, hoists, steel mills, and printing presses.

SUMMARY OF IMPORTANT POINTS

1. Two general types of induction motors are: (a) squirrel-cage and (b) the wound-rotor.

Construction

2. The stator has three phase windings placed 120 electrical degrees apart. They may be connected in Wye or delta.
3. The rotor winding consists of parallel conductors embedded in the rotor core surface. These conductors are short-circuited with two end rings on the shaft. There are no brushes and the induced currents in these bars are from transformer action.

Operation

4. The stator produces a rotating magnetic field, with a synchronous speed $S = 60 \times f$/pole pairs or $\eta = 2\pi/P$.
5. Interchanging any two of the three live wires will reverse the direction of rotation.
6. The attraction between the stator field and the induced rotor field causes the rotor to rotate and follow the stator rotating field.
7. The rotor always lags slightly behind the

synchronous speed. This slip is about 2 to 5 percent of rated speed and is necessary for rotor induction.

8. The frequency of the induced rotor voltage is f_R = pole pairs × slip/60.

9. The slip increases slightly for an increase in load and this will cause the induced rotor current to increase to satisfy the new load.

10. The motor has relatively constant speed. It will develop maximum breakdown torque at between 10 and 25 percent slip. The torque equation $T = K\phi_s I_R \cos \underline{/\theta_R}$ shows torque varies almost directly with the rotor current.

11. At starting the slip is 100 percent; torque is low because of the high rotor reactance. The power factor is also low and lagging.

12. Output watts = 2π torque × r/min/60 = torque x rad/s.

13. The stator windings may be each constructed in two sections which can then be connected for operation at one of two different voltages.

14. The motor will run but not start on a single-phase voltage. Its torque will be reduced.

15. Like the dc motor, its power losses are made up of I^2R, core (hysteresis and eddy currents) and friction and windage losses. The effective rotor resistance may be determined from the "blocked rotor test."

16. The fixed losses are determined at no-load and rated voltage conditions. Fixed losses = watts at no-load−Cu losses at no-load.

17. The efficiency is low at light loads but improves considerably with load.

Ratings and Identification

18. Temperature rating depends on its winding insulation. The temperature rating is given as the temperature rise above the ambient.

19. The motor is designed with various types of enclosures: open, drip-proof, explosion proof, etc.

20. A letter classication is given based on the motor's starting and rated currents and torques. CEMA B has low rotor resistance, fair starting torque, and low slip. CEMA C has a dual cage, good starting torque, and low slip. CEMA D has high rotor resistance, excellent starting torque, and a greater slip.

21. The motor is not generally suited for variable speed applications. Speed may be varied by changes in (a) supply frequency, (b) stator poles, and (c) rotor resistance.

The Wound-Rotor Motor

22. This motor has the same stator design as the cage motor. However, its rotor has three phase windings connected in Wye. The three winding ends are brought out through slip rings and brushes to an external resistance controller.

23. The resistance controller is used to vary the motor's speed. The motor has excellent starting torque but poor speed regulation and low efficiency.

ESSAY AND PROBLEM QUESTIONS

1. (a) List three advantages of using an ac induction motor rather than a dc motor.
   (b) List four uses of a polyphase induction motor.
   (c) List the three main parts of an ac induction motor.

2. Name the two general groups of three-phase induction motors. What do these two groups of motors have in common?

3. Describe the construction of (a) the stator and (b) the rotor for a three-phase squirrel-cage motor.

4. (a) Explain how the stator of a three-phase induction motor produces a rotating magnetic field. (b) How can

the direction of this rotating field be reversed?

5. (a) Explain what is meant by the synchronous speed of an induction motor.
   (b) Name the two conditions which determine this speed.

6. Explain how induction motor action produces torque in an induction motor.

7. Which of the windings in an induction motor may be described as (a) the primary and (b) the secondary winding?

8. (a) Describe the terms: (i) rotor revolutions slip (ii) percent slip, and (iii) rotor frequency.
   (b) Why is rotor slip necessary in an induction motor?

9. A six-pole, three-phase, 240-V, 60-Hz induction motor has a rated full-load speed of 1140 r/min (119.32 rad/s). Determine:
   (a) synchronous speed
   (b) rotor slip
   (c) rotor percent slip
   (d) rotor frequency

10. (a) What is the rotor frequency for the motor in Q. 9 at the instant of startup?
    (b) what effect does the rotor frequency have on the rotor?

11. A 240-V, 60-Hz, three-phase, 12-pole squirrel-cage motor has a full load speed of 560 r/min (58.6 rad/s). Determine:
    (a) the synchronous speed
    (b) rotor slip at startup
    (c) rotor slip at rated load
    (d) percent slip at startup
    (e) percent slip at rated load
    (f) rotor frequency at startup
    (g) rotor frequency at rated load

12. The load on a three-phase squirrel cage motor was increased from no-load to rated full-load. Explain the rotor action to meet this increase in load.

13. Draw characteristic curves to illustrate: (a) torque versus rotor current and (b) torque versus rotor slip for a squirrel-cage induc-

tion motor.

14. Name two rotor properties which will affect the amount of torque produced by an induction motor for a given increase in slip.

15. If a standard squirrel-cage motor is loaded much beyond rated full-load, explain the effect on the following:
    (a) rotor impedance
    (b) rotor power factor
    (c) the increase in rotor current for a given increase in slip, and
    (d) motor torque

16. (a) Explain why the starting torque for the standard induction squirrel-cage motor may be considerably less than its rated full-load torque.
    (b) What rotor change will improve this starting torque?

17. Explain why the in-rush starting current for an induction motor may be considerably greater than its rated full-load current.

18. Explain why the torque of an induction motor will vary as the square of the applied voltage for a given rotor slip.

19. Describe the effects that a low rotor resistance would have on the following properties of squirrel-cage induction motor:
    (a) starting torque
    (b) locked rotor current,
    (c) slip and
    (d) efficiency.

20. Describe the effects that a high-resistance rotor would have on the following properties of a squirrel-cage induction motor:
    (a) starting torque
    (b) locked rotor current
    (c) slip and
    (d) efficiency

21. If a six-pole, 36 kW motor operates at 1150 r/min (120.36 rad/s), determine its output torque.

22. A three-phase, 60 Hz, four-pole, 240-V squirrel-cage induction motor draws a line current of 48 A at full-load. The power factor is 0.8 lag and the efficiency is

85 percent. The slip is 3 percent. Determine the following at rated load:
(a) synchronous speed
(b) rotor speed
(c) input power
(d) output power
(e) torque

23. Draw the speed versus output power characteristic curve for a typical standard squirrel-cage motor and describe the reason for this speed characteristics from no-load to full load.

24. The stator magnetic circuit in an induction motor includes the air gap, which separates the stator core from the rotor core. What effect does this air gap have on (a) the stator magnetic reluctance (b) The motor power factor?

25. Explain (a) why the standard squirrel-cage motor has a low lagging power factor at startup and at no-load and (b) why the motor's power factor will increase as the motor's load is increased.

26. (a) Name the different power losses in an induction motor.
    (b) Which of these losses are affected by changes in load and which are affected by changes in speed?

27. In a blocked rotor test of a three-phase induction motor, a line voltage of 70 V cause a rated current of 16.5 A. The readings of the two wattmeters in the circuit were $W_1 = 700$ W and $W_2 = -2W$. Determine (a) the effective phase resistance of this motor and (b) the total copper losses at a load $I_L = 16.5$ A.

28. When the motor in Q. 27 was operated at no-load and under its rated voltage of $V_L = 240$ V, the line current was measured as $I_L = 6.5$ A, and the wattmeters W_1 and W_2 read 670 W and − 250 W, respectively. Determine the motor's fixed losses.

29. When the motor in Q. 27 and Q. 28 was operated under rated load and voltage conditions, the following readings were taken: $I_L = 16.5$ A, $W_1 = 3550$ W, and $W_2 = 1890$ W. Determine (a) the motor's efficiency and (b) the motor's power factor.

30. Sketch the power factor and efficiency characteristic curves for a typical three-phase squirrel-cage induction motor.

31. Which induction motor will have the higher power factor?
    (a) no-load or full-load
    (b) low speed or high speed
    (c) 8 kW or 50 kW

32. Which induction motor will be generally more efficient?
    (a) no-load or full-load
    (b) low speed or high speed
    (c) 8 kW or 50 kW

33. The following results were obtained from a prony brake test of a three-phase induction motor: $V_L = 240$ V, $W_1 = 1680$ W, $I_L = 14.5$ A, $W_2 = 1830$ W, and rotor speed = 1760 r/min (184.21 rad/s). If the prony brake arm was 0.5 m, and the total force on the arm was 30 N, determine: (a) torque output in N.m, (b) the output power, (c) the efficiency, and (d) the power factor.

34. A. six-pole, three-phase induction motor is rated at 240 V, 32.5 A. The full-load efficiency is 80 percent, the power factor 0.85 lag, and the slip is 3 percent. Determine: (a) the rotor speed, (b) the output power, and (c) the total losses.

35. The name plate on a three-phase induction motor indicates a rating of 240/480 V and 18.5/9.3 A and a Wye-connected stator.
    (a) draw a diagram to illustrate the stator Wye-connection, and stator coil sections for this motor
    (b) draw a diagram to show the terminals in the terminal box on this motor.
    (c) show the terminal connections for 480-V operation
    (d) show the terminal connections for 240-V operation

36. Explain (a) why a three-phase induction

motor may not start when energized from a single-phase supply. (b) Why the motor will continue to run if one of its line wires is disconnected while the motor is running.

37. Describe the following terms:
    (a) temperature rise continuous duty
    (b) open-type motor
    (c) close-type motor

38. Explain the purpose of the CEMA letter code stamped in the name plate of an induction motor,

39. (a) Briefly describe the operating characteristics for an induction motor coded CEMA B.
    (b) Name six applications for this type of motor.

40. Explain why the double squirrel-cage induction motor will have: (a) good starting torque, (b) relatively low starting current, and (c) relatively small percent slip.

41. Briefly describe the rotor construction and operating characteristic for a squirrel-cage induction motor coded CEMA D.

42. Explain how the speed of a squirrel-cage motor may be controlled.

43. Describe the construction of a wound-rotor induction motor.

44. The external speed controller to a wound rotor motor is set at its maximum resistance. What effect will this have on (a) the motor's starting characteristics and (b) the motor's operating characteristics.

45. Why is the wound-rotor induction motor suited for heavy loads requiring frequent start-stop operations and long starting periods?

46. List the advantages and disadvantages of the wound-rotor induction motor compared to the squirrel-cage induction motor.

47. Describe the speed regulation and efficiency of a wound-rotor motor: (a) with maximum controller resistance in the rotor circuit and (b) with minimum resistance in the rotor circuit.

CHAPTER
EIGHTEEN

SYNCHRONOUS MOTORS AND SELF-SYNCHRONOUS DEVICES

Synchronous motors and self-synchronous devices are used in many industrial applications. A synchronous motor is one that moves in synchronism with the rotating magnetic field that is produced by its stator windings. Most industrial synchronous motors are three-phase motors. However, small single-phase synchronous motors are also manufactured and these are very common in various timing devices.

The construction of the three-phase synchronous motor is very similar to that of an ac generator. In fact, just as a dc generator may be operated as a dc motor, an ac generator may be operated as a synchronous motor. This motor not only has a constant speed from no-load to full-load, but its dc rotor field can also be adjusted to produce a wide range of lagging and leading power factor values. The single-phase synchronous motor, on the other hand, does not use any dc rotor excitation and,

hence, does not have this power factor characteristic. This motor will be described later in this chapter.

THREE-PHASE SYNCHRONOUS MOTORS

CONSTRUCTION

The synchronous motor is almost identical with the corresponding ac generator, and its main elements consists of the stator (armature), end shields and brush assembly, rotor (field), and amortisseur (damper) windings.

This motor may have either a revolving armature or a revolving field, although most synchronous motors are of the revolving-field type. The stationary armature or stator is similar to that of the three-phase induction motor. It consists of three phase coils mounted 120° apart. These coils are connected in either

Figure 18-1 The synchronous motor rotor. *Courtesy: John Dubiel*

delta or Wye (star), and mounted on the stator frame to form an even number of stator poles. The leads for the stator windings are brought out to a terminal box on the frame and marked T_1, T_2, and T_3.

The rotor of the synchronous motor contains both the field and amortisseur windings. The field windings are usually wound to form salient-field poles, with alternate polarity, and of the same number as the stator poles. The field circuit must be energized from a dc source, and its leads (F_1 and F_2) are brought to the terminal box via two slip rings and brushes.

The synchronous motor, as will be shown later, is not by itself self-starting. In order to operate, the motor's speed must first be brought up to synchronous or near synchronous speed. Therefore, most synchronous motors are fitted with a starting winding called an amortisseur winding. This winding is a form of squirrel-cage winding. It consists of copper bars embedded in the laminated rotor core of each of the field poles, and brazed to two end rings, one on each side of the rotor.

The end shields are fitted with bearings. They support the rotor assembly and are bolted to the stator frame. The brush assemblies are necessary in order to supply dc current to the rotor field circuit.

PRINCIPLE OF OPERATION

When a three-phase voltage is applied to the stator (armature) windings of a synchronous motor, it produces a rotating magnetic field in the same way as in a three-phase induction motor. The speed of this rotating field will be in synchronism with the supply current and depends only on the supply frequency and the number of stator poles. This synchronous speed may be expressed either in radians per

Table 18-1

FREQUENCY	SPEED IN η (RAD)S FOR VARIOUS NUMBER OF POLES							
Hz	2	4	6	8	10	12	16	20
25	50 π	25 π	16.67 π	12.5 π	10 π	8.33 π	6.25 π	5 π
50	100 π	50 π	33.33 π	25 π	20 π	16.67 π	12.5 π	10 π
60	120 π	60 π	40 π	30 π	24 π	20 π	15 π	12 π

second or in revolutions per minute, as shown below (note: 1 revolution equals 2π radians).

$$\eta = \frac{2\pi f}{P} \qquad (18\text{-}1(a))$$

where η = speed in rad/s
and P = number of pole pairs per phase

$$S = \frac{f \times 60}{P} \qquad (18\text{-}1(b))$$

where S = speed in r/min
and P = number of pole pairs per phase

Table 18-1 shows the pole-speed relation for three different frequencies.

When this rotating magnetic field cuts the damper (amortisseur) cage winding, it produces a starting torque. Therefore, this synchronous motor starts up as a squirrel-cage induction motor, and its speed will increase to a value that is slightly less than synchronous speed.

If the rotor field windings are now excited from an external dc supply, fixed poles of alternate polarity will be set up in the rotor core, as shown in Fig. 18-2. These field poles will be attracted to unlike poles from the rotating stator field, and the rotor will be pulled into synchronism or locked into position with the rotating magnetic field. The rotor will now be rotating along at synchronous speed. Any variation in the field current now will produce no further changes in the motor's speed.

Note that the dc rotor field is applied only after the rotor has speeded up to about 95 percent of synchronous speed. The dc field will not produce any starting torque. This is because, with the stator field rotating at synchronous speed and the rotor at a standstill, the rotor poles are first attracted in one direction and then in the other, resulting in zero torque on the rotor. The rotor dc poles, however, are required for it to pull from near synchronous speed into synchronism. The initial starting torque for the synchronous motor described was set up by the amortisseur winding. In fact, the rotor may be started and accelerated to near synchronous speed by other means, such as by another small auxiliary motor mounted on the main motor shaft.

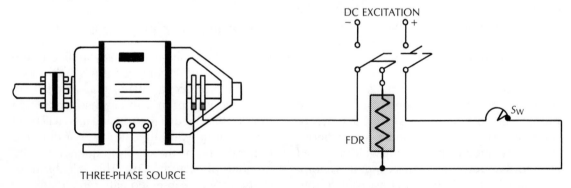

Figure 18-2 Operating principle of a synchronous motor with field discharge circuit.

STARTING THE SYNCHRONOUS MOTOR

It was already explained that the synchronous motor is not self-starting and that some type of secondary starting method must be employed. Also, the dc field excitation is applied only after the rotor has been accelerated to near synchronous speed. It should be obvious that during starting and while the motor is building up its speed, a voltage will be induced in the rotor field winding by the rotating stator field. This induced ac voltage can be dangerously high because of the many turns in the rotor field coils.

For safety reasons, therefore, when starting this motor, the field circuit is usually short-circuited through a resistance to limit this induced voltage. The induced ac current which flows through the field circuit will now be limited by the high reactance of this circuit. When the motor has speeded up to near synchronism, the field discharge resistor is removed at the same time the dc field excitation current is applied. A switch, as shown in Fig. 18-2, may be used to add or remove the field discharge resistor (FDR) from the field circuit. At synchronous speed, the rotor slip will be zero and, hence, no voltage will be induced in the field coils.

The rotor of a synchronous motor is pulled into synchronism because of the attractive force between the rotor field poles and the stator poles. The amount of torque that a motor will exert to pull into synchronism is called the *pull-in torque*. If the motor's load exceeds the motor's pull-in torque, it is obvious that the motor will not be able to pull into synchronism. The pull-in torque developed by a synchronous motor will depend on its design. These motors may be designed to have a pull-in torque as high as 150 percent of rated full-load torque.

Generally, the starting current for synchronous motors is less than that for induction motors of the same speed and power rating. It is not uncommon practice for using full-rated voltage when starting synchronous motors with power rating as high as 10 kW. Full-voltage starting circuitry is generally simpler and less expensive than the circuitry for reduced voltage starting. Reduced voltage starting for ac motors using auto transformers or reactors will be discussed in Chapter Twenty-One.

DC FIELD EXCITATION

The excitation current for the field circuit may be supplied from a separate dc exciter such as from a small dc generator mounted on the motor shaft. This system of excitation requires the use of brushes and slip rings.

The advent of solid state technology, however, has now made possible the brushless synchronous motor. The excitation system on this type of motor uses an exciter, and a bridge rectifier circuit mounted right on the rotor spider. The rectifier's output is connected directly to the field circuit and, hence, the need for brushes and slip rings is eliminated. Special control circuitry on the rotor is used to automatically switch the field circuit on at the proper time to bring about rotor synchronism. The main disadvantage of the brushless system is that the field current is not variable and also the added circuitry makes these motors more expensive.

LOAD EFFECT ON A SYNCHRONOUS MOTOR

When the mechanical load on either a dc motor or an ac induction motor is increased, the motor speed decreases. This decrease in speed reduces the motor's counter voltage which allows additional current to be drawn from the supply to meet the increase in load. In a synchronous motor, however, the rotor is locked in synchronism with the rotating magnetic field and so there will no change in motor speed for an increase in load. Therefore, the decrease in counter voltage, and the resulting increase in the supply current to carry an increase in load, must be caused by some other changes in the motor's operating properties.

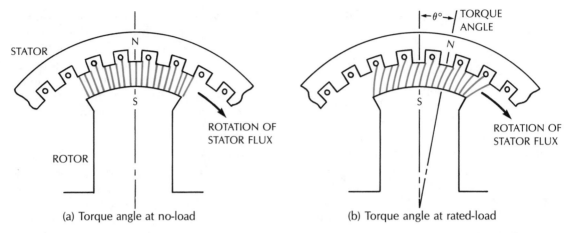

(a) Torque angle at no-load   (b) Torque angle at rated-load

Figure 18-3 Relative positions of the stator poles and the rotor field poles at: (a) no-load and (b) full-load conditions.

The change in the counter voltage of a synchronous motor for a change in load is caused by a shift of the rotor poles with respect to the attracting stator poles. As shown in Fig. 18-3(a), when the motor is operated at no-load, the centre lines of the two poles coincide (if losses are neglected). However, when the mechanical load is increased, there is a backward shift of the rotor pole relative to the stator pole, as shown in Fig. 18-3(b). Note that there is no change in rotor speed; there is only a shift of the rotor poles opposite to the direction of motor rotation. It is this phase displacement between the rotor poles relative to the stator poles that causes the necessary change in the stator current to satisfy the change in the motor's load. This angular displacement between a rotor and stator pole is called the *torque angle*.

The effect of this torque angle can also be illustrated by the phasor diagrams of Fig. 18-4. When the synchronous motor operates at no-load, the torque angle is practically zero, and the motor's counter voltage, (V_g), is equal and opposite to the applied line voltage, (V_L). However, when the load is increased, the phase position of the counter voltage drops behind the no-load position by the torque angle ($\alpha°$), as shown in Fig. 18-4(b). Note the

increase in load changes only the phase position of V_g; its magnitude remains unchanged for the same applied rotor field current.

The effective stator voltage (V_e) will be the phasor resultant of the counter voltage, (V_g), and the applied terminal voltage, (V_L). This resultant stator voltage causes the necessary stator current to flow. Any change in the load will change the resultant voltage, thus changing the corresponding stator load current. Since the stator windings have a high inductance, the stator current will always lag the effective voltage, (V_e), by almost 90°, as shown in Fig. 18-4(b).

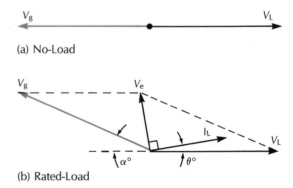

(a) No-Load

(b) Rated-Load

Figure 18-4 Phasor diagrams of a synchronous motor showing the applied and counter voltages for two different loads and the same field excitation.

If the load placed on a synchronous motor is made too high, the rotor will pull out of synchronism and it may either stall or, if it has an amortisseur winding, it may continue to run, but as an induction motor. The maximum torque that a synchronous motor will develop without losing synchronism is called the *pull-out torque*. The pull-out torque, like the pull-in torque, of a synchronous motor will depend on its design. This torque can be made as high as 300 percent of the motor's full-load torque. A decrease in the field excitation current will also reduce this torque. Synchronous motors are usually designed to have torque characteristics for definite applications.

It can also be reasoned from the phasor diagrams that the increase in load for a constant field current will also cause the motor's power factor to increase in a lag direction. The power factor angle, $\underline{/\theta}$, for the given load and field current is shown in Fig. 18-4(b). From Fig. 18-4, the input power to a three-phase synchronous motor may be expressed as:

$$\text{input power} = \sqrt{3}\, V_L I_L \cos \underline{/\theta} \qquad (18\text{-}2)$$

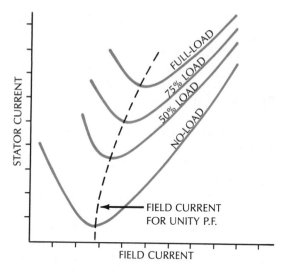

Figure 18-5 Typical V-curve characteristics of a synchronous motor.

The effects of load and field current variations on a motor are also illustrated by the V-shaped curves shown in Fig. 18-5.

The speed-torque-current characteristics of a high-speed synchronous motor are illustrated in Fig. 18-6. Note that when the motor has achieved synchronous speed, this speed will remain constant from no-load to the load

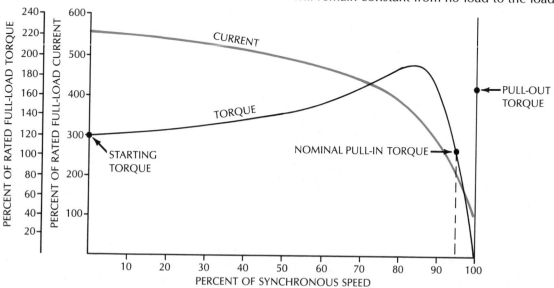

Figure 18-6 The speed-current and speed-torque characteristics of a typical high-speed synchronous motor.

at which the motor will pull-out of synchronism.

A sudden change in load may also cause a *hunting* effect in the operation of a synchronous motor. Each time there is a change in load there will be change in the torque angle between the rotor and stator fields. This may cause the rotor field poles to oscillate slightly backward and forward until the exact torque angle is attained. This oscillating effect is called hunting.

The hunting effect in a synchronous motor is lessened by the amortisseur windings in the pole tips of the rotor. When the motor is hunting, it will momentarily slow down and speed up as the field poles oscillate back and forth. This change in speed compared to the rotating magnetic field will cause current to be induced in the amortisseur windings. The effect of this induced current will be to cancel out any change in the rotor's speed. Thus, the amortisseur windings also serve to dampen any hunting effect in the operation of this motor.

POWER FACTOR

An outstanding characteristic of the synchronous motor is that its power factor, for a constant load, can be varied over a wide range from leading to lagging values by simply adjusting the dc field excitation current. An under-excited field will produce a lagging power factor and an over-excited field will produce a leading power factor. This property increases the industrial uses for this type of motor.

In the previous section, it was shown that the motor's power factor angle will increase in a lagging direction when the motor's mechanical load is increased and the field current is kept constant. This change in power factor, as shown in Fig. 18-4, is brought about by a change in the motor's torque angle. Under the constant field current condition, the magnitude of the counter voltage, V_g, remained the same and only its phase position changed with a change in load.

If the load on a synchronous motor is now kept constant and the field excitation changed, the motor's torque angle will remain the same but the magnitude of the counter voltage will be altered, as shown in Fig. 18-7. The field current may be adjusted to supply either too little of the motor's magnetizing current (lagging P.F.), the exact amount of magnetizing current (unity P.F.), or too much magnetizing current (leading P.F.). Note that the field variation affects only the power factor and not the motor's speed.

Figure 18-7(a) shows the results for a field current adjusted to supply the exact amount of magnetizing current required by the motor.

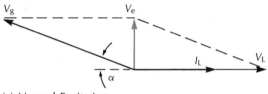

(a) Normal Excitation

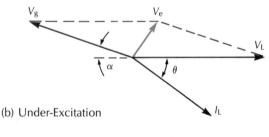

(b) Under-Excitation

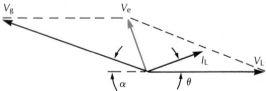

(c) Over-Excitation

Figure 18-7 Phasor diagrams of a synchronous motor with a constant load but with the field current adjusted for: (a) unity power factor, (b) lag power factor, and (c) lead power factor.

Under this condition, the motor's power factor will be unity. The field current required for unity power factor is called the *normal excitation current*.

Figure 18-7(b) shows the results for a field adjusted to supply less than the normal field current. Under this condition, the counter voltage, V_g, is reduced. Therefore, the ac stator voltage must now provide some of the required magnetizing current. As a result, the motor will have a lagging power factor.

When the field current is increased beyond the normal value, as shown in Fig. 18-7(c), the counter voltage, V_g, increases so that there is now an excess of magnetizing current. This excess magnetizing current shows up at the ac stator voltage supply, and thus the stator voltage must now provide a negative magnetizing current component to cancel this out. As a result, the stator current will lead the stator supply voltage, and the motor will now have a lead power factor.

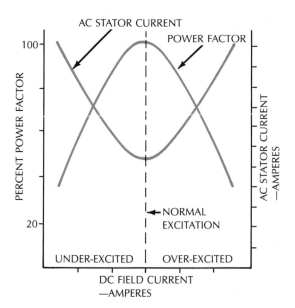

Figure 18-8 The characteristic curves for the stator current and power factor of a synchronous motor under a constant load and with a variation of its field excitation current.

Figure 18-8 compares the power factor and stator current characteristics for a constant motor load and a variation of the excitation field current. Note that with the normal field current, the power factor is unity and the stator current is at its minimum value. As the field current is decreased below the normal value, the power factor becomes lagging, and as the field current is increased beyond the normal value, the power factor becomes leading. When the motor power factor decreases in either a lagging or leading direction, the stator current increases. The stator current must increase in order to maintain the same output power at a reduced power factor.

$$\text{power} = \sqrt{3} \; V_L I_L \cos \underline{/\theta}$$

INDUSTRIAL APPLICATIONS OF SYNCHRONOUS MOTORS

Synchronous motors are used for constant speed applications in sizes of 15 kW and larger. They are commonly used for driving large air and gas compressors, because these devices must be driven at a constant speed in order to maintain a constant output and maximum efficiency. They are also used to drive large pumps, blowers, fans, pulverizers, and dc generators. The development of the brushless synchronous motor has also now made possible its use where formerly only induction motors were considered.

Another outstanding use of synchronous motors is their use for correcting the power factor of a lagging system. When used for this purpose, the synchronous motor is often called a *synchronous condenser* or a *synchronous capacitor*.

POWER FACTOR CORRECTION

It was shown that when the power factor of a system decreases from unity, the supply current must increase in order to maintain the same system power. However, since the supply current to a system must usually not exceed the system's rated value, it follows that operating a system at a low power factor may also limit the power that the system may deliver. For example, a system power of 2 kW at unity power factor and a rated current of 10 A will reduce to 1.2 kW for the same current and a 0.60 power factor.

The increase in current caused by a low power factor will also increase the system's voltage (IR) and power (I^2R) losses. Therefore, a low power factor may reduce a system's efficiency and cause poor voltage regulation in devices such as generators and transformers.

The advantages of operating a system at unity or near unity power factor should now be obvious. In fact, electric utilities will often include a penalty clause in the contract of companies operating below a certain power factor value.

Most industrial systems use induction motors and thus they will have a lag system power factor. In order to gain the benefits of a high power factor, synchronous motors are often installed in these systems, as shown in Fig. 18-9. When these motors are operated with their field over-excited, they will compensate for the lagging reactive power (VARs) taken by the induction motors. The following examples will illustrate how a synchronous condenser can be used to improve the overall power factor of a system.

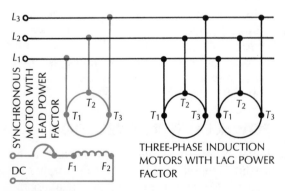

Figure 18-9 A synchronous motor used to correct a system power factor.

Example 18-1: An industrial system, as shown in Fig. 18-9, uses two three-phase induction motors, one motor is rated at 30 A and 0.80 lag power factor and the second motor is rated at 40 A and 0.85 power factor. The supply voltage to the system is 240 V. A synchronous motor, rated at 25 A and 0.65 lead power factor is used to improve the overall system power factor. Determine:

(a) the true, apparent, and reactive power for each of the two induction motors,
(b) the total true, apparent, and reactive power supplied to the two induction motors,
(c) the power factor with only the two induction motors in operation,
(d) the true, apparent, and reactive power for the synchronous motor,
(e) the total system true, apparent, and reactive power when all three motors are in operation.
(f) the overall system power factor,
(g) the supply current with only the two induction motors in operation,
(h) the supply current with all three motors in operation.

Solutions:

(a) *Motor #1*:
(I_L = 30 A, P.F. = 0.8 lag, $\underline{/\theta_1}$ = 36.870)

$$\text{true power} = \sqrt{3}\ V_L\ I_L\cos\underline{/\theta_1}$$
$$= 1.732 \times 240 \times 30 \times 0.8$$
$$= 9964.80\ \text{W}$$

$$\text{apparent power} = \sqrt{3}\ V_L \cdot I_L$$
$$= 1.732 \times 240 \times 30$$
$$= 12\ 456.0\ \text{VA}$$

$$\text{reactive power} = \sqrt{3}\ V_L I_L \sin\underline{/\theta}$$
$$= 1.732 \times 240 \times 30 \times 0.6$$
$$= 7473.6\ \text{VARs}$$

Motor #2:
(I_L = 40 A, P.F. = 0.85 lag, $\underline{/\theta_2}$ = 31.79°)

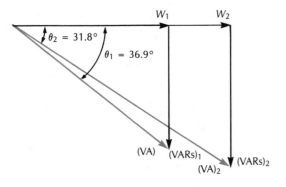

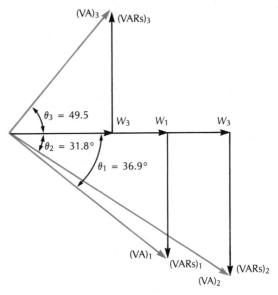

Figure 18-10(a) Phasor power triangle for the two induction motors.

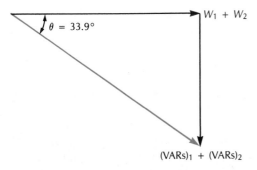

Figure 18-11(a) Phasor power triangles for the three motors in the system.

Figure 18-10(b) Phasor addition of the two induction power triangles.

$$\text{true power} = \sqrt{3}\ V_L I_L \cos\ /\underline{\theta_2}$$
$$= 1.732 \times 240 \times 40 \times 0.85$$
$$= 14\ 116.8\ \text{W}$$

$$\text{apparent power} = \sqrt{3}\ V_L I_L$$
$$= 1.732 \times 240 \times 40$$
$$= 16\ 608.0\ \text{VA}$$

$$\text{reactive power} = \sqrt{3}\ V_L I_L \sin\ /\underline{\theta}$$
$$= 1.732 \times 240 \times 40 \times 0.53$$
$$= 8802.20\ \text{VARs}$$

(b) With only the two induction motors on line: total true power $= W_1 + W_2$
    (see Fig. 18-9)
    $= 9964.8 + 14\ 116.8$
    $= 24\ 081.6\ \text{W}$

total reactive power $= (\text{VARs})_1 + (\text{VARs})_2$
    (see Fig. 18-10)
    $= 7473.6 + 8802.2$
    $= 16\ 275.8\ \text{VARs}$

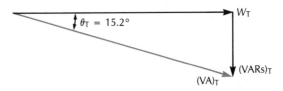

Figure 18-11(b) Phasor addition of the three power triangles.

Total apparent power from the phasor diagrams of Fig. 18-9:

$$= \sqrt{[W_1 + W_2]^2 + [(\text{VARs})_1 + (\text{VARs})_2]^2}$$
$$= \sqrt{(24\ 081.6)^2 + (16\ 275.8)^2}$$
$$= \sqrt{8.45 \times 10^8}$$
$$= 29\ 065.9\ \text{VA}$$

(c) Power factor with only the two induction motors on the line:

$$\text{power factor} = \frac{\text{total W}}{\text{total VA}}$$
$$= \frac{24\ 081.6}{29\ 065.9}$$
$$= 0.83\ \text{lag or 83\% lag}$$

(d) Power to the synchronous motor ($I = 25$ A, P.F. $= 0.65$, $/\underline{\theta} = 49.5°$):

true power $= \sqrt{3} \ V_L I_L \cos \angle\theta_3$
$= 1.732 \times 240 \times 25 \times 0.65$
$= 6747$ W

apparent power $= \sqrt{3} \ V_L I_L$
$= 1.732 \times 240 \times 25$
$= 10\ 380$ VA

reactive power $= -\sqrt{3} \ V_L I_L \sin \angle\theta$
$= -1.732 \times 240 \times 25 \times 0.76$
$= -7888.8$ VARs

(Note: reactive power of the synchronous motor is leading; therefore, it is negative.)

(e)   Total system power (see Fig. 18-11):

total true power $= W_1 + W_2 + W_3$
$= 9964.8 + 14\ 116.8$
$\qquad + 6747$
$= 30\ 828.6$ W

total reactive power
$= (VARs)_1 + (VARs)_2 - (VARs)_3$
$= 7473.6 + 8802.2 - 7888.8$
$= 8387$ VARs

(Note: net reactive power is still inductive, but it is greatly reduced.)

total apparent power
$= \sqrt{(W_T)^2 + (VARs)_T^2}$
$= \sqrt{(30\ 828.6)^2 + (8387)^2}$
$= \sqrt{1.02 \times 10^9}$
$= 31\ 949.1$ VA

(f)   Overall system power factor:

$$P.F. = \frac{\text{total watts } (W_T)}{\text{total voltamperes } (VA_T)}$$
$$= \frac{30\ 828.6}{31\ 949.1}$$
$$= 0.965 \text{ lag}$$

(Note: the insertion of the over-excited synchronous motor has improved the system power factor from 0.83 lag to 0.925 lag.)

(g)   The supply current with only the two induction motors on the line:

$$I_L = \frac{\text{VA for two motors}}{\sqrt{3} \ V_L}$$
$$= \frac{29\ 065.9}{1.732 \times 240}$$
$$= 70.0 \text{ A (for the two motors)}$$

(h)   The supply current with all three motors connected into the system.

$$I_{LT} = \frac{VA_T}{\sqrt{3} \ V_L}$$
$$= \frac{31\ 949.1}{1.732 \times 240}$$
$$= 76.95 \text{ A}$$

Example 18-2: The electrical load of an industrial system is 800 kVA at a power factor of 0.75 lag. The system is operated from a 4800-volt supply. Another motor is to be added to the system to carry an additional load of 200 kW. Determine the following if the new motor is (a) an induction motor with 0.90 lag power factor and (b) a synchronous motor with a 0.85 lead power factor:

Solutions:

(a)   the new kVA load of the system (b) the new system power factor, and (c) the supply current.

From Fig. 18-12(a) with an original load of 800 kVA at 0.75 lag power factor:
$\angle\theta_1 = \cos^{-1} 0.75$
$\qquad = 41.4°$

original load (kW) $= $ kVA $\cos \angle\theta_1$
$= 800 \times 0.75$
$= 600$ kW

original load (kVARs) $= $ kVA $\sin \angle\theta_1$
$= 800 \times 0.66$
$= 528$ kVARs (lagging)

From Fig. 18-12(a), the additional load of 200 kW using an induction motor with an 0.9 lag power factor, $\angle\theta_2 = 25.84°$:

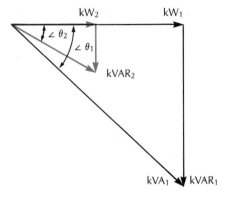

Figure 18-12(a) Power phasor triangles for original load and for the new induction motor.

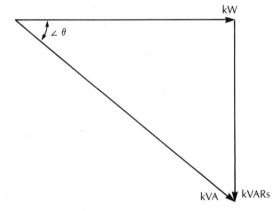

Figure 18-12(b) Resultant power triangle when the induction motor is added to the system.

$$\text{additional load (kVA)} = \frac{kW}{\text{power factor}}$$
$$= \frac{200}{0.9}$$
$$= 222.22 \text{ kVA}$$

additional load (kVARs)
$$= kVA \times \sin \angle\theta_2$$
$$= 222.22 \times 0.44$$
$$= 97.78 \text{ kVARs (lagging)}$$

From Fig. 18-12(b), the resultant system power when the induction motor is added to the system:

new net load kW
$$= kW_1 + kW_2$$
$$= 600 + 200$$
$$= 800 \text{ kW}$$

new net load kVARs
$$= kVAR_1 + kVAR_2$$
$$= 528 + 97.78$$
$$= 625.78 \text{ kVARs (lagging)}$$

new net kVA
$$= \sqrt{(kW^2 + (kVARs)^2}$$
$$= \sqrt{(800)^2 + (625.78)^2}$$
$$= \sqrt{1\ 301\ 600.6}$$
$$= 1015.68 \text{ kVA}$$

new system P.F.
$$= \frac{kW}{kVA}$$
$$= \frac{800}{1015.68}$$
$$= 0.79 \text{ lag}$$

new load current
$$= \frac{kVA \times 1000}{\sqrt{3}\ V_L}$$
$$= \frac{1015.68 \times 1000}{1.732 \times 4800}$$
$$= 122.31 \text{ A}$$

(b) From Fig. 18-13(a), the additional load of 200 kW, using a synchronous motor with a 0.85 lead power factor ($\angle\theta_3 = 31.79$):

additional load (kVA)
$$= \frac{kW}{\text{power factor}}$$
$$= \frac{200}{0.85}$$
$$= 235.29 \text{ kVA}$$

additional load (kVARs)
$$= - kVA \times \sin\angle\theta_3$$
$$= -235.29 \times 0.53$$
$$= -124.7 \text{ kVARs (leading)}$$

From Fig. 18-13(b) (see page 368), the resultant system power when the synchronous motor is added to the system:

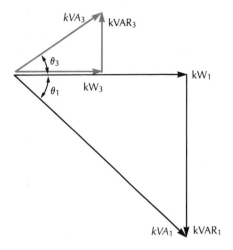

Figure 18-13(a) Power phasor triangles for the original load and for the new synchronous motor.

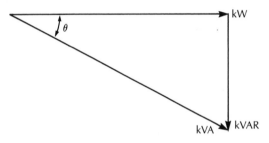

Figure 18-13(b) Resultant power phasor triangle when the synchronous motor is added to the system.

new net kW
$$= kW_1 + kW_2$$
$$= 600 + 200$$
$$= 800 \text{ kW (same as before)}$$

new net kVAR
$$= kVAR_1 - kVAR_3$$
$$= 528 - 124.7$$
$$= 403.7 \text{ kVARs (lagging)}$$

new net kVA
$$= \sqrt{(kW)^2 + (kVARs)^2}$$
$$= \sqrt{(800)^2 + (403.7)^2}$$
$$= \sqrt{802\,973.7}$$
$$= 896.09 \text{ kVA}$$

new system P.F.
$$= \frac{kW}{kVA}$$
$$= \frac{800}{896.09}$$
$$= 0.89 \text{ (lagging)}$$

new load current
$$= \frac{kVA \times 1000}{\sqrt{3}\,V_L}$$
$$= \frac{896.09 \times 1000}{1.732 \times 4800}$$
$$= 107.91 \text{ A}$$

Note that when the synchronous motor was substituted for the induction motor, the load kW remained the same; however, the net VARs from the supply decreased. Therefore, as shown by the results, the use of the synchronous motor with a leading power factor improves the overall system power factor and reduces the net supply current.

MOTOR RATINGS AND EFFICIENCY

The name plate data on a synchronous motor is very much the same as that given for an ac generator except for the output power rating. On ac generators, the output power is specified in kVA, whereas on motors, this is replaced with a kW rating.

The name plate data of a synchronous motor will also include a power factor rating. Most synchronous motors are rated at unity power factor, 90 percent lead, or 80 percent lead power factor. While a lead power factor synchronous motor will provide more power factor correction, it must be designed to carry larger currents. Therefore, because of the heavier windings that are required, lead power factor synchronous motors are usually more expensive than unity power factor motors.

A unity power factor motor may also be operated at a leading power factor, but the power output will have to be lower than the rated value. This property is illustrated in

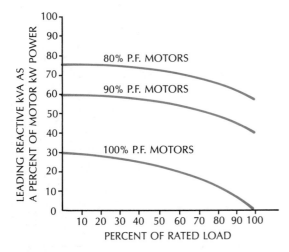

Figure 18-14 Typical leading kVA of a high-speed synchronous motor operating at different loads and with rated field excitation.

Fig. 18-14 for three different power factor motors. When the power factor is made more leading, the current taken by the motor will also increase. Since this current must not exceed the motor's rating, it will be necessary to lower the mechanical load on the motor in order to limit the current to the motor's rated value.

The efficiency of a three-phase synchronous motor is slightly higher than the efficiencies of induction motors of the same speed and power ratings. The losses of a synchronous motor are identical with those of an ac generator. These losses consist of the stray power losses (which comprise the friction, windage, and iron losses), and the copper losses in the stator and in the separately excited field windings.

SELF-SYNCHRONOUS DEVICES

Self-synchronous devices are also called *selsyn* devices. These devices can be electrically interconnected to provide synchronized control between different points in a system. They are used in numerous industrial control and indicating systems. Some applications for self-synchronous devices are: to transmit control

information between the bridge and engine room of a ship; to indicate at some distant point the position of a generator rheostat or a transformer tap connection; to indicate wind direction and velocity; to indicate the angular position of devices such as swing bridges, gates and valves; to control the roll height in steel rolling mills; and to maintain synchronism between several motors such as in large paper presses.

PRINCIPLE OF OPERATION

Self-synchronous devices are very similar in construction to small synchronous motors. The stator of a selsyn unit contains a three-phase winding similar to that of a synchronous motor or generator, and the rotor has two salient poles complete with slip rings. A simple selsyn indicator system will have at least two self-synchronous motors, one acting as a generator or transmitter, and the other as a motor or receiver.

The connections for a single self-synchronous system are shown in Fig. 18-15 on page 370. The rotor or primary circuits of the two units are connected to an external single-phase ac supply, and the stator or secondary circuit of the transmitter is connected to the stator circuit of the receiver. Note that the system requires only a single-phase ac excitation.

When the rotor or primary circuits are energized, a voltage is induced in the stator or secondary circuits by transformer action. The value of the voltage induced in each winding of the three-phase stator in both the transmitter and receiver will depend on the angular positions of their rotor. It is this induced stator voltage that causes the self-synchronous action in the system.

When the rotor of the transmitter is in the same relative position as the rotor of the receiver, the voltage induced in the two stators will be equal and opposite, and no current will flow between the stators. However, when the rotors are at different positions, the voltages

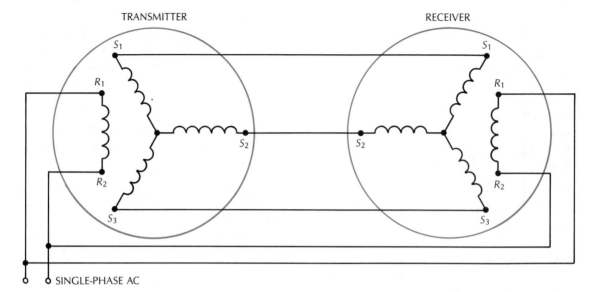

Figure 18-15 A simple self-synchronous circuit.

induced in the stators are no longer balanced and a current will now flow through the secondary circuits of both the transmitter and receiver. This current immediately sets up torque in both the transmitter and receiver. The torque in the transmitter acts to oppose any change to the transmitter rotor setting, while the torque in the receiver causes the receiver rotor to turn. The receiver rotor will turn to a new position where the two stator voltages are again balanced. Therefore, as the transmitter rotor is turned, either manually or mechanically, the receiver rotor follows in exactly the same direction and at the same speed.

SELSYN POWER DRIVES AND DIFFERENTIAL DEVICES

Two variations to the selsyn-indicating device described above are the self-synchronous motor power drives and the self-synchronous differential devices. Self-synchronous motor drives are used in applications where more than one motor must work together in synchronism, and differential selsyns are used to modify the electrical angle transmitted by a transmitter selsyn.

The *self-synchronous power drive motor* resembles a three-phase wound-rotor induction motor. The rotor has three windings usually connected in Wye, and brought out through three slip rings. Likewise, the stator has three windings connected in Wye.

In a power drive, the rotor circuits are excited from a three-phase supply, and the stator or secondary circuits are interconnected. The motors in the system must first be synchronized by applying a single-phase supply to the rotors, and then the three-phase power is applied. If the motors are not synchronized, the unbalanced induced stator voltage may cause the motors to start as induction motors. Once the motors are synchronized and then started, the motors in the drive system will all rotate at the same speed.

The *differential selsyn* is very similar in construction to the power drive selsyn motor described above. It resembles a miniature wound-rotor induction motor. A simple differential system will contain a selsyn transmitter and receiver and a differential selsyn, as shown in Fig. 18-16.

The electrical connection of the differential selsyn, however, is different from that of the self-synchronous power drive motor. In fact,

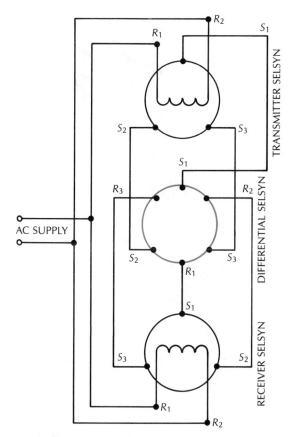

Figure 18-16 The differential selsyn connections.

selsyns is held fixed and a second is rotated, the third selsyn will rotate through the same angle. This direction of rotation may be reversed by simply interchanging any pair of leads to either the rotor or stator of the differential unit. If two of the selsyns are rotated simultaneously, the third will rotate through an angle equal to the algebraic sum of their movements. Therefore, the differential selsyn may be used to modify the electrical angle transmitted by the transmitter unit. The receiver selsyn can be made to respond to either the sum or the difference of the angular movements applied to the transmitter and the differential selsyns. This movement will depend simply on the physical direction of rotation of the two input rotors and also on the phase rotation of the windings.

SMALL SINGLE-PHASE SYNCHRONOUS MOTORS

The single-phase synchronous motor bears little, if any, resemblance to the three-phase synchronous motor. In fact, the single-phase synchronous motor falls into a special class of low wattage (10-25 watts) motors. One of the differences between the single-phase and three-phase synchronous motor is that, unlike the three-phase motor, the single-phase synchronous motor does not use any dc rotor excitation. As a result, the small synchronous motor is not an equal opposite to the ac generator as is the case for the three-phase synchronous motor. Because of its relatively small size and constant speed characteristics, the single-phase synchronous motor is widely used in various timing devices and recording instruments.

Like the three-phase motor, the stator of the small synchronous motor also produces a rotating magnetic field. And as stated before, the synchronous speed of this field equals:

$$\eta = \frac{4\pi f}{P}$$

where η = speed in rad/s
P = number of poles

the differential selsyn is really a static single-phase transformer with no three-phase voltages and currents present. Whereas the rotor windings from the transmitter and receiver selsyns are directly excited from a single-phase ac supply, the same as for a selsyn indicator system, there are no direct supply connections to the differential selsyn. Instead, the three-phase differential stator windings are connected to the transmitter stator windings, and the three-phase differential rotor windings are connected to the three-phase receiver stator windings. In the differential selsyn, unlike the transmitter and receiver selsyns, the stator acts as the primary and the rotor acts as the secondary.

The primary of the differential unit will have the same voltage distribution as the secondary of the excited transmitter. If one of the three

However, the stator of the small synchronous motor must usually produce its rotating field from single-phase current. To produce this rotating field effect from single-phase ac, most small synchronous motors are designed with shaded pole stators.

In a shaded pole stator, a section of each of the stator main poles is (shaded) encircled by a heavy shorting conductor ring. The flux induced in this shorting ring combines with the main pole flux to produce the rotating stator field. The method of producing a rotating magnetic field from single-phase ac current will be described in more detail in the chapter on single-phase ac motors.

There are two general types of small synchronous motors classified according to the type of rotor used: (a) the reluctance-type synchronous motor which uses a modification of the squirrel-cage rotor, and (b) the hysteresis synchronous motor which uses a rotor made of a permanent magnetic alloy. Because of their different operating characteristics each type will be discussed separately.

THE RELUCTANCE SYNCHRONOUS MOTOR

The reluctance synchronous motor uses a special type of magnetic squirrel-cage rotor, as shown in Fig. 18-17. The rotor contains equally spaced areas of high reluctance. This may be done by designing notches in the rotor periphery. The number of notches will correspond to the number of stator poles. The sections between the notches on the rotor will become salient rotor poles because of their relatively low reluctance to the stator flux.

The reluctance motor starts and accelerates like a regular squirrel-cage motor, but as it approaches the synchronous speed of the field, a critical point is reached where the rotor will snap into synchronism. This pull-in point is attained when the rotor slip becomes low enough that the stator flux can now sufficiently magnetize the low reluctance rotor sections. The rotor poles thus formed are now

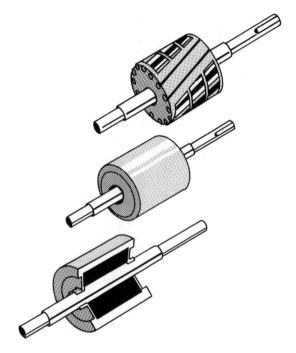

Figure 18-17 Typical reluctance (above) and hysteresis (below) synchronous motor rotors.
Courtesy: Bodine Electric Co.

attracted to the stator poles.

This motor will adjust its torque angle for a change in load in a similar way to that described for the three-phase synchronous

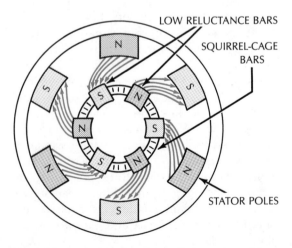

LOW RELUCTANCE BARS

SQUIRREL-CAGE BARS

STATOR POLES

Figure 18-18 Attraction between stator poles and the magnetized low reluctance poles.

motor. If the applied load at starting is too great, the motor may not pull in to synchronous speed, or if the motor is already running, it may pull out of synchronism. It will then operate in the squirrel-cage mode and its operation may even become rough and nonuniform.

THE HYSTERESIS SYNCHRONOUS MOTOR

A good example of the hysteresis synchronous motor is the General Electric or Warren Clock Motor. This motor has a shaded pole stator design similar to the reluctance motor. However, the differences in its performance to the reluctance motor are associated with its rotor design. The rotor of this motor is made of two or more hardened cobalt steel disks with soft-iron cross bars, as shown in Fig. 18-19. These disks are pressed on to the rotor shaft. Another type of rotor design used is a hardened cast cobalt steel cylinder securely mounted to the shaft with a nonmagnetic support. The cobalt steel material has good magnetic retentivity and is highly permeable to a magnetic field.

When the stator is energized with single-phase ac, the rotating stator field induces a current in the rotor cross bars and the motor starts as an induction motor. However, at the same time as the motor develops speed, a comparatively large hysteresis loss occurs in

COBALT STEEL DISKS PRESSED ON SHAFT

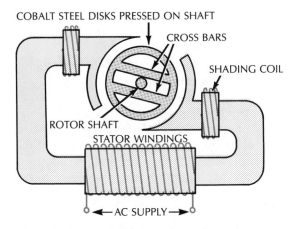

Figure 18-19 A typical hysteresis synchronous motor.

the disk rotor, creating a larger torque than is available from a conventional motor. Therefore, this motor also starts on the hysteresis principle. The motor will now accelerate at a relatively constant rate towards synchronous speed.

The hysteresis effect is produced as follows: when a magnetic material is placed in a magnetic field, the electrons of the material's atoms align themselves with the field flux and, thus, the material itself becomes magnetized. If the material or the field is rotated, then the electrons of the material must also turn in such a way as to maintain their alignment with the external field. This change in position of the electrons with respect to the position of the material requires energy and this energy is called hysteresis energy.

Because the hysteresis loss in the hardened cobalt steel rotor is quite high, a considerable amount of energy is expended in this rotor. As a result, a large amount of operating torque is produced.

Instead of the permanently fixed poles found in the reluctance rotor, the hysteresis rotor poles are "induced" by the rotating stator fields. During the acceleration period, the stator field will be rotating faster than the rotor, and the poles which this field induces in the rotor bars will shift around the rotor's periphery. When the rotor's speed reaches the synchronous speed of the field, the rotor poles will take up fixed positions, and at this point the rotor will lock into synchronism with the rotating stator field.

Hysteresis synchronous motors, with their more gradual acceleration characteristics, can pull into synchronism any load that is within their capacity to start and accelerate. This type of motor will maintain a constant speed very effectively, even with erratic load changes. Because of these operating properties, hysteresis motors are well known for their use in high-fidelity tape recorders and phono turntables. They are also widely used in clock motors and numerous timing devices.

SUMMARY OF IMPORTANT POINTS

Construction

1. A synchronous motor is the exact opposite of the ac generator.
2. It has the same stator structure as the induction motor which produces a rotating field:
   $S = 60 \times f/\text{pole pairs or } (\eta = 4\pi/P)$
3. The rotor has a dc field winding and a damper or cage-starter winding. Therefore, the motor starts on the induction principle.

Operation

4. The dc excitation does not produce any starting torque. It produces rotor poles which lock in with the stator poles. An over-excited field will change the motor's power factor from lagging to leading. This property is useful for system power factor correction.
5. A load change displaces the motor's counter voltage by a backward shift of the rotor poles with respect to the stator poles. This displacement is the torque angle.
6. Its name plate data is similar to that for the alternator except output power is given in kW instead of kVA, and also a power factor rating is included.
7. A load beyond the pull-out torque will cause the motor to pull out of synchronism.

Self-synchronous devices

8. Self-synchronous devices are also called selsyn devices. The stator contains a three-phase winding and the rotor has two salient poles.
9. Between a transmitter and a receiver selsyn, the stators are interconnected; whereas, the rotors are supplied with single-phase ac. When the rotors are unaligned, unequal induced stator voltages are set up and the receiver rotor turns towards alignment.
10. Other selsyn systems are the power drive and differential system.

Single-phase synchronous motors

11. These small synchronous motors require no rotor dc excitation. Their stators use a shaded pole design to produce a rotating field.
12. Since the reluctance motor uses a special type of magnetic cage rotor, this motor starts on the induction principle.
13. The hysteresis motor has a rotor made of several hardened steel disks with cross bars. The Warren or G.E. clock motor is an example of this motor.
14. The hysteresis loss in the rotor is responsible for the motor's torque. Near-rated speed fixed rotor poles are induced and the rotor will pull into synchronism.

ESSAY AND PROBLEM QUESTIONS

1. List the factors which determine the speed of a synchronous motor.
2. Briefly describe the construction of the stator and the rotor of a three-phase synchronous motor.
3. Describe the operation of a three-phase synchronous motor from start to run.
4. Compare the operation of a synchronous motor to that of a squirrel-cage induction motor.
5. What is an amortisseur winding and what are its functions in a synchronous motor?
6. Describe how a synchronous motor adjusts its electrical input for an increase in mechanical load.
7. What is the purpose of the field discharge resistor used in the dc rotor field?
8. A three-phase 60-Hz synchronous motor with 12 poles is used to drive an ac generator with 20 poles. Determine (a) the synchronous speed of the motor and (b) the frequency of the ac generator's output.
9. List two methods for starting a three-phase synchronous motor.
10. List three methods for exciting the dc field of a synchronous motor.
11. Describe the following: (a) torque angle,

(b) pull-in torque, and (c) pull-out torque.

12. With the use of a phasor diagram, show the effect on motor power factor for an increase in load and a constant dc rotor excitation.

13. Describe what is meant by hunting and how this effect is reduced in a synchronous motor.

14. Draw a V-curve to show the effect of motor power factor on stator line current.

15. (a) How can the rotation of a synchronous motor be reversed?
    (b) When is a synchronous motor normally excited?

16. With the aid of phasor diagrams, describe the effect on motor power factor when the rotor field is (a) under-excited and (b) over-excited.

17. List five applications for the synchronous motor.

18. An industrial system has two, three-phase 208-V motors. One motor is a squirrel-cage induction motor which requires 40 A at a 0.75 lag power factor. The second motor is a synchronous motor which requires 30 A at 0.8 lag power factor.
    (a) Determine the following for the squirrel-cage motor:
        (i) the true power
        (ii) the reactive power
        (iii) the apparent power
        (iv) sketch the phasor diagram
    (b) Determine the following for the synchronous motor:
        (i) the true power
        (ii) the reactive power
        (iii) the apparent power
        (iv) sketch the phasor diagram
    (c) Determine the following for the industrial system:
        (i) the total true power
        (ii) the total reactive power
        (iii) the total apparent power
        (iv) the system's power factor
        (v) the total line current
        (vi) sketch the phasor diagram

19. What precaution must be taken when operating a synchronous motor at a more leading power factor than its rated power factor?

20. Why is a low system power factor undesirable?

21. An industrial load of 800 kVA at a power factor of 0.7 lag is supplied from a three-phase, 2400-V feeder line. A synchronous motor operating at 300 kVA and 0.85 lead power factor is installed to carry an additional system load. Determine the following:
    (a) the total system load in kW
    (b) the total system reactive power in kvars
    (c) the total system apparent power in kVA
    (d) the new system power factor

22. Compare the name plate data of a synchronous motor with that of a three-phase ac generator.

23. What are some of the uses of self-synchronous devices?

24. Illustrate the electrical connections of a simple indicator selsyn system and then describe its operation.

25. Compare a power drive motor from a power drive synchronous system with the receiver selsyn in a synchronous indicator system.

26. (a) Draw a synchronous circuit to include a transmitter, receiver, and a differential selsyn.
    (b) Describe the operation of the differential selsyn.

27. Describe how the stator in a small single-phase synchronous motor produces its rotating magnetic field.

28. Briefly describe the rotor construction in
    (a) a reluctance synchronous motor
    (b) a hysteresis synchronous motor

29. (a) What is a hysteresis motor?
    (b) List some applications for this motor.

30. Compare the operation of the reluctance synchronous motor with that of the Warren clock motor.

CHAPTER
NINETEEN _____

SINGLE-PHASE AC MOTORS

The importance of the study of single-phase motors is evident from their numerous industrial and domestic applications. Single-phase motors form a sizeable part of the total number of electric motors manufactured in this country. These motors are generally referred to as low-power (fractional horsepower) motors. They usually have power ratings of less than 600 W, although motors with power ratings as high as 2 kW are also available.

Single-phase motors were one of the first types developed for use on alternating voltage and current. Today, they are manufactured in a large number of types to suit various applications. These motors may be classified into three general categories: (a) induction motors, (b) commutator motors, and (c) synchronous motors.

Single-phase induction motors are further classified according to the method used for starting them. The three general types of induction motors are: the split-phase, the repulsion-induction, and the shaded-pole induction motors. Commutator motors are so-called because they use commutator and brush assemblies. The two general types of

commutator motors are the repulsion and the series motors. Synchronous motors turn at synchronous speed. These motors were described in the previous chapter.

THE SINGLE-PHASE INDUCTION MOTOR PRINCIPLE

The basic principle of operation of the single-phase induction motor is similar to that of the three-phase induction motor. Like the three-phase motor, the single-phase induction motor will have a rotating stator field and a cage rotor. The rotor will turn as its induced poles are attracted to the rotating stator field. The main difference is, however, that the single-phase induction motor must produce its rotating stator field from a single-phase voltage supply.

It was shown in the previous chapter that if a three-phase induction motor was single-phased while it was operating, for example, by disconnecting one of the three line wires to the motor, the motor would continue to run as a single-phase motor. If the motor was at rest, it would not start on a single phase. It would not start because a single-phase voltage does not produce a rotating magnetic field, as does a three-phase voltage. A single-phase voltage will produce only a pulsating field.

The cage rotor of a single-phase induction motor is very similar to that of a three-phase induction motor. When this cage rotor is placed in a single-phase magnetic field, currents will be induced in the rotor bars by transformer action. These currents will set up rotor poles that will be in the same direct line with the stator poles. Since these two sets of poles pulsate but are also in alignment, the cage rotor might vibrate violently but would have no tendency to rotate. Therefore, the single-phase induction motor will not start under a single-phase pulsating stator field and some special starting means must be employed.

PRODUCING A ROTATING FIELD FROM TWO 90° OUT-OF-PHASE FIELDS

It was demonstrated in the previous chapter that a rotating stator field will provide the necessary starting and accelerating torque for the cage rotor. Several methods of producing a rotating field from a single-phase supply have been developed, and single-phase induction motors are classified according to these starting methods.

One method of producing a rotating magnetic field from a single-phase supply is by using a second set of windings on the stator core. This winding, called the *start winding*, is connected in parallel with the main stator windings, but it is displaced 90 electrical degrees from it. If the two windings are made the same, then when energized from a single-

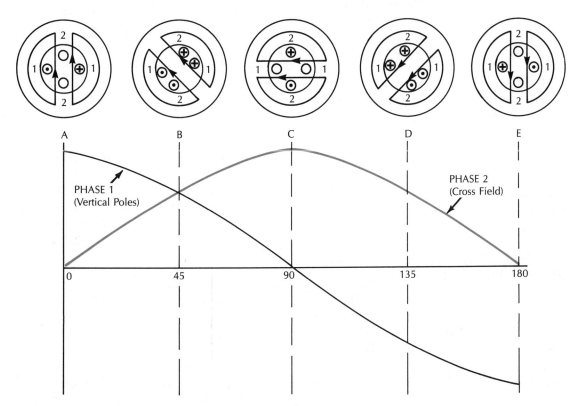

Figure 19-1 Producing a rotating field from two 90° out-of-phase fields.

phase voltage, the two fields will act in unison and the combined field will alternate but will not revolve. However, if one of the windings is designed to have a high impedance compared to the other, the two coil currents and, hence, their magnetic fields can be made to be as much as 90° out-of-phase. This is called *phase splitting* and the motor is called a split-phase motor. The two out-of-phase fields would combine to produce a rotating magnetic field as described below.

Figure 19-1 shows two 90° out-of-phase pulsating fields. Phase 1 is produced by the vertical stator poles, and phase 2 by the horizontal poles. The result for only one-half cycle is given and the second half of the cycle is left as an exercise for the student. The diagram shows the combined field effect at 45° increments.

At position "A," only phase 1 is producing flux, and the net flux direction will be in a vertical direction, as shown. At instant "B," 45° later, both phases are producing flux and the net flux direction will have also rotated 45°. At position "C," the maximum flux is now in a horizontal direction because only phase 2 is producing flux. At instant "D," the flux from phase 1 is building up again but in a new direction, while that from phase 2 is now decreasing. Therefore, the net flux at this instant will be as shown. At position "E," the maximum flux is just the opposite to what it was at instant "A." It should now be evident that the two out-of-phase fields are combining to produce a net rotating field effect.

THE EFFECT OF THE GENERATED ROTOR VOLTAGE

Once the cage rotor starts to rotate, there will be an additional voltage induced in the rotor. The first voltage, described previously, was caused by transformer action. The second voltage is induced by generator action, as the cage rotor cuts through the stator flux. This generated voltage will vary in-phase with the stator current and flux. However, because the

rotor has a low resistance and a high inductance, the resulting rotor current will lag the induced voltage by almost 90°, as shown in Fig. 19-2.

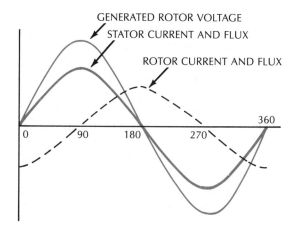

Figure 19-2 Phase relations of the stator current and flux, the generated voltage, and the rotor current and flux.

The field produced by this generated rotor voltage will thus lag the stator field by about 90°. This field will be similar to the field produced by the auxiliary stator winding and is called a *cross field*. Because the cross field and the main stator field are 90° out-of-phase, they will combine, as described before, to produce a rotating magnetic field. Therefore, once the single-phase induction motor is started, the motor's operation can be maintained with only a single stator field. The rotation will continue because of this new rotating magnetic field.

Since the cross field is produced by generator action, it follows that the strength of this field will be proportional to the rotor speed. However, at rated speed, the maximum strength attained by the cross field will always be less than that of the stator field. This, as may be recalled, is due to the inherent rotor slip in an induction motor. This difference in field strength means that the resulting rotating field will be irregular and not as constant as that produced by a polyphase motor. Therefore, the torque produced by a single-phase induction motor is pulsating. This is the reason why these motors must often be set on spring or rubber mounts to reduce possible motor vibration and noise.

SPLIT-PHASE INDUCTION MOTORS

Split-phase motors are perhaps the most widely used, relatively constant-speed ac motors employed for operating domestic appliances and for a variety of industrial applications. The simple construction, low cost, coupled with good efficiency, fair starting torque, and relatively good output for a given frame size has made the split-phase induction motor today's general purpose motor.

The name *split-phase*, as described previously, is taken from the "phase-splitting" method used for starting these single-phase induction devices. There are three general types of split-phase motors; (a) the resistance-start, (b) the capacitor-start, and (c) the permanent capacitor motors. These three types of split-phase motors will be described later in separate sections.

The stator of a split-phase motor is equipped with a main winding and an auxiliary start winding connected in parallel. The two windings are wound on the same stator slots, but displaced 90 electrical degrees in space. The stator produces a rotating field, and the synchronous speed for this rotating magnetic field is determined in the same way as that for the three-phase induction motor. The direction of this rotating field and thus the rotation of the motor may be reversed by simply interchanging the leads to either one of the two stator coils.

The rotor of the split-phase motor has the same cage design as that of the three-phase induction motor. The student may recall that the rotor bars in most cage rotors are skewed, and that the size and shape of the bars have a

demonstrable effect on the motor's speed-torque characteristics.

Split-phase motors may also be designed for dual voltage 115-V or 230-V operation. In a dual voltage motor, the main stator coil has two sections and their four leads are labelled T_1, T_2, T_3, and T_4. The start winding is usually a single 115-V coil and its leads are labelled T_5 and T_6. For 115-V operation, the two sections of the main coil are connected in parallel with T_1 connected to T_3, and T_2 connected to T_4. For 230-V operation, the main coil sections are connected in-series with the lead T_2 connected to T_3. The connections are shown in Fig. 19-3.

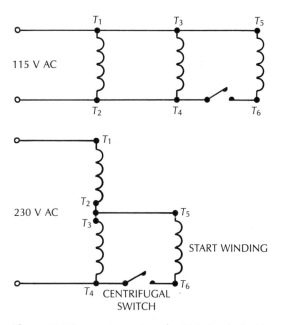

Figure 19-3 Stator connections for 115-V and 230-V operation.

THE RESISTANCE-START SPLIT-PHASE MOTOR

The resistance-start motor is commonly called a split-phase motor even though there are two other types of split-phase motors. In this motor, the start winding is designed to have a higher resistance and a lower reactance than the main stator winding. This is done by constructing the auxiliary winding with a smaller conductor and with fewer turns than the main winding. Also, the main winding is surrounded by more iron by placing it deeper into the stator slots.

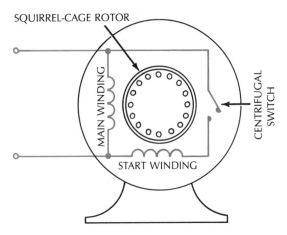

Figure 19-4(a) The resistance-start split-phase motor.

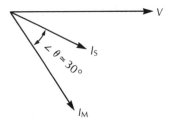

Figure 19-4(b) Phase relations of the stator's two coil currents.

It should be obvious that the stator current would "split," as shown in Fig. 19-4. The current component, I_S, through the high-resistance start winding will be more nearly in phase with the supply voltage, V, than the current component, I_M, through the main stator winding. Therefore, these two currents will differ in time phase and, as explained previously, their magnetic fields will combine to produce a rotating magnetic field.

Because the high-resistance start winding

will have some inductance and the main winding some resistance, the phase angle, $\underline{/\theta}$, between the two coil currents will be small. This angle is only about 30°, as shown in Fig. 19-4, and as a result, the rotating field produced is weak. Also, because of the difference in the coil's impedance, the magnitude of the two currents will be unequal. Therefore, the starting torque produced by a resistance-start motor will be both weak and nonuniform.

The split-phase motor is also fitted with some means of disconnecting the auxiliary winding from the supply after the motor has attained about 70 to 80 percent of synchronous speed. It was previously explained that once this motor is started, its auxiliary winding may be disconnected, and the motor will continue to operate with only its main run winding. The start winding is disconnected from the supply, after the motor has been started, to protect it against overheating and burning out. The high resistance in this auxiliary winding will produce an excessive amount of heat if the winding is left in the circuit for too long a period.

Figure 19-5 Single-phase cage rotor with centrifugal switch.

Courtesy: Franklin Electric.

The *centrifugal switch* mechanism, shown in Fig. 19-5, is the method most commonly used for disconnecting the start winding from the voltage supply. The switch is normally closed, and contains a stationary and a rotating part. It is mounted on the inside of the motor. When the motor reaches a predetermined speed, the switch is activated by centrifugal force action.

The *electromagnetic relay* is another device used for disconnecting the auxiliary winding after the motor has been started. It is commonly used where the motor is sealed, such as in a refrigerator. The switch is normally open and is mounted on the outside of the motor. At the instant of starting, the motor will probably draw five to ten times its normal current, with the largest component flowing through the run winding. This high-current component is used to energize the relay and the contacts will close, connecting the start winding into the circuit. When the motor reaches the proper speed, the current in the main winding will have decreased and the relay is de-energized, opening the contacts to the start winding.

The operation of the split-phase motor is dependent on the proper operation of these switching devices. If the start winding is kept connected to the supply voltage for too long a time, the winding will overheat and burn. On the other hand if, at the instant of starting, the auxiliary winding is held open for any reason, the motor will fail to start. If the motor does not start but simply gives a humming sound, then this indicates that one of the windings is open. The operation of the centrifugal switch or the electromagnetic relay can be detected by a clicking sound coming from the device soon after the motor is started or just before the motor comes to a stop.

The split-phase motor has very good speed regulation with a full-load slip of about 4 to 6 percent. However, as explained, the starting torque for the resistance-start motor is small and its starting current is relatively high. For these reasons, the resistance motor is most commonly used only in sizes ranging from approximately 25 W to 350 W (0.03 to 0.5 hp). This motor is widely used in applications with easily started loads. Some common applica-

tions are for driving fans, grinders, washing machines, and woodworking tools.

THE CAPACITOR-START SPLIT-PHASE MOTOR

The capacitor-start (CS) induction motor is another type of split-phase motor. It is practically the same as the resistance-start split-phase motor, except that a capacitor is connected in series with the start windings. The capacitor is usually mounted on top of the motor's frame, although it may be mounted at any convenient location.

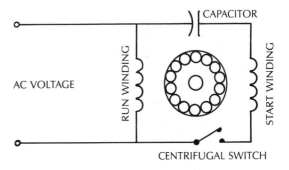

Figure 19-6(a) The schematic diagram of a capacitor-start motor.

Once the capacitor-start motor has accelerated to about 75 percent of synchronous speed, the auxiliary start winding with its series-connected capacitor is disconnected from the circuit by the centrifugal switch. The motor will continue to operate but with only its single-phase main stator field. Therefore, the capacitor is used only to modify the starting characteristics of the motor, and its operating properties will remain the same as that of the resistance-start motor.

As shown in Fig. 19-6(b), with the proper size of capacitor, the auxiliary coil current can be made to lead the main coil current by 90 electrical degrees. Since the starting torque of the motor is directly proportional to the sine of the angle between the stator two coil currents, it follows then that the starting torque of the

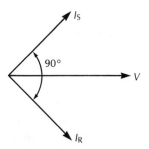

Figure 19-6(b) Phase relation of the stator two coil currents.

capacitor-start motor could be about two and a half times greater than that of the split-phase motor with only a 30 electrical degrees phase displacement. It should also be obvious from the phasor diagram that the starting power factor will be improved, and thus the motor's starting surge current will be reduced.

Because of the improved starting characteristics, the capacitor-start motor is slowly replacing the resistance-start motor, especially in applications involving hard-starting loads. The motor is commonly used in refrigerator compressors and various machine tool systems. It is usually manufactured in sizes from about 100 to 600 W, but larger sizes are also available.

The capacitor-start motor is normally fitted with an electrolytic-type capacitor. This type of ac, nonpolarized capacitor consists of two sheets of aluminum foil separated by an electrolyte and is designed for only intermittent service. Therefore, excessive start-and-stop operation of the motor or prolonged connection of the capacitor to the supply will probably cause the capacitor to fail. If the capacitor is short-circuited, the start-winding current will increase and the winding may burn out.

THE CAPACITOR-START, CAPACITOR-RUN MOTOR

The capacitor-start, capacitor-run motor is similar to the capacitor-start motor except that the motor is designed to operate with the auxiliary winding and its series capacitor per-

manently connected to the supply voltage. This type of motor is also called a *permanent split-phase capacitor motor* (PSC) or simply a *capacitor motor*. The capacitance in series with the auxiliary winding may have one fixed value or it may have one value for starting and another for running. These two types of permanent capacitor motors are described below.

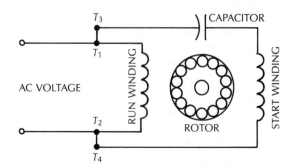

Figure 19-7 A permanent capacitor motor using a single capacitor.

The capacitor-motor shown in Fig. 19-7 uses a single fixed capacitance for both starting and running. The motor will have no centrifugal switch because the start and run windings are energized at all times. Also, because of this permanent connection of the two stator windings, the rotating stator field will closely resemble that of a two-phase voltage supply. Therefore, the field will be more constant and as a result, the motor will have a greater efficiency and will operate more quietly and smoother than either the resistance-start or capacitor-start split-phase motors.

The type of capacitor used on these motors is the oil-type rather than the electrolytic-type as used on capacitor-start motors. Unlike the electrolytic capacitor, the oil-type capacitor is rated for continuous duty. It consists of aluminum sheets separated by an oil-impregnated paper dielectic. This type of capacitor is larger in physical size and is more costly than the electrolytic type capacitor.

Another version of the permanent capacitor motor is the "two capacitor" motor, shown in

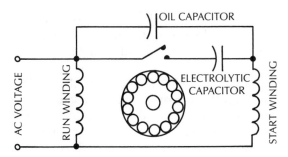

Figure 19-8 A permanent capacitor motor using two capacitors.

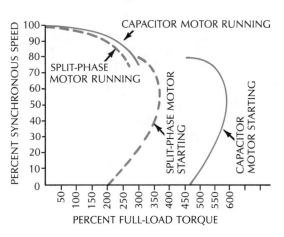

Figure 19-9 Typical speed-torque curves for split-phase motors.

Fig. 19-8. This motor uses an electrolytic capacitor in addition to the oil-type capacitor. The two capacitors are connected in parallel and in series with the auxiliary winding. When the motor reaches a predetermined speed, the electrolytic capacitor is disconnected by a centrifugal or relay switch and the motor will now run with only the oil-type capacitor in its circuit.

The use of the two capacitors produces an improvement of the motor-starting characteristics, while at the same time it preserves the efficiency and quiet operation of the permanent capacitor motor. Using a capacitance above the rated value will normally increase the starting torque, but at the expense of the motor's running performance. This is because the rotor speed affects both the magnitude and time phase of the current in the capacitive winding. The current in the auxiliary capacitive winding is lowest at the instant of starting and highest at rated rotor speed. Therefore, the motor's starting and running characteristics may be optimized by using a higher than rated capacitance when the motor is started and "rated" capacitance when the motor is running.

Figure 19-9 shows the typical speed-torque curves for a capacitor start (CS) motor and a permanent capacitor (PSC) motor. Since the oil-type capacitor has a quite low capacitance (approximately 15 μf) compared to the ones used on the capacitor-start motors, the starting torque of the single capacitor PSC motor is

relatively low, about 100 percent of rated torque. However, as explained, when the second capacitor is added, the starting torque of the PSC motor will be much greater.

Although the PSC motor is generally more expensive, it will have a better efficiency, a higher starting torque, and a quieter operation than the CS or the split-phase motor. It will also have a better operating power factor and will cause less line voltage disturbance when it is started. Another advantage of the PSC motor is that, unlike the CS or split-phase motor which may overheat under frequent start/stop operation, this motor has a better capability for this kind of application. Permanent split-phase capacitor motors are used in such systems as oil burners, air conditioners, fans, etc.

THE SHADED-POLE MOTOR

The shaded-pole motor is another type of single-phase ac motor using a rotating stator field. However, unlike the split-phase stator design, the rotating field of this motor is produced by using a shading coil in each of the stator poles. The small single-phase synchronous motor described in the previous chapter was also shown to use a shaded-pole stator construction.

A shading coil is a low-resistance short-cir-

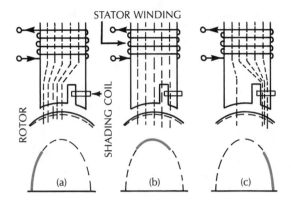

Figure 19-10 The rotating field effect in a shaded-pole stator.

cuited copper loop which is placed around a part of each stator pole. Figure 19-10 shows one pole of a motor fitted with a shading coil. Note that the shading loop is not connected to the supply, and that only the main stator winding is directly energized. However, the changing flux produced by the main winding will induce a current flow in the shaded ring by transformer action.

Figure 19-10 illustrates how a rotating field effect is set up by a shaded-pole stator. The effect for one-half cycle of voltage on a single-stator pole only is shown. In Fig. 19-10(a), since the current in the main winding is increasing, a current will be induced in the shading coil. This induced current then sets up a flux which opposes the changing flux that created it (Lenz's law). Therefore, as shown, the greater part of the flux will pass through the left section of the pole during this part of the ac voltage cycle.

In Fig. 19-10(b), the rate of change of the current is very small across the top of the sine wave. Thus, there will be no induced shaded coil flux during this part of the ac cycle and, as a result, the main field flux is now evenly distributed over the entire pole face, as shown.

In part (c), the ac voltage is now decreasing rapidly toward zero, and this will again cause an induced current to flow in the shading coil.

Since the shading-coil flux tends to oppose the decrease (change) of the main flux, the flux will now be concentrated over the shaded-pole area, as shown.

For the negative half of the ac voltage cycle (not shown in the diagram), the polarity of the pole is reversed, but the flux will again shift across the pole face in the same direction. Therefore, the effect of the shading ring causes a rotating field effect, and the direction of this rotating field will be from the nonshaded section to the shaded section of a pole face.

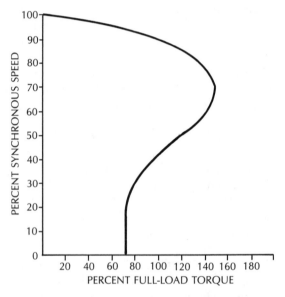

Figure 19-11 Typical characteristic curve for a 5-W shaded-pole motor.

The field produced by a shaded-pole stator is relatively weak, so that this type of motor will have very little starting torque and low efficiency. However, the motor has no centrifugal switch and is very rugged and reliable. The slip of this motor will also increase as the voltage is decreased and this property is often used in controlling the speed of the motor. The shaded-pole motor is generally manufactured only in small sizes ranging from about 5 to 40 W, and it is commonly used in devices such as small fans and blowers.

UNIVERSAL MOTORS

It may be recalled that the direction of rotation of a dc shunt or series motor is unaffected when the polarity of the supply voltage is reversed. It is, therefore, reasonable to expect that either motor will also operate on single-phased ac voltage. However, the operation of a dc shunt motor on an ac supply is impractical because the high inductance of the shunt field will cause the field current to lag the armature current by almost 90°. As a result, the shunt motor will develop very little torque if it is operated on ac voltage. The series motor, on the other hand, with some changes to reduce the effect of hysteresis, eddy currents, and series-field impedance, will operate satisfactorily on either dc or single-phase ac voltage. This type of series motor which may be operated on either dc or ac voltage is called a *universal motor*.

Unlike the shunt motor, the current through the series field and the armature of a series motor is the same, and thus the series-field flux and armature flux will be in phase. Hence, the torque developed by a series motor can be about the same for either dc or an equivalent amount of ac voltage. However, when a dc series motor is operated on ac voltage, its series-field impedance is high, and thus the current and the resulting torque will be small. The field reactance of the motor will also cause sparking and low motor power factor.

In the series motor, the adverse effects caused by a high field impedance, iron losses, eddy currents, and hysteresis are minimized by constructing the motor with few field turns and with a low reluctance magnetic circuit. The field is well laminated and is operated at low flux densities. The number of armature conductors and commutator segments is also increased.

The effect of armature reaction is reduced by the use of compensating windings placed in the stator pole faces. The compensating winding may be connected in series with the

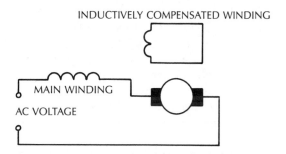

Figure 19-12(a) A conductively compensated universal motor.

Figure 19-12(b) An inductively compensated universal motor.

field and armature, in which case the motor is said to be *conductively compensated*, or the motor may be *inductively compensated* by short-circuiting the compensating winding on itself. These two connections are shown in Fig. 19-12. Since the inductive compensation works by transformer action, it is obvious that the inductively compensated motor is designed for only ac voltage operation.

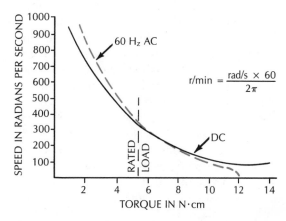

Figure 19-13 Speed-torque characteristics of a low-power universal motor.

The universal motor or the ac series motor has approximately the same operating characteristics as the dc series motor. The starting torque of the motor is high and the speed may increase to an excessively high value at light loads. For this reason, these motors are usually permanently connected to the devices being driven. The speed of the series motor may also be regulated by adding resistances in series with the motor. A common application for large ac series motors is in traction devices. Universal or ac-dc series motors are usually designed with smaller power ratings and are commonly used in devices such as sewing machines, vacuum cleaners, projectors, business machines, and various portable power tools.

The direction of rotation of any series motor can be reversed by reversing the current flow through either the armature or the field circuit. Universal motors, however, are very sensitive to brush position. These motors are usually wound for operation in only one direction, and reversing the rotation may cause severe sparking at the brushes.

THE REPULSION MOTOR

The repulsion motor is another type of commutator motor. Like the series motor, it has a commutator and brush assembly. The repulsion motor, however, operates on the principle of a repulsion torque developed between the stator and rotor poles. The stator winding is connected directly to the single-phase ac voltage and the motor winding is similar to that of a dc motor. The brushes, however, are not connected to the voltage, but instead are shorted to each other. Hence, the current in the armature is induced by transformer action. There are three-types of repulsion motors: (a) the repulsion motor, (b) the repulsion-start, induction-run motor, and (c) the repulsion-induction motor.

The positioning of the brushes is critical to the operation of repulsion motors. If the

brushes are placed so that the line connecting them is along the stator pole axis, then the net rotor torque produced will be zero. When the brush axis is the same as the pole axis, the

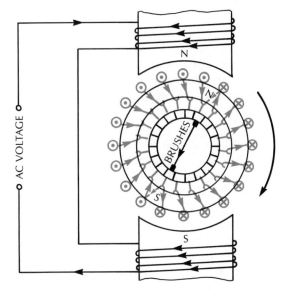

(a) Brush position for clockwise rotation

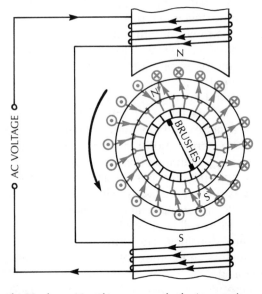

(b) Brush position for counter-clockwise rotation

Figure 19-14 Simplified schematic of a repulsion motor.

poles induced in the rotor are not only of the same polarity as adjacent stator poles, but they are also in alignment with the stator poles. Hence, the repulsion torque in either direction is equal and cancels out each other.

In repulsion motors, the brushes are placed so that their axes make an angle of about 20° with the neutral axis of the field poles. The position of the brushes is shown in Fig. 19-14. Since the induced rotor poles are set up in the vicinity of the brushes, their axes will be displaced from the stator poles. Therefore, because the induced rotor poles have the same polarity as the adjacent field poles and are also slightly displaced from them, the rotor poles will be repelled, causing the armature to rotate. The motor derives its name from this action.

It should be evident now from Fig. 19-14 that the direction of rotation of the repulsion motor will be in the same direction that the brushes are shifted, relative to the neutral field axis. Therefore, when the brushes are shifted in a clockwise direction, as in Fig. 19-14(a), the motor will rotate in a clockwise direction, and when the brushes are shifted in a counter-clockwise direction, as shown in Fig. 19-14(b), the motor will rotate in a counter-clockwise direction.

The repulsion motor has operating characteristics very similar to those of the series motor. It has a high starting torque and a high speed at light loads. The repulsion motor principle is mostly used for providing the starting torques for some types of induction motors, as described below.

THE REPULSION-START INDUCTION-RUN MOTOR

It was shown in a previous section that the single-phase cage induction motor is in itself not self-starting. One method described for starting this type of motor is the use of a start winding in the stator. The fields from the start and run windings combined to produce the required rotating field to start the single-phase cage motor. The repulsion principle is another method which may be used to start a single-phase induction motor.

The repulsion-start induction-run motor is basically a repulsion motor in construction. However, at a predetermined speed, the rotor winding is short-circuited to give the equivalent of a squirrel-cage winding. Therefore, this type of motor will start as a repulsion motor with its accompanying high torque, but runs as an induction motor with its constant-speed characteristics. After the commutator has been short-circuited, the brushes no longer carry any current and may be lifted from the commutator.

These motors are more complex than the cage-induction motors. Their efficiency and operating torque are also less than those of a cage-induction motor of comparative size. As a result they are not commonly used.

REPULSION-INDUCTION MOTORS

The repulsion-induction motor is another type of commutator motor. The rotor of this motor contains a squirrel-cage winding in addition to the repulsion winding. The motor has a commutator and brush assembly but no centrifugal or automatic-type switches. Hence, this motor will start on the repulsion principle and operate on the combined repulsion and induction principles.

In the repulsion-induction motor, the starting characteristics of the repulsion motor and the constant-speed characteristics of the induction motor are obtained. However, at no-load, the speed of the motor is slightly above synchronous speed. Also, at full-load speed and up to about the maximum running torque point, the motor torque is greater than the combined torque from the cage and commutated windings. The torque-speed characteristics are similar to those of a dc compound motor. Typical applications for this motor are in refrigerator compressors, lathes, and boring mills.

SUMMARY OF IMPORTANT POINTS

Induction motors

1. Three types of single-phase induction motors are (a) split-phase, (b) the shaded pole, and (c) repulsion-induction.

2. The split-phase motor has a cage rotor, and the stator has a high inductance main winding and a high resistance start winding. The two paralleled stator windings produces a rotating field from the single-phase ac voltage.

3. To bring the phase shift between the stator two-coil currents closer to 90°, a capacitor is connected in series with the start winding. This strengthens the rotating field and so improves the starting torque.

4. Reversing either one of the two stator coils will reverse the motor's rotation. The main coil is sometimes designed with two sections for dual voltage operation.

5. After startup, the start winding is disconnected from the supply voltage by a centrifugal switch mechanism. In some capacitor motors, the capacitor and start winding remains connected in the circuit.

6. The shaded-pole motor obtains its rotating stator field by using a shading coil in each of the stator poles. The field produced is weak, but no centrifugal switch is needed.

7. The shaded-pole motor rotates in the direction from the nonshaded pole section to the shaded section.

Commutator motors

8. The universal motor is an ac-dc type motor. It is very similar to the dc series motor.

9. The repulsion motor is another type of commutator motor. Torque is set up by the repulsion between the stator and rotor poles.

10. To develop the repulsion torque, the brushes must be displaced by about 20° from the pole axis.

11. Two types of repulsion motors are the repulsion-start induction run and the repulsion-induction motor. These motors will have the high starting torque of repulsion motors and the constant speed of induction motors.

12. To reverse rotation, the brushes must be moved 20° to the other side of the pole axis.

REVIEW QUESTIONS

1. Name the three general groups into which single-phase motors may be classified.

2. Name three types of single-phase induction motors.

3. Name two types of single-phase commutator motors.

4. Compare the basic construction of the three-phase induction motor with that of a single-phase induction motor.

5. Briefly explain why the basic single-phase induction motor is not self-starting.

6. Describe the origin of the term "split-phase" as used to describe the split-phase induction motor.

7. Briefly describe the relative positions and the impedance properties of the two stator windings in a single-phase induction motor.

8. Describe with the aid of diagrams how two 90° out-of-phase alternating fields will combine to produce a rotating field.

9. (a) Name the electrical property responsible for each of the two voltages set up in the rotor of an operating single-phase induction motor.

   (b) Which of the two rotor fields is produced only while the motor is rotating, and which one will combine with the main stator field to produce a rotating field effect?

10. Explain why the single-phase induction motor, once it is started, will continue to operate with only its main stator winding energized.
11. Why must single-phase induction motors often be mounted on spring or rubber mounts?
12. Name the three general types of split-phase induction motors.
13. (a) How may the rotation of a split-phase motor be reversed?
    (b) What is the synchronous speed of a six-pole, 240-V, 60-Hz split-phase motor?
14. Draw diagrams to illustrate the connections of a 120/240-V split-phase motor for (a) 120-V and (b) 240-V operation.
15. Why is the starting torque of the resistance-start split-phase motor (a) relatively weak? and (b) nonuniform?
16. (a) Briefly describe the function and operation of the centrifugal switch mechanism used in a split-phase induction motor.
    (b) What effect will an open centrifugal switch have on starting this motor?
    (c) What is another device which is sometimes used in place of the centrifugal switch mechanism?
17. (a) Explain why the starting torque of a capacitor-start motor is better than that of the resistance-start split-phase motor.
    (b) Name two other starting characteristics that are improved.
18. (a) Explain what may happen if for some reason the centrifugal switch in a capacitor-start motor fails to open.
    (b) Explain what effects excessive start-and-stop operation will probably have on a capacitor-start motor.
19. Compare (a) the construction and (b) the operating characteristics of the capacitor-start with the permanent capacitor motor.
20. Show the connections for (a) the resistance-start and (b) capacitor-start induction motor.
21. (a) Describe the capacitor types and connections of the two capacitors in a two-capacitor motor.
    (b) What are the advantages of using two capacitors?
22. List three applications for each of the three types of split-phase induction motors.
23. Briefly explain how the shaded-pole motor produces its rotating stator field.
24. (a) In what direction does the rotating field in a shaded-pole motor rotate?
    (b) How can the speed of this motor be controlled?
25. A four-pole, single-phase induction motor operates on 240 V, 60 Hz, and runs at 182.21 rad/s. Determine: (a) the synchronous speed and (b) the percent slip.
26. Briefly explain why it is impractical to operate a dc shunt motor on single-phase ac voltage.
27. Why is a low-power ac series motor called a universal motor?
28. What are the differences between the dc series motor and the ac series motor?
29. Briefly describe two methods used for reducing armature reaction in ac series motors.
30. List four application for universal motors.
31. Briefly describe the repulsion motor principle.
32. How can the rotation of a repulsion motor be reversed?
33. Explain the operation of a repulsion-start induction-run motor.
34. Explain the difference between a repulsion-start induction-run motor and a repulsion-induction motor.
35. Compare the operating characteristics of a repulsion motor with those of a repulsion induction motor.

CHAPTER
TWENTY

AC MEASURING INSTRUMENTS

The d'Arsonval moving-coil that was discussed in Chapter Six is basically a dc meter movement and, when used in an ac circuit, the meter will not give a proper reading. As the alternating current goes positive for the first half cycle, the moving coil would turn in one direction, then during the negative half cycle, the moving coil would turn in the opposite direction. Thus, when measuring ordinary 60-Hz line current, the meter movement would be unable to follow the reversing current fast enough (120 reversals per second), and would vibrate back and forth at the zero position. Since the positive and negative alternations cancel each other, the alternating current average position would be zero. For large currents, this vibration may damage the pointer.

To measure ac, the meter used must produce an up-scale pointer deflection, regardless of the polarity of the alternating current. This type of deflection may be accomplished by one of the following methods: (a) rectifier type, (b) electromagnetic type, and (c) thermal type.

RECTIFIER-TYPE METER

In the rectifier-type instrument, a dc meter movement is used to measure ac. A device called a rectifier changes the ac to dc before it is applied to the meter movement. The rectifier is usually a semiconductor device called a diode, as shown in Fig. 20-1.

(a) Junction diode rectifier

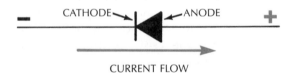

CURRENT FLOW

(b) Rectifier symbol

Figure 20-1 The junction diode.

The *junction diode* is made from semiconductor materials such as germanium (Ge) or silicon (Si). It offers a very high opposition to current flow in one direction, but a very low opposition to current flow in the opposite direction. The current always flows from the cathode end to the anode end, as shown in Fig. 20-1(b) (opposite the direction that the arrow is pointing). The cathode end is marked with a black band, as in Fig. 20-1(a). When an ac wave is applied to the rectifier either the positive alternation or negative alternation will be allowed to pass through, depending which way the rectifier is connected into the circuit, as shown in Fig. 20-2.

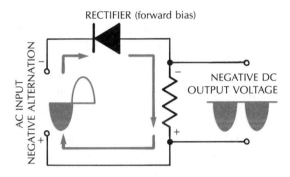

(a) Rectifier producing a negative output.

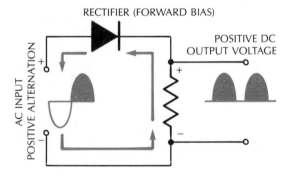

(b) Rectifier producing a positive output.

Figure 20-2 Rectifier changing ac to dc.

There are two basic rectifier circuits: the half-wave and the full-wave. In the *half-wave rectifier circuit*, one pulse of the ac input wave passes through the meter movement, while the opposite alternation is bypassed around the meter movement by the shunt rectifier, Fig. 20-3.

The current that flows through the meter movement, as shown, is a pulsating dc. These pulses are all of the same polarity, causing an up-scale deflection of the meter. If the frequency of the current being measured is lower

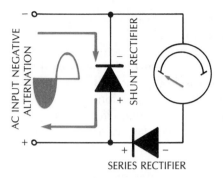

(a) Negative input pulse.

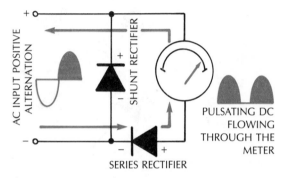

(b) Positive input pulse.

Figure 20-3 Half-wave rectifier meter.

than 10 Hz, the pointer will fluctuate as the pulsating dc rises and falls, making it difficult to read the meter. At the higher frequency currents, the pointer cannot move rapidly enough to return to zero between pulses; thus it will indicate an average value of the current pulses. The average current for one pulse of a sine wave is 0.637 of the peak value. Since the next alternation is bypassed by the rectifier, its average value is zero. Thus, the average value for a complete sine wave is the sum of both alternations divided by two, (0.637/2 = 0.318), which is 0.318 times the peak value. The meter will therefore show a reading that represents 0.318 of the peak value.

However, because the effective, or root-mean-square (rms) values are more useful, ac meters are calibrated for measuring rms values. A rectifier instrument correctly indicates rms values only for the waveshape for which it is calibrated.

In a *full-wave rectifier meter*, the rectifiers are connected so that both halves of the ac current must follow paths that flow through the meter in the same direction. This is accomplished by using a *bridge circuit*, using four rectifiers, as shown in Fig. 20-4.

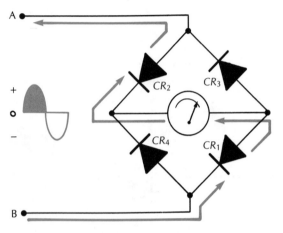

(a) Current flow on positive alternation.

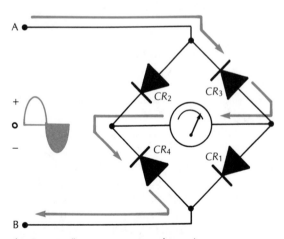

(b) Current flow on negative alternation.

Figure 20-4 Bridge rectifier meter circuit.

The four rectifiers are used in conjunction with one meter movement. During the positive alternation of the input sine wave (Fig. 20-4(a)), current flows from terminal B through

rectifier CR_1, through the meter movement, then through rectifier CR_2, and back to terminal A. During the negative alternation of the input sine wave (Fig. 20-4(b)), current flows from terminal A, through rectifier CR_3, through the meter movement, then through rectifier CR_4, and back to terminal B. Notice that even though the input current changes direction, the current flows through the meter in the same direction. The four rectifiers convert both halves of the input ac sine wave into a pulsating dc across the meter movement, and are said to be providing full-wave rectification. There will be two dc pulses through the meter for each ac input cycle being measured.

The average value of the current flow through the meter movement is twice as great in the full-wave rectifier arrangement than it was in the half-wave circuit, because both halves of the input ac sine wave flow through the meter movement. Thus, the average current may be calculated by multiplying 0.637 by the peak value. However, as stated before, the ac meter scale is calibrated in effective or rms values which are equal to 0.707 times the peak current being measured.

Example 20-1: If the peak value is 200 V, (a) what is the average value, (b) what is the rms value at this reading?

(a) average value = 0.637 × peak value
$$= 0.637 \times 200$$
$$= 127.4 \text{ V}$$

(b) rms value = 0.707 × peak value
$$= 0.707 \times 200$$
$$= 141.4 \text{ V}$$

The scale used in an ac rectifier-type meter is linear, that is, the values on the scale are equally spaced. The accuracy of most rectifier-type meters is ±5 percent.

ELECTROMAGNETIC-TYPE METERS

The electromagnetic-type meters are capable of responding directly to the ac input. Basically, when two iron bars are placed inside an electromagnetic coil, and the coil is energized, the two bars become magnetized with the same polarity; thus they repel each other. The magnetic polarity is maintained constant although the current reverses. This effect is utilized in moving-vane meters to measure electric current. These moving iron meters are commonly referred to as a radial-vane or concentric-vane-type meter. The moving-iron vane instruments are widely used for the measurement of alternating currents and voltages.

RADIAL-VANE METER MOVEMENT

Two soft iron vanes, one stationary and one securely attached to the central shaft of the moving element, with a pointer attached, are placed inside a coil, Fig. 20-5.

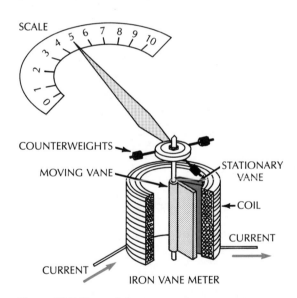

Figure 20-5 The radial-vane meter movement.

When current flows through the coil, an electromagnetic field is set up around the coil. The two iron (stationary and moving) vanes are magnetized, with the same polarity. Therefore, the vanes will repel each other. As the moveable vane is repelled by the stationary vane, the pointer moves across the scale. The amount of movement is dependent on the amount of current flow in the coil. A spiral spring provides a restraining force.

When the moving iron-vane meter is to be used as a voltmeter, the field coil is wound with many turns of fine wire. This will produce a strong field but with only a small current flow.

When used as an ammeter, the moving iron-vane field coil is wound with relatively few turns of wire heavy enough to carry the rated current. As in most ac meters, the scale is calibrated in effective values (rms), and its accuracy is about ±0.5 percent.

CONCENTRIC-VANE METERS

The differences between the radial-vane and concentric-vane meters are the shape of the vanes and their placement with respect to each other inside the coil. In the concentric-vane, shown in Fig. 20-6, the moveable vane is attached to the central shaft of the moving semicircular element, which is placed inside another semicircular vane that is stationary. The operation of the concentric-vane meter is similar to the radial-vane meter described earlier except that the outside vane has one tapered edge, which provides a nonuniform magnetic field between the two vanes. Hence, the resulting force will not be the same in both

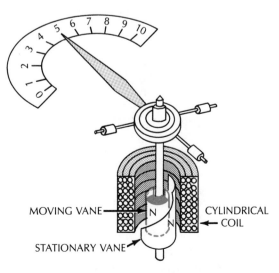

Figure 20-6 Concentric-vane meter movement.

vanes. The wider end of the outside vane conducts a greater number of lines of force than the narrower end. Thus, the wider end of the vane produces a stronger magnetic force than the narrow end. The result is a stronger repulsion force between the wide end of the fixed vane and moving vane than at the tapered end of the fixed vane. Therefore, the inner moving vane will rotate in the direction of the tapered end of the fixed vane causing the meter pointer to deflect up scale.

Moving iron-vane meters are primarily used to measure alternating current, although they may be used to measure direct current as well. For dc measurement, however, an error is introduced by the residual magnetism that is produced in the vanes by the direct current. When used on ac circuits, the iron-vane meters have an accuracy of ±5 percent. They cannot provide accurate readings over a wide range of frequencies, as the reactance of the coil increases when the frequency increases.

In general, iron-vane meters require more current to produce full-scale deflection than is required by the d'Arsonval moving-coil meter because of the reluctance of the magnetic circuit. Therefore, iron-vane meters are seldom used in low-power circuits.

Moving-iron vane meters have a scale that is nonlinear. The values on the meter's scale are not equally spaced; the numbers at the low end of the scale are crowded, and further apart at the high end of the scale. The deflection of the pointer in iron-vane meters increases with the square of the current. For example, if a current flow of 5 mA causes the pointer to deflect a distance of 2 cm on the meter's scale, then a current of 10 mA (doubled) will produce about 8 cm (four times) of a deflection.

When the current is doubled, this in turn doubles the strength of the magnetic field about each vane, causing the combined repulsion forces of the two vanes to become four times stronger. Thus, this meter will have a nonlinear or a square-law scale.

THERMOCOUPLE METERS

In Chapter One, you learned how thermoelectricity was produced. (Basically, the thermocouple meter utilizes the thermal and electromagnetic effects.) The thermocouple meter consists of a thermocouple and a moving-coil d'Arsonval meter movement, as shown in Fig. 20-7. The thermocouple generates the current that is required to drive the meter movement.

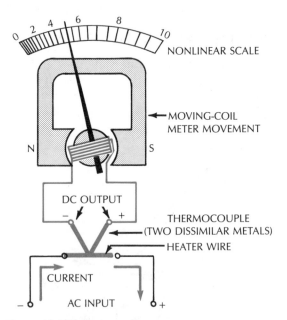

Figure 20-7 Thermocouple meter.

The thermocouple is made up of two dissimilar metal strips or wires joined together at one end. When their junction is heated, a potential difference is produced across the opposite ends of the thermocouple.

In this meter movement a short heater wire is attached to the junction of the thermocouple, as shown in Fig. 20-7. The junction is heated by a current flowing through the heater wire. The input current can be either ac or dc because the heating effect is not affected by the direction of current flow. The heat produced by the heater wire causes the thermocouple to generate a dc output voltage. The voltage causes a current to flow through the moving-coil meter movement. The pointer turns in proportion to the amount of heat generated by the heater element. Thus, a higher input current will produce more heat at the junction, resulting in a higher dc output voltage from the thermocouple; this in turn will produce a higher current flow through the meter movement, and a greater pointer deflection.

The amount of heat produced by the flow of current through the heater element is proportional to the square of the heating current ($P = I^2R$). Because the output dc voltage is proportional to the temperature, the amount of deflection in the moving-coil meter movement is proportional to the square of the current flowing through the heater element. Therefore, thermocouple meters, like iron-vane meters, have a *square-law* or nonlinear scale, Fig. 20-7.

Thermocouple meters usually have an accuracy of ±2 percent. The meter movement that is used with a thermocouple should have a low resistance to match the low resistance of the thermocouple. The moving coil must have a high sensitivity, since it must produce a deflection with a small generated voltage.

The thermocouple meter can measure alternating currents over a very wide range of frequencies, extending into the radio frequency range, several thousand megahertz. These meters may be used to measure direct current as well, if their scales are properly calibrated. This is because these meters will respond to the amount of heat that an ac or dc input voltage can produce.

ELECTRODYNAMOMETER MOVEMENT

The electrodynamometer movement utilizes the same basic operating principles as the permanent-magnet moving-coil meter movement, that is, the deflection of the moving coil is obtained from the interaction of two mag-

netic fields. In the electrodynamometer, the permanent magnets are replaced by two stationary coils, sometimes called the field coils, Fig. 20-8 and Fig. 20-9.

These stationary (current) coils are made up of a few turns of heavy wire and are connected in series, while the moving (voltage) coil consists of many turns of fine wire that are connected in parallel. The moving coil is similar to the type used in the moving-coil meter movement, and is mounted inside the two field coils. The moving coil is fastened to the central shaft so that it may rotate inside the two stationary coils.

When current flows through the stationary and moveable coils, two magnetic fields are set up and interact with each other, which cause the moving coil and its attached pointer to deflect up scale.

If the current flow through the dynamometer is reversed, the polarity of both the stationary and moveable coils are reversed; thus the deflection of the moving coil is always in the same direction, regardless of the direction of current low through the coils. That is why the

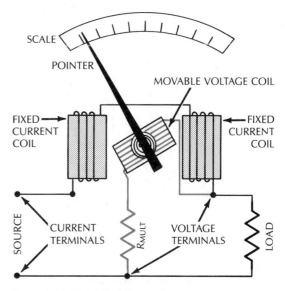

Figure 20-8 Simplified electrodynamometer wattmeter circuit.

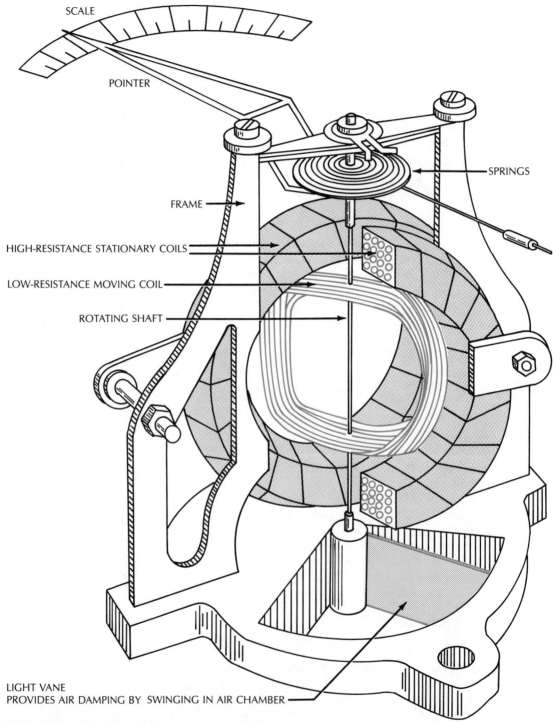

SCALE

POINTER

SPRINGS

FRAME

HIGH-RESISTANCE STATIONARY COILS

LOW-RESISTANCE MOVING COIL

ROTATING SHAFT

LIGHT VANE
PROVIDES AIR DAMPING BY SWINGING IN AIR CHAMBER

Figure 20-9 Modern electrodynamometer movement.

electrodynamometer may be used for ac or dc measurements.

Because of the oscillating effect of alternating current, ac meter movements are usually equipped with a damping mechanism. In Fig. 20-9 damping is accomplished by the use of an aluminum vane, which is attached to the central shaft and fits closely in an enclosed box. When the moving coil turns, the vane moves, forcing the air from one side of the vane to the other, thus producing a damping effect.

By changing the connections between the field coils, and moving coils, and by adding resistors, these meters may be used to measure volts, amperes, watts, and power factor. Since the moveable coil is designed to carry small currents, the dynamometer's range as a voltmeter or ammeter is limited to low values. Higher ranges can be attained by use of current or potential transformers. These transformers have been covered in Chapter Fourteen.

POWER MEASUREMENTS

True power consumed by an electric circuit is measured by an instrument called a *wattmeter*. Since power is a function of both voltage and current, the wattmeter measures the effective values of the voltage and current in a circuit, computes the amount of power consumed, and indicates this value on a calibrated scale. Thus, a wattmeter must have two elements built into the meter, as shown in the electrodynamometer in Fig. 20-8. The moveable coil, which has a multiplier resistance connected in series, is connected across (parallel) the line, and the stationary coils are connected in series with the circuit (load).

When the line current flows through the current (stationary) coils, a magnetic field proportional to the current is set up around the coils. When a voltage is applied to the potential (moving) coil, the current flowing through it is proportional to the line voltage. The inter-

action of the two magnetic fields causes the moving coil to deflect. The turning force acting on the moving coil at any instant is proportional to the product of the instantaneous values of the load voltage and the line current. Thus, the meter indicates the actual power dissipated in the circuit.

The current and potential coils have specific current and voltage ratings. Therefore, when using a wattmeter you must be very careful not to exceed these voltage and current ratings. Exceeding these ratings will damage the coils. A wattmeter is always distinctly rated, that is, the voltage and current ratings are marked on the instrument.

Figure 20-10 shows how a wattmeter is connected in a (a) single-phase circuit, (b) a three-wire single-phase circuit, (c) a three-wire three-phase circuit, and (d) a four-wire three-phase circuit. This was also described in the chapter on three-phase connections.

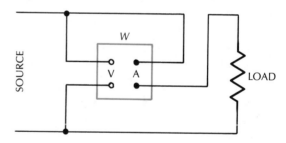

Figure 20-10(a) Single-phase circuit.

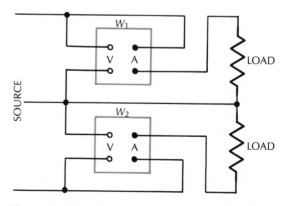

Figure 20-10(b) Three-wire single-phase circuit.

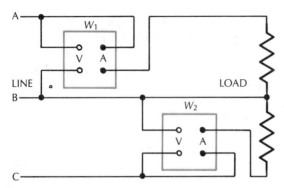

Figure 20-10(c) Three-wire three-phase circuit.

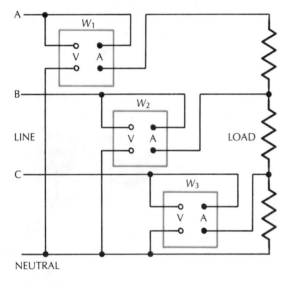

Figure 20-10(d) Four-wire three-phase circuit.

ENERGY MEASUREMENTS IN AC CIRCUITS

The induction watthour meter is an instrument that is used for measuring and recording electrical energy consumed in an ac circuit. Electrical energy is sold by the kilowatthour. One kilowatthour is the quantity of electrical energy used when power is consumed for one hour at the rate of one kilowatt. Thus, the watthour meter must take both quantities of power and time into consideration. The modern kilowatthour meter is shown in Fig. 20-11.

Figure 20-11 Kilowatt hour meter.
Courtesy: Ferranti-Packard Transformers Ltd.

The watthour meter is basically an induction motor, consisting of an aluminum disk, a moving magnetic field, damping magnets, current and potential coils, a registering mechanism of dials, and a system of gears. A simplified diagram of an induction watthour meter is shown in Fig. 20-12.

The rotating element (aluminum disk) is mounted on a vertical spindle, which has a

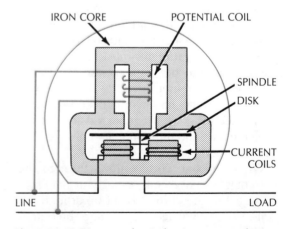

Figure 20-12 Diagram of an induction-type watthour meter.

worm gear at one end. As the gears turn the dials indicate the amount of energy passing through the meter. The rotation of the aluminum disk is accomplished by an electromagnet, which consists of a potential coil and current coils. The potential coil is connected across the load and is wound with many turns of fine wire. It induces an eddy current in the rotating aluminum disk. The eddy current produces a magnetic field which reacts with the magnetic field produced by the current coils to produce a driving torque on the disk. The current coils, connected in series with the load, are wound with a few turns of heavy wire, since they must carry the full line current. The speed of rotation of the aluminum disk is proportional to the product of the amperes (in the current coils) times the volts (across the potential coil). The total electrical energy that is consumed by the load is proportional to the number of revolutions made by the disk during a given time period.

A small copper disk (shading disk) or coil (shading coil) is placed in the air gap under the potential coil (not shown in diagram), to produce a forward torque large enough to counteract any friction produced by the rotating aluminum disk.

Two damping magnets restrain the aluminum disk from racing at a high speed. They produce a counter torque that opposes the turning of the aluminum disk.

This counter torque is produced when the aluminum disk rotates in the magnetic field established by the damping magnets. The eddy currents, in turn, produce a magnetic field that reacts with the field of the damping magnets, causing a restraining action that is proportional to the speed of the disk. The faster the disk rotates, the greater the induced eddy currents, and the greater the restraining action.

In a three-phase system, electrical energy may be measured by means of two single-phase watthour meters, as shown in Fig. 20-10. In a poly-phase system, the number of

watthour meters required is one less than the number of conductors used in the circuit.

A three-phase watthour meter is basically two single-phase watthour meters that have the moving elements connected to the same spindle.

POWER-FACTOR METERS

As was indicated in Chapter Ten, the power factor of a circuit is the ratio of the true power to apparent power, or the cosine of the phase angle between the current and voltage in the circuit. In a pure resistive circuit, where the voltage and current are in phase, the power factor has a value of one (*unity*). In a pure inductive circuit, the current *lags* the voltage by 90°, the resulting power factor is zero (lagging). Whereas in a pure capacitive circuit, the current *leads* the voltage by 90°, the resulting power factor has a value of zero (leading).

The electrodynamometer-type power-factor meter is shown schematically in Fig. 20-13.

The power-factor meter consists of a pair of coils, *A* and *B*, fixed at right angles to each

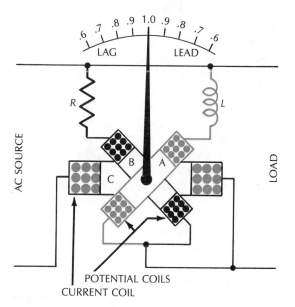

Figure 20-13 Schematic electrodynamometer type of power-factor meter.

other. These coils are connected across the line and are moveable potential coils. Coil A is connected in series through a high inductance (L), while Coil B is connected in series with a high noninductive resistance (R). The two series combinations are joined together in parallel and connected across the line. Coils A and B are mounted on a spindle so that the assembly, together with the attached pointer, is free to rotate through an angle of 90°. The currents in these two coils are 90° out of phase, and thus produce a rotating magnetic field.

The meter has a third coil, C, which is stationary and is mounted at right angles to both coils A and B. It is connected in series with the load, and thus it carries the line current.

At unity power factor, the current in coil B (connected to the resistance) is in phase with the line voltage, while the current in coil A lags the line voltage. The magnetic field produced by the fixed coil C will exert a torque on moveable coil B, causing the pointer to move up scale to unity power factor. For power factors other than unity, the current of one potential coil will always be in phase with the line current, while the current of the other potential coil will be out of phase with the line current. The pointer will receive corresponding deflections on the calibrated scale, indicating the new power factor. The scale may be calibrated in terms of the phase angle or power factor.

Power-factor meters are designed to measure the power factor at low power line frequencies (60 Hz). Phase measurements at frequencies higher than 60 Hz can be accurately attained by special electronic instruments.

THE CLAMP-ON VOLT-AMMETER

The clamp-on volt-ammeter is one of the most commonly used instruments for electrical testing in industrial operations. It is used to measure ac voltage and ac current without the problem of opening a circuit. Figure 20-14

Figure 20-14 Digital clamp-on volt-ammeter.

Courtesy: Amprobe Instruments.

shows a photograph of a clamp-on volt-ammeter, and Fig. 20-15 shows a schematic diagram of a basic clamp-on meter.

The line current is measured through the use of a split-iron core that can be opened and closed around a current-carrying conductor. An alternating current produces a magnetic field that induces an alternative current in the coil. The split-iron core contains a current transformer, which has a secondary winding suitably divided by a multiple-range series shunt for several ranges of current. This secondary current is rectified (converted to dc) for the permanent-magnet moving-coil meter.

For use as a voltmeter, the test leads are connected across the circuit to be measured. These leads make a direct connection to the rectifier, with multiplier resistances in series to produce the desired range.

Since transformer action is required for the operation of the clamp-on volt-ammeter, only ac current can be measured. Clamp-on volt-ammeters are useful in measuring high ac currents. These meters are very sensitive and strong external or stray magnetic fields should be avoided.

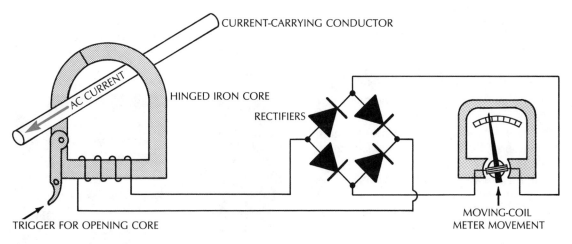

Figure 20-15 A basic clamp-on meter circuit.

MULTITESTERS

The multitester is a multipurpose instrument designed to measure both ac and dc current, voltage, and resistance. Some multitesters include a decibel (dB) scale. There are basically three types of multitesters: (a) the *volt-ohm meter* (VOM), (b) the *field-effect transistor* (FET)-VOM which has a built-in transistorized amplifier that increases its sensitivity, and (c) the *digital multimeter* (DMM), an amplified multitester in which the readout is a digital display in the form of a light emitting diode (LED) or liquid crystal display (LCD).

Most multimeters use a single multiposition switch which selects both the function and range. The volt-ohmmeter, shown in Fig. 20-16, is an analog display meter which makes use of the d'Arsonval meter movement.

The VOM may have one or more batteries that supply power for resistance measurements, one battery for all the low ranges, and another battery for the high resistance range.

The differences among the three types of multitesters referred to earlier are their sensitivity. In Chapter Six, it was pointed out that the higher the sensitivity rating of the meter, the higher is its input resistance, and the less loading effect the multitester will have on circuits being tested. This is very important in

Figure 20-16 The volt-ohmmeter (analog).
Courtesy: Amprobe Instruments.

semiconductor circuits with very low dc voltages and currents.

The sensitivity of the VOM is rated on the basis of ohms-per-volt; thus, the higher the ohms-per-volt rating, the more sensitive the meter movement. Most VOMs that do not use built-in amplifiers have a rating of 1000 to

100 000 ohms-per-volt for dc volts, and 1000 to 5000 ohms-per-volt for ac volts. The ohms-per-volt rating on the ac function is always lower than on the dc function, due to the losses that occur in the rectifier circuit. However, multitesters with built-in amplifiers have much higher sensitivity ratings. The FET-VOMs and the DMM have an input resistance of 10 to 20 megohms, thus reducing any loading effect to a minimum value.

The built-in amplifier also may have a negative feedback circuit. This minimizes the effects caused by components that change in value due to age or environmental conditions which normally would affect the accuracy of the meter. Figure 20-17 shows a FET volt-ohm-meter.

Figure 20-17 FET volt-ohmmeter (analog).
Courtesy: B & K Precision.

This meter uses two types of battery supply, one battery (D cell) for resistance measure-

ments and one or two batteries (9 V) for powering the FET amplifier circuit. Both the VOM and the FET-VOM make use of the analog display meter movement.

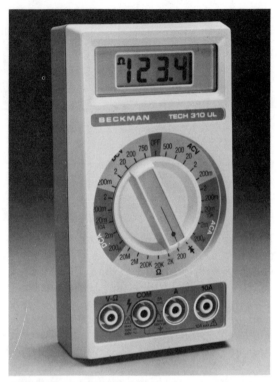

Figure 20-18 Digital multimeter.
Courtesy: B & K Precision.

The digital multimeter (DMM) with its digital display is easy to read, thus reducing possibilities of human error. The digits are displayed with units, decimal point, and polarity. Figure 20-18 shows a typical DMM. It offers many advantages such as greater speed, increased accuracy, reduction of operator errors, and the ability to make automatic measurements.

The circuitry of the DMM is very complex. A microprocessor converts the voltage being measured to a series of pulses. These pulses, in turn, trigger an oscillator on and off. The oscillator produces an output signal that is fed to a counting circuit, which converts the signal to a numerical display. When the DMM is being used on the ac voltage mode, the circuit is calibrated to display the rms value of a pure sine wave.

Some DMMs have no range switches, but instead have five function push buttons. The only basic decision required by the operator is whether voltage, current, or resistance will be measured, and if it will be ac or dc. This type of meter is referred to as an *auto-ranging digital multimeter*. The range selection is made by the microprocessor, which senses the input level and selects the range that will provide the greatest display resolution. It gives a range command that changes both the range and the decimal point automatically. Thus, in the auto-ranging mode, it can perform measurements of a volt one minute, and then measure several hundred volts the next minute. The DMM automatically rounds off the reading to three places.

Other features usually found in the DMM are automatic polarity indication, zero adjust, protection against overloads, low battery indicators, and "continuity testing," with a buzzer sound appearing when continuity occurs in a circuit.

FREQUENCY METERS

Alternating voltages are usually generated at a particular frequency. A frequency meter is an instrument that will measure or indicate the frequency of a particular ac voltage in an electric circuit.

Until recently, there were several types of frequency meters: the vibrating-reed frequency indicator, moving-pointer frequency instrument, the moving-disk frequency meter, and others that involve special bridge circuits that produce shifting magnetic fields.

The advent of the integrated circuit has resulted in digital display frequency counters that are extremely fast, accurate, and cover a wide frequency spectrum, with no moving parts. Figure 20-19 shows a frequency counter that is compact. The circuitry of the frequency

Figure 20-19 Frequency counter.
Courtesy: Triplett Corp.

counter is too complex and beyond the scope of this book.

SUMMARY OF IMPORTANT POINTS

1. An ac meter that contains a d'Arsonval meter movement must convert the input ac to dc for its operation.
2. An ac measuring instrument produces an up-scale deflection by the following methods: thermal, electromagnetic, and rectification.
3. A junction diode (rectifier) changes ac to dc before it is applied to the meter movement.
4. The current in a diode always flows from cathode to anode.
5. The half-wave rectifier circuit produces one pulse per cycle. The full-wave and bridge rectifier circuits produce two dc pulses per cycle. The bridge rectifier is commonly used in ac meters.
6. Ac rectifier-type meters respond to average values, but are calibrated for measuring rms values.
7. Electromagnetic iron-vane meters are of the radial, concentric, or plunger-vane type. They require more current to produce FSD than the d'Arsonval moving coil.
8. Iron-vane meters cannot provide accurate readings over a wide range of frequencies, as the reactance of the coil changes when the frequency increases.
9. A thermocouple (made up of two dissimi-

lar metals joined together) generates the current that is required to drive a d'Arsonval meter movement.

10. The amount of heat produced by the flow of current through the heater element of a thermocouple meter is proportional to the square-law scale.

11. The thermocouple meter will respond to the amount of heat that an ac or dc input voltage can produce.

12. In an electrodynamometer movement, the deflection of the moving coil is obtained by the interaction of two magnetic fields, a stationary field (current coils), and a moving coil (potential coils).

13. The electrodynamometer may be used to measure ac or dc watts (true power).

14. An induction watthour meter is an instrument that is used for measuring and recording electrical energy consumed in an ac circuit.

15. Power factor may be indicated by an electrodynamometer-type meter.

16. The clamp-on volt-ammeter is used to measure ac voltages and currents without the problem of opening the circuit.

17. There are three types of multitesters: the volt-ohmmeter (VOM), the field effect transistor (FET)-VOM, and the digital multimeter (DMM). The VOM and FET-VOM are analog display meters, while the DMM has a digital readout. Both the FET-VOM and the DMM have built-in amplifiers which makes them very sensitive. The DMM has a microprocessor which performs many automatic decisions (functions).

18. Frequency meters are instruments that indicate the frequency of a particular ac voltage in an electric circuit.

REVIEW QUESTIONS

1. Name three ways that the deflection of a d'Arsonval moving coil may be accomplished for ac measurement.

2. Describe how a permanent-magnet moving-coil meter can be used for measuring alternating current.

3. Describe the principle of the moving iron-vane meter.

4. What currents do moving iron-vane meters measure?

5. Describe an outstanding feature of the electrodynamometer.

6. Name two types of moving iron-vane instruments.

7. Why is a rectifier necessary in the clamp-on meter?

8. Why are electrical meters damped?

9. Describe how a twisting torque is produced in an electrodynamometer.

10. What currents are electrodynamometer meters designed to measure?

11. Describe how a clamp-on meter operates.

12. What is a multimeter used for?

13. Describe how an electrodynamometer is connected in a circuit to measure electrical power.

14. How many single-phase wattmeters are required to measure power in a three-wire three-phase circuit?

15. Describe how the driving torque is developed in an induction watthour meter.

16. Describe the construction of the electrodynamometer-type power-factor meter and explain its principle of operation.

17. Explain how kilowatt hour meter connections in a circuit compare to the connections of a wattmeter.

18. Why are ac meters less sensitive than dc meters?

19. Describe how the fixed and moving coils are energized in an electrodynamometer wattmeter.

20. What is a thermocouple meter designed to measure?

21. Describe the basic principle of operation of a thermocouple meter.

22. Describe what effects hysteresis, eddy currents have on the operation of ac moving-coil meters that have iron cores.

23. Explain why a power-factor meter (an electrodynamometer-type) is suitable for measuring power factor at a specific frequency.
24. What is meant by an analog meter and a digital meter?
25. Describe how an FET-VOM differs from an ordinary multimeter.
26. Describe some of the distinguishing features between a FET-VOM and a DMM.
27. What is a frequency meter?
28. What are potential and current transformers used for?
29. List the advantages of a DMM over other meters.

CHAPTER
TWENTY-ONE

AC MOTOR CONTROL

Alternating current motor controls are classified by the methods used in starting: across-the-line-starter (full-voltage starter) and reduced-voltage starters.

If an ac induction motor is started by connecting it across the line, it will draw a current from four to six times greater than its normal full-load running current. Alternating current motors are constructed to withstand a high starting current and, unlike dc motors, they don't have commutators nor brushes which can be damaged by the excessive sparking that is created by the large currents.

When large ac motors are connected across-the-line, the high starting current may produce great stress, which could damage the machinery driven by the motor. The high starting current will also cause large line voltage fluctuations that may affect the operation of other motors connected to the same line. Under these conditions, the starting voltage should be increased in stages as the motor accelerates to full running speed. Thus, some form of control is necessary to start the ac motor.

MOTOR CONTROLLERS

Motor controllers must be designed to fit the needs of the machine to which they are con-

nected. Thus, they may be classified into four categories, according to the functions they perform:
1. Starting—with provision to limit the inrush current.
2. Reversing.
3. Running—motor speed control.
4. Stopping—braking action.
The motor may be controlled manually by an operator, automatically, or by some combination of the two systems, which could be called semi-automatic.

MANUAL CONTROLLER

A motor may be controlled manually by an operator who will activate a device such as a switch, drum controller, or a face plate-type starter. The most common type of manual starter is the simple **on** and **off** toggle switch.

Both the manual and automatic controllers may have built-in protection against overloads, over-speeding and over-heating.

Manual starters are used on low-power single-phase or three-phase motors of 746 W or less.

One disadvantage of a manual starter in some applications is that the contacts remain closed in the **on** position if a power failure occurs. Thus, the motor will automatically start when the power is restored. This is not a *safe* condition, because any machinery attached to the motor could be damaged when the line voltage returns. In certain applications such as air fans, pumps, etc., this is desireable as these motors run continuously. Figure 21-1 shows a diagram of a three-phase manual starter with overload heater protection.

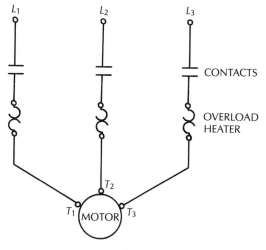

Figure 21-1 Three-phase manual starter with overload heater protection.

All motors require overload protection. An overload relay is usually built into each controller with the exception of the reversing drum switch, where it is not practical mechanically to build it into the switch. When a reversing drum switch is used in a control circuit, a magnetic starter with overload protection is generally placed in the line ahead of it. Drum switches were covered in Chapter Nine, dc motor control.

SEMI-AUTOMATIC CONTROLLER

In semi-automatic controllers, the operator must push a start button which completes the circuit. The ensuing current flow activates a magnetic contactor which connects the motor across-the-line. Many electric circuits utilize this method of control.

AUTOMATIC CONTROLLER

Automatic controllers are operated by pilot devices such as a float, pressure switch, thermostat, or time clock that react to certain conditions. For example, a float in a sump pump will close a switch when the water level rises, activating a pump automatically. There are a varied number of pilot devices available, and each one reacts to a different condition that will open or close a switch. Figure 21-2 shows an automatic pressure switch with a manual starter controlling a motor.

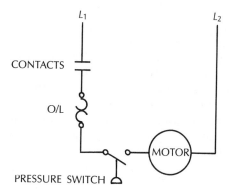

Figure 21-2 Automatic pressure switch with manual starter.

Another example of automatic control found in every home is the heating and air-conditioning system. The thermostat is a pilot device that closes a switch when the temperature drops to activate the furnace, or in the case of an air conditioner, when the temperature rises, the compressor is energized.

PUSH BUTTONS

Push buttons are the most common pilot devices used in control circuits. They are mechanically operated, and have a varied set of contact arrangements. The *single-break* (snap-action type) is similar in operation to a common knife switch and is used in small motors of approximately 249 W. It provides a maintained contact. The other type of push button is a momentary *double-break* contact that may be either *normally open* (N.O.) or *normally closed* (N.C.).

The normally open (N.O.) double-break contact makes contact while the pushbutton is being depressed. The normally closed (N.C.) double-break contact opens when the pushbutton is depressed.

Both the single-break and double-break contacts may be arranged by *poles* and also have a number of *throws*. A pole is a single or a set of contacts that control an individual circuit; throws refer to the number of closed contact positions allocated to each pole on a particular switch.

Some pilot control devices are made up of a combination of normally open and normally closed contacts. By connecting the proper set of contacts, a normally open or a normally closed contact can be obtained.

Pushbuttons are usually mounted adjacent to a controller or some distance away from it, and perform such operating conditions as start, stop, reverse, forward, slow, and fast.

CONTACTORS

Magnetic contactors are classified in two categories: one that operates on ac current and the other on dc current. The basic operation of the dc contactor was described in Chapter Nine, DC Motor Control. However there is a difference between the two contactors and that is in the design of the magnetic structure.

In the dc contactor, the heating effect of the current flowing through the coil causes the iron frame (core) of the contactor to heat up. Whereas, in an ac contactor the iron core heats up from the copper loss of the coil and the flip-flopping of the molecules in the iron core.

With a dc contactor, the strength of the magnetic field is dependent on the coil resistance and the applied voltage. However, in the ac contactor, in addition to the above factors, the strength of the magnetic field is also

dependent on the reluctance of the iron core, the number of turns in the coil, and the frequency of the applied voltage.

To reduce the iron core losses and eddy-current effects, the iron core is laminated. As a result, the laminations may vibrate (hum) whenever the contactor is operating. This hum is found in all ac magnets where the iron core is laminated, and it is caused by the alternating magnetic field.

The holding coil on an ac magnetic contactor is excited by an alternating current that is varying; thus, the magnetic pulling force is also varying. Each time that the ac current drops and reverses from positive to negative, the magnetic pulling force does not have enough strength to pull the contacts shut. This tends to cause the contactor to become noisy, a condition that is known as *chattering*.

SHADING COIL

To overcome this chattering effect, a small auxiliary coil, called a *shading coil*, is embedded around a portion of the iron core tip (pole face). This shading coil produces an out-of-phase magnetic field which overcomes the tendency of the iron core to open with each reversal of the magnetic field (Fig. 21-3)

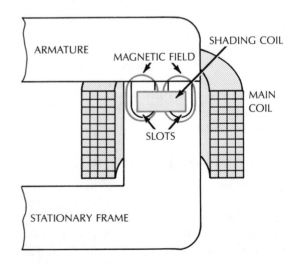

Figure 21-3 Shading coil.

The shading coil may be a short-circuited single turn, or several turns of wire, or just a copper strap. Because of the resistance and reactance, the current induced in this coil produces a magnetic field that is out of phase with the main field. This field is sufficiently strong, maintaining the pulling force on the armature as the main magnetic field passes through the zero point when the ac current reverses.

The dc magnetic contactor does not require a shading coil or laminated core as found in the ac device.

Magnetic contactors (starters and controllers) provide a safe, convenient way of starting and stopping motors from a remote position. They are used where across-the-line (full-voltage) starting can be safely utilized, and may be controlled by various types of pilot devices. Magnetic starters are manufactured in various sizes and power ratings, (eg., size 00,10 A—size 8,1300 A). A two-pole starter is used on single-phase circuits, while three-pole starters are used with motors operating on three-phase, three-wire ac systems, shown in Fig. 21-4.

When a short-circuit condition occurs in a motor, the current will rise to a high value and destroy the magnetic starter and the other components on that line. This type of damage may be prevented by adding fuses in the lines ahead of the magnetic starter. The amperage ratings of the fuses are about three times the full-load current rating of the motor. Thus, the fuses will not open the circuit during the startup period when the motor draws more current than it does at full-load. Momentary overload conditions are controlled by the overload relays.

SEQUENCE OF OPERATION

The sequence of operation of the control and load circuit of the starter shown in Fig. 21-4 is as follows: when the startbutton is pressed,

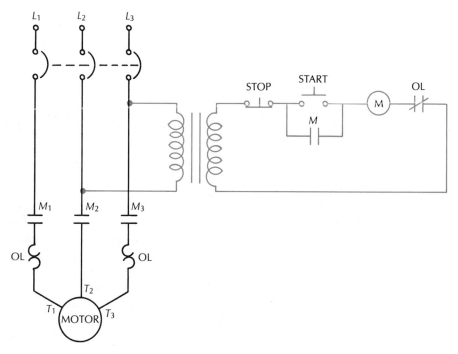

Figure 21-4 Across-the-line starter used in a three-phase circuit.

current flows from line L_3 through the stopbutton, the startbutton, through coil M to L_2. Contactor M becomes energized closing the contacts M_1, M_2, and M_3 in the power circuit, applying full voltage to the motor. At the same time, the sealing contact M in parallel with the startbutton closes. Coil M remains energized after the pushbutton is released because current flow continues to flow in the control circuit around the start button through the sealing contact M. When the normally closed stopbutton is pressed, no current flows in the control circuit. Coil M is de-energized, contacts M, M_1, M_2, M_3 open, removing power from the motor and control circuit. The motor will not restart when the pushbutton is released. The most complicated control diagrams can be analyzed by utilizing this method.

CONTROL CIRCUIT WIRING

There are two methods of wiring a control circuit; the two-wire control which is a no-voltage, low-voltage release; and the three-wire control which provides no-voltage or low-voltage protection.

TWO-WIRE CONTROL

In this control system, a magnetic starter will become de-energized when the line voltage drops to a low value or zero volts. Figure 21-5 shows a two-wire control circuit utilizing a float switch as a pilot device. Two wires connect the pilot device to the magnetic starter. The magnetic starter is energized when the line voltage is restored because the contacts on the pilot device remain closed (single contact). It is not necessary for an operator to restart the motor. Thus, this control circuit lends itself for such installations as exhaust fans, sump pumps, etc. However, it could present a safety hazard to the operator, machine or work in progress when the machine suddenly starts up as the line voltage is restored. Contact M is not utilized in this circuit. Although two-wire control provides low-voltage-release, it does not provide low-

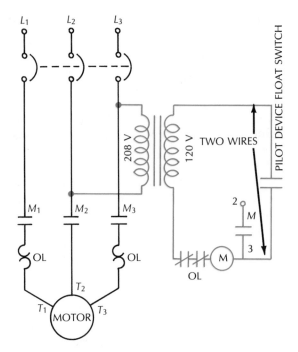

Figure 21-5 Two-wire control.

voltage protection. The two-wire control is not suitable in providing protection from the effects of low voltage. A three-wire control, which provides the desired protection, should be utilized.

THREE-WIRE CONTROL

The no-voltage (low-voltage) protection control system has three wires connecting the pilot device to the magnetic starter, as shown in Fig. 21-6.

When the line voltage drops to a low value or the power fails, the magnetic starter is de-energized (drop-out), stopping the motor. The sealing contacts of the "start" button will open. When the line voltage is restored, the motor will not start until the operator presses the "start" button. The operator and machine are protected against unforseen startups. In no-voltage protection, the control coil (M) is maintained through a double contact pushbutton station.

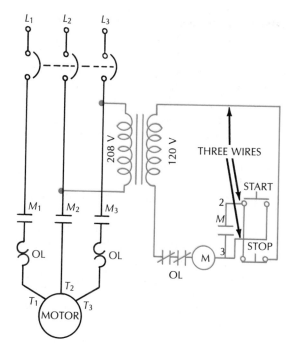

Figure 21-6 Three-wire control.

To restart the motor, the startbutton is depressed, current flows from Line One (L_1) through the normally closed stopbutton, the depressed "start" button, control coil (M), overload (normally closed) contacts to Line Two (L_2). Coil M is energized, closing contacts M, M_1, M_2 and M_3. Contact M provides a path for current flow when the startbutton is released. To stop the motor, coil M is de-energized when the "stop" button is pressed, opening contacts M, M_1, M_2, M_3.

In comparing the manual and magnetic starters the following points should be noted:

MANUALLY OPERATED MOTOR STARTER

1. May be operated only from the starter location.
2. Automatically restarts the motor when power is restored after a failure.
3. Provides overload protection in all cases except in the reversing drum switch.

MAGNETICALLY OPERATED MOTOR STARTER

1. May be operated from more than one location, local or remote.
2. With three-wire control circuit, momentary pushbuttons may be used. Restart of the motor is impossible when power is restored after a failure. The operator must restart the motor.
3. With two-wire control automatic pilot devices are utilized. They permit automatic restart of the motor when the power is restored after a failure.
4. Overload protection is provided in all cases.

RELAYS

A relay is a magnetic auxiliary device used to control one or more circuits such as starter coils, small motors, or light loads rather than operate power circuits. Contactors, on the other hand, carry heavy motor currents.

VOLTAGE AND CURRENT RELAYS

A *voltage relay* is designed to operate at a certain voltage, and is rated by the current-carrying capacity of the contacts. Magnetic contactors, on the other hand, are rated in watts or kilowatts depending on the size of the motor that they are controlling.

A *current relay* is a current-sensitive device and is activated when a certain value of current flows through the control coil. Current relays are designed to operate on overcurrent or undercurrent conditions.

Relays may differ in appearance and construction from one manufacturer to another. A complete description on the operation of relays was covered in Chapter Nine on dc motor control.

SOLID STATE RELAYS

Solid state relays have been in use for a number of years and have proved themselves to the point that they are replacing the electromechanical relay in many applications.

There are several disadvantages to conventional electromechanical relays:

1. The electromechanical relay contains moving parts which eventually will break down. For example, an electromechanical relay when operated under rated load conditions may have a life of 1 000 000 operations. The solid state relay, on the other hand, when operated within rated conditions at the same load could have a life of ten times, or 10 million operations.

2. In the electromechanical relay, the contacts wear out due to sparking that takes place as the contacts open under load. In solid state relays, there is no arcing; thus there is no erosion of the contacts.

3. The electromechanical relay cannot open and close the contacts at a high rate of speed that can be achieved by a solid state relay.

4. Solid state relays have no moving parts; thus they are completely noiseless when they operate.

5. Electromechanical relays have moving parts that have a degree of elasticity. These moving parts are affected by shock and vibration forces; thus an open contact may close or a closed contact may open for a microsecond, causing an unsafe condition. Solid state relays having no moving parts can be operated in an environment where vibration is persistant.

6. Solid state relays are completely encapsulated; thus no hazardous arcing is generated. In an electromechanical relay, a high temperature arc is generated as the contacts open. This could result in a fire if the surrounding environment contains gas or dust.

7. An electromechanical relay requires a greater input power to activate it and maintain the load switch in the energized position. The lower input power requirements of solid state relays makes them compatible with low power integrated logic circuitry.

DISADVANTAGES OF SOLID STATE RELAYS

1. Solid state relays are susceptible to transients in the power line. These transients may be large enough to turn a semiconductor device in the solid state relay into a conducting mode for a brief interval of time, or even permanently damage the semiconductor. An electromechanical relay with its higher input power requirements would be unaffected by these transients.

2. Because the contacts are physically separated when an electromechanical relay is open, there is no electrical connection between the control and load circuits. In the solid state relay, a certain amount of current leakage exists in the semiconductor device when it is in the **off** mode. Thus, some potential difference remains in the load circuit, presenting a safety hazard. This hazardous condition may be overcome by incorporating special disconnecting devices, which of course increase the cost of the solid state relay.

An ac solid state relay is shown in Fig. 21-7.

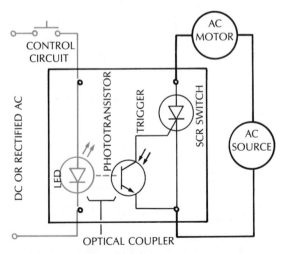

Figure 21-7 Alternating current solid state relay.

Solid state relays use semiconductor devices such as the *silicon-controlled rectifier* (SCR) or *triac* to act as an electronic switch.

Most ac solid state relays contain three main

components: a light-emitting diode (LED), a phototransistor, and the switching element (SCR or triac) in the load-circuit. When an LED and phototransistor are used together, they are called *opto-couplers*.

When the voltage is applied to the control circuit, current flows through the LED, causing it to light up. The light strikes the sensitive surface of the phototransistor, causing the internal resistance to drop to a low value. The phototransistor conducts heavily and triggers the gate of the SCR, causing it to conduct. This action of the SCR or triac is like the closing of a switch which allows current to flow through the load from the ac source.

Solid state relays are constructed with normally open or normally closed contacts.

The LED and phototransistor are physically separated (*opto-isolation*), insuring that there is no electrical connection between the control circuit and load circuit.

Solid state relays may incorporate a time delay along with the operation of a switching device. These devices are called *time-delay relays* (TDR). In the control circuit of Fig. 21-8 when the pushbutton is pressed, power is applied to the internal circuits of the TDR. The TDR circuit is activated and immediately initiates the time delay. At the end of the time delay, the relay switches its contacts to the **on** state, applying power to the motor from the ac source.

Time-delay relays may provide a time delay of one millisecond to several hundred hours.

TDRs may be used in various applications such as mixing chemicals, bulk-lot batching.

1. Mixing chemicals: where several chemicals are poured into a basic mixing tank. The TDR opens the valve through a solenoid for a prescribed period of time to supply the correct amount of formula constituents. Figure 21-8 shows a simple arrangement whereby the control signal is applied simultaneously to all five batches of chemicals. After the flow of the longest delay has stopped (batch 2), the mixer motor will be activated for a set period of time.

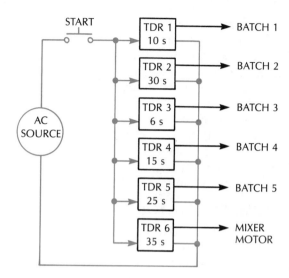

Figure 21-8 Time-delay relay.

2. Limiting of surge currents by sequential delay: such as in a large auditorium that contains large banks of lamps. The line voltage would drop drastically if all lamps were turned on at the same time. By turning on a few banks at a time, large current surges are avoided.

The theory of operation of the SCR and triac will be covered in Chapter Twenty-Two, industrial electronics. Other relays such as overload protective devices were covered in Chapter Nine on dc motor control.

REDUCED-VOLTAGE STARTING

Large three-phase induction motors are never started across the line.

Reduced-voltage starters are used to reduce the inrush current, prevent excessive line voltage fluctuations, and limit the torque which otherwise could damage the driven machinery.

Automatic reduced-voltage starters may use one or more across-the-line motor starting contactors. Once the operator presses the START button, the rest of the operation is completed by the controller. At start, the motor is connected to a low-voltage circuit, then as it attains rated speed, full voltage and current are applied. The length of time it takes to complete the starting sequence is automatically controlled by a timing relay (TR). The voltage may be reduced by various means.

The most common reduced-voltage starting methods are:

1. primary-resistor starters
2. auto-transformer starters
3. part winding starters
4. Wye-delta starters
5. solid state starters

PRIMARY-RESISTOR STARTERS

By connecting a resistor in each line leading to the motor, a voltage drop across each resistor produces a reduced voltage at the motor. The motor starting speed and current has been reduced. At the same time, the timing mecha-

nism is activated. After a set period of time has elapsed during which the motor speed increases, the timer closes its contacts, short-circuiting the resistors, and applying full voltage to the motor.

An elementary wiring diagram of an automatic reduced-voltage starter is shown in Fig. 21-9. When the startbutton is pressed, a complete circuit is established from L_2 to L_3. Coil M is energized, closing the main power contacts M and the sealing or maintaining contact M in the control circuit. The motor is now connected to the line through the starting resistors and is energized. At the same time, a timing relay TR is activated. A voltage drop occurs in the resistors, resulting in a reduced voltage being applied to the motor.

As the motor accelerates, the voltage drop

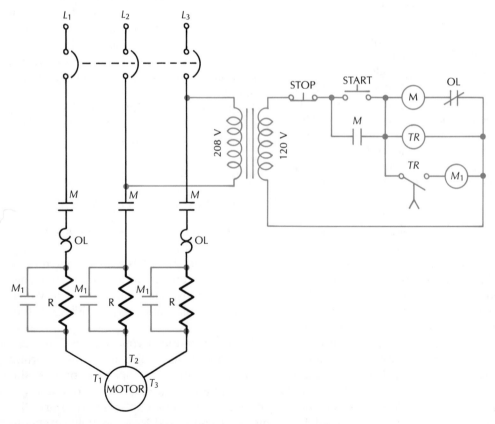

Figure 21-9 Elementary diagram of an automatic reduced-voltage starter.

across each resistor decreases gradually due to the reduction in line current. At the same time, the terminal voltage at the motor increases.

After a predetermined period of acceleration, the time-delay contact *TR* closes the circuit to coil M_1. This energizes the accelerating contactor coil M_1, which causes contact M_1 to close. The resistors are shorted, and the motor is connected across the full line voltage.

To obtain smoother acceleration at startup, starters with adjustable resistance (three or more points) are utilized. The starting resistance is cut out in several steps, increasing the current to the motor at each step.

This controller has overload and undervoltage protection, a large load current will trip the overload relay, and a low voltage or no voltage will de-energize the main contactors M_1.

AUTO-TRANSFORMER STARTER

Step-down auto-transformers are sometimes used to reduce the starting voltage in ac squirrel-cage motors. Figure 21-10 shows an auto-transformer starter connected to a three-phase motor.

Two auto-transformers provided with taps are connected in open-delta. These taps provide 55, 65, 75 percent of the line voltage during the acceleration period. Because of the lower starting voltage the motor draws less current and develops a lower torque than it would at full line voltage. A tap is preselected to match the starting torque of a given motor with an applied load. An adjustable control relay, *TR*, has time-opening (*TO*) and time-closing (*TC*) contacts that control the transfer of voltage from a reduced level to full line

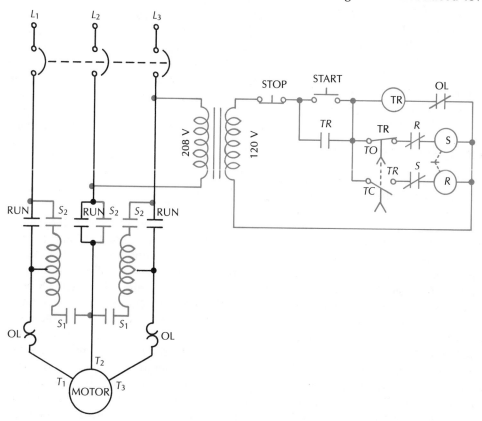

Figure 21-10 Auto-transformer starter (closed transition).

voltage. Contactors *RUN* and S_1 are mechanically interlocked to prevent both from closing at the same time. When the startbutton is pressed, the timing relay *TR* is energized. The normally open sealing contacts *TR* across the startbutton are closed, energizing the starting coil *S*, closing contacts S_1 and S_2. Contacts S_1 and S_2 apply a full voltage across the auto-transformers and a partial voltage across the motor which is connected to preselected taps on the auto-transformer. The *RUN* contacts cannot close at this moment because of the mechanical interlock. After a preset time delay, during which the motor accelerates, the timing relay *TO* opens and contacts *TC* close. Coil *S* is de-energized, opening contacts S_1 and S_2, coil *R* is energized simultaneously, closing contacts *RUN*. The motor is now transferred from the auto-transformer and connected across the full line voltage. Auto-transformer starters provide a voltage to the motor at all times, during the transition period from reduced voltage to full line voltage. Thus, it is referred to as "closed-transition" starting.

Auto-transformer starters develop the greatest starting torque for every ampere of current produced in the circuit.

PART-WINDING MOTOR STARTERS

Part-winding and Wye-delta starters require motors specifically designed for these control circuits. These motors are similar to squirrel-cage motors except that they have two identical sets of starter coils that are Wye (star) or delta connected. The motor terminals are joined to form a Wye connection. During startup, full voltage is applied to one set of windings, and in the *RUN* position, the two sets of windings are connected in parallel.

During the starting operation the locked-rotor current is approximately 65 percent of the value when both windings are connected in parallel, and the motor develops about 45 percent starting torque. Thus, the starting torque produced is low, and may be inadequate to start or accelerate a heavy load.

The part-winding starter has some obvious advantages. It does not require auxiliary voltage-reducing devices such as resistors, transformers or reactors, and only half-size contactors are necessary. Thus, it is less expensive than other reduced-voltage starters.

STAR (WYE)-DELTA MOTOR STARTERS

As the name implies, in this method the controller connects the motor in Wye (star) during the starting period and then in delta after the motor has accelerated to the *RUN* mode. Thus, the motor used with this starter must have its windings delta connected.

The line current and starting torque are considerably lower in value when the motor windings are connected in Wye (star) than when connected in delta.

The line current in a Wye is one-third of that in the delta connection. The winding current in Wye is 1.73 times the winding current in delta. The voltage across each winding is equal to the line voltage divided by 1.73. Thus, if the line voltage is 208 V, the voltage across each winding will be 208/1.73 or 120 V. This value is approximately 58 percent of the line voltage during the starting period. Because of this, it is important that the overload relays must be rated on the basis of winding current. Figure 21-11 shows a Wye-delta starter, open transition.

The operation of the above circuit depends upon first closing the contacts *M* and *S*, connecting the motor winding in a Wye circuit. The motor accelerates to full speed. After a predetermined time interval, contacts *S* open first, and then contacts *D* close, connecting the motor windings in delta. The motor proceeds to run normally at rated voltage. It is important that contactors *S* and *D* are not permitted to operate simultaneously, as a short-circuit could occur. Therefore, both mechanical and electrical interlocking is provided. During the interval between contacts *S* opening and contacts *D* closing, there is no

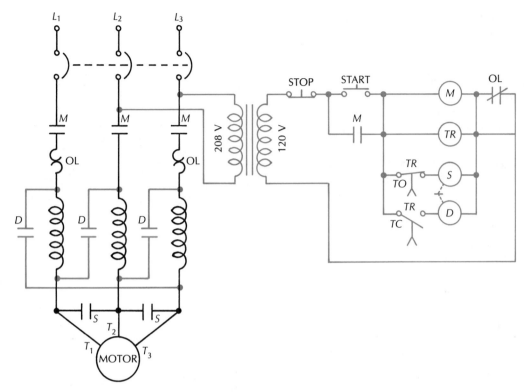

Figure 21-11 Wye (star)-delta starter (open transition).

electrical power applied to the motor; this is called "open transition."

Pressing the startbutton will energize coil M, which closes sealing contacts M, activating relays TR, S, and D. Contacts M and S close simultaneously, applying power to the motor and connecting the three windings into a Wye circuit. When the TR relay times out, contact TR-TO opens first, de-energizing contactor S, which opens contacts S. The relay TR-TC closes, energizing contactor D which closes the three D contacts connecting the motor windings to run in delta. The voltage across each winding in a delta connection equals the line voltage.

This starter is used for starting motors with loads that require a long acceleration period, such as in fans, blowers, air-conditioning pumps, etc. It does not have the flexibility found in other types of reduced-voltage starters.

SOLID STATE STARTERS

As seen earlier in this chapter in the discussion of solid state relays, it is possible to start an electric motor electronically. The starting inrush current is regulated by solid state components to produce a smooth acceleration from zero to full voltage. Whereas, in the electromechanical motor control devices, acceleration is achieved in steps from reduced voltage to full voltage. This smooth acceleration achieved by solid state ac motor starters is known as "soft starting."

Solid state ac motor starters utilize silicon-controlled rectifiers (SCR). (SCRs are discussed in detail in Chapter Twenty-Two.)

An SCR is similar to a diode, as it can conduct current in one direction only and blocks the flow of current in the other direction. Unlike the diode, it will not conduct until it is turned on or triggered by a signal (voltage) called a *gate pulse*.

Once the SCR is triggered into conduction, the gate loses control. The SCR will continue to conduct until the current flow drops to zero at the end of each half cycle. With the addition of a rectification circuit to supply dc power for the shunt field, this circuit may be used to control a dc motor.

The triggering circuit controls the exact point at which the pulse is applied to the gate. Triggering can take place at electrical angles from 0-360° (firing angle) during each half-cycle of applied voltage. In a half-wave control circuit, triggering takes place only during the positive half-cycle of the applied voltage. In Fig. 21-12, the SCR circuit is used to control a single-phase motor, with triggering taking place during the positive and negative alternation of the ac voltage. One SCR controls the positive alternation while the other controls the negative alternation. This is known as full-wave motor control.

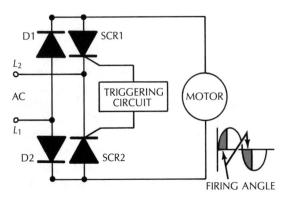

Figure 21-12 SCR motor control circuit (single-phase, full wave).

Figure 21-13 shows a diagram of a solid state control used to soft start a three-phase induction motor. When the motor is started, it accelerates smoothly under the control of the SCRs until it attains operating speed. Then the SCRs are turned **off** and the *RUN* contacts close, applying full voltage to the motor.

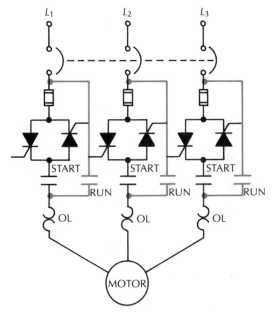

Figure 21-13 Solid state starter for a three-phase induction motor.

REVERSING CONTROLLERS

Some industrial applications such as conveyors, hoists, elevators, etc., require a motor starter which can reverse the direction of rotation of a motor when a pushbutton is depressed. This may be accomplished by utilizing a reversing magnetic starter (Fig. 21-14). The reversing starter contains two contactors, one for the forward and one for reverse operation. The reversing starter is able to change the direction of rotation of a three-phase motor by interchanging any two of the three line leads. In a single-phase motor, the direction of rotation can be changed by reversing one of the field windings.

Single-phase and three-phase motors may be reversed manually with a drum controller. A magnetic starter is normally connected ahead of the drum controller so that the magnetic starter will not energize until the drum contacts are closed during startup or the magnetic starter must be de-energized before the drum contacts open on reversing. Thus, the mag-

Figure 21-14 Reversing starter.
Courtesy: Allen-Bradley Canada Ltd.

netic starter opens the load circuit, and the drum controller switches the lines while there is no power in the lines. Drum controllers were discussed in Chapter Nine on dc motor control.

INTERLOCKS

An interlocking system is built in to prevent accidental reversal of rotation while the motor is running in one direction. Accidental reversal will cause a short-circuit. There are three basic methods of interlocking.

MECHANICAL INTERLOCK

When the motor is running in one direction, a mechanical interlock "jambs" the mechanism of the second contactor, preventing it from closing. If the forward contacts are closed, then the reversing contacts cannot close, due to the interlock, and vice versa. Figure 21-15(a) illustrates a mechanical interlock with a revers-

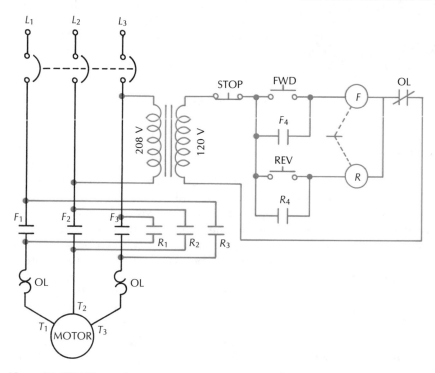

Figure 21-15(a) Three-phase motor reversing starter with mechanical interlock.

ing starter as used in controlling a three-phase motor, while Fig. 21-15(b) shows a typical three-phase motor control panel.

Figure 21-15(b) Three-phase motor control panel.
Courtesy: Franklin Electric.

A mechanical interlock is shown as a broken line between coils F and R. When the *forward* pushbutton is pressed, contactor coil F is energized, closing the sealing contact F_4 and the forward contacts F_1, F_2, F_3. The motor now rotates in the forward direction. Coil R is prevented from closing by means of the mechanical interlock. The *stop* pushbutton must be pressed before the motor can be reversed. If the *reverse* pushbutton is pressed, contactor coil R is energized, closing the sealing contact

R_4 and the reverse contactors R_1, R_2, R_3. Coil F is prevented from closing by means of the mechanical interlock. Two of the line leads, L_1 and L_3, are interchanged, and the motor now rotates in the reverse direction.

PUSH BUTTON INTERLOCK

Push button interlock is an electrical method that utilizes double-circuit push buttons that have a N.O. and N.C. contact, and were discussed in Chapter Nine, dc motor control.

AUXILIARY CONTACT INTERLOCK

In this method, interlock is achieved by utilizing N.C. auxiliary contacts on both the forward and reverse contactors, as shown in Fig. 21-16.

When the *forward* push button is pressed, contactor coil F is energized, closing the sealing contact F_4 and the forward contacts F_1, F_2, F_3. The N.C. contact F_5 opens, preventing the reverse contactor R from being energized and closing. After the motor has been stopped, pressing the *reverse* push button will energize contactor R. Sealing contact R_4 and reverse contacts R_1, R_2, R_3 will close, and N.C. contact R_5 will open to disconnect the forward contactor F from the circuit.

SEQUENCE CONTROL

Sequence control is necessary in many industrial applications such as conveyor systems or in a machine tool that is equipped with several operating motor drives and a coolant pump motor. The coolant pump must be operating before the machine itself can be safely activated. The motor-driven machinery requires a systematic sequence of motor startup. Starters are connected in such a manner that one cannot be energized until the preceding has been energized first. This method of startup is accomplished with the use of interlocking circuits. For example, an auxiliary contact in one

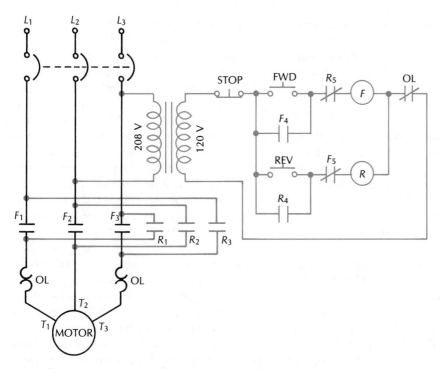

Figure 21-16 Auxiliary contact interlock.

starter may be placed in series with the control circuit of the succeeding starter. Thus, when the first starter is activated, it closes contacts which are in series with the control circuit of the second starter, allowing the second motor to be started. Figure 21-17 illustrates a sequence control circuit.

Coil M_p may be energized independently by pressing the start push button. Coil M_1 cannot be energized until coil M_p is energized first, which closes M_p on the second step of the ladder diagram. To activate coil M_2, contacts M_p and M_1 on the third step must be closed. This means that coils M_p and M_1 must be energized before coil M_2 can be energized. In the last step, coil M_3 cannot be energized unless either coils M_1 or M_2 are energized first. Each coil can be turned off individually. To turn off all coils, the stop push button on step one must be pressed.

An automatic sequence control circuit

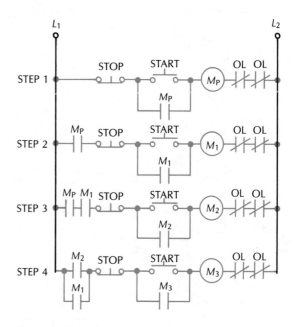

Figure 21-17 Sequence control circuit.

could be arranged where a series of motors may be started automatically with only one start-stop control station. Various pilot devices are used to activate motors in an automatic sequence.

JOGGING

In some industrial applications, a machine may have to be moved in small increments. This is accomplished by a control system called *jogging* (inching). Jogging is defined as the repeated closure of the circuit in quick succession which applies short intervals of power to the motor, producing small movements in the driven machine. By pressing the jog push button, the magnetic starter is energized, and when the job push button is released, the motor stops.

Jogging may be accomplished by several methods: (1) a selector switch, (2) a push button, and (3) a push button with a jog relay:

1. When a selector switch is used in a jogging circuit, a control station with start, stop, and selector functions must be utilized. The startbutton is used to jog or run the motor,

depending on whether the selector switch is in the jog or run position.

2. Figure 21-18 shows an across-the-line starter connected to a start-jog-stop push button station, using a selector push button. When the jog push button is pressed, coil M is energized, closing the M contactors. The motor is connected to the power source and starts turning. The sealing contact M does not maintain the circuit around the jog push button because as soon as the jog push button is released, the main contactor coil M is de-energized, opening all the M contacts. the control circuit is now de-energized. The motor will run only when the jog push button is held down. By pressing the jog push button repeatedly, this action will cause the motor to start and stop, causing the driven machinery to "inch" its way to the desired position.

3. Jogging circuit using a control relay is illustrated in Fig. 21-19. Pressing the start push button will cause the control relay R to be energized, closing contacts R_1 and R_2. R_2 completes the circuit to energize coil M,

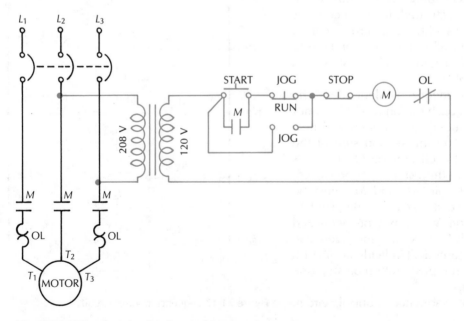

Figure 21-18 Jogging circuit using pushbuttons.

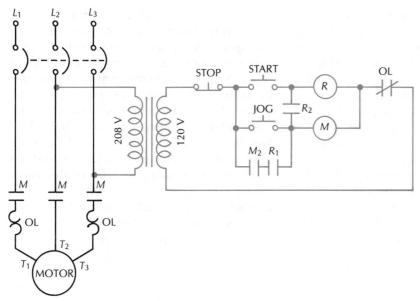

Figure 21-19 Jog control using relays.

closing contacts M and M_2. Contacts M_2 and R_1 complete the sealing circuit for the holding coil M when the start push button is released. If the jog push button is pressed while the motor is stopped, an electric circuit is completed through the holding coil M, as long as contact is maintained. The motor is activated performing the jogging function.

BRAKING INDUCTION MOTORS

The various methods used in motor deceleration (braking) were discussed in Chapter Nine on dc motor control.

MULTISPEED MOTORS

The speed of a three-phase squirrel-cage motor operated on a constant frequency source can be changed by switching the external connections. By reconnecting the stator windings, a different number of poles can be formed in the motor, resulting in a limited number of different speeds. The number of poles is either half or twice the number of original poles. A multispeed motor may use separate windings, or one winding that may be switched to produce two or four different speeds. Each winding is centre-tapped and all three centretaps are joined together, producing a parallel Wye circuit, high-speed connection.

Special magnetic starters are available for the purpose of changing the motor connections to obtain the different speeds.

Since the multispeed motors do not have a continuously variable speed, they are used extensively where multispeeds are definitely required, as in ventilation systems, conveyors, machine tools, washing machines.

There are other types of multispeed motors available that may be required to produce a constant torque, variable torque, or a constant power output.

WOUND-ROTOR MOTOR CONTROL

A wound-rotor motor is basically an induction motor with a rotor winding connected to three slip rings. That is why it is sometimes referred to as a slip-ring motor. Connected to the slip rings through carbon brushes is a motor controller in the form of a Wye-connected variable resistance. A high resistance is connected to the slip rings during the starting period, resulting in high starting torque. As the motor accelerates the resistance is gradually reduced by the controller. As high starting torque is developed at startup, and as the motor accelerates, the torque is reduced. When the motor reaches full speed, the slip rings are short-circuited, cutting all the external resistance out of the circuit. At this point, the wound-rotor motor is running under the same

conditions as a squirrel-cage motor.

Because the rotor-circuit resistance will change in value as the motor is rotating, it may cause the motor to increase or decrease its speed. Some form of control must be provided for adjusting the speed of the wound-rotor motor. These controls may be manual or automatic. Manual control is obtained by a drum switching device which may provide as much as five steps of resistance. Automatic systems short-circuit resistance sections by the use of timing, current, or frequency relays. Figure 21-20 shows a diagram of an automatic wound-rotor motor acceleration system.

When the start pushbutton is pressed, coil

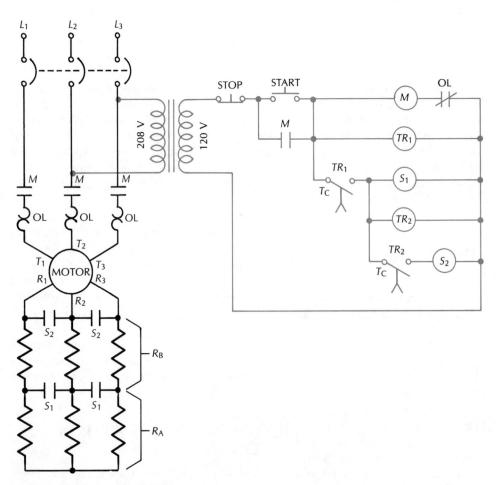

Figure 21-20 Automatic wound-rotor starting system.

M is energized, the sealing contact M closes, and the main contacts M connect the motor to the three-phase line voltage with full resistance in the secondary circuit. The motor starts to accelerate. The sealing contact M completes the circuit to timing coil TR_1. As the timing relay TR_1 is energized, it actuates the N.O. contacts of TR_1, timed to close (TC). After a preset period of time, contacts TR_1 close, energizing coil S_1, which closes contacts S_1, shunting out resistors R_A in the rotor circuit. The motor accelerates as resistors R_A are shunted out of the circuit. Coil TR_2 is energized, and after a preset time period the N.O. contacts TR_2 (TC) close, energizing coil S_2, which closes contacts S_2, shunting resistors R_B out of the circuit. The motor accelerates to its maximum speed.

WOUND-MOTOR DRIVES

In the conventional method of speed control used in wound-rotor motors, inserting rotor resistance is inefficient since power is dissipated in the form of heat.

Electronic circuits control the external resistance used in wound-rotor motors during startup and running conditions.

Figure 21-21 shows a block diagram of a solid state system that regulates the speed of a three-phase wound-rotor induction motor by controlling the rotor resistance.

The three-phase rotor current is rectified and filtered. A pure dc current flows through the resistor, R. An SCR (thyristor) is connected in parallel with the resistor and acts as a high-frequency switch. The value of the resistor is changed by increasing or decreasing the duration of the pulse.

Rotor slip power inherent in the wound-rotor induction motor is actually wasted power being dissipated in the form of heat. This slip frequency power generated in the rotor of the motor can be recovered by converting it back to ac (inverter) and returning it to the line, improving the power factor of the motor. This type of drive is appropriate for applications where a wound-rotor motor is operated at less than rated speed.

SYNCHRONOUS MOTOR CONTROL

A synchronous motor is a three-phase motor that maintains a constant speed as the load varies from no-load to full-load. Because of this constant-speed characteristic, synchronous motors are perfect machines for heavy loads that need to be driven continuously in such applications as pumps and compressors.

The synchronous motor and induction motor are very similar in the principle of operation and stator construction. The difference is in the rotor construction. In the synchronous motor, the rotor must have the same number of poles as the stator if proper synchronization is to take place. The synchronous motor has a special starting winding; thus, it is always

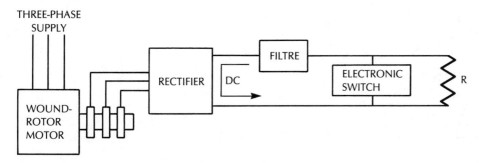

Figure 21-21 Induction-motor speed control by electronic circuits.

started the same way as an induction motor. The starting method used may be any one of the six starting methods that we studied earlier in this chapter. During the starting period, the motor is accelerated to a value approximately 95 to 98 percent of its synchronous speed. A timing device delays synchronization until the proper speed is reached. Synchronization takes place after a dc voltage is applied to the rotor. Figure 21-22 illustrates a control system for starting and synchronizing a synchronous motor.

A synchronous motor circuit must control the following operations:

1. During startup and acceleration period, the control circuit must prevent the field windings from being excited by the dc voltage.

2. During the period of acceleration, a field discharge resistor must be connected to the rotor circuit. The purpose of this resistor is to dissipate power generated in the field coil and prevent damage to the rotor windings. The N.C. contact F completes the short-circuit.

3. As the rotor approaches synchronous speed, a timing relay TR(TC) is timed to close, energizing contactor F which opens the N.C. contact F. This action disconnects the field discharge resistor and simultaneously applies the dc excitation to the rotor windings (field coil). The field of the rotor winding is now synchronized (locked) in step with the revolving field of the stator windings.

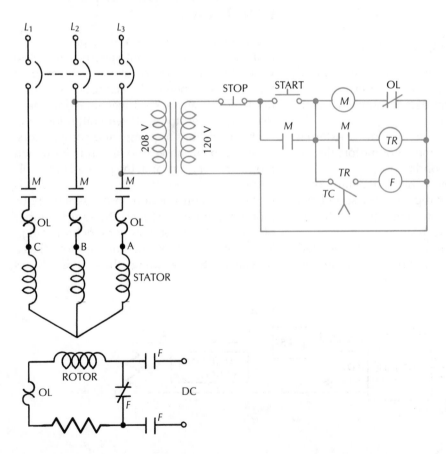

Figure 21-22 Synchronous motor control.

If the rotor fails to synchronize, a protective device senses the induced current in the rotor winding and disconnects the motor from the power source.

STATIC MOTOR CONTROL

Static logic is the basis of today's digital computer. The term *static control* means that there are no moving parts or contacts in the control device. It is ideally suited to applications where the environmental conditions (dirt, oil, moisture) make the operation of conventional relay control unreliable, and where high speed and repetitive operation is required. In applications where a system cycles at a slow rate, a low production piece of equipment, or the control functions are mainly motor starters, then a static control system may not be suitable. A static control system can operate at frequencies higher than 10 kHz.

Transistorized static control is a system of control that provides a method of electrically controlling machines and processes with logic elements. Logic elements accomplish the same job in static control that relays do in a conventional control system with the following advantages:

1. The response time (switching speed) is much faster than ordinary relays.
2. They have no moving parts to wear out as in relays, resulting in long life and reliability.
3. They are considerably smaller in size, equipment is lighter weight.
4. Lower power requirements.
5. They can operate in adverse environments where conditions exist which can cause malfunctions in conventional relays.
6. As modern industrial equipment becomes more and more complex and requires a greater number of relay functions, the chances of breakdowns is increased. With static control, the number of breakdowns is reduced.

These same basic logic functions that are used in static control are also contained in digital computers. Computer logic will be dealt with in greater detail in Chapter Twenty-Two, Industrial Electronics.

INVERTERS

An inverter is an electronic circuit that is used to convert dc power to ac power of arbitrary frequency. Modern inverters utilizing solid state components are highly efficient and economical. Inverter systems are used to provide power for variable frequency ac motors, computers, and stand-by power supplies.

An inverter is basically a switching circuit that turns a dc power source **on** and **off** by a solid state switching device such as a transistor or SCR (thyristor). The inverter is basically a square-wave generator. When the **on** or **off** time is varied (switching rate), then the output voltage and frequency of the inverter is varied.

The basic inverter circuit shown in Fig. 21-23 is a single-phase parallel inverter circuit. In

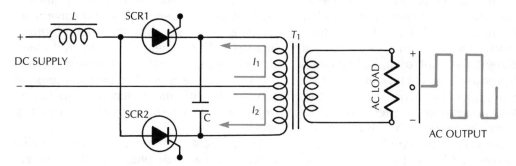

Figure 21-23 Basic inverter circuit.

this circuit, the two SCRs are alternately turned **on**. When SCR1 conducts, current I_1 flows through the transformer T in the direction shown, and when SCR2 conducts, current I_2 flows, as shown. The source voltage is alternately applied to one-half of the transformer winding and then the other, producing an alternating square-wave at the load. This output can be filtered to a sine wave. The frequency of the ac voltage produced is controlled by the voltage pulses (gating pulses) applied to the gate of the SCR. These pulses are usually generated by an electronic oscillator or multivibrator circuit.

Once an SCR starts conducting, the gate can no longer control the anode current, when used in dc circuits. A special circuit, called a commutating circuit (capacitor C charging action), turns the SCR off at the proper times.

VARIABLE FREQUENCY SPEED CONTROL

In many applications, a motor may have to operate at speeds other than for which it was designed. For example, in a conveyor system that requires several motors, it is necessary to vary the speed of these ac motors simultaneously. The most common method where the speed of all motors in a system may be changed at one time is one in which the frequency of the applied voltage is varied. This type of motor is highly efficient and performs with a high degree of accuracy.

Some drives may convert electrical energy directly into mechanical energy, while other drives may have several steps of conversion. Though each conversion that takes place is highly efficient, the result of multiple conversions can produce a lower efficiency.

Induction motors and synchronous motors receive their power from inverters. The motor speed control is obtained by means of frequency control. From the synchronous speed formula derived in Chapter Eighteen, it can be shown that if all the components in the formula are held constant except for frequency,

the result will be a change in motor speed (radians/minute). An increase in frequency will increase the motor speed and, conversely, decreasing the frequency of the applied voltage will result in a lower motor speed. Figure 21-24 shows a diagram of a variable frequency ac motor speed control.

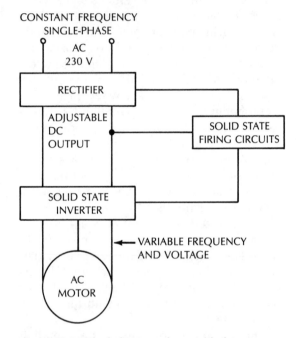

Figure 21-24 Block diagram of a variable frequency ac motor speed control.

There is one problem associated with variable frequency drives. When the frequency of the applied voltage is varied, the inductive reactance (X_L) of the stator winding varies as well. At low frequencies, the inductive reactance is low, and thus the motor current increases. This increased current will cause the windings to heat up. Therefore, drives must reduce the applied voltage in a constant proportion with the reduced frequency in order to maintain the current at a constant level.

In Fig. 21-24 the induction motor is being powered by a single-phase ac 230-V source. The 230-V single-phase ac is rectified to a direct current. The dc is fed to an inverter, that converts the dc to a three-phase ac power with a variable frequency. This output can be used to operate a conventional three-phase squirrel-cage induction motor.

Variable frequency drives may be provided with such special features as dynamic braking, reversing, power-factor correction, over-current, and high-low voltage protection.

Wider ranges of speed control may be attained by using electronic (microprocessor) variable speed drives. The microprocessor can produce a high pulse rate (per cycle) and control the size as well as the time interval of the pulse.

PROGRAMMABLE CONTROLLERS

Electromechanical switches are being made obsolete by electronic *programmable controllers*. A programmable controller (PC) is a digitally operating electronic apparatus that uses a programmable memory for the internal storage of instructions that implement specific functions such as logic, sequencing, timing, counting, and arithmetic to control machines and processes.

Unlike the computer, the programmable controller is designed specifically to interact directly with the industrial environment. It does not require as high a degree of programming knowledge as the computer.

The PC replaces a conventional relay system, and it can be used to control processes that couldn't be controlled by relays. The PC can actuate sequential steps in the function of any automated system. It can monitor a wide variety of inputs such as switches, pushbuttons, sensing devices, or outputs of other microbased devices. The PC also has a number of outputs, each capable of driving a relay, contactor, indicator light, buzzer, solenoid, motor, or any other solid state device.

The programming instructions can be easily written in relay ladder language. Each element in a conventional relay control system is represented by an identical symbol on the CRT (video) screen programming panel.

The program specifies the type of element used (solenoid, timer, etc.) and the exact location of each element in the circuit. Figure 21-25 shows a photo of a typical PC programming console.

Figure 21-26, shown on page 430, illustrates

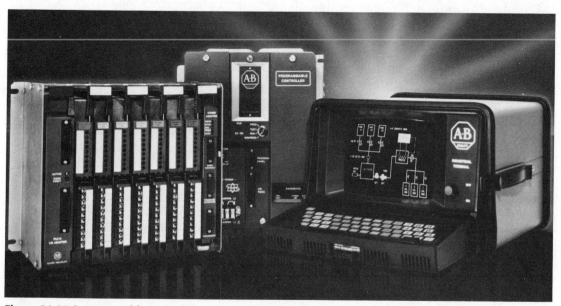

Figure 21-25 Programmable controller. *Courtesy: Allan-Bradley Canada.*

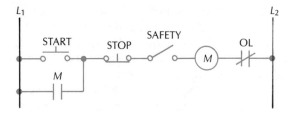

Figure 21-26 Relay control circuit.

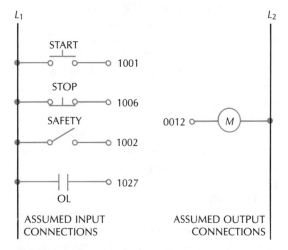

ASSUMED INPUT
CONNECTIONS

ASSUMED OUTPUT
CONNECTIONS

Figure 21-27 Working wiring diagram.

a typical control (ladder) diagram used in relay circuits. Figure 21-27 shows the same control elements in an assumed wiring diagram connected to input circuits in the input-output (I/O) configuration, with assigned output number 12, which will operate an external device. The resultant PC logic is shown on the video screen in Fig. 21-25 of the programming console. Each rung of the control diagram may be shown on the display at one time for purposes of monitoring output status, debugging the system, or editing the program. Once a program has been utilized and the process or machine has been updated, it is not necessary to rewire the machine as with relays or static control; the entire program can be changed in a short time by writing a new program.

The PC can gather and store data such as number of parts produced, machine downtime, rejects, tool changes, operating efficien-cies, and diagnostic monitoring, etc.

Diagnostic monitoring consists of logic checks preprogrammed to detect, identify, and report equipment faults on the video screen as they occur. Some of these faults do not necessarily stop the machine or process cycle by a few milliseconds longer than it should. Relay control systems cannot perform diagnostic monitoring.

SUMMARY OF IMPORTANT POINTS

1. Large ac motors should be started at a reduced voltage to reduce the high start-ing currents, which could cause large line voltage fluctuations and produce a great stress on the driven machinery.
2. Motor controllers are classified into four categories according to the functions they perform: (1) starting, (2) reversing, (3) run-ning, (4) stopping.
3. Motor control can be manual, semi-auto-matic, or automatic.
4. Magnetic contactors provide a safe, con-venient way of starting and stopping motors from a remote location. They are used where across-the-line starting can be safely utilized.
5. There are two methods of wiring a control circuit: (1) two-wire control, (2) three-wire control.
6. A relay is a magnetic auxiliary device used to control one or more circuits.
7. Solid state relays contain no moving parts, operate at a high speed, have no contacts to burn or wear out, do not produce haz-ardous arcing, have lower power require-ments. However, they are susceptible to transients, and have a certain amount of current leakage when in the **off** mode.
8. Reduced-voltage starters are used to reduce the inrush current, preventing excessive line voltage fluctuations, and limit the starting torque, which could damage the driven machinery.

9. Solid state motor starters regulate the inrush current and produce a smooth acceleration, electronically.

10. Interlocks prevent accidental reversal of rotation while the motor is running in one direction. This is accomplished by mechanical interlock, pushbutton interlock, and auxiliary contact interlock.

11. Sequence control is the control of a system that is equipped with several operating motor drives.

12. Jogging is the repeated closure of the circuit in quick succession, which applies short intervals of power to the motor, producing small movements in a driven machine.

13. A multispeed motor may use separate windings, or one winding that may be switched to produce two or four different speeds.

14. Variable frequency speed control is highly efficient and extremely accurate.

15. Programmable controller is a digitally operating electronic device. It uses a programmable memory to store instructions that initiate specific functions to control machines.

REVIEW QUESTIONS

1. Alternating current controls are classified by the methods used in starting; what are these methods?

2. What detrimental effects will be produced by the high starting current when ac motors are connected across-the-line?

3. What is meant by the term "pilot device?"

4. Motor controllers are classified into four categories according to the functions they perform. Name the four functions.

5. Describe the disadvantage of a manual starter in some applications.

6. List two applications where a manual starter would definitely be desirable.

7. Describe what is meant by "normally open" and "normally closed" contacts.

8. Explain why N.O. and N.C. contacts cannot be closed simultaneously.

9. Describe the difference between a manual, semi-automatic, and automatic control system.

10. State two advantages that the magnetic starter has over the manual starter.

11. Define the term "interlock."

12. Name the three types of interlocks.

13. Draw the control circuit only for each of the three interlocks in Q. 12.

14. Why are interlocks required on reversing starters?

15. Design (draw) a circuit for a single-phase motor where: (a) the control circuit will provide low-voltage release protection, (b) the motor must be protected from overloads, (c) the motor must have forward and reverse controls from one location, as well as stopbutton, (d) the starter must have both mechanical and pushbutton interlock.

16. Name the two categories that classify magnetic contactors, and describe the differences in the design between the two contactors.

17. Name the two methods of wiring control circuits, and explain the difference between the two.

18. Describe what a relay is, and explain what the difference is between a voltage and current relay.

19. List seven disadvantages of electromechanical relays.

20. Give two disadvantages of a solid state relay.

21. What are the three component parts of a solid state relay?

22. List the five common methods of reduced-voltage starting.

23. What are two advantages of reduced-voltage starting?

24. What main disadvantage is there to reduced-voltage starting?

25. What is the purpose of an overload relay?

26. How is the direction of rotation of a three-

phase motor reversed?

27. Why is it important that a motor be allowed to come to rest before reversing the rotation?
28. Describe how different speeds may be obtained in a dual-speed induction motor.
29. Explain what is meant by the term "jogging."
30. Draw and label fully an elementary diagram of an auto-transformer-type motor starter connected to a three-phase induction motor. Explain the operation of this circuit.
31. Give one advantage and one disadvantage of an auto-transformer starter.
32. Describe how the speed of a wound-rotor motor may be varied. What are the disadvantages of this method of speed control?
33. Draw an elementary diagram of a reversing controller that may be started in either direction with pushbuttons, equipped with electrical interlock.
34. What method is commonly used to start synchronous motors?
35. Explain the basic principle of controlling the speed of ac induction motors by varying the frequency of the applied voltage.
36. Why must the voltage be reduced when the frequency is decreased in a variable frequency speed controller?
37. List the advantages of static motor control.
38. Give two disadvantages of static motor control.
39. Give two advantages that programmable controllers have over a relay and static motor control system.

CHAPTER
TWENTY-TWO

INDUSTRIAL AND COMPUTER ELECTRONICS

The study of electricity and of electronics cannot be completely separated. For example, the control circuits used on many industrial electrical systems are actually electronic circuits. It should be obvious that the electronic field is extremely broad and is rapidly expanding and changing. These changes have resulted from the new demands of the world around us. Today, vacuum tubes have been largely replaced by solid state components, and computers and robots are becoming common everyday devices. These developments are among many that have opened up countless new areas of study for the student.

This field of industrial electronics is so wide that it will be possible to describe only a few of the components and their general applications here. This chapter will introduce the student to some of the more commonly used solid state components and to some of the fundamentals of analog and digital systems. It is hoped that this introduction will encourage the student to further readings and study in this fascinating field of ever-expanding opportunities.

SEMICONDUCTOR DEVICES

Semiconductor devices are also called solid state devices. These include diodes, transistors, SCRs, ICs and many others. Solid state devices are manufactured almost exclusively from the semiconductor materials silicon and germanium. Semiconductor materials are neither good conductors nor good insulators. However, their electrical characteristics may be altered so that they can act either to allow an easy current flow or to block a current flow.

The electrical characteristics of silicon and germanium may be altered by adding a controlled amount of impurities during the manufacturing process. This process is called *doping*.

When an N-type impurity such as arsenic or antimony is added to the semiconductor material, it produces loosely bound electrons in the material. The reason for this is because four of the five valence electrons from the dopant atoms combine with the four valence electrons of the semiconductor atoms to form covalent bonds. The fifth dopant valance electron does not become part of a covalent bond. This negative charge is held loosely in the valence shell and can easily be removed. The doped material is called a N-type material because of this relatively free negative charge. Note that the atoms of the semiconductor and dopant materials are neutral atoms. As such, there are actually no surplus of electrons in the doped material.

Another type of impurity such as indium or gallium may be used as the dopant. However atoms of these impurities have only three valance electrons. The formation of covalent bonds leaves a space in the dopant atom's valence shell. This space in the dopant's valence shell, in which there is no electron, is called a *hole* and can readily accept an electron. Since a hole can readily accept an electron, the hole acts like a positive ion, and for this reason the doped material is called a P-type material. Note again that the atoms of the semiconductor and dopant materials are neutral atoms. As such, there are actually no surplus or shortage of electrons in the doped material.

Because of these two possible arrangements of the valence electrons in a doped material, conduction through a N-type material is considered as conduction by negative charges, and conduction through a P-type material is considered as conduction by holes or positive charges. The actual conductivity of the material, however, will depend on the amount of doping.

THE PN JUNCTION

The properties of solid state components are due largely to the electrical characteristics at the junction between a P-type and N-type material. The PN junction is formed by fusing the two types of material in special ovens. The types and applications of such PN junctions seem endless because of the many ways in which the characteristics can be controlled by the amount of doping, type of material, method of activation, shape, and other factors.

When no external voltage is applied across the junction, some of the loosely held electrons from the N-type material will diffuse across the junction to the P-type material. These electrons will "fill" some of the "holes" in the P-type material, making these atoms into negative ions. The electrons from the N-type material that have diffused across the

junction leave behind them positive ions. Therefore a small voltage will exist across the junction (approximately 0.5 V) when silicon is used). This barrier voltage prevents further diffusion of charges across the junction, and the junction is now said to be in a state of equilibrium.

When an external voltage is connected across the junction with the negative terminal connected to the N-type material and the positive terminal connected to the P-type material, as shown in Fig. 22-1(a), a current will flow through the circuit, and the junction is said to be *forward-biased*. The negative charge of the voltage supply will repel the negative charges through the N-type material toward the junction. Similarly, the positive charge of the voltage supply will attract electrons from across the junction. This action causes a free flow of charges across the junction. Also, the junction will have a low resistance when it is forward-biased.

The effect is quite different when the N-type material is connected to the positive terminal and the P-type material to the negative

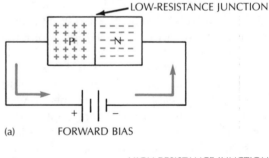

(a)   FORWARD BIAS

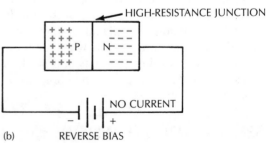

(b)   REVERSE BIAS

Figure 22-1 The PN junction: (a) forward-biased and (b) reverse-biased.

terminal of the voltage supply. With this reverse connection, charges will be drawn away from the junction. Consequently, there is no flow of charges across the junction, and the junction then has a very high resistance. The junction is now said to be *reverse-biased*, as shown in Fig. 22-1(b).

It was shown that when the PN junction is forward-biased, the junction has a low resistance, resulting in a high-current flow through the circuit. When it is reverse-biased, the junction resistance is very high and little or no current flows. Therefore, the PN junction will allow a current to flow in only one direction and, hence, can act as a diode or rectifier.

The response of a typical PN diode is shown in Fig. 22-2. When the forward bias voltage exceeds the barrier voltage, the forward current begins to rise. An increase in the forward bias will produce an increase in the junction current. When the diode is reverse-biased, the junction resistance is very high. There may be a small reverse current present in the diode due to "minority carriers" in the P and N materials. (The discussion of "minority carriers" is beyond the scope of this text.) At low operating temperatures, this current is usually considered insignificant. The current will also change very little for an increase in the reverse bias voltage but only up to the diode ava-

lanche breakdown value. At this voltage, called the *peak inverse voltage* (PIV), there is an abrupt increase in the junction current. Beyond this PIV value, a complete breakdown of the electron bond structure occurs, and the reverse current rises sharply.

RECTIFIER CIRCUITS

It was shown in the previous section that the PN junction will act as a undirectional current device and, hence, can be used as a diode or rectifier. The symbol for a diode and its forward current path is shown in Fig. 22-3.

Figure 22-3 The semiconductor diode.

Although both silicon and germanium diodes are used as rectifiers, the silicon diodes are more widely used because they can be operated at higher current and voltage levels. They can also withstand higher temperatures, and are small, highly efficient, and shock resistant.

A very important industrial use for the silicon diode is its use in rectifier circuits for converting ac voltage into dc voltage. One type of rectifier circuit is the full-wave bridge rectifier circuit shown in Fig. 22-4, on page 436. For analysis, consider the input half-cycle in which point A is negative with respect to point B. The diodes D_2 and D_4 are forward-biased and the current flow will be from point A through D_2, the load R, D_4, and back to point B. On the other input half-cycle, point B will now be negative with respect to point A, and the diodes D_3 and D_1 are forward-biased. The current flow is now from point B through D_3, the load R, D_1 and back to point A.

It is obvious that during each half-cycle, two diodes conduct in series, the other two being reverse-biased, and the current pulses

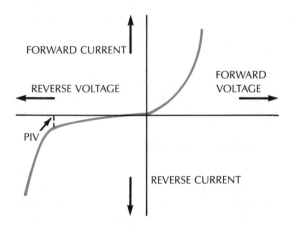

Figure 22-2 Forward and reverse current and voltage values for a diode.

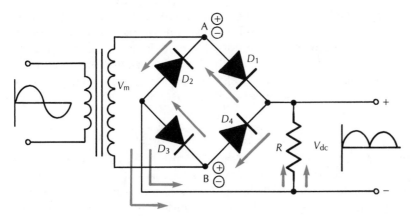

Figure 22-4 The full-wave bridge rectifier.

through the load are always in the same direction, as shown in Fig. 22-4. Study of the bridge circuit will also show that the reverse bias applied to each diode is equal to the peak secondary transformer voltage, V_m. Thus, the minimum PIV rating for the diodes used in the bridge circuit must be that of the peak secondary coil voltage, V_m.

Most amplifiers and electronic equipment require relatively constant dc voltage and current with a minimum amount of ripple. The ripple represents the ac pulsations in the rectifier dc output. These pulsations are usually undesirable. For example, if they are not sufficiently filtered out, they may cause an audible hum in audio circuits. The ripple content in a rectifier output is reduced by connecting a filter circuit to the rectifier output. One example of a filter circuit uses two capacitors and an inductor, as shown by the π-filter in Fig. 22-5.

The charging and discharging properties of the capacitor are responsible for maintaining a more constant dc output. When the input voltage exceeds the capacitor voltage, the capacitor will charge; when the input voltage drops lower than the capacitor voltage, the capacitor will discharge, thus maintaining a high voltage for a longer period of time than if no capacitors were used. Also, as shown in Fig. 22-5, the diodes will now conduct for only a short interval of high current. It is often said

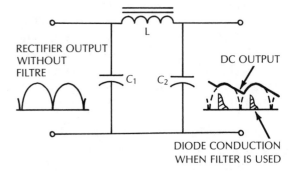

Figure 22-5 The π-filter circuit.

that the major effect of C_1 is on regulation and that of C_2 is on ripple.

The series inductor will also represent a series impedance to the ripple component in the rectifier output and, thus, it too will act as a filter. The inductor may be viewed as storing magnetic energy when the current is above the average value, and releasing this energy when the current falls below the average, thus maintaining a more constant output.

THE ZENER DIODE

Another type of diode often used as a voltage regulator is the *Zener diode*. Unlike the rectifier diode described previously, this diode depends on its reverse-biased properties. When the reverse voltage applied to the PN junction of a Zener diode is increased, the

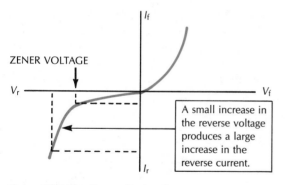

Figure 22-6 The Zener diode effect.

reverse current initially remains small and relatively constant. However, beyond a point called the Zener level, the current will increase by a very large amount for a small increase in the reverse voltage, as shown in Fig. 22-6.

In Fig. 22-6, it was shown that the Zener current can increase by a considerable amount for a small increase in the reverse-bias voltage. That is, the diode's dynamic resistance will decrease as the Zener current is increased. The use of this Zener property is illustrated in the voltage regulator shown in Fig. 22-7.

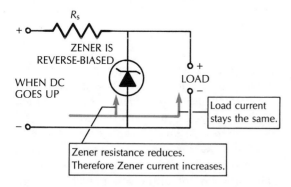

Figure 22-7 A voltage regulator using a Zener diode.

The Zener diode is connected in parallel with the operating load, and a limiting resistor is placed ahead in series, as shown in the diagram. The value of this resistance must be selected to fix the operating regulator voltage at a desired level, and also to prevent the power rating of the diode from being exceeded.

In the regulator circuit shown in Fig. 22-7, when the supply voltage increases, the Zener current will increase. However, as the Zener current increases, its resistance also decreases, and it will allow still more current to flow through it. This increase in current will increase the voltage drop across the limiting resistor and, hence, the voltage output to the load will remain almost constant.

If the supply voltage decreases, the Zener resistance will now increase. Therefore, the load current and voltage will again remain almost constant. It should be now evident that the Zener action in the regulator circuit will keep the load voltage constant even though the supply voltage may be changing.

THE LIGHT-EMITTING DIODE (LED)

The light-emitting diode is another family of semiconductor devices. They have the unique property of radiating light. Actually, all semiconductor devices possess some ability to convert a current into visible or infrared light. However, the light radiation from the semiconductors' silicon and germanium is extremely small and thus they are not used for this purpose. The material which made possible the construction of LEDs is the semiconductor material gallium arsenide. This material can be stimulated to emit light by an electron beam, a bright light source, or by an electric current injected through it.

When the PN junction of a LED is forward-biased, the electrons injected into the N-material must first move to a higher energy level, called the *conduction band*, in order to cross the junction voltage barrier. When these electrons cross the barrier, they then drop back to holes at a lower energy level. The energy converted in this process appears as light or radiated energy.

The voltage-current characteristic of a light-emitting diode is very similar to that of most junction diodes. As shown in Fig. 22-6, the forward current increases rapidly, with the voltage remaining nearly constant. The limit to the forward current is usually between 60 to 100 mA, without heat sinking. Also, a reverse voltage of up to about two volts will usually not cause any damage.

The characteristics of the LED lend themselves to numerous applications. For example, because of its linear current-light relationship, the LED is used in circuits such as in AM voice communications and optical potentiometers. Another useful property of this diode is its fast response time which makes it suitable for high-frequency optical systems. The LED is also very sturdy and has virtually an unlimited operating life. It is commonly used as indicators and in readout displays. Also, it is manufactured to emit infrared radiation. This property makes it suitable for use in some laser and detection systems.

The description of the many applications for the LED is beyond the scope of this text, and only three example circuits are given here. In Fig. 22-8, the LED is used as an indicator, and it

shows how the output may be varied from an off state to full brightness. The resistance R_S is used to limit the diode current to its safe level. The value of R_S may be determined from the equation

$$R_S = V_i - \frac{V_L}{I} \qquad (22\text{-}2)$$

where V_i is the source voltage
$\quad V_L$ is the LED forward voltage
$\quad I$ is the desired maximum current

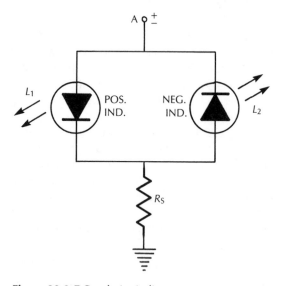

Figure 22-9 DC polarity indicator.

Figure 22-9 shows a simple but very useful polarity indicator. Since the LED possesses the same rectification properties of conventional diodes, only the forward-biased diode will light when a voltage is applied to the circuit. Therefore, when terminal A is positive, diode L_1 will emit light, and when terminal A is negative, L_1 will be off and diode L_2 will emit light.

A very common use for the visible LED is in the seven-segment numeric readout circuit, as shown in Fig. 22-10. The display incorporates seven separate rows of LEDs to permit the generation of digits 0 to 9.

As indicated in Fig. 22-10, the input signal to the display circuit is usually in binary-coded-

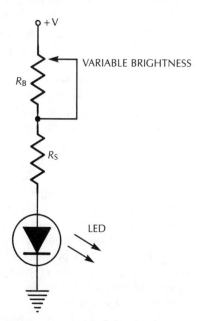

Figure 22-8 A simple LED circuit.

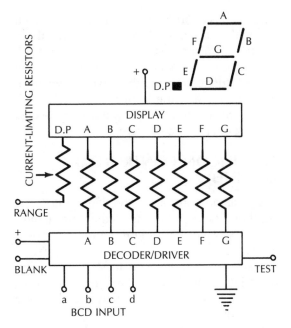

Figure 22-10 A basic BDC seven-segment circuit diagram.

decimal (BCD) format. This form of input signal requires a decoder-driver circuit to activate the appropriate segments in the display. The TTL 7447 is a popular decoder IC. Terms such as BCD and decoder will be described more fully later in this chapter. As shown in the circuit, this decoder contains provisions for both blanking all the segments and for testing the individual segments for proper operation. The BCD input combination and respective decoder output for each display 0 through 9 is given in the truth table of Table 22-1.

THE THERMISTOR

The thermistor is a type of temperature-to-resistance transducer. It is a relatively simple two-terminal device made from special mixtures of semiconductors such as magnesium, manganese, nickel, and oxides of cobalt. Its name is obtained from the phrase *therm*ally sensitive res*istor*. The thermistor has a wide

Table 22-1 Truth table for seven-segment display

BCD INPUT				OUTPUT TO SEGMENTS							DECIMAL DISPLAY
d	c	b	a	A	B	C	D	E	F	G	
0	0	0	0	0	0	0	0	0	0	1	0
0	0	0	1	1	0	0	1	1	1	1	1
0	0	1	0	0	0	1	0	0	1	0	2
0	0	1	1	0	0	0	0	1	1	0	3
0	1	0	0	1	0	0	1	1	0	0	4
0	1	0	1	0	1	0	0	1	0	0	5
0	1	1	0	1	1	0	0	0	0	0	6
0	1	1	1	0	0	0	1	1	1	1	7
1	0	0	0	0	0	0	0	0	0	0	8
1	0	0	1	0	0	0	0	1	0	0	9

range of applications, extending from temperature measurement and control to automatic regulation of electrical signals.

The effect of temperature on resistance can be a nuisance for the stability of certain circuits, and often elaborate steps are required to correct this problem. For this reason, conventional resistors are deliberately made to have a temperature coefficient of resistance as close to zero as possible. There are many other circuit applications, however, in which operation may be modified or controlled or certain measurements performed if the circuit's resistance can be made to vary with temperature in a predictable way. The thermistor is especially suited for this purpose because its resistance will always vary by a known amount for a given change in temperature. Unlike pure metals, however, most thermistors have *negative temperature coefficients* (NTC).

The circuit of Fig. 22-11 illustrates one of the many industrial applications for the thermistor. The circuit shown is a modified thermistor bridge thermometer and may be operated with either dc or ac voltages. The thermistor, R_3, is one arm of the bridge, and thus makes the resistance of that arm temperature-sensitive.

The bridge may be balanced at any desired temperature value by means of potentiometer,

R_4, which is another arm of the bridge circuit. When the bridge is balanced (null state), there will be no current through the coil or relay, L_1, and the relay is not activated. However, if the temperature changes, either up or down, from the preset value, the bridge becomes unbalanced and a current will flow through the coil, causing the relay to operate.

THE SILICON-CONTROLLED RECTIFIER

The silicon-controlled rectifier (SCR) is another member of the semiconductor family. It is a four-layer PNPN component, as shown in Fig. 22-12. It has normal blocking characteristics, since there is at least one reverse-biased junction in either direction. The use of the SCR has helped to simplify greatly many types of industrial control circuits.

The PNPN structure has three junctions, as shown in Fig. 22-12(a). When a forward voltage is supplied to the anode and cathode terminals, the junctions J_1 and J_3 are forward-biased, but junction J_2 is reversed-biased. If the terminal voltage is reversed, the junctions J_1 and J_3 become reverse-biased, and the SCR will exhibit the normal characteristics of reverse-biased diodes. Therefore, the SCR will normally not conduct in either direction.

It was shown that when the SCR is forward-biased, the junction J_2 provides the control of

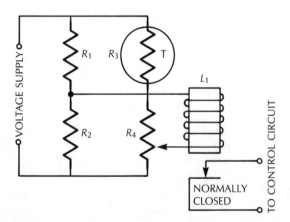

Figure 22-11 A thermistor bridge circuit for temperature control.

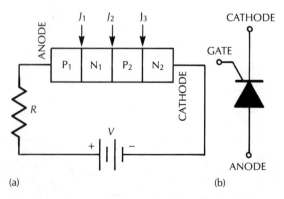

Figure 22-12(a) The SCR, PNPN, structure.
(b) Symbol for the SCR.

current through the device. There are two means of erasing this blocking characteristic and allowing a forward current to flow. One method is by increasing the forward voltage until avalanche breakdown occurs. This method, however, is usually not employed. Instead, the operating voltage is set well below the forward avalanche level and triggering is achieved by applying a gate current to the P-material between junctions J_2 and J_3. The anode, cathode, and gate terminals of an SCR are shown in Fig. 22-12(b).

The gating control of the SCR accounts for the usefulness of this device. The SCR requires only a short pulse to trigger it into conduction. That is, once the SCR is triggered, it continues to conduct even with the gate signal removed. Therefore, the SCR is like a latching switch. To turn the SCR off and return it to gate control, the forward-applied voltage must be turned off or reversed. The SCR will also turn off if the forward current drops below its holding limit.

The switching, latching, high current capacity and other properties of the SCR make it useful for many control applications. The complete description of these circuits, however, is beyond the scope of this text, and only two relatively simple but useful circuits are given below.

The circuit shown in Fig. 22-13 may be used to control the speed of universal brush-type motors found in devices such as food blenders, drills, and sabre saws. It is not suitable for controlling the speed of other types of motors. The operating speed is set through the use of an SCR. The speed may be varied from zero to approximately 80 percent of full speed by the potentiometer R_2. For 100 percent full speed, switch S_2 is turned on. This will circumvent the SCR control.

The setting of R_2 controls the amount of time the SCR conducts during each cycle of applied voltage. During one pulse of the ac cycle, the SCR is forward-biased, and will conduct when the appropriate gate signal is applied. During the other half of the cycle, the SCR is turned off because it is now reverse-biased. Hence, the SCR will only conduct dur-

---

PARTS

R_1—2.5 kΩ
R_2—500 Ω, 2 W, POT
R_3—1.5 kΩ, 0.5 W
D_1—Motorola HEP 162
D_2—Motorola HEP 162
SCR—Motorola HEP 302
S_1 and S_2—5 A rated

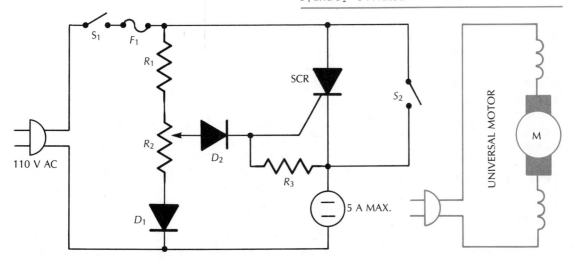

Figure 22-13 Speed control for a power tool equipped with a series universal motor.

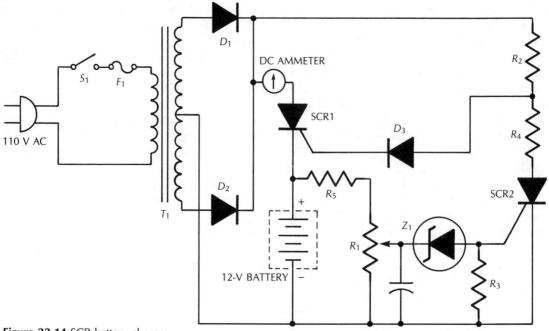

Figure 22-14 SCR battery charger.

ing a set interval of the forward ac voltage pulses.

The circuit shown in Fig. 22-14 may be used to charge a 12-V battery from a 120-V voltage line. When the battery is low, SCR1 is fired through D_3 and R_2, and a charging current flows to the battery. As the battery's voltage reaches the desired level set by R_1, the Zener diode, Z_1, which acts as a switch, will turn on SCR2. The current through SCR2 will now develop a voltage across R_2, removing the gate voltage to SCR1, and thus cutting off the charging current. If the battery voltage again drops below the preset value, the Zener will block, turning off SCR2. This will allow SCR1 to trigger back on and recharge the battery.

JUNCTION TRANSISTORS

Transistors are manufactured either as PNP or NPN devices and are made from either silicon or germanium. They have three terminals called the *collector*, *emitter*, and *base*, as shown in Fig. 22-15(a) and 2-15(b). Generally,

PARTS

T_1—Transformer, 24 V, centre-tapped
R_2 and R_4—27 Ω, 5 W
R_5—47 Ω, 1 W
R_3—1 kV, 0.5 W
R_1—500 Ω, 2 W, POT
C_1—100 μF, 25 V, electrolyte
D_1 and D_2—IR-20HB20
D_3—IR-5A4D
Z_1—1 W, 8.4 V
    IR-Z1108
SCR1—IR-SCR-03
SCR2—IR-SCR-01

silicon transistors have higher current capability than germanium transistors.

The major difference between the PNP and NPN transistor is in the method of biasing used and the resulting current direction through the transistor. For example, in Fig. 22-15(a), when the collector is made negative with respect to the emitter and a negative signal is applied to the base, the PNP transistor conducts and a current will flow *from the collector to the*

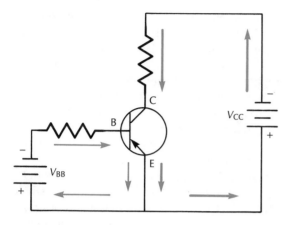

Figure 22-15(a) PNP common-emitter circuit.

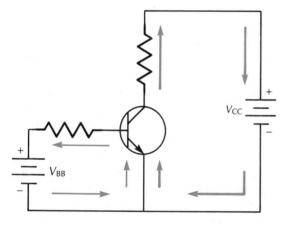

Figure 22-15(b) NPN common-emitter circuit.

emitter. If the base signal is reduced to zero or becomes positive, then the PNP transistor will not conduct. In the NPN circuit of Fig. 22-15(b), the collector must be made positive with respect to the emitter, and the base signal must also be positive in order for the NPN transistor to conduct. Note that the resulting current flow is now *from the emitter to the collector*. Simple tests may be made with an ohmmeter to determine whether a transistor is made from silicon or germanium, and whether it's a PNP or NPN transistor. These tests are left as research exercises for the student.

The circuits shown in Figs. 22-15(a) and 22-15(b) are called common-emitter circuits because in each circuit the emitter is common

to both the collector and base. Other configurations such as the common-base and the common-collector circuits are also possible. In the common-base circuit, the base is the common terminal. Similarly, in the common-collector circuit, the collector is made the common terminal.

Although transistors are enclosed in various sizes and shapes, they may be grouped into two basic types, those designed to amplify a signal from **off** through to a certain value and those designed to switch a signal to either **on** or **off**. Some transistors can also do both, amplify and switch. Transistors are also rated as either the small-signal type or the power type. Power transistors have larger physical enclosures and are mounted on some type of metal plate to help remove the excess heat produced.

Amplifying transistors are used in circuits such as radios and stereos where a weak signal must be amplified in order to be heard over a speaker. These types of systems are discussed in numerous electronics texts and will not be discussed here. Switching transistors are used in timing, counting, and computer circuits. In these circuits, the signal is either present or is not present. There is no "in-between." Some of these applications will be discussed later in this chapter.

THE FIELD-EFFECT TRANSISTOR

The field-effect transistor, or FET as it is often called, may resemble the junction transistor in outward appearance because it also has three terminals. However, the actual construction and principle of operation of the FET are very different from that of the transistor. The properties of the FET also make it more suitable than the transistor for many applications.

The FET as shown Fig. 22-16, for example, is made from a small cylinder of N-type material which acts as the channel. To this a small collar of P-type material is attached around the side of the channel to form a PN junction. The P-material forms the "gate" terminal, and the leads to either side of the channel form the "source" and "drain" terminals. The gate, source, and drain terminals of the FET may be compared to the base, emitter, and collector, respectively, of the transistor.

The FET made in this way is called a junction field-effect transistor, or JFET. When a P-type material is used for the channel, the FET is called a P-channel JFET; when an N-type material is used for the channel, the FET is called an N-channel JFET. The operation of the P-channel and N-channel JFETs are very much the same, with only their polarities being reversed. The symbol for the N-channel JFET is shown in Fig. 22-16.

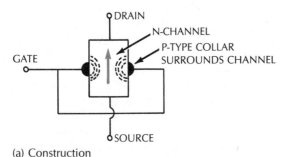

(a) Construction

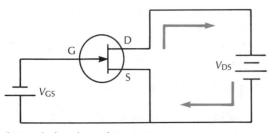

(b) Symbol and simple circuit

Figure 22-16 N-channel JFET.

Unlike the junction transistor in which the collector current is controlled by a forward-biased base current, the channel current of a JFET is controlled by a reverse-biased gate voltage. In a JFET, the channel current flows from source to drain. If the gate voltage, V_{GS}, is zero or forward-biased, then the channel current will be limited only by the applied voltage, V_{DS}, and the total circuit resistance. How-

ever, when the gate is reverse-biased, as shown in Fig. 22-16, the channel conduction is now under the control of the gate voltage, V_{GS}.

The reverse-gate bias creates a depletion region in the channel adjacent to the gate material. An increase in the reverse bias will increase the depletion region, thus increasing the channel resistance.

At a certain value of reverse bias, the channel conduction will be completely cut off. Therefore, the JFET may be used as a switch, just as is the transistors. Also, since a small change in the reverse-gate bias will produce a relatively large and proportional change in the channel current for a constant V_{DS}, the JFET may also be used as an amplifier. A third application of the JFET, as an inverter device, should also be evident. It was shown that when the gate bias is zero, the channel current is high, and when the gate reverse bias is increased, the channel current is reduced to zero.

An important property of the JFET is that the channel current is controlled by a voltage rather than by a current, as in the junction transistor. This is because the JFET has a very high input impedance. As a result, the JFET has a more quiet operating characteristic than the transistor and, also, its power consumption is lower. The JFET, however, is not completely perfect because there is still some leakage gate current. This leakage current may alter the JFET's operating characteristics and may even cause it to burn out.

THE MOSFET

MOSFET is the abbreviation for metal-oxide semiconductor field-effect transistor. It is a modified JFET and is also called an *insulated gate FET or IGFET*. As the latter name implies, the gate of this FET is insulated from the channel by a dielectric. The insulating layer is a thin film of silicon dioxide, as shown in Fig. 22-17.

The construction of the MOSFET is achieved by a diffusion process. As shown in Figure 22-17(a), the N-type semiconductor is

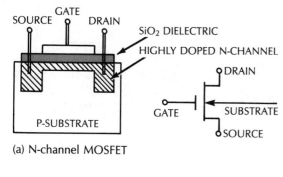

(a) N-channel MOSFET

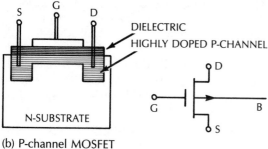

(b) P-channel MOSFET

Figure 22-17 Construction and symbols for the N-channel and P-channel MOSFETS.

diffused into a P-type substrate to form the channel, drain, and source. The channel deposit is made thinner than that for the drain and source terminals. Next, a thin layer of a dielectric (silicon dioxide) is grown on the channel. On top of this, a metal oxide layer is deposited, and this layer forms the gate. This arrangements produces a N-channel MOSFET. The P-channel MOSFET is constructed in a similar way, but with a N-type substrate and P-type channel.

The gate voltage controls the channel current by either depleting the charge carriers in the channel or by enriching them. For example, when the gate voltage is made positive in a N-channel MOSFET, electrons from the highly doped ends are drawn into the thin channel. This enriches the channel and thus reduces its resistance. When the gate voltage is made negative, the opposite charge movement occurs in the channel and its resistance increases.

This variation in channel for changes in the gate potential produces a significant change in channel current similar to the gate action in the JFET. However, there are two advantages. Since the gate is insulated from the channel, there is very little leakage current, and this reduces any energy loss in the input circuit. Secondly, as described, the MOSFET will respond to both positive and negative gate polarities, whereas the JFET will respond to only one bias polarity.

One important precaution must be observed when working with MOSFET components. The dielectric forms a small capacitor, with the gate and channel acting as the plates. Any sudden current spikes, from handling, for example, can damage the capacitor and, hence, the MOSFET. For this reason, MOSFETs are shipped with their leads shorted.

ENHANCEMENT MOSFET
There is a whole series of MOSFETs. One type is called the enhancement-mode MOSFET. The enhancement MOSFET has the advantage that it requires a relatively strong gate bias before it will allow channel current to flow. Since there is now no leakage current, the energy expended when the device is not conducting is practically zero and, as a result, this MOSFET is highly energy-efficient.

The channel of the enhancement MOSFET is not continuous. Therefore, a voltage simply connected across the source and drain terminals will not produce any channel current. However, when a sufficiently high-gate bias is added, the channel continuity is induced by electrons drawn from the source, drain, and substrate. As a result, a current will now flow through the channel. For the N-channel MOSFET, a channel current will flow only for a positive-gate bias and not for a negative bias. This condition is the reverse for the P-channel MOSFET.

A very useful switching device is obtained when a P- and N-channel enhancement-mode MOSFET are combined. The new device is often called a *complementary metal-oxide semiconductor* (CMOS). It is used extensively in digital circuits such as in calculators, watches, and computers.

INTEGRATED CIRCUITS

An integrated circuit, or IC, as it is often called, is a circuit containing components such as transistors, resistors, and capacitors, made very small and enclosed in a compact package. The circuitry enclosed may be just a simple amplifier or a complete system. Today, ICs are produced in the millions. These tiny packages of electronic marvel are now used, in varying degrees, in almost all areas of our society.

A complete discussion on the fabrication, types and various applications for the IC is beyond the scope of this text. Instead, student attention is directed to the general shapes, symbols, and a few applications of certain common ICs.

There are three major integrated circuit packages. These are the TO-5, the surface

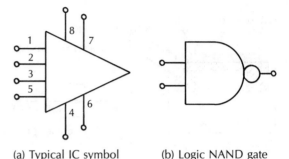

(a) Typical IC symbol   (b) Logic NAND gate

Figure 22-19 The most commonly used symbol for the IC is the triangular shape shown. Special symbols are used for logic ICs.

supported, and the dual-in-line types, as illustrated in Fig. 22-19. These ICs may either be soldered directly to PC boards or they may be socket-mounted.

Since the actual circuitry incorporated into an IC may be very complex, much time and space may be required to draw the entire circuit. As a result, it is common practice to use symbols to represent ICs. Once the operating properties of the IC are known, it becomes important only to recognize its external terminals. Figure 22-19 shows the symbols for two ICs. The triangular representation shown is the most commonly used symbol. Special symbols are used for logic devices, as shown in the example for the logic NAND gate. Some circuit applications for the IC may be found in the following sections on computer and logic systems.

ANALOG VERSUS DIGITAL SYSTEMS

It may be said that the development of IC technology made possible the widespread growth of computer and robotic systems in our modern society. Today, numerous computer devices utilizing several different ICs may be found in almost every aspect of our work and play. Some applications for computers are to solve complex mathematical problems, to prepare bills and make out

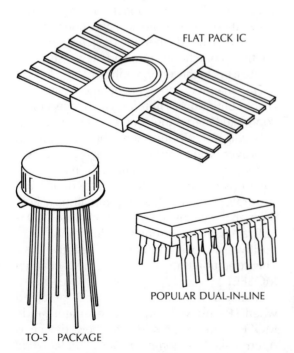

FLAT PACK IC

POPULAR DUAL-IN-LINE

TO-5 PACKAGE

Figure 22-18 The ICD is made in several shapes.

paycheques, to analyze and guide the function of a rocket, to predict the weather, to automatically operate machines, and to explore the relations in various science processes.

Computers can be classified into two general groups by the way in which they represent information. One type is called *analog* computers. In this type of computer, computation is performed by a continuous measurement of a physical quantity analogous to the quantities in the problem under consideration. The second group is called *digital* computers. In this type of computer, computation is performed by a numerical counting of discrete data (digital).

ANALOG DEVICES

Although the analog computer is not as widely used as the digital computer, there are certain purposes for which the analog system is more convenient. For example, the mercury thermometer, for measuring temperature, is an analog device. In this instrument, the distance along its stem is analogous or proportional to the temperature. The variable resistor or rheostat is another example of a common analog device. It may be seen that an analog system always mirrors the relations in the actual system. It is also possible to construct an analog computer to solve very complex problems, and the computer itself may contain only a few special devices.

Some of the disadvantages of analog systems, however, are their limitations for storage of data, speed, and accuracy. For example, in an electronic analog computer, variable quantities are represented to scale by means of electric voltages. Since the voltage is varied smoothly over its range, there are no discrete settings and the accuracy can easily be affected.

The detailed construction and operation of analog computers is beyond the scope of this text. However, there are two circuits that are worth a brief review because of their impor-

tance to both analog and digital computers. These two systems are the *differential* and *operational amplifier*.

THE DIFFERENTIAL AMPLIFIER

The fact that the differential amplifier is the key circuit of most ICs is reason enough to appreciate its importance. Figure 22-20 shows that the circuit is basically a direct emitter-coupled configuration. This means that few or no coupling capacitors are required between

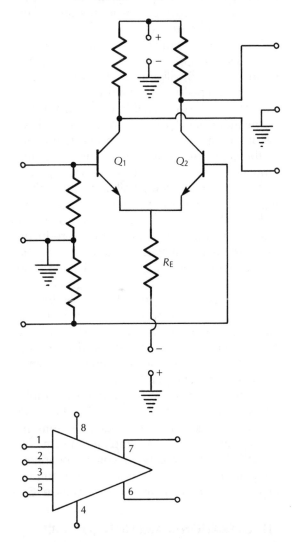

Figure 22-20 The basic differential amplifer and the typical triangle representation.

stages; this property saves space. The circuit also has fine stability and good ability to reject unwanted components.

The differential amplifier amplifies the difference voltage between the two base inputs. Customarily, the desired signal is applied to one base input while the other is grounded.

The outputs at the two collector terminals will then have the same magnitude but are out-of-phase. An alternative input will be to apply opposite polarity signals to the two bases. These two types of inputs are called *differential-mode inputs*.

For example, in the differential-mode operation, a signal applied to the base of transistor Q_1 in Fig. 22-20 will appear at both the common-emitter resistor, R_E, and the collector output of Q_1. Since the signal across R_E also serves as the input signal to transistor Q_2, the collector output of Q_2 will have the opposite polarity to the collector output of Q_1. Therefore, the amplifier, in effect, will act as a phase splitter.

If the two bases are paralleled so that their input signals have the same magnitude and polarity, the net signal across the common-emitter resistor will be greatly reduced because the two signals at this resistor will be subtractive. When perfectly balanced, the amplifier will now act like a bridge and the output from collector to collector will drop to zero. This type of in-phase input signal is called common-mode input, and it usually appears in the form of undesired hum and interference. Thus, the differential amplifier will reject this kind of interference.

Another property of the differential amplifier is its excellent dc stability. This is because any voltage variations will act as common-mode signals, which the amplifier in turn will reject. This dc stability to voltage changes, temperature, etc., is important, for example, in the construction of multistage circuits in ICs.

THE OPERATIONAL AMPLIFIER (OP AMP)

The operational amplifier, also called an "op Amp," is a special linear IC which is used in many analog, as well as digital, applications. It is used, for example, in circuits for waveform generations and shaping, signal amplification, instrumentation, impedance transformation, analog-digital converters, and to perform mathematical operations that include summations, subtractions, differentiation, and integration.

The op Amp is a multistage IC that includes circuits such as the differential amplifier, voltage regulation, and level shifting. The actual circuitry is beyond the scope of this text and, instead, only the often-used symbol is shown in Fig. 22-21.

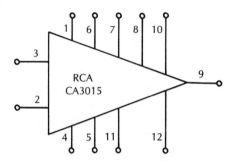

Figure 22-21 Operational amplifier IC symbol.

The op Amp uses a type of feedback that gives the amplifier high stability and characteristics that can be related back to the feedback network. Some op Amps also have high dc and ac voltage gain, and their common-mode rejection is excellent. Most operational amplifiers have push-pull inputs and single-ended outputs. An advantage of the difference input to an op Amp is shown by the inverting and noninverting circuits in Fig. 22-22. When the signal is applied to one of the differential inputs, as shown, the output is inverted. If the signal is transferred to the other input, the output retains the same polarity as the input. This means that the input can be selected to produce either an out-of-phase or in-phase output. The simple op Amp circuits for subtraction and summation are also illustrated in Fig. 22-22.

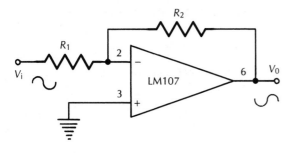

(a) Inverting

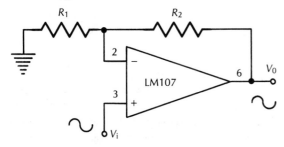

(b) Noninverting

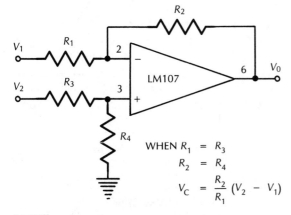

WHEN $R_1 = R_3$

$R_2 = R_4$

$V_C = \dfrac{R_2}{R_1}(V_2 - V_1)$

(c) Difference

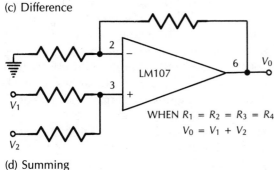

WHEN $R_1 = R_2 = R_3 = R_4$

$V_0 = V_1 + V_2$

(d) Summing

Figure 22-22 Basic op amp applications.

DIGITAL PRINCIPLES AND DEVICES

The development of microelectronics and IC technology has brought about a revolution in digital computers. The student is certainly aware of the great impact that digital devices and digital computers are making in industry and other institutions. Digital circuits may now be found in automobiles and toys. Digital controllers are being used to control complete sections in industrial systems, and computerized robots are being designed to perform numerous industrial tasks. Digital technology, therefore, is an important area of study which may lead to many job opportunities.

Digital devices operate by counting or adding discrete items such as gear teeth, holes punched in paper, or electrical pulses. Unlike analog systems, digital devices operate by converting any problem into arithmetic, specifically addition. For example, multiplication is performed by a series of additions. Therefore, digital operation involves controlled arithmetic. All problems must be broken down into arithmetical steps, and then these steps must be completed in some sequence.

THE BINARY NUMBER SYSTEM

Since digital systems deals with numbers, it is important that the student understand the types of number systems that are used. Although most people are accustomed to the decimal or base 10 system of numbers, there are other systems of numbers which are better suited for digital applications. The binary or base 2 system is the most popular, because it has only two discrete coefficients (1 and 0) which correspond to the two stable states (**on** and **off**) for most electronic components. The **on** state of a circuit is represented by 1 and the **off** state by 0.

A close look at the two systems of counting shown on page 450 will indicate that counting in the binary system is as easy as counting in the decimal system. The major difference

Table 22-2 Decimal and binary systems

BASE 10 (DECIMAL)	BASE 2 (BINARY)
0	0000
1	0001
2	0010
3	0011
4	0100
5	0101
6	0110
7	0111
8	1000
9	1001
10	1010

being, that whereas the "carry" in the decimal system occurs after 9, the "carry" in the binary system occurs after 1. Also, as shown, the binary representation of a number uses a greater number of characters or binary digits (bits). Note too, that a binary number such as 101 is not read as one hundred and one, but rather as one-zero-one.

$$496 = \frac{10^2 \mid 10^1 \mid 10^0}{4 \mid 9 \mid 6} \quad \text{Decimal (Base 10)}$$

$$1010 = \frac{2^3 \mid 2^2 \mid 2^1 \mid 2^0}{1 \mid 0 \mid 1 \mid 0} \quad \text{Binary (Base 2)}$$

The two numbers above show how numbers are "weighted." In the decimal system, each column's "weight" increases by a factor of 10, because 10 is base. Similarly, in the binary system, each column's "weight" increases by a factor of 2, because 2 is the base. From this reasoning, it should be an easy matter to understand how a binary number may now be converted into its decimal equivalent and vice versa. The example below illustrates these two conversion methods.

BINARY TO DECIMAL CONVERSION

$$101011 = (1 \times 2^5) + (0 \times 2^4) + (1 \times 2^3) + (0 \times 2^2)$$
$$+ (1 \times 2^1) + (1 \times 2^0)$$
$$= 32 + 0 + 8 + 0 + 2 + 1$$
$$= 43$$

DECIMAL TO BINARY CONVERSION

$$53 = 53$$
$$\frac{-32}{21} \longrightarrow (1) \times 2^5$$
$$\frac{-16}{5} \longrightarrow (1) \times 2^4$$
$$\frac{-0}{5} \longrightarrow (0) \times 2^3$$
$$\frac{-4}{1} \longrightarrow (1) \times 2^2$$
$$\frac{-0}{1} \longrightarrow (0) \times 2^1$$
$$\frac{-1}{} \longrightarrow (1) \times 2^0$$
$$= 110101$$

The process for addition, subtraction, multiplication, and division of binary numbers uses a similar set of rules as for that of base 10 numbers. This investigation is left as a research exercise for the student. Instead, a brief description of another binary number system is presented.

BINARY CODED DECIMAL

The binary coded decimal system, also called the BCD code, is a modification of the binary system. This code makes it easier to recognize decimal numbers because it uses a 4-bit binary code for each decimal digit in a number.

DECIMAL NO.	4	2	6
BCD CODE	0100	0010	0110

As shown in the example above, the decimal number 426 uses three 4-bit binary modules. Since each decimal digit may range only from 0 to 9, it follows that each 4-bit binary module will use only the binary codes 0000 through 1001. Therefore, in this system there are some binary combinations for each of the 4-bit module that is not used. The BCD code is sometimes called the 8-4-2-1 code because this is the weight assigned to each of the four bits in a 4-bit module.

DIGITAL LOGIC DEVICES

The term *digital* refers to the fact that a number in a digital device is represented by one and only one combination of binary digits, and the term *logic* is a science dealing with the rules of thought and reasoning. Digital logic, therefore, is based on binary mathematics used to predict the result of combining various binary propositions such as true and false statements. One form of logic is called Boolean algebra, named after its originator, George Boole.

Digital systems use logic mathematics. In fact, a characteristic of digital systems is the simplicity of the basic components. For example, in a computer there might be thousands of electronic circuits and components but only five to ten different types of circuits. These few circuits are the result of the simplicity of binary logic operations.

There are five fundamental operations called NOT, AND, OR, NAND, and NOR. These operations may be generated in several ways using different switching (**on-off**) circuits and components. The actual electronic circuits are not discussed here, but rather their logic symbols, shown in Fig. 22-23, will be used. These logic devices are also called *logic gates*. Combinations of these gates are used to perform the different operations in digital computer systems.

THE NOT GATE

The NOT gate is a simple *inverter*. If the input is *high* (1), then the output will be *low* (0); if the input is *low* (0), then the output will be *high* (1). In logic mathematics, a line above the input or output is used to indicate inversion. For example, if $A = 1$, then $\overline{A} = \overline{1} = 0$, and if $A = 0$, then $\overline{A} = \overline{0} = 1$.

THE AND GATE

The AND gate will have a high output only when all its inputs are high. The logic equation, Boolean rules, and the truth table for the AND logic are given below. The logic symbol is given in Fig. 22-23. Note that in logic mathematics, the "dot" represents the AND logic. The truth table is a simple means of displaying

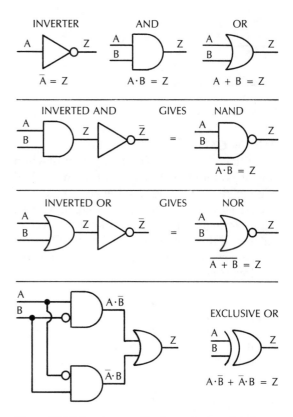

Figure 22-23 Fundamental logic symbols and their logic equations.

the output for different input combinations. For two inputs (A and B) there will be ($2^2 = 4$) four possible input combinations, as shown. If the device has n inputs, then there will be 2^n possible input combinations.

AND LOGIC EQUATION

$A \cdot B = Z$
(Reads A AND B)

BOOLEAN RULES

$0 \cdot 0 = 0$
$1 \cdot 0 = 0$
$0 \cdot 1 = 0$
$1 \cdot 1 = 1$

TRUTH TABLE

B	A 0	1
0	0	0
1	0	1

THE OR GATE

The OR logic gate will give a high output when any of its input is high. This is sometimes called the *inclusive OR* logic gate because it will also give a HI output when a combination of its input is high. The logic equation, Boolean rules, and the truth table for the inclusive OR gate are given below. Note that "+" represents the logic OR.

OR LOGIC EQUATION

$A + B = Z$
(Reads A OR B)

BOOLEAN RULES

$0 + 0 = 0$
$1 + 0 = 1$
$0 + 1 = 1$
$1 + 1 = 1$

TRUTH TABLE

B	A 0	1
0	0	1
1	1	1

A variation to the inclusive OR gate is the *exclusive OR* gate, abbreviated XOR.

The XOR gate will give a high output only when one of its input is high. Therefore, as shown below, the output will be low for a combination of high inputs. The output for different combinations can be easily verified using the Boolean rules for the AND, OR, and INVERT logic gates.

XOR LOGIC EQUATION

$\overline{A} \cdot B + A \cdot \overline{B} = Z$
(Reads NOT A AND B
OR A AND NOT B)

BOOLEAN RULES

$(\overline{0} \cdot 0) + (0 \cdot \overline{0}) = 0$
$(\overline{1} \cdot 0) + (1 \cdot \overline{0}) = 1$
$(\overline{0} \cdot 1) + (0 \cdot \overline{1}) = 1$
$(\overline{1} \cdot 1) + (1 \cdot \overline{1}) = 0$

TRUTH TABLE

B	A 0	1
0	0	1
1	1	0

THE NAND GATE

The NAND gate is the inversion of the AND gate. It is formed by combining the AND gate with a NOT (inverter) gate. This gate will produce a low output only when all its inputs are high. An important property of the NAND gate is that it can also be used as a building block for all other combination logic functions. This property will be illustrated in a later section.

The symbol for the NAND gate is given in Fig. 22-23. Note how the bubble at the end of the symbol is used to indicate the attached inverter. Also, the logic equation, Boolean rules, and truth table for a 2-input NAND gate are given below.

NAND LOGIC EQUATION

$\overline{A \cdot B} = Z$

BOOLEAN RULES

$\overline{0 \cdot 0} = 1$
$\overline{0 \cdot 1} = 1$
$\overline{1 \cdot 0} = 1$
$\overline{1 \cdot 1} = 0$

TRUTH TABLE

B	A	
	0	1
0	1	1
1	1	0

THE NOR GATE

The NOR gate is formed in a similar manner as the NAND gate. It is an inverted OR gate. This gate will produce a high output only when all its inputs are low. Like the NAND gate, the NOR gate can also be used as a building block for all other combination logic function. The symbol for the NOR gate is shown in Fig. 22-24, and the Boolean rules and truth tables are given below.

NOR LOGIC EQUATION

$$\overline{A + B + Z}$$

BOOLEAN RULES

$$\overline{0 + 0} = 1$$
$$\overline{1 + 0} = 0$$
$$\overline{0 + 1} = 0$$
$$\overline{1 + 1} = 0$$

TRUTH TABLE

B	A	
	0	1
0	1	0
1	0	0

The flip-flop (or bistable multivibrator) is called a memory or latch circuit because it will maintain its logic outputs determined by a set of inputs even after these inputs are removed. The flip-flop has two stable outputs (Q and $\overline{Q}$) which are always the opposite of each other. For example, for one set of inputs, the outputs may be Q = 1 and $\overline{Q}$ = 0, and when triggered by another set of inputs, the flip-flop will latch on to the inverse output state, Q = 0 and $\overline{Q}$ = 1.

When working with memory circuits, time becomes an important factor. It is important that the flip-flop change states at the right time. Also, it may be necessary to coordinate the overall action of the many flip-flops generally used in the device. To accomplish this, a square wave signal, called the *clock*, is sent to each flip-flop. The other inputs such as the S and R inputs are now called the conditioning lines. These inputs decide what the output of the flip-flop will be at the next pulse.

THE FLIP-FLOP—MEMORY CIRCUITS

Computers usually require the storage of large amounts of data in a readily available form. Devices used include paper tape, punched cards, magnetic cores, magnetic tape and disks, as well as electronic circuits. This section, however, will describe only how logic gates are combined to form electronic memory circuits, called *flip-flops*. These circuits are commonly used in computers for storing binary digit information which is used in intermediate high-speed operations. For example, flip-flops are used for storing the result of an arithmetic operation which is to be used in another operation.

Several different types of flip-flops have been developed. Some of these are the S-R (set-reset), J-K (S-R with no input restriction), D (delay), and T (toggle) types. Figure 22-24 shows the circuits for NAND and NOR S-R flip-flops, and Fig. 22-25 shows the J-K flip-flop.

THE S-R FLIP-FLOP

The S-R flip-flop is perhaps the most simple type of memory circuit. In the timing circuit and truth table given for the NAND S-R flip-flop circuit in Fig. 22-24, when the S (set) input is 1, the Q output will also latch to 1 if the R (reset) input is 0. The output, Q, will remain high even if the S input is now changed to low. However, if the R input is now made 1 and the S input made 0, the output, Q, will then latch

R	S	Q	Comment
0	0	*	Race
0	1	1	Set
1	0	0	Reset
1	1	NC	No change

R	S	Q	Comment
0	0	NC	No change
0	1	1	Set
1	0	0	Reset
1	1	*	Race

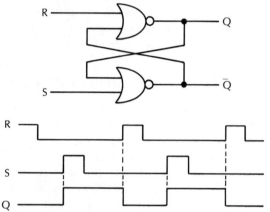

Figure 22-24 The NAND and NOR S-R flip-flops with their truth tables and timing diagrams.

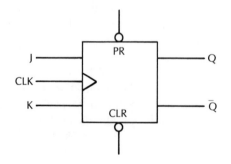

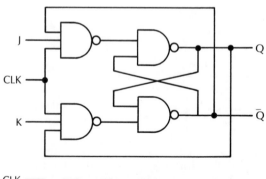

CLK	J	K	Q	Remarks
	0	0	*	Disable
	0	1	0	
	1	0	1	
	1	1	*	Toggle

Figure 22-25 The J-K flip-flop, symbol, and truth table.

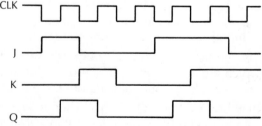

to 0. Q will now stay low even when input, R, is returned to low. The table does not show the output of $\overline{Q}$, but recall that when Q = 1, $\overline{Q}$= 0, and vice versa.

The S-R flip-flop will operate satisfactorily for the above stated input combination; however, the output for the NAND S-R flip-flop becomes unpredictable if both inputs (S and R) are simultaneously made 0.

This condition is called *racing*, and is undesirable. For the NOR S-R flip-flop, the racing condition exists when both inputs (S and R) are 1 instead of 0, as shown in Fig. 22-24.

THE J-K FLIP-FLOP

THE J-K flip-flop is ideal for use in building counters. Figure 22-25 shows a simple J-K flip-flop. Because of the double inversion through the two NAND gates, this flip-flop is enabled only when the CLK input is high. Also, if both J and K inputs are low, the two input gates will be disabled and the flip-flop will be inactive under any (1 or 0) input CLK signals.

The J-K flip-flop is set when J = 1, K = 0, and the CLK is high. If $\overline{Q}$ and input J are both high, a high CLK signal will set the flip-flop by making Q = 1 and sending $\overline{Q}$ = 0. The flip-flop will now hold on to this set state, except under the reset condition. For reset, J=0 and K=1. When the CLK is now made high, the output resets. It becomes Q=0 and $\overline{Q}$ = 1.

The J-K flip-flop may toggle when both J and K inputs are high. Each time the CLK input switches from low to high, the flip-flop will set or reset, depending on the current state of the output. This toggling effect can cause an oscillating or racing condition in the basic J-K flip-flop. For this reason, the J-K flip-flop is usually edge-triggered. In edge triggering, the flip-flop is made active only on the forward rise of the CLK pulse. This increases the propagation time between pulses and thus reduce the possibility of racing.

COMBINATIONAL LOGIC

Logic gates are widely used in industry for constructing numerous computer components and control circuits. For example, the flip-flop, which was formed by combining certain gates, is the basis in the design of such computer components as counters, shift registers, adders, decoders, and memory. In order to analyze and simplify logic circuits, a knowledge of Boolean algebra and such topics as Karnaugh mapping is often very useful. In this section, some additional Boolean rules will be summarized. (The Boolean rules for the basic logic gates were already stated in a previous section.) Some simple circuit applications will also be given. However, Karnaugh mapping is left as research reading for the student.

SUMMARY OF ADDITIONAL BOOLEAN RULES

(1) Involution: $\overline{\overline{A}} = A$
(2) Special properties: $1 + A = 1$
$$0 \cdot A = 0$$
$$0 + A = A$$
$$1 \cdot A = A$$
$$\overline{A} + A = 1$$
$$\overline{A} \cdot A = 0$$
(3) De Morgan's law:
$$\overline{A + B + C} = \overline{A} \cdot \overline{B} \cdot \overline{C}$$
$$\overline{A \cdot B \cdot C \cdot} = \overline{A} + \overline{B} + \overline{C}$$
$$(\overline{A \cdot B}) \cdot (C + \overline{D}) = ([A + \overline{B} + (\overline{C} \cdot D)])$$
(4) Indempotent laws: $A + A = A$
$$A \cdot A = A$$
(5) Absorption laws: $A + A \cdot B = A$
$$A \cdot (A + B) = A$$
(6) Commutative laws: $A + B = B + A$
$$A \cdot B = B \cdot A$$
(7) Associative laws: $(A + B) + C = (C + B) + A$
$$(A \cdot B) \cdot C = (B \cdot C) \cdot A$$
(8) Distributive laws: $(A \cdot B) + (A \cdot C) = A \cdot (B + C)$
$$(A + B) \cdot (A + C) = A + B \cdot C$$

Example circuit 1:
In this example, the task is to generate the NOR logic using only NAND gates. By using

De Morgan's rule, above, it is possible to change the NOR logic equation to its NAND logic equivalent.

De Morgan's rules states that two or more expressions all joined by the same operations (AND or OR) may be changed by: (a) complementing every expression, (b) complementing the whole group, and (c) changing the operation. Therefore, the NOR equation may be changed, as shown below. The result can also be easily verified by constructing the truth table. The final equivalent circuit is shown in Fig. 22-26 (b) and (c).

$$\overline{A + B} = Z \quad \Rightarrow \quad \overline{\overline{A} \cdot \overline{B}} = \overline{Z}$$
$$\overline{A} \cdot \overline{B} = Z$$

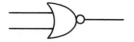

(a) NOR

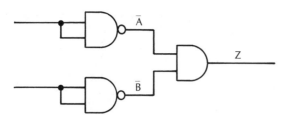

(b) Equivalent Circuit

(c) Simplified (Bubbled AND)

Figure 22-26 (a) NOR gate.
(b) Boolean equivalent NOR gate.
(c) Simplified equivalent called the bubbled AND gate.

Example circuit 2:
In digital design, it is often convenient to first construct the truth table for a required system. Once this is done, the next step is to write the logic equation for the system. This can be done by the *sum-of-products* method. In this

method, the fundamental products of each high output in the truth table is ORed. The final step is then to transform this equation into an equivalent logic circuit.

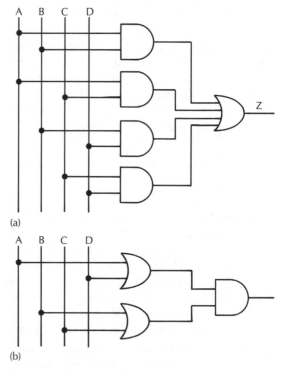

(a)

(b)

Figure 22-27 (a) Circuit for A*B + A*C + B*D + C*D.
(b) Simplified equivalent circuit (A + D) * (B + C).

Suppose the logic equation derived from the truth table of a required system was : $A \cdot B + A \cdot C + B \cdot D + C \cdot D = Z$. The logic circuit can now be easily constructed using four AND and one OR gates, as shown in Fig. 22-27(a). However, although this circuit will perform the required logic, it is not necessarily the simplest possible logic circuit.

In this example, the task is to simplify this circuit using Boolean rules. The original logic equation is rewritten below, and then it is simplified by factoring and rearranging terms.

$$A \cdot B + A \cdot C + B \cdot D + C \cdot D = Z$$
$$A \cdot (B + C) + D \cdot (B + C) = Z$$

$(A + D) \cdot (B + C) = Z$

The circuit for the simplified logic equation is shown in Fig. 22-27(b). Note that it uses fewer components. Also, it has only six input leads compared to twelve for the original circuit. Derive the truth table for each of the two equations and compare them. Do this by plugging in the different combinations of 0 and 1 for the inputs A, B, C, and D. The results should indicate identical truth tables.

Example circuit 3:
In this example, logic gates are used to perform the sealing action in a simple across-the-line starter control system. The task is to start the motor by momentarily pressing the start push button and to stop it by momentarily pressing the stop push button.

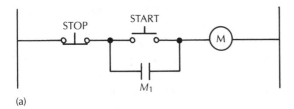

(a)

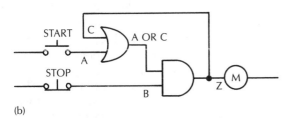

(b)

Figure 22-28 (a) Across-the-line starter using contactors.
(b) Across-the-line starter using logic gates.

Figures 22-28 (a) and (b) show the two possible designs for the across-the-line starter. An analysis of the OR-AND gate logic combination will show that it performs the same start-stop action as the magnetic starter.

MICROCOMPUTER SYSTEMS

The microcomputer technology revolution was made possible by the ability of engineers to fabricate very complex electronic circuits on integrated circuit substrates. These devices are now being used in a wide range of applications such as in grocery scales, tests equipment, display controls, cameras, TV games, industrial machine controls, and in automobiles and airplanes. The growth of possible application for microcomputers is nearly limitless.

A detailed discussion of the microcomputer is beyond the scope of this text. In fact, there are now numerous texts that are solely devoted to this area of study. The brief description presented here should dispel some of the mystery and provide a simple and general understanding of these systems. Also, the student should feel more confident in doing any additional research and experimentation in this fascinating field.

THE COMPUTER SYSTEM

The basic sections of a computer are shown in Fig. 22-29. They are: (a) a memory, (b) a control unit, (c) an arithmetic-logic unit (ALU), (d) input and output (I/0) devices, and (e) a number of registers.

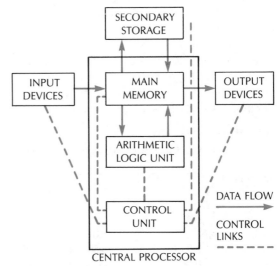

Figure 22-29 The basic sections of a computer.

The control unit, the ALU, and a number of registers are usually formed on a single chip. This chip, therefore, forms the heart of any computer system and is often called the microprocessor unit (MPU). These MPU chips may have as many as 5000 electronic components organized into thousands of gates. These gates are then organized into controllers, registers, counters, etc. Two widely used microprocessors are the Intel 8080 and the Motorola 6800.

PROGRAMMING THE MPU

In order to make use of the tremendous properties of a MPU, it must be programmed with a set of step-by-step instructions. The instructions that a MPU understands are called its *instruction set*. Some common MPU instructions are summarized in Table 22-3.

Abbreviated instructions like those listed below are called mnemonics (memory aids). They are useful in computer work because they describe the operation being executed. Computer programs using mnemonics are called *assembly language programs*.

In order to load these instructions and other data words into the microcomputer, some kind of binary code, which the MPU can interpret, is required. This code tells the MPU what instruction to perform. It is called the *operation code* (op code). Some examples of the op code are listed in Table 22-3. Computer instructions using strings of 0s and 1s are called *machine language programs*. Machine language programs may also be written in hexadecimal instead of the binary number system.

A SIMPLE PROGRAM

Table 22-4, on opposite page, on opposite page, is an example of how an MPU may be programmed to solve the arithmetic problem $16 + 24 - 12$.

Table 22-4 shows the assembly language and the two machine language representation for solving this arithmetic problem. Note that in each of the programs the instructions are stored ahead of the data. Also, note that the program information is stored starting with address memory location "0." The "H" which appears after any of the listed numbers is a reminder that it is a hexadecimal number.

The machine language program shown in Table 22-4 is the direct translation of the assembly language version. The student should also note that the contents column of the programs contains two bits of information, namely: an instruction field and an address field. For example, in the statement SUB 7H, the "SUB" part (or 0010 in machine language) is the instruction field, and the "7H" (or 0111 in machine language) is the address field. Note, too, that the OUT and HLT do not require any address field information.

In the assembly language program, the data (decimal numbers 16, 24, and 12) are stored in addresses 5H, 6H, and 7H, respectively. The corresponding hexadecimal numbers for the decimal numbers 16, 24, and 12, as shown, are 10H, 18H, and CH. Therefore, the first instruction, LDA 5H, means load the accumulator with the contents of memory address 5H. The content of address 5H is given in data as 10H.

The second instruction, ADD 6H, will now add the contents of memory location 6H (which is 18H) to the content already in the accumulator. Similarly, the instruction, SUB 7H, will reduce the content of the accumula-

Table 22-3

MNEMONIC	OPERATION	OP CODE
LDA	Load RAM data into accumulator	0000
ADD	Add RAM data to accumulator	0001
SUB	Subtract RAM data from accumulator	0010
OUT	Load accumulator data into output register	1110
HLT	Stop processing	1111

Table 22-4

| ASSEMBLY LANGUAGE | | MACHINE LANGUAGE | | | |
| | | | BINARY | | HEXADECIMAL |
ADDRESS	CONTENTS	ADDRESS	CONTENTS	ADDRESS	CONTENTS
0H	LDA 5H	0000	0000 0101	0H	05H
IH	ADD 6H	0001	0001 0110	1H	16H
2H	SUB 7H	0010	0010 0111	2H	27H
3H	OUT	0011	1110 XXXX	3H	EXH
4H	HLT	0100	1111 XXXX	4H	FXH
5H	10H	0101	0001 0000	5H	10H
6H	18H	0110	0001 1000	6H	18H
7H	CH	0111	0000 1100	7H	0CH

tor by the content stored in location 7H. The OUT instruction will now transmit the accumulator content to an output register. From this register, the result can now be displayed, printed, etc. Finally, the HLT command stops the MPU's processing.

THE CONTROL OF FETCH CYCLE
When the MPU is executing a program, the order and rate at which information is fetched, executed, and stored are very important functions within the MPU. These functions are carried out by the *control unit* within the MPU. The control unit contains a ring counter, which generates control words at clocked intervals (timing states.) The total timing states, called the *machine cycle*, are divided into two sections, a *fetch cycle* and an *execution cycle*. These timing states occur very rapidly and are measured in nanoseconds.

During the fetch cycle, the address in the program counter (PC) is transferred to the address register. This is called the *address timing state*. In the next timing state, the PC is incremented with the new address for the next fetch cycle. The PC is part of the control unit. It keeps track of the order at which addressed instructions are to be executed. In the final timing state of the fetch cycle, the addressed instruction pointed to in the RAM is now transferred to the instruction register.

In the execution cycle, data transfer during the different timing states depends on the program instruction being executed. Although the number of timing states remains constant, the actual active states and the type of registers used can be different. This is briefly shown in the comparison for the LDA and ADD instructions below.

If at the end of the fetch cycle, the instruction is LDA 5H, then during the first timing state of the instruction cycle, the instruction field, LDA, is decoded, and the address field, 5H, is stored in the memory address register. In the next timing state, the data pointed to in the RAM by this address are now loaded into the accumulator. All the other timing states in this cycle are now inactive.

If the program instruction is ADD 7H instead of LDA 5H, then during the first timing state, the instruction field, ADD, is again decoded and the address field, 7H, is stored in the memory address register. In the next timing state, however, the data pointed to in the RAM by this address are transferred to an *adder register*. During the third timing state, the contents of the adder and accumulator registers are now combined, and the new result stored in the accumulator.

MICROPROGRAMMING
The foregoing discussion on the machine cycle indicated that the controller sends out a control word during each timing state. These

control words tell the computer what to do. These small-step instructions are called micro instructions and are permanently stored in the ROM (read-only memory) of the MPU. Note the difference between this and RAM (random access memory). Program instructions and data entered into the MPU are stored in RAM. Therefore, information can both be written into and read from the RAM. However, information can usually only be read from the ROM.

In order to differentiate between these small-step instructions (micro instructions) and actual programming instructions such as LDA, ADD, SUB, etc., the term macro instruction is used to describe the latter. A macro instruction such as ADD will involve at least three micro instructions. The micro instruction in the ROM, which is needed for the execution of a macro instruction, is accessed by applying the appropriate address for that micro instruction. The controller sets up the sequence of micro instruction addresses.

INTERFACING

The MPU chip, when properly programmed, is capable of doing numerous calculations at very fast speed. However, it will be a practically useless device unless it can be interfaced with the outside world to both receive operating data and to provide readable outputs. Every use that the MPU may be put to involves some kind of input and output signals. Interfacing is so important to the MPU that some interfacing circuitry is often part of the MPU.

The interfacing unit provides the boundary link between the MPU and the outside world. It provides the go-between which will respond to data being sent to the MPU and data coming from the MPU. Three reasons why the MPU cannot be connected directly to the outside devices are: (a) the power of the MPU busses is too small to drive most external equipment, (b) the output device may be analog, whereas the MPU is digital, and (c) the MPU itself must be isolated from the output devices so that it can communicate with various circuits, not just one. Examples of interfacing are almost limitless. A few examples are described below.

The simple calculator and the microcomputer are two common examples which use interface chips to connect, for example, their input keyboards and their output displays to their MPUs. Of course, the microcomputer may use additional interface circuits to connect to such other devices as printers and industrial measuring circuits. In the case of the keyboard, the interface chip must decode the input signal to the language understandable by the MPU. For the display screen, the interface chip must convert the output from the MPU back into readable characters. It must also increase the power to drive the video screen.

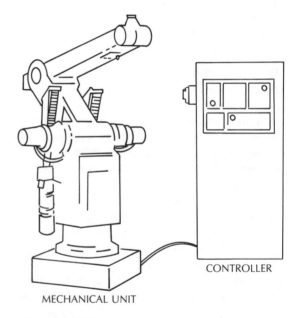

Figure 22-30 A robotic system.

A robotic system with a programmable controller is another example which uses some kind of interface circuits. The controller itself is digital, whereas the operation of the robot may contain both analog and digital signals. The programmable controller is a specially designed computer. The operator uses it to program the robot with a particular pattern of motions. The entered program is stored in the computer's memory so that the sequence of instructions can be repeated over and over by the controller. The computer monitors and controls the input/output signals between the robot's operations and the control system.

MPUs can be designed to handle input/output signals in different ways. For example, some MPUs operate on a 8-bit code, while others may use a 16-bit code. An 8-bit code will require a 8-line data bus and, thus, the interface connector will need to have 8 pins in order to communicate with the MPU.

It is also possible to simplify the interface connection by using serial data output. Only one data line is now required instead of 8 lines. However, the MPU must now be instructed to send its 8-bit code, one bit at a time. This takes up more MPU time, but it makes the interface cheaper. When more than one similar circuit is involved, for example, as in the number of 7-segment displays in a calculator, a type of parallel arrangement in the interface is also possible.

Any peripheral device may be interfaced to a MPU using the appropriate combinational logic circuit. To simplify the design of these logic interface circuits, special *peripheral interface chips*, called PIAs, have been designed. One such PIA is the Motorola 6820. This PIA is digital and is compatible with many 8-bit MPUs. The 6820 PIA also has its own set of registers. These registers allow the PIA to control the flow of data. It can send information or receive information from the MPU, while at the same time it is either receiving information from or sending information to the peripheral.

Many peripheral applications are also non-digital. Some examples are a pen recorder, temperature measurement, and the operation of a robot. These devices will require analog interfacing. Among other functions, analog PIAs must convert the analog signals from the peripheral to digital signals before they can be presented to the MPU. Similarly, digital signals from the MPU must be reconverted to analog signals before being sent to peripheral device. Analog PIAs, therefore, must contain both analog-to-digital converters (ADC) and digital-to-analog converters (DAC). The op Amp described in an earlier section is an important circuit in the design of converters.

SUMMARY OF IMPORTANT POINTS

Semiconductor devices

1. A doping process is used to produce P- and N- type materials.
2. The PN diode will allow current flow only when it is forward-biased.
3. A filter circuit is used to remove the ac ripple from a dc rectifier output.
4. The Zener diode depends on its reverse-biased properties. A common application for this diode is in voltage regulators.
5. The LED will emit light when a current flows through it.
6. The thermistor is a type of temperature-to-resistance transducer. They usually have negative temperature coefficients.
7. The SCR is a four-layer PNPN component. When forward-biased, it will trigger into conduction only when its gate signal is applied.
8. The main difference between PNP and NPN transistors is in their method of biasing. They are used in both amplifying and switching circuits.
9. The FET has three terminals: gate, source and drain. Unlike the transistor, it is controlled by a reverse-biased gate voltage rather than a forward-biased base current.
10. The MOSFET is an insulated gate FET. There are various MOSFET devices.
11. Three major IC packages are TO-5, the surface-supported, and the dual-in-line types.

Computer devices

12. An analog computer operates by taking continuous measurements of the physical

quantities analogous to the quantities in the problem.

13. In digital computers, computation is performed by numerical counting of discrete data.

14. The differential amplifier amplifies the difference voltage between two inputs. It is a direct emitter-couple circuit and forms a key part of most ICs.

15. The op Amp is a linear IC used in many analog and digital applications such as in converters, integrators, and summing circuits.

16. Binary digits (bits) have an increasing "weight value" of 0,2,4,8,16, etc.

17. In the BCD code, each module is used to represent numbers between 0 (0000) and 9 (1001).

18. The six basic logic gates are: NOT, AND, OR, NAND, NOR, and XOR.

19. Either the NANDs or NORs may be combined to perform any logic.

20. A knowledge of Boolean rules is useful in simplifying some logic circuits.

21. The basic circuit in counters, adders, memory, and other computer components is the flip-flop. The flip-flop will latch on to one of two possible stable states. Some types of flip-flops are the S-R, clocked D, and the J-K.

22. The MPU is the heart of any microcomputer system. The main parts of a MPU are the registers, memory, the control unit, and the ALU.

23. The instruction set (op codes) for a MPU may be written in either assembly or machine language. However, it can only be entered into the MPU in machine language form.

24. Two important sequences in a computer control cycle are the fetch and execution cycles.

25. The micro instructions for each op code (macro instruction) are stored in the ROM.

26. The MPU is interfaced to the outside world through some kind of interface chips (PIAs).

REVIEW AND PROBLEM QUESTIONS

1. Briefly describe how the electrical properties of silicon are modified to produce (a) a P-type semiconductor material and (b) a N-type semiconductor material.

2. Name four factors that affect the amount of conduction across a PN function.

3. Briefly describe how the small barrier potential that is present across a PN junction is formed.

4. Describe the conduction of (a) a forward-biased PN junction and (b) a reverse-biased PN junction.

5. Describe the conduction properties of a diode as its reverse bias voltage is increased from zero to its PIV value.

6. Draw a bridge rectifier circuit and briefly describe its operation.

7. Describe the purpose and operation of a filter circuit connected to a rectifier output.

8. Briefly describe the operation of a Zener diode and give an example of where it is used.

9. (a) Briefly describe how a LED emits light and (b) list five industrial applications for this device.

10. A LED with a forward voltage of 2 V is to be connected across a 12-V supply. Determine the series resistance required to limit the diode current to 150 MA.

11. (a) Draw a circuit to show how LEDs may be used as a voltage polarity indicator.
    (b) Briefly describe the operation of your circuit.

12. (a) Describe the operating properties of thermistors and, (b) list four industrial applications.

13. Construct a circuit to show how a thermistor may be used to turn off another circuit if the temperature were to vary from a set value. Briefly describe the operation of your circuit.

14. (a) Describe the junction characteristics of a SCR when it is forward-biased.

(b) Explain how this device is triggered into conducting and how it is turned off.

(c) List three applications for this device.

15. Draw a circuit and briefly describe how a SCR may be used to control the speed of an electric drill.

16. Draw a circuit and briefly describe how SCRs may be used to charge a battery from a 120-V ac supply.

17. (a) Identify the three leads of a transistor.

(b) Illustrate the biasing used for a PNP and for a NPN transistor.

(c) Briefly describe when the above transistors will conduct and when they will not.

18. Describe a simple test for determining (a) whether a transistor is made from silicon or germanium and (b) whether it is PNP or NPN.

19. Give two general type of grouping for transistors.

20. (a) Briefly describe the construction of the N-channel JFET, and, (b) describe how the device's channel current is controlled.

21. Compare the operating characteristics of the JFET with that of the junction transistor.

22. Briefly describe the construction of a MOSFET device.

23. (a) What is an IC? (b) Draw the commonly used symbol for an IC. (c) What are the three major methods of packaging ICs?

24. Describe how information is represented in (a) an analog device and (b) a digital device. (c) Give two examples of these two devices.

25. (a) Draw a circuit for a basic differential amplifier.

(b) Explain the general circuit operation under (i) differential mode input and (ii) common mode input.

26. (a) List five types of circuits or systems which may use operational amplifiers.

(b) List four important properties of this device.

27. Convert each of the following decimal number into their binary equivalent: (a) 8 (b) 15 (c) 22 (d) 27 (e) 5 (f) 7 (g) 9 (h) 11.

28. Define (a) the term "bit" and (b) state the "weighted value" for each of the first four columns in a binary number.

29. Convert each of the following binary numbers to their decimal equivalent: (a) 1010 (b) 1110 (c) 10111 (d) 0110

30. Give an example to illustrate (a) the addition of two binary numbers, (b) the subtraction of two binary numbers, and (c) the multiplication of two binary numbers.

31. (a) Describe the properties of the BCD number system.

(b) Give the BCD equivalent for the decimal numbers 27 635, and 909

32. Explain why the BCD code is also called the 8421 code.

33. Give (a) the symbol, (b) logic equation, and (c) the truth table for each of the five fundamental logic gates.

34. (a) List the gates that can be used as inverters. (b) Compare the properties of the OR and XOR gates. (c) Draw the symbol and write the equation for the XOR gate.

35. Draw the truth table and the circuit for each of the following equations:

(a) $A \cdot B + C \cdot D = Z$

(b) $A + B \cdot C + D = Z$

36. Construct the following systems using only NAND gates:

(a) $A + B = Z$ (b) $A \cdot B + C \cdot D = Z$

(c) $A + B \cdot C + D = Z$

37. Construct the following systems using only NOR gates:

(a) $A \cdot B = Z$ (b) $A \cdot B + C \cdot D = Z$

(c) $A + B \cdot C + D = Z$

38. List six types of devices that are used for data storage in computers

39. (a) Name three properties of the flip-flop circuit.

(b) Explain why a clock input signal is applied to all flip-flops in a computer device.

40. Draw the circuit and briefly describe the operation for:

(a) the S-R flip-flop and (b) the J-K flip-flop.

41. List five computer logic devices that are formed from the combination of logic gates.

42. (a) Draw the circuit for $\overline{A + B} = Z$ using only NAND gates.
    (b) Draw the circuit for $A \cdot B = Z$ using only NOR gates.

43. Determine the NAND equivalent equation for:
    (a) $(A \cdot B) + C = Z$
    (b) $A + B + C = Z$
    (c) $(A \cdot B) + (C \cdot D) = Z$
    (d) $(\overline{A \cdot B}) + \overline{C} = Z$

44. Write the Boolean equation for the following truth tables:

(a)

A	B	Z
0	0	0
0	1	1
1	0	1
1	1	0

(b)

A	B	C	Z
0	0	0	0
0	0	1	0
0	1	0	1
0	1	1	0
1	0	0	0
1	0	1	1
1	1	0	0
1	1	1	0

(c)

A	B	C	Z
0	0	0	0
0	0	1	1
0	1	0	1
0	1	1	0
1	0	0	1
1	0	1	0
1	1	0	0
1	1	1	0

45. Construct the logic circuit for each of the three truth tables above.

46. Draw the truth table for each of the given logic expressions:
    (a) $A \cdot \overline{B} \cdot C = Z$
    (b) $A + B \cdot C = Z$
    (c) $A \cdot \overline{B} + \overline{A} \cdot B = Z$
    (d) $A \cdot \overline{B} \cdot C + \overline{A} \cdot B \cdot \overline{C} + A \cdot \overline{B} \cdot \overline{C} = Z$

47. (a) Draw the logic circuit for the expression $\overline{A} \cdot B + A \cdot \overline{B} + A \cdot B = Z$
    (b) Simplify the above expression and redraw the circuit.

48. (a) Construct and describe the operation of a simple start-stop circuit using logic gates.
    (b) Write the logic equation and draw the truth table for the start-stop circuit.

49. List the major sections found in (a) the MPU device and (b) the microcomputer.

50. (a) Name two commonly used MPUs.
    (b) What is called the "instruction set" for a MPU?
    (c) Give three mnenomic codes used in assembly language and state their meaning and their possible op code.

51. Complete the chart opposite. Note the letter H at the end of a number signifies that the number is hexadecimal.

52. The content column in the binary machine language section in the chart opposite contains two fields of numbers. What are the names given to these two fields?

53. Describe the properties of the timing states generated by the control unit of the MPU.

54. The timing states in a machine cycle are divided into two sections. Name these two sections and briefly describe what occurs during these timing cycles.

55. What are micro instructions and where are they stored?

56. (a) Differentiate between the terms (i) ROM and (ii) RAM.

(b) Differentiate between the computer terms (i) micro instructions and (ii) macro instruction.

57. (a) What is the purpose of interfacing in computer systems?

(b) List three reasons why a MPU usually cannot be directly connected to outside devices.

58. (a) List four examples of interfacing used in computer systems.

(b) What are the properties of a serial interface system?

59. (a) Briefly describe the functions that must be performed by an analog PIA system?

BINARY		HEXADECIMAL	
ADDRESS	CONTENT	ADDRESS	CONTENT
0001	0001 0110	1H	
0011	1110 xxxx		
		4H	FH
011			0CH

APPENDIX A

STRUCTURE OF ELECTRON SHELLS

ATOMIC NUMBER	ELEMENT	SYMBOL	NUMBER OF SHELLS	STRUCTURE OF ELECTRON SHELLS						
				1ST K	2ND L	3RD M	4TH N	5TH O	6TH P	7TH Q
1	Hydrogen	H	1	1						
2	Helium	He	1	2						
3	Lithium	Li	2	2	1					
4	Beryllium	Be	2	2	2					
5	Boron	B	2	2	3					
6	Carbon	C	2	2	4					
7	Nitrogen	N	2	2	5					
8	Oxygen	O	2	2	6					
9	Fluorine	F	2	2	7					
10	Neon	Ne	2	2	8					
11	Sodium	Na	3	2	8	1				
12	Magnesium	Mg	3	2	8	2				
13	Aluminum	Al	3	2	8	3				
14	Silicon	Si	3	2	8	4				
15	Phosphorus	P	3	2	8	5				
16	Sulphur	S	3	2	8	6				
17	Chlorine	Cl	3	2	8	7				
18	Argon	A	3	2	8	8				
19	Potassium	K	4	2	8	8	1			

No.	Element	Symbol					
20	Calcium	Ca	4	2	8	8	2
21	Scandium	Sc	4	2	8	9	2
22	Titanium	Ti	4	2	8	10	2
23	Vanadium	V	4	2	8	11	2
24	Chromium	Cr	4	2	8	13	1
25	Manganese	Mn	4	2	8	13	2
26	Iron	Fe	4	2	8	14	2
27	Cobalt	Co	4	2	8	15	2
28	Nickel	Ni	4	2	8	16	2
29	Copper	Cu	4	2	8	18	1
30	Zinc	Zn	4	2	8	18	2
31	Gallium	Ga	4	2	8	18	3
32	Germanium	Ge	4	2	8	18	4
33	Arsenic	As	4	2	8	18	5
34	Selenium	Se	4	2	8	18	6
35	Bromine	Br	4	2	8	18	7
36	Krypton	Kr	4	2	8	18	8
37	Rubidium	Rb	5	2	8	18	8	1
38	Strontium	Sr	5	2	8	18	8	2
39	Yttrium	Y	5	2	8	18	9	2
40	Zirconium	Zr	5	2	8	18	10	2
41	Niobium	Nb	5	2	8	18	12	1
42	Molybdenum	Mo	5	2	8	18	13	1
43	Technetium	Tc	5	2	8	18	14	1
44	Ruthenium	Ru	5	2	8	18	15	1
45	Rhodium	Rh	5	2	8	18	16	1
46	Palladium	Pd	5	2	8	18	18	0
47	Silver	Ag	5	2	8	18	18	1
48	Cadmium	Cd	5	2	8	18	18	2
49	Indium	In	5	2	8	18	18	3
50	Tin	Sn	5	2	8	18	18	4
51	Antimony	Sb	5	2	8	18	18	5
52	Tellurium	Te	5	2	8	18	18	6

(Continued)

ATOMIC NUMBER	ELEMENT	SYMBOL	NUMBER OF SHELLS	STRUCTURE OF ELECTRON SHELLS						
				1ST K	2ND L	3RD M	4TH N	5TH O	6TH P	7TH Q
53	Iodine	I	5	2	8	18	18	7		
54	Xenon	Xe	5	2	8	18	18	8		
55	Cesium	Cs	6	2	8	18	18	8	1	
56	Barium	Ba	6	2	8	18	18	8	2	
57	Lanthanum	La	6	2	8	18	18	9	2	
58	Cerium	Ce	6	2	8	18	19	9	2	
59	Praseodymium	Pr	6	2	8	19	20	9	2	
60	Neodymium	Nd	6	2	8	19	21	9	2	
61	Promethium	Pm	6	2	8	18	22	9	2	
62	Samarium	Sm	6	2	8	18	23	9	2	
63	Europium	Eu	6	2	8	18	24	9	2	
64	Gadolinium	Gd	6	2	8	18	25	9	2	
65	Terbium	Tb	6	2	8	18	26	9	2	
66	Dysprosium	Dy	6	2	8	18	27	9	2	
67	Holmium	Ho	6	2	8	18	28	9	2	
68	Erbium	Er	6	2	8	18	29	9	2	
69	Thulium	Tm	6	2	8	18	30	9	2	
70	Ytterbium	Yb	6	2	8	18	31	9	2	
71	Lutetium	Lu	6	2	8	18	32	9	2	
72	Hafnium	Hf	6	2	8	18	32	10	2	
73	Tantalum	Ta	6	2	8	18	32	11	2	
74	Tungsten	W	6	2	8	18	32	12	2	
75	Rhenium	Re	6	2	8	18	32	13	2	
76	Osmium	Os	6	2	8	18	32	14	2	
77	Iridium	Ir	6	2	8	18	32	15	2	
78	Platinum	Pt	6	2	8	18	32	16	2	
79	Gold	Au	6	2	8	18	32	18	1	
80	Mercury	Hg	6	2	8	18	32	18	2	
81	Thallium	Tl	6	2	8	18	32	18	3	

Atomic Number	Element	Symbol								
82	Lead	Pb	6	2	8	18	32	18	4	
83	Bismuth	Bi	6	2	8	18	32	18	5	
84	Polonium	Po	6	2	8	18	32	18	6	
85	Astatine	At	7	2	8	18	32	18	7	
86	Radon	Rn	7	2	8	18	32	18	8	
87	Francium	Fr	8	2	8	18	32	18	8	1
88	Radium	Ra	8	2	8	18	32	18	8	2
89	Actinium	Ac	8	2	8	18	32	18	9	2
90	Thorium	Th	8	2	8	18	32	19	9	2
91	Protactinium	Ps	8	2	8	18	32	20	9	2
92	Uranium	U	8	2	8	18	32	21	9	2
93	Neptunium	Np	8	2	8	18	32	22	9	2
94	Plutonium	Pu	8	2	8	18	32	23	9	2
95	Americium	Am	8	2	8	18	32	24	9	2
96	Curium	Cm	8	2	8	18	32	25	9	2
97	Berkelium	Bk	8	2	8	18	32	26	9	2
98	Californium	Cf	8	2	8	18	32	27	9	2
99	Einsteinium	E	8	2	8	18	32	28	9	2
100	Fermium	Fm	8	2	8	18	32	29	9	2
101	Mendelevium	Mv	8	2	8	18	32	30	9	2
102	Novelium	No	8	2	8	18	32	31	9	2
103	Lawrencium	Lw	8	2	8	18	32	32	9	2

APPENDIX B

COLOUR CODING OF RESISTORS

The colour of the first band indicates the first digit. The colour of the second band indicates the second digit. The third band indicates the number of zeros to add, or it can represent a decimal multiplier if the colour silver or gold appear in the third band. The fourth band indicates the tolerance. (A fifth band may appear on some resistors—this merely indicates that this resistor can be used in specialized equipment such as computers, missiles, etc.)

Colour coding of Resistors

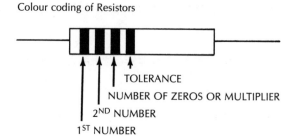

Colour Values for Resistors

COLOUR	DIGIT	MULTIPLIER	TOLERANCE
Black	0	1	
Brown	1	10	
Red	2	10^2	
Orange	3	10^3	
Yellow	4	10^4	
Green	5	10^5	
Blue	6	10^6	
Violet	7	10^7	
Grey	8	10^8	
White	9	10^9	
Gold		0.1	5%
Silver		0.01	10%
No Colour			20%

APPENDIX C

THE J OPERATOR

It was pointed out in Chapter Eleven that a phasor such as a voltage of 100 V with a lagging angle of 30° may be written in *polar form* as 100 V $\underline{/-30°}$. This phasor, as shown in Fig. A-1, may also be expressed in *rectangular form* as $86.6 - j\,50$.

$$100 \text{ V } \underline{/-30°} = 86.6 - j\,50 \text{ V}$$

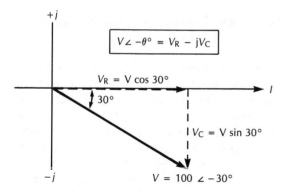

Figure A-1 Rectangular coordinates for a lagging phasor.

The phasor $(86.6 - j\,50)$ volts indicates that the phasor has an in-phase component of 86.6 V and a 90° lag component of 50 V. It should be obvious that the $(-j)$ factor is used to indicate a lag angle of 90°. Similarly, as shown in Fig. A-2, a phasor written in polar form as 200 V $\underline{/60°}$ may be expressed in rectangular form as $100 + j\,173.2$ V. The $(+j)$ factor is used to indicate a lead angle of 90°.

When working with phasors in rectangular coordinate form, the j operator follows the same set of rules as for the system of complex

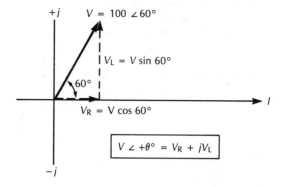

Figure A-2 Rectangular coordinates for a leading phasor.

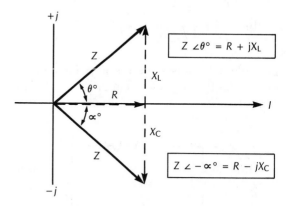

Figure A-3 Inductive and capacitive impedance triangles.

algebra. By definition, $j = \sqrt{-1}$, the in-phase component is represented by the "real" axis, and the 90° out-of-phase component is represented by the "imaginary" or j axis. The use of the j operator can sometimes simplify the analysis of many ac circuits. The following expressions will assist the student in obtaining accurate solutions.

1. (a) $j = \sqrt{-1}$
   (b) $j^2 = (\sqrt{-1})^2 = -1$
   (c) $j^3 = j^2 \times j = -j$
   (d) $j^4 = j^2 \times j^2 = 1$
   (e) $j^5 = j^4 \times j = j$

2. (a) $V\underline{/\theta°} = V\cos\theta° + jV\sin\theta°$
   $= V_R + jV_L$
   (b) $V\underline{/-\theta°} = V\cos\theta° - jV\sin\theta°$
   $= V_R - jV_c$

3. (a) $V_R + jV_c = \sqrt{(V_R)^2 + (V_L)^2}\underline{/\tan^{-1}\left(\dfrac{V_L}{V_R}\right)}$
   (b) $V_R - jV_c = \sqrt{(V_R)^2 + (V_c)^2}\underline{/-\tan^{-1}\left(\dfrac{V_c}{V_R}\right)}$

4. (a) $V_1\underline{/\theta°} \times V_2\underline{/\propto°} = V_1 \times V_2\underline{/\theta° + \propto°}$
   (b) $\dfrac{V_1\underline{/\theta°}}{V_2\underline{/\propto°}} = \dfrac{V_1}{V_2}\underline{/\theta° - \propto°}$

5. (a) $\dfrac{1}{jX} = \dfrac{1}{X\underline{/90°}} = \dfrac{1}{X}\underline{/-90°} = -j\dfrac{1}{X}$
   (b) $\dfrac{1}{-jX} = \dfrac{1}{X\underline{/-90°}} = \dfrac{1}{X}\underline{/90°} = +j\dfrac{1}{X}$

Example A-1: Determine the impedance and current for the series *R-C* circuit shown in Fig A-4.

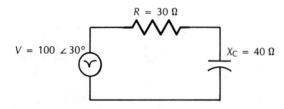

Figure A-4(a) Series *R-C* circuit.

$Z = R - jX_c$ (from impedance triangle)
$= 30 - j40$
$= \sqrt{(30)^2 + (40)^2}\underline{/-\tan^{-1}\dfrac{40}{30}}$

$= 50\ \Omega\ \underline{/-53.1°}$
(Z lags the real axis by 53.1°)

$I = \dfrac{V}{Z} = \dfrac{100\underline{/30°}}{50\underline{/-53.1°}}$
$= 2\ A\ \underline{/83.1°}$
(*I* leads the real axis by 83.1°)

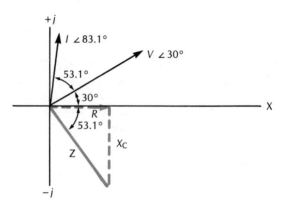

Figure A-4(b) Phasor diagram. (Note all directions relative to the X-axis.)

The phasor diagram in Fig. A-4(b) shows the quantities for the total voltage and current in the circuit. Note that the angular difference between *V* and *I* is 53.1°. This is the circuit phase angle, and it is also always the angle associated with the circuit impedance. For this circuit, *I* leads *V* by 53.1°. The impedance triangle is also shown on this diagram, but note that they are actually not phasor quantities.

Example A-2: Determine the impedance and total current for the curcuit shown in Fig. A-5.

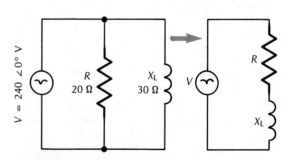

Figure A-5(a) Parallel *R-L* circuit with its series equivalent circuit.

According to Kirchhoff's current law, the circuit current, I_T, is equal to the phasor sum of the branch current, I_R and I_L.

$$I_R = \frac{V}{R} = \frac{240\ \underline{/0^\circ}}{20\ \underline{/0^\circ}} = 12\ A\ \underline{/0^\circ}$$

$$I_L = \frac{V}{jX_L} = \frac{240\ \underline{/0^\circ}}{30\ \underline{/90^\circ}} = 8\ A\ \underline{/-90}$$

$$\begin{aligned}I_T &= I_R + I_L \\ &= 12\ \underline{/0^\circ} + 8\ \underline{/-90^\circ} \\ &= 12 - j\ 8\ A\ \text{(rectangular form)}\end{aligned}$$

or

$$\begin{aligned}I_T &= \sqrt{(I_R)^2 + (I_L)^2}\ \underline{/-\tan^{-1}\frac{I_L}{I_R}} \\ &= \sqrt{(12)^2 + (8)^2}\ \underline{/-\tan^{-1}\frac{8}{12}} \\ &= 14.4\ A\ \underline{/-33.7^\circ}\ \text{(polar form)}\end{aligned}$$

$$\begin{aligned}Z &= \frac{V}{I_T} = \frac{240\ \underline{/0^\circ}}{14.4\ \underline{/-33.7^\circ}} \\ &= 16.6\ \Omega\ \underline{/33.7^\circ}\ \text{(polar form)}\end{aligned}$$

$$\begin{aligned}Z &= 16.6 \cos 33.7^\circ + j\ 16.6 \sin 33.7^\circ \\ &= 16.6\ (0.83) + j\ 16.6\ (0.55) \\ &= 13.8 + j\ 19.2\ \Omega\ \text{(rect. form)}\end{aligned}$$

or

$$\frac{1}{Z} = \frac{1}{R} + \frac{1}{jX_L}$$

$$\begin{aligned}Z &= \frac{R \times (jX_L)}{R + jX_L} \\ &= \frac{(20) \times (j\ 30)}{20 + j\ 30} \\ &= \frac{j\ 600}{20 + j\ 30} \\ &= \frac{600\ \underline{/90^\circ}}{36.1\ \underline{/56.3^\circ}} \\ &= 16.6\ \Omega\ \underline{/33.7^\circ}\ \text{(polar form)} \\ &= 13.8 + j\ 9.2\ \Omega\ \text{(rect. form)}\end{aligned}$$

Note that with the use of the j operator, either the circuit current or the circuit impedance may be determined first. The rectangular

form for the circuit impedance also indicates that the series R-L equivalent circuit may now be easily drawn, as shown in Fig. A-5. The phasor diagram for this circuit is shown in Fig. A-5(b)

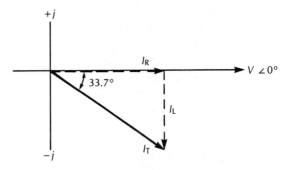

Figure A-5(b) Phasor diagram.

Example A-3: Determine the circuit impedance and circuit currents for the series-parallel R-L-C circuit, shown in Fig. A-6

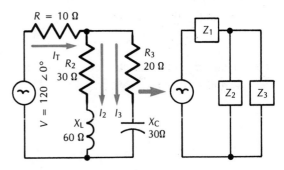

Figure A-6(a) Series-parallel R-L-C circuit and simplified circuit.

(a) $Z_1 = R = 10\ \underline{/0^\circ}\ \Omega$
(b) $Z_2 = R_2 + j\ X_L$
   $= 30 + j\ 60\ \Omega$
   $= 67.1\ \underline{/63.4^\circ}\ \Omega$
(c) $Z_3 = R - jX_c$
   $= 20 - j\ 30\ \Omega$
   $= 36.1\ \underline{/-56.3^\circ}\ \Omega$

The simplified circuit with the Z_1, Z_2, and Z_3 is shown in Fig. A-6(a).

(d) $Z_T = Z_1 + \dfrac{Z_2 \times Z_3}{Z_2 + Z_3}$

    (note that Z_2 and Z_3 are in parallel)

$= 10. \underline{/0°} + \dfrac{(67.1 \underline{/63.4°})(36.1 \underline{/-56.3°})}{(30 + j\,60) + (20 - j\,30)}$

$= 10 \underline{/0°} + \dfrac{2422.3 \underline{/7.1°}}{50 + j\,30}$

$= 10 \underline{/0°} + \dfrac{2422.3 \underline{/7.1°}}{58.3 \underline{/31°}}$

$= 10 \underline{/0°} + 41.5 \underline{/-23.9°}$

$= (10 + j\,0) + (37.9 - j\,16.8)$

$= 47.9 - j\,16.8 \ \Omega$

or

    $= 50.8 \underline{/-19.3°} \ \Omega$

(e) $I_T = \dfrac{V}{Z_T} = \dfrac{120 \underline{/0°}}{50.8 \underline{/-19.3°}}$

    $= 2.4 \underline{/19.3°} \ A$

(f) $V_1 = I_T Z_1$

    $= (2.4 \underline{/19.3°})(10 \underline{/0°})$

    $= 24 \underline{/19.3°} \ V$

(g) From (d), the parallel combination of Z_2 and Z_3 is equal to $41.5 \underline{/-23.9°} \ \Omega$.

$Z_{23} = 41.5 \underline{/-23.9°} \ \Omega$

$V_2 = V_3 = I_T Z_{23}$

    $= (2.4 \underline{/19.3°})(41.5 \underline{/-23.9°})$

    $= 99.6 \underline{/-4.6°} \ V$

(h) $I_2 = \dfrac{V_2}{Z_2} = \dfrac{99.6 \underline{/-4.6°}}{67.1 \underline{/63.4°}}$

    $= 1.5 \underline{/-68°} \ A$

(i) $I_3 = \dfrac{V_3}{Z_3} = \dfrac{99.6 \underline{/-4.6°}}{36.1 \underline{/-56.3°}}$

    $= 2.8 \underline{/51.7°} \ A$

The phasor diagram for this circuit is shown in Fig. A-6(b).

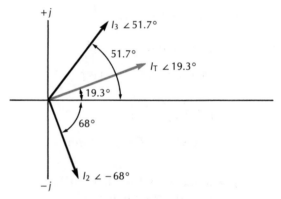

Figure A-6(b) Phasor diagram.

APPENDIX D

NATURAL TRIGONOMETRIC FUNCTIONS

Natural Trigonometric Functions

DEGREES	SINE	COSINE	TANGENT	DEGREES	SINE	COSINE	TANGENT
0	0.000	1.000	0.000	46	0.719	0.695	1.03
1	0.017	1.000	0.017	47	0.731	0.682	1.07
2	0.035	0.999	0.035	48	0.743	0.669	1.11
3	0.052	0.999	0.052	49	0.755	0.656	1.15
4	0.070	0.998	0.070	50	0.766	0.643	1.19
5	0.087	0.996	0.087	51	0.777	0.629	1.23
6	0.105	0.995	0.105	52	0.788	0.616	1.28
7	0.122	0.993	0.123	53	0.799	0.602	1.33
8	0.139	0.990	0.141	54	0.809	0.588	1.38
9	0.156	0.988	0.158	55	0.819	0.574	1.43
10	0.174	0.985	0.176	56	0.829	0.559	1.48
11	0.191	0.982	0.194	57	0.839	0.545	1.54
12	0.208	0.978	0.213	58	0.848	0.530	1.60
13	0.225	0.974	0.231	59	0.857	0.515	1.66
14	0.242	0.970	0.249	60	0.866	0.500	1.73
15	0.259	0.966	0.268	61	0.875	0.485	1.80
16	0.276	0.961	0.287	62	0.883	0.470	1.88
17	0.292	0.956	0.306	63	0.891	0.454	1.96
18	0.309	0.951	0.325	64	0.899	0.438	2.05
19	0.326	0.946	0.344	65	0.906	0.423	2.14
20	0.342	0.940	0.364	66	0.914	0.407	2.25

(Continued)

DEGREES	SINE	COSINE	TANGENT	DEGREES	SINE	COSINE	TANGENT
21	0.358	0.934	0.384	67	0.920	0.391	2.36
22	0.375	0.927	0.404	68	0.927	0.375	2.48
23	0.391	0.920	0.424	69	0.934	0.358	2.61
24	0.407	0.914	0.445	70	0.940	0.342	2.75
25	0.423	0.906	0.466	71	0.946	0.326	2.90
26	0.438	0.899	0.488	72	0.951	0.309	3.08
27	0.454	0.891	0.510	73	0.956	0.292	3.27
28	0.470	0.883	0.532	74	0.961	0.276	3.49
29	0.485	0.875	0.554	75	0.966	0.259	3.73
30	0.500	0.866	0.577	76	0.970	0.242	4.01
31	0.515	0.857	0.601	77	0.974	0.225	4.33
32	0.530	0.848	0.625	78	0.978	0.208	4.70
33	0.545	0.839	0.649	79	0.982	0.191	5.14
34	0.559	0.829	0.674	80	0.985	0.174	5.67
35	0.574	0.819	0.700	81	0.988	0.156	6.31
36	0.588	0.809	0.726	82	0.990	0.139	7.11
37	0.602	0.799	0.754	83	0.993	0.122	8.14
38	0.616	0.788	0.781	84	0.995	0.105	9.51
39	0.629	0.777	0.810	85	0.996	0.087	11.4
40	0.643	0.766	0.839	86	0.998	0.070	14.3
41	0.656	0.755	0.869	87	0.999	0.052	19.1
42	0.669	0.743	0.900	88	0.999	0.035	28.6
43	0.682	0.731	0.932	89	1.000	0.017	57.3
44	0.695	0.719	0.966	90	1.000	0.000	∞
45	0.707	0.707	1.000				

INDEX

AC Sine wave, 109, 111
AC generators (alternators)
  amortisseur winding, 328, 329
  armature reaction, 322
  construction, 314
  cylindrical rotor, 318
  delta connection, 316
  efficiency, 329
  exciter, 318,320
  field discharge, 319
  hunting, 328
  induced voltage, 315, 317, 320
  load power factor, 322
  parallel operation, 324
  phase sequence, 325
  phasor diagrams, 322
  power losses, 325, 329
  rotor, 318
  salient rotor, 318
  slip rings, 314
  stator, 314
  synchronous operation, 321,
    325
  ventilation, 319
  voltage regulation, 323
  Wye connection, 316
AC power, 196
AC resistance
  dielectric loss, 198
  eddy currents, 198
  hysteresis, 198
  pure, 197
  skin effect, 198
AC resistive circuits, 196
AC voltage, 182, 184
AC voltage
  advantages, 182
  average value, 190
  effective (rms) value, 189
  instantaneous value, 188
  peak value, 187
AC instruments
  electro magnetic type, 392
  rectifier type, 390
  thermal type, 394
AC motor controls, 406
Across-the-line starter, 164, 165
Additive polarity, 264

Adjustable voltage sped control,
  157
Admittance, 253, 255
Alkaline, 23
Alnico, 59
Alternating current, 7
Alternator-ac generator, 111, 184
Amalgamation, 22
Ambient temperature, 345
Ammeter, 14
  basic properties, 90
  multirange, 93
  sensitivity, 89
  shunt, 90
Ampere, 7, 14
Ampere-hour, 30
Ampere-turns, 267
Analog device, 446
Analog display, 402
AND gate, 390
Anode, 221
Anti-plugging relay, 171
Antimony, 433
Apparent power, 235, 243
Arc barriers, 167
Arc chute, 167
Armature, 109, 115, 116
  self-induced voltage, 124
  coil induction, 124
Armature reaction, 123, 124
  causes, 123
  correction, 125
  effects, 124
Armature resistance, 123
Armature core, 116
Armature winding, 116
  drum windings, 116
  frog leg windings, 117
  lap windings, 118
  wave windings, 117
    ring wound, 116
    lap, 141
    wave, 141
Arsenic, 433
Artificial magnets, 58
Assembly language, 458
  address field, 458
  instruction field, 458
Atomic
  mass, 2

  number, 2,4
  shells, 2, 4, 5
Attracting magnetic force, 64
Auto-couplers, 413
Auto-ranging meters, 403
Automatic reversing control, 169
  electrical (push button), 169
  mechanical interlock, 169
Automatic controllers, 407
Automatic motor control, 161
  pilot control devices, 161
  primary control devices, 161
Autotransformer, 274
Autotransformer starters, 415,
Avalanche breakdown, 435
Average value, 391, 392
Ayerston shunt, 94

Basic motor principle, 137
Battery, definition, 18
Binary coded decimal (BCD),
  450
Binary number system, 449
Bistable multivibrator, 453
Bits, 449
Block rotor test, 343
Blowout coil, 167
Boolean algebra, 455
Braking
  electrical, 175
  mechanical, 176
Braking force, 88
Breakdown torque, 340
Bridge circuit, 49, 103, 392, 436,
  440
Bridge rectifier
  full-wave, 198
  half-wave, 198
Brush
  construction, 115
  positioning, 115
Buildup action, 128

Canadian Electrical
    Manufacturers Association
    (CEMA), 346
Canadian Standards Association
    (CSA), 264
Capacitive, vars, 242, 250
Capacitive circuits

V, I, and Q waveforms, 232
  impedance, 228, 234
  phase relationship, 231
  power factor, 236
  reactive power, 232
Capacitor potential transformer, 274
Capacitor start motor, 380
Capacitors, 219
  capacitance, 219, 221
  charge stored, 222
  charging current, 221, 223
  counter voltage, 221, 228
  discharging current, 221, 223
  in parallel, 229
  in series, 228
  ratings, 222
  reactance, 228
Cathode, 390
Cells
  alkaline manganese, 23
  lead acid, 28
  lithium, 25
  magnesium, 25
  mercury, 23
  nickel cadmium, 31
  parallel, 27
  primary, 18
  secondary, 18
  series, 27
  series parallel, 27
  silver oxide, 24
  voltaic, 19
  zinc carbon, 21
Centrifugal switch, 379
Ceramic magnet, 59
Charges, electric, 2, 3, 4
Circuit breaker, 167
Clamp-on voltmeter, 400
Clock signal, 455
Cobalt steel, 373
Coercive force, 73
Collector of transistor, 442
Combination logic, 455
Commutating axis, 112, 113
Commutation, 113, 139, 140
Commutator, 111, 116, 139
Commutator
  action, 113
  motor, 386

Compensating winding, 126, 144
Complementary metal, oxide
  semiconductor (CMOS), 445
Complex circuits, 48, 49
Compound, 2
Compound generator, 129
  cumulative, 130
  differential, 130
  load characteristics, 130
  long shunt, 130
  regulation, 130
  short shunt, 130
Compound motor, 147
  cumulative, 150
  differential, 150
  long shunt, 146, 150
  short shunt, 150
Computer, 457
  ALU, 457
  analog, 446
  control unit, 457
  digital, 446
  execution cycle, 458
  fetch cycle, 459
  I/O devices, 457
  machine cycle, 458
  machine language, 459
  macro instruction, 460
  memory, 457
  micro instruction set, 460
  peripheral devices, 460
  PIAs, 460
  program counter, 459
  RAM, 458
  registers, 457
  ring counter, 459
  ROM, 459
  serial date output, 461
Concentric vane meter, 393
Condensors, 219
Conductance, 13, 68, 253, 255
Conductively compensated, 385
Conductors, 5, 13
Consequent poles, 63
Contact resistance, 95
Contactor
  clapper, 166
  solenoid, 166
Control circuit, 164, 410

Control cycle, 459
Control relay, 165
Controllers
  above and below normal, 159
  above normal, 159
  automatic, 161
  drum, 160
  enclosures, 159
  face plate, 159
  operation, 159
Converters
  analog to digital (ADC), 460
  digital to analog (DAC), 460
Copper power losses, 131
Core lamination, 116
Core losses, 131, 314, 318
Core saturation, 122
Cosine function, 187
Coulomb, 6, 14
Coulomb, Charles, 4
Counter torque, 142
  voltage, 144, 149, 203, 204, 265
Counter voltage controller, 172
Cross magnetic field, 378
Cunife, 59
Current
  ac, 7
  dc, 7
  flow, 6
Current limit acceleration
  controller, 173
Current transformer
  bar type, 272
  window type, 272
  wound type, 272
Cycle, 184, 185
Cylindrical roter, 318

D'Arsonval meter, 87, 88, 390
Damper winding, 362
Damping, 89
De Morgan's rule, 453
Definite time accelerating
  starter, 172
  electrical, 186
  mechanical, 186
Degree of compounding, 130
Delta connection, 288
Delta-delta connection, 297

Delta-Wye circuit conversions, 49, 50
Delta-Wye connection, 304
Depolarizing, 21, 22
Diagrams
  schematic, 162
  wiring, 162, 165
Dielectric
  constant, 220, 222
  loss, 198
  material, 220, 222
  strength, 220, 222
Differential amplifiers, 447, 448
Differential generator, 130
Differential motor, 151
Differential selsyn, 370
Digital logic, 450
Digital multimeter (DMM), 402
Digital devices, 446, 449
Diode, 390
Diomagnetic, 59
Direct current, 7, 189
Domains, 60
Doping, 433
Drain terminal, 444
Drum controller, 160
  windings, 116
Dry cell, 21
Dynamic braking, 175
Dynamic resistance, 437

Eddy-current loss, 116, 151
Eddy currents, 89, 198
Effective resistance, 211
Effective value (RMS), 189, 391, 393
Efficiency, 44, 130, 147, 329
Electrical charge, 6
Electrical
  diagrams, 161
  symbols, 162
Electrical braking
  dynamic, 175
  plugging, 176
  regenerative, 175
Electrical degrees, 186, 315
Electrical inertia, 202
Electrical junction (node), 40
Electrical rotation, 186
Electrodes, 19

Electrodynamometer, 87
Electrodynamometer
  movement, 395
Electrolyte, 19
Electrolytic capacitor, 220
Electromagnet, 63
Electromagnetic field, 63
Electromagnetic induction, 79, 109
Electromagnetic meters, 87, 392
Electromotive force, 7, 66
Electron
  shells, 4
  valence, 4
Electrostatic field, 220
Element, 1, Appendix A
Emitter terminal, 442
Energy, 44, 45
Energy, kinetic, 1
Energy potential, 1
Energy measurement in ac, 398
Enhancement MosFET, 445
Exclusive or grate (XOR), 451
Exponential, constant "e", 225
Exponential equations, 225

Face plate starters, 156
Farad, 220, 222
Faraday's law, 79, 80, 202
Feedback, 448
Ferrites, 59
Ferromagnetic, 59
Fetch cycle, 459
Field Effect Transistor (FET), 443
Field excitation, 119
Field discharge resistor, 360
Field effect transistor
  volt-ohmmeter
  (FET-VOM), 402
Field poles, 120, 185
Filter circuit, 436
Fixed power lossess, 131
Flemings left-hand rule, 138
Flip-flop, 453
  edge triggering, 455
  JK, 455
  NAND (SR), 453
  NOR (SR), 453
  racing, 455
  toggling, 455

Flux density, 67
Flux flow, 66
Forward bias, 434
Four-point starters, 157
Fractional horsepower, 376
Free electrons, 6
Frequencey, 185
Frequency meter, 403
Friction lossess, 147
Fundamental generator
  equation, 317
Fuses, 167

Gallium, 434
Gallium arsenide, 437
Galvonameter, 87
Gate terminal 443
Generation action, 144
Generator
  efficiency, 130
  equation, 121
  field poles, 120
  left-hand rule, 81
  load characteristics, 127, 130
  magnetization curve, 121
  power losses, 131
  voltage losses, 122
  voltage regulation, 126, 127
Generator type
  compound, 120, 129
  separately excited, 120, 127
  series, 120, 129
  shunt self-excited, 120, 127
Germanium, 390

Henry, 202
Hexadecimal system, 449
Holding coil, 157
Horsepower, 341
Hunting, 328, 362
Hydrogen
  density, 320
  heat conductivity, 319
Hydrometer, 31
Hysteresis
  loop, 73
  loss, 73, 116, 146, 198
Hysteresis synchronous motor, 373

Indium, 434
Induced current, 79
Induced voltage, 79, 80, 315, 317, 320, 322
Inductance definition, 203, 204
    effect of frequency on, 204
    in ac circuit, 204
    in dc current, 203
    in parallel, 207
    in series, 206
Inductance mutual, 205
Inductance voltage regulators, 279
Induction
    electromagnetic , 78, 79
    electrostatic, 78
    magnetic, 78
Induction motor (single phase), 376, 378
Induction watthour meter, 398
Induction wattmeter, 398
Inductive reactance, 205
Inductive VARs, 242
Inductively compensated, 385
Inductors, 202
Inrush starting current, 155, 167
Instantaneous value, 188
Instrument transformer
    current, 271
    voltage, 273
Insulated gate FET (IGFET), 445
Insulators, 6
Integrated circuit, 446
Interfacing, 460
Interlocking
    electrical, 170
    mechanical, 170
Interlocking controls, 419
Interlocks, 169
Internal resistance, 90
Internal losses
    power, 46
    voltage, 46
Internal resistance, 21
Interpoles, 126, 143, 144
Inverter gate, 451
Inverters, 427
Ions, 5
Ionic coduction, 26
Ionization, 5

Iron losses, 147

J Operator, 212, 235
JFET, 443
Jogging, 422
Joules, 44
Junction diode, 390
Junction transistor, 442

Karnaugh mapping, 455
Kilowatt hour, 45, 398
Kinetic energy, 1
Kirchhoff's laws, 370, 40 144

Ladder diagrams, 429
Laminated core, 116
Lap winding, 118, 141
Latch circuit, 453
Lead acid
    cell, 28
    charging, 31
    construction, 28
    maintenance, 31
    operation, 28
    rating, 30
    safety precautions, 33
    testing, 30
Left-hand generator rule, 110
Left-hand rule, coil, 65
    straight wire, 64
Lenz's law, 82, 202, 337
Light emitting diode (LED), 437
Line diagram, 162
Line losses
    power, 46
    voltage, 46
Line power losses, 261
Line voltage, 286, 289
Linear scale, 102, 392
Lithium cell, 23
Load characteristics, 127-130
Load effect on motor speed, 145, 148
Loading effect of voltmeters, 100
Local action, 22
Locked rotor torque, 343
Lockout contactor controller, 174
Lodestone, 58
Logic functions, 451
    AND, 451

NAND, 451
NOR, 451
OR, 451
XOR, 451
Low voltage protection, 158

MOSFET (metal oxide semiconductor), 445
Magnesium cell, 25
Magnetic characteristics, 61
    circuit, 65, 115
    coil, 64
    dipole, 60
    domain, 60
    field, 61
    flux, 61, 66
    force, 62, 68
    induction, 62
    intensity, 68
    laws, 62
    memory, 84
    poles, 61, 59
    relay, 83
    shielding, 63
Magnetic contactor, 165
    chattering, 408
    shading coil, 408
Magnetic curves, 69
Magnetic noise, 334
Magnetic overload relay, 167
    instanteous trip, 168
    time delay trip, 168
Magnetic starter, 155
Magnetism, theory of, 59
Magnetite, 58
Magnetizing current, 265, 266, 363
Magnetization curve, 121
Magnetizing VARs, 242
Magneto, 119
Magnetomotive force, 66
Manganin, 91
Manual starting rheostat, 158
Matter, 1
Mechanical brakes, 160
Mechanical degrees, 186, 315
Mechanical power losses, 131, 325, 329
Megger, 105
Memory circuit, 460

Mercury cell, 23
Meters
  electrodynamometer, 87
  internal resistance, 90
  moving coil, 87, 88
  moving iron vane, 87
  multiplier resistor, 97
  parallax, 90
  sensitivity, 89
  shunt, 90
Mica, 116
Microprocessor (MPU), 458
  instruction set, 458
  Intel 8080, 458
  machine cycle, 459
  machine language, 458
  Motorolla 6800, 458
Minority carriers, 435
Mnemomics, 458
Molecule, 2
Motor classification, 146
  compound, 146, 150
  controllers, 155
  counter voltage, 144
  efficiency, 147
  field effect, 145, 148
  load effect, 145, 148
  losses, 146
  natural speed, 144
  rating, 146
  series, 146, 149
  shunt, 146, 148
  speed regulation, 145
  starters, 155
  starting requirements, 155
  torque, 138
Motor armature current
  running, 157
  starting, 155, 172
Motor action principle, 337
Motor controllers, 406
  automatic, 407
  manual, 406
  semi automatic, 407
Motor enclosures, 345
Moving coil regulator, 280
Moving iron vane meter, 393
Multiple source circuits, 51
Multirange meter, 93, 96, 101
Multispeed motors, 423

Multitesters (multimeter), 401, 402
Mutual induction, 205 265

N-type impurity, 433
NAND logic gate, 452
Natural magnetism, 58
Negative temperature
    coefficient, 169, 440
Negative feedback, 402
Neutral plane, 112, 138, 143
Neutral wire, 286
Neutrons, 2
Newtons, 44
Nichrome, 91
Nickel cadmium, 31
Nonlinear scale, 101, 394, 395
Normal excitation current, 363
Nucleous, 2
Null state, 440

Ohm's law, 14, 38
Ohm's-per-volt rating, 99
Ohmmeter, 15, 101
Op code, 458
Open circuit voltage, 46
Open-delta connection, 307
Operational amplifier, 448
Optical potentiometers, 438
Oscillation, 89
Overload protection
  circuit breakers, 167
  fuses, 167
  overload relays, 168
P-type material, 434
Parallax, 90
Parallel circuit, 39
  current, 57
  resistance, 40
  voltage, 37
Parallel operation of alternators, 324
Paramagnetic, 59
Part winding starters, 416
Peak inverse voltage (PIV), 435
Peak value, 187
Period, 185
Peripheral interface adapter
    (PIA), 461
Permanent magnet, 58

Permanent split-phase motor, 379
Permeability, 58, 68
Phase angle, 191, 196, 214
Phase relationship
  in phase, 191
  inductive circuits, 208
  lagging, 191
  leading, 192
  out-of-phase, 191, 192
Phase rotation, 286
Phase splitting, 378
Phasor addition
  component method, 194
  triangular addition, 193
Phasors, properties, 192, 193
Piezoelectric crystal, 109
Pigtail, 115
Pilot control device, 161, 407
Plugging, 176
PN junction, 434
Polar expression, 196
Polarity indicator, 438
Polarization, 20, 24
Positive temperature coefficient, 169
Potential
  chemical action, 9
  difference, 9, 10
  energy, 1
  friction, 9
  heat, 10
  light, 10
  magnetism, 10
  pressure, 9
Potential transformers, 273
Power, 15
  calculation, 44, 45
  in a pure inductance, 209, 210
  losses, 47
  reactive, 210, 211
  transfer, 47, 48
  true power, 209, 210
Power factor, 214
  capacitive, 236
  correction, 249
  lag, 242
  lead, 242
  correction, 363
  synchronous motor, 362

three-phase system, 292
Power factor meters, 399
Power grid, 324, 325
Power losses-alternators 147, 325
  copper (I^2R), 131
  core, 131
  eddy currents, 116, 131
  friction, 131
  hysteresis, 116, 131
  mechanical, 131
Power measurement
  three wattmeter method, 296, 397
  two wattmeter method, 294
Primary control devices, 161
Primary cells, 18
Primary resistor starters, 414
Programmable controller, 429, 461
Programming
  assembly language, 458
  machine language, 458
  mnemonics, 458
  Op code, 458
Prony brake, 142
Proton, 2
Pull-in torque, 359
Pull-out torque, 361
Pulsating dc, 391
Push-pull inputs, 448
Push button switches, 407
Pythagorean relationship, 194

Q factor of a coil, 211, 243

RLC (parallel circuit)
  apparent power, 241, 248
  impedance, 241, 251
  net reaction effects, 241, 258
  phase angle, 242, 247
  power factor, 242, 247, 249
  reactive power, 242, 249
  resonance, 244, 248
  true power, 242, 247
RLC (series-parallel circuit),
  series RL to parallel, 253
RLC (series-paralled circuit),
  susceptance (Siemens),
  253, 255

Radial-vane meter, 393
Radians per second, 120, 316
Radio frequency, 395
Reactance, 228
Reactive power, 211, 232
Rectifier circuit, 435, 436
Rectifier-type meter, 39, 390, 392
Reduced voltage starters
  autotransformer type, 176, 413, 415
  counter voltage, 172
  current limit method, 173
  definite time acceleration, 172
  lockout contactors, 174
  part winding type, 416
  primary resistor type, 414
  solid state type, 417
  voltage drop acceleration, 174
  Wye-delta type, 416
Reed relay, 83, 84
Regenerative braking, 175
Registers, 457
Regulators, 279
Relative density, 30
Relays, 411
Reluctance, 58, 66, 67
Reluctance synchronous motor, 372
Repelling magnetic force, 62
Repulsion, induction motor, 387
Repulsion motor, 386
Repulsion start induction run, 387
Residual magnetism, 58, 128
Resistance
  colour code, Appendix B
  definition, 11
  factors affecting, 11
  measurement, 15
Resistance losses,
  power, 131
  voltage, 123
Resistivity, 12
Resonant frequency, 244, 248
Retentivity, 73
Reverse bias, 434
Reversing controllers, 418
Revolution per minute, 185, 120
Right-hand motor rule, 138

Ring shunt, 94
Ring-wound winding, 116
Ripple voltage, 114, 436
Root-mean-square (RMS), 189, 391, 393
Rotating magnetic field, 334, 377
Rotational power loss, 147
Rotor, cylindrical, 318
Rotor, field discharge, 319
Rotor, resistance, 319
Rotor, salient pole, 318
Runaway action, 148

Safety precautions, 33
Saturation, 69, 122
Scalar quantity, 192
Schematic diagram, 162
Scientific notation, 8
Sealing circuit, 164
Secondary cell, 18
Self-differential devices, 370
Self-induced voltage, 124
Self-induction, 125, 265
Self-power drives, 370
Self-selsyn devices, 370
Self-synchronous device, 369
Self-regulating, 126
Semiconductors, 5, 390
Semiconductor devices, 433
Sensitivity, 401
  ammeter, 89
  voltmeter, 96
Separately excited generator, 120, 127
Sequence control circuit, 409, 420
Serial data output, 461
Series dc motor starters, 157, 158
Series motor, 146, 149
Series circuit, 15, 36, 39
Series generator, 120, 129
Series multiplier, 90, 91
Series transformer, 271
Seven segment display, 437
Shaded-pole motor, 383
Shading coil, 408
Shunt motor, 146, 148
Shunt resistor, 98
Siemens, 13

Silicon chip, 390
Silicon control rectifer (SCR),
    440, 413, 417, 427, 441
Silver oxide, 24
Sine wave, 111, 184
Single-phase motor, 377
Skin effect, 198
Slidewire bridge, 104
Slip rings, 109
Solar cell, 109
Solder pot relay, 168
Solenoid, 168
Solid state devices, 433
Solid state motor control, 179
Solid state relay, 411
Solid state starters, 412, 417
Specific resistance, 12
Speed controllers, 159, 161
Speed load characteristics, 145,
    150
Speed regulation, 145
Split rings, 111
Split-phase induction motors,
    379, 381, 385
Square-law scale, 394
Squirrel cage, motor, 333–348
Star connections, 246
Starters, automatic, 161
    four point, 157
    manual, 156
    three point, 156
Static motor control, 427
Stator
    armature reaction, 322
    coil reactance, 321
    induced voltage, 315
    load power factor, 322
    synchronous generator, 321
    voltage losses, 321, 322, 329
Storage cell, 28
Stray power loss, 329
Subtractive polarity, 264
Superposition principle, 52
Synchronous condenser, 366
Synchronous motor (1 phase)
    construction, 371
    hysteresis type, 373
    reluctance type, 372
    shaded pole stator, 372
Synchronous motor (3 phase)

construction, 356
dc field excitation, 359
efficiency, 368
field current effect, 360
hunting effects, 362
industrial applications, 363
load effects on, 359
normal excitation current, 363
power factor, 362
power factor correction, 363
principal of operation, 357
pull-in torque, 359
pull-out torque, 361
ratings, 368
starting properties, 359
torque angle, 360
Synchronous motor control, 425
Synchronous operation, 321, 325
Synchroscope, 326, 327

Tangent function, 187
Tap changing transformer, 276
Taut-band suspension, 89
Temperature coefficient, 13, 91
Temporary magnet, 59
Terminal voltage, 46
Thermal overload relay, 168
Thermistor, 439
Thermocouple, 10, 109, 394
Thermocouple meters, 394
Three-wire control, 410
Three-phase circuits
    advantages, 286
    apparent power, 292
    delta connections, 288
    line voltage, 286
    phase rotation, 286
    phase voltage, 286
    power factor, 289
    true power, 292
    Wye connection, 286
Three phase motor, squirrel
    cage, 333
    wound rotor, 349
Three-phase transformer, 280
Three-point starters, 156
Three-phase motors,
    synchronous, 356
Time constant, 224, 226
Time delay relay, 168

Torque, 88, 138, 142
Torque angle, 360
Torque calculations, 341
Torque load characteristics, 140,
    150
Transformer
    construction, 262
    core type, 262
    counter voltage, 265
    definition, 261
    efficiency, 261, 268, 269
    H-type, 263
    impedance voltage, 270
    instrument transformers, 271
    current, 271
    instrument transformers
    voltage, 273
    losses (copper), 261, 270
    losses (iron), 269, 270
    mutual induction, 265
    open circuit test, 270
    operation, 265
    phasor diagrams, 265, 266
    polarity, 264
    power loss, 267, 275
    shell type, 263
    short-circuit test, 269
    turns ratio, current, 267
    turns ratio, voltage, 266
    under load, 266
Transformer connections
    delta-delta, 297
    delta-Wye, 304
    open-delta, 307
    Wye-delta, 305
    Wye-Wye, 302
Transistor, 442
Triac, 412
Trigonometric functions, 186-187
True power, 209, 210
Two-wire control, 410

Universal motor, 385

Valence, electron, 4
    shell, 4
Variable frequency speed
    control, 428
VARs, 211, 232
Vector, 192